Snake Venom Metalloproteinases

Special Issue Editors

Jay Fox
José María Gutiérrez

MDPI

Special Issue Editors

Jay Fox
University of Virginia
USA

José María Gutiérrez
Universidad de Costa Rica
Costa Rica

Editorial Office
MDPI AG
St. Alban-Anlage 66
Basel, Switzerland

This edition is a reprint of the Special Issue published online in the open access journal *Toxins* (ISSN 2072-6651) from 2016–2017 (available at: http://www.mdpi.com/journal/toxins/special_issues/snake-venom-metalloproteinases).

For citation purposes, cite each article independently as indicated on the article page online and as indicated below:

Author 1; Author 2; Author 3 etc. Article title. *Journal Name*. **Year**. Article number/page range.

Cover photo courtesy of Carlos Andrés Bravo Vega.

ISBN 978-3-03842-426-0 (Pbk)
ISBN 978-3-03842-427-7 (PDF)

Table of Contents

Section 2: Original Research

About the Guest Editors

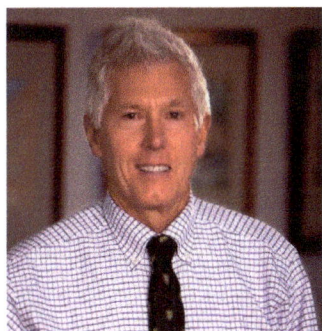

Jay William Fox graduated from Monmouth College with a B.A degree with an emphasis on biology, chemistry and philosophy. During the course of his undergraduate studies, he developed a keen interest in biochemistry and decided to pursue an advanced degree in that subject. He matriculated into the Biochemistry Ph.D. Program at Colorado State University. During his second year in the program, he joined the laboratory of Professor A. T. Tu. Fox's dear friend and future long-time collaborator, Professor Jon Bjarnason, was also a student in Tu's laboratory, which was focused on protein chemistry and protein structure and function. All of these features were a strong match with Professor Fox's interests. The topic of these areas of study in Tu's laboratory was snake venom toxins and Fox's first assignment as a graduate student in the laboratory was to travel to Asia to collect sea snake venoms from which he was to then isolate and characterize their major neurotoxins. After a successful collection of venom in Thailand and India Fox isolated the major neurotoxin from the sea snake *Hydrophis hardwickii* (*Lapemis Hardwickii*). He went on to characterize the toxin and determine its primary sequence. While in Tu's lab, Fox studied other snake toxins including myotoxins and hemorrhagic toxins. Upon graduation, Professor Fox joined the group of Dr. Marshall Elzinga at Brookhaven National Laboratory where he continued to develop his skills in applying advanced technologies to protein characterization and structural studies. From Brookhaven, he then joined the group of the noted peptide chemist Professor John Stewart at the University of Colorado School of Medicine under a NIH post-doctoral fellowship. While in Professor Stewart's lab, Fox worked on the synthesis of adrenocorticotropic hormone analogs as well as the synthesis of peptide analogs of snake neurotoxins. From Denver, Professor Fox accepted his first faculty position at the University of Virginia School of Medicine. In his early years, he undertook many different areas of study but the study of toxins was always in the mix. Over the years, Fox's group isolated and characterized a number of snake venom metalloproteinases (SVMPs), determined their primary structures and provided convincing biochemical evidence that these SVMPs could effectively cleave extracellular matrix to produce hemorrhage. Fox's group was also one of the first to clone and determine the cDNA sequences of the SVMPs as well as perform mass spectrometry-based proteomics on venoms. In addition to venom work, the Fox laboratory has been very active in the field of cancer biology, focusing on the microenvironment and host–tumor interactions. Over his time at the University of Virginia, Fox rose to the rank of Professor of Microbiology, Immunology and Cancer Biology and has served as the Assistant Dean and Associate Dean for Research at the School of Medicine as well as currently serving as the Director for Research Infrastructure and the Associate Director of the University of Virginia Cancer Center. He has over 200 publications in peer-reviewed journals and has been awarded several patents. Professor Fox has served on a number of editorial boards and NIH Study Sections as well as serving on external advisory committees to a variety of universities and organizations. He is on the Science Policy Committee of the Federation of American Societies for Experimental Biology and has served on FASEB's Executive Board. Professor Fox is a former president of the Association of Biomolecular Resource Facilities as well the recipient of the Outstanding Service Award from the association. Currently, Professor Fox serves as President of the International Society on Toxinology.

José María Gutiérrez (San José, Costa Rica, 1954) obtained a B.Sc. degree in Microbiology and Clinical Chemistry at the University of Costa Rica (1977) and a PhD in Physiological Sciences at Oklahoma State University, USA (1984). In 1975, he started working as a research assistant at Instituto Clodomiro Picado and, after 1977, became a researcher at this institute. Since 1984, he teaches at the School of Microbiology of the University of Costa Rica at graduate and undergraduate levels. Ha has taught courses in Immunology, Biochemistry, Cellular Pathology and Research Methods. His research interests have focused on the biochemical characterization of snake venoms and toxins, including proteomics studies, and on the mechanisms of action of toxins, particularly of snake venom metalloproteinases and phospholipases A_2 responsible for the local tissue damage induced by viperid venoms. He has been involved in the study of the preclinical neutralizing ability of antivenoms, and in the development of novel antivenoms for various regions of the world, including Latin America, Africa, Papua New Guinea and Sri Lanka. Gutiérrez has been interested in the history of science and in the social implications of science and technology, particularly in developing countries, and has been involved in extension programs to improve the prevention and management of snakebite envenomings. His research work has resulted in over 450 publications in specialized journals and books. He was Director of Instituto Clodomiro Picado and Head of the Research Division of this institute, and has been consultant to the World Health Organization in antivenoms. Gutiérrez coordinated a network of public laboratories in Latin America devoted to the manufacture and quality control of antivenoms, and is a member of the Board of Directors of the Global Snakebite Initiative. He has received a number of awards and recognitions, such as the National Award of Science (Costa Rica), the Sven Brohult Award (International Foundation for Science) and the Redi Award (International Society on Toxinology).

Preface to "Snake Venom Metalloproteinases"

A simple review of PubMed for "venom hemorrhage" shows that one of the first scientific indications that snake venom can produce hemorrhage appeared in *the Journal of Experimental Medicine* in 1909 where it was demonstrated that intravenous injection of *Crotalus atrox* venom in rabbits gives rise to glomerular lesions including the presence of hemorrhage and exudate in the kidney [1]. And even earlier, in 1894 and 1896, de Lacerda, and Mitchel and Reichert, respectively, had described macroscopic hemorrhagic lesions in animals after application of viperid venoms [2,3]. Similarly, a search using the terms "venom coagulopathy" identified a work from 1949 showing that snake venoms could alter the prothrombin time in normal blood [4]. Thus, from the earliest of formally published scientific efforts, there was a clear demonstration that some snake venoms could significantly influence the victims' normal blood coagulation as well as giving rise to hemorrhage. These pathologies began to be attributed to specific metalloproteinases in the snake venoms as early as the 1950s when investigators were becoming successful in isolating proteins from the venom which they characterized as metalloproteinases and their proteinase activity correlated to their pathological activities. One such example is evidenced by the work of Maeno and colleagues who isolated a hemorrhagic metalloproteinase, H⊕-proteinase, from *Trimeresurus flavoviridis* [5]. This work and many subsequent studies clearly highlighted the major role of snake venom metalloproteinases (SVMPs) in the pathophysiology of a snake bite [6,7].

In this book, we have attempted to provide the reader with a solid, scientific review of a number of the key functional and structural characteristics associated with SVMPs as well as to describe some aspects of this family of venom proteinases which are still not fully understood. We begin with a brief discussion about the field of SVMP investigation by ourselves (Understanding the Snake Venom Metalloproteinases: An Interview with Jay Fox and José María Gutiérrez), which serves as an important preface for students and seasoned investigators alike interested in the field. This is followed by a historical review, by Drs. Giebeler and Zigrino, of the family of proteins termed "A Disintegrin and Metalloproteinase" (ADAMs), that are orthologs of the SVMPs (A Disintegrin and Metalloproteinase (ADAM): Historical Overview of Their Functions). Dr. Takeda follows this with a comprehensive examination of the structural features of the SVMPs that support the manifold biological activities associated with the class of venom toxins (ADAMs and ADAMTs family proteins and snake venom metalloproteinase: a structural overview).

As venomic studies have well described, the family of SVMPs is quite diverse, both in terms of their size, structure and biological activities. In the chapter by Dr. Moura-da-Silva and colleagues (Processing of snake Venom Metalloproteinases: Generation of Toxin Diversity and Enzyme Inactivation), a rich description is provided on how post-translational processing of precursors of the SVMPs found in venoms contributes to the diversity of structure and function of the SVMP family.

The next section of the collection focuses more on the functional aspects of the SVMPs, notably the pathophysiologies of hemorrhage and coagulopathies, often observed in envenomations. An overarching discussion of the action of SVMPs on the extracellular matrix is provided by Dr. Gutierrez and colleagues (A Comprehensive View of the Structural and Functional Alternations of Extracellular Matrix by Snake Venom Metalloproteinases (SVMPs): Novel Perspectives on the Pathophysiology of Envenoming), where novel aspects of the action of these enzymes on matrix components are discussed. This is followed by a more in-depth look at hemorrhage induced by SVMPs in venom, through a historical review on how our understanding on the pathogenesis of this effect has developed over time (Hemorrhage Caused by Snake Venom Metalloproteinases: A Journey of Discovery and Understanding).

The next section by Drs. Kini and Koh discusses the role of SVMPs in coagulopathies commonly observed in envenomation by many snakes, as well as the structural features associated with the SVMPs that play a role in coagulopathy. This chapter provides a close examination of not only venom-induced coagulopathy itself but also on how SVMPs impact fibrinolysis and platelet aggregation, key components in blood coagulation (Metalloproteinases Affecting Blood Coagulation, Fibrinolysis and Platelet Aggregation from Snake Venoms: Definition and Nomenclature of Interaction Sites).

In the yin and yang of Toxinology, we must not only consider how natural toxins such as the SVMPs function but also the nature of their naturally occurring inhibitors. Dr. Bastos and colleagues provide an informative review of naturally occurring inhibitors to the SVMPs as well as insight into how they function to abrogate the activity of these toxins. Further discussion is presented as to how knowledge of these inhibitors can inform the field to promote the development of inhibitors of the SVMPs for therapeutic applications (Natural Inhibitors of Snake Venom Metalloendopeptidases: History and Current Challenges).

The review section of the book ends with the exciting observations that have been derived from snake genomics with the advent of a number of next-generation sequencing protocols and instruments. Dr. Kerkkamp and colleagues provide the reader with excellent examples of toxinological and evolutionary insights gained from snake genomic analyses and leave one with a keen sense of enthusiasm for a rejuvenation of the studies of SVMPs based on insights provided by genomic analyses of venomous snakes (Snake Genome Sequencing: Results and Future Prospects).

The next section of this book is comprised of novel research findings in the field of SVMPs, and highlights some of the current critical topics that engage investigators in Toxinology. This begins with a report by Dr. Camacho and colleagues on the isolation and characterization of a catalytically inactive PII SVMP. This work underscores how often venom proteins go unobserved due to a lack of appropriate tests, and suggests that enzymatically inactive SVMP homologues may be more frequent in venoms than it was previously thought (Novel Catalytically-Inactive PII Metalloproteinases from a Viperid Snake Venom with Substitutions in the Canonical Zinc-Binding Motif). This is followed by what may become a seminal work in understanding an in-direct role SVMPs may play in envenomation pathophysiology. In this work, the authors describe how wound exudate from a viperid envenomation can contribute to vascular permeability in part through a DAMPs/TLR-4 mediated pathway. This discovery could lead to a new understanding of venom-induced hemoconcentration, edema and a number of other systemic effects of envenomation (Viperid Envenomation Wound Exudate Contributes to Increased Vascular Permeability via a DAMPs/TLR-4 Mediated Pathway). The following work by Dr. Yee and colleagues describes the presence of natural peptide inhibitors of SVMPs found in the venom of the Myanmar Russell's viper using both standard protein purification techniques as well as venom transcriptomics (Snake Venom Metalloproteinases and Their Peptide Inhibitors from Myanmar Russell's Viper Venom). The final contribution to the section of novel studies describes a genomic sequencing study of *Echis ocellatus*. In this investigation Drs. Sanz and Calvete demonstrate how genomic sequencing can inform on venom toxin evolution based on the genomic organization observed from SVMP genes (Insights into the Evolution of a Snake Venom Multi-Gene Family from the Genomic Organization of *Echis ocellatus* SVMP Genes).

In summary, we hope that this collection of reviews and novel scientific reports will provide both students and established investigators in the field of Toxinology and beyond with a solid foundation and understanding of the field of SVMPs studies, and how these impact not only our realm of Toxinology but other scientific fields as well. Also, it is our hope that those who read this collection will generate new ideas and potential questions about the SVMPs and then seek to further the field with their own studies on this intriguing and fascinating family of venom proteins.

Jay W. Fox and José María Gutiérrez
Guest Editors

References:

[1] Pearce, R.M. An experimental glomerular lesion caused by venom (*Crotalus adamanteus*). *J. Exp. Med.* **1909**, *11*, 532–541.
[2] de Lacerda, J.B. *Leçons sur le Venin des Serpents du Brésil*; Lombaerts: Rio de Janeiro, Brasil, 1884.
[3] Mitchel, S.W.; Reichert, E.T. *Researches upon the Venoms of Poisonous Serpents*; Smithsonian Institution: Washington, DC, USA, 1886.
[4] Macht, D.I. Influence of snake venoms on prothrombin time of normal and hemophilic blood. *Fed. Proc.* **1949**, *5*, 69.

[5] Maeno, H.; Morimura, M.; Mitsuhashi, S.; Sawai, Y.; Okonogi, T. Studies on Habu Snake Venom. 2b. Further purification and enzymic and biological activities of Ha-proteinases. *Japan. J. Microbiol.* **1959**, *3*, 277–284.

[6] Fox, J.W.; Serrano, S.M.T. Structural considerations of the snake venom metalloproteinases, key members of the M12 reprolysin family of metalloproteinases. *Toxicon* **2005**, *45*, 969–985.

[7] Gutiérrez, J.M.; Rucavado, A. Snake venom metalloproteinases: their role in the pathogenesis of local tissue damage. *Biochimie* **2000**, *82*, 841–850.

toxins

MDPI

Editorial

Understanding the Snake Venom Metalloproteinases: An Interview with Jay Fox and José María Gutiérrez

Jay W. Fox [1] and José María Gutiérrez [2]

[1] Department of Microbiology, Immunology and Cancer Biology, University of Virginia School of Medicine, P.O. Box 800734, Charlottesville, VA 22908, USA; jwf8x@virginia.edu
[2] Instituto Clodomiro Picado, Facultad de Microbiología Universidad de Costa Rica, San José 11501-2060, Costa Rica; jose.gutierrez@ucr.ac.cr

Interview by Chao Xiao (Managing Editor, *Toxins* Editorial Office)
Received: 3 January 2017; Accepted: 11 January 2017; Published: 16 January 2017

Abstract: Jay W. Fox and José María Gutiérrez recently finished editing a Special Issue on the topic "Snake Venom Metalloproteinases" in *Toxins*. The Special Issue covers a wide range of topics, including the molecular evolution and structure of snake venom metalloproteinases (SVMPs), the mechanisms involved in the generation of diversity of SVMPs, the mechanism of action of SVMPs, and their role in the pathophysiology of envenomings, with implications for improving the therapy of envenomings. In this interview, we discussed with Jay W. Fox and José María Gutiérrez their research on the SVMPs and their perspectives on the future trends and challenges for studying snake venoms.

Jay Fox is a Professor of Microbiology, Immunology, and Cancer Biology, at the University of Virginia School of Medicine and an Associate Director of the UVA Cancer Center. Dr. Fox currently is engaged in research on carcinogenesis in women with dense breasts focusing on the interaction of stroma and breast epithelium. He is also interested in the secondary or indirect effects of viper envenomation focusing on the roles of venom and host generated damage-associated molecular pattern molecules (DAMPs) in the pathophysiology of snake bites. Dr. Fox directs the Office of Research Core Administration and oversees the operation of 15-shared resource core facilities employing approximately 60 faculty and staff. Dr. Fox teaches courses to both medical and graduate students on cancer biology and also teaches a course on research ethics. He has served as the President of the Association of Biomolecular Resource Facilities, was a member of the Federation of Associations of Experimental Biology, and is currently serving as the President of the International Society on Toxinology. Dr. Fox participated on numerous NIH Study Panels and sits on the External Advisory Committees for two National Cancer Institute Designated Cancer Centers. Outside of work, he enjoys being a Scoutmaster for Troop 37 in Charlottesville, Virginia and sailing and oyster ranching at his home on the Chesapeake Bay (Figure 1).

José María Gutiérrez is a Professor at the University of Costa Rica, where he performs research on snake venoms and antivenoms at the Instituto Clodomiro Picado and teaches Immunology, Research Methods, Cellular Pathology, and Biochemistry at the School of Microbiology. Dr. Gutiérrez's main research interests are related to the composition and mechanism of action of snake venom toxins, particularly regarding metalloproteinases and phospholipases A_2 responsible for the drastic local tissue damage characteristic of viperid snakebite envenomings. Dr. Gutiérrez is also involved in the development of novel antivenoms for various regions of the world and in the preclinical evaluation of antivenom efficacy, as well as in the search for novel inhibitory compounds that could be used to treat envenomings. Dr. Gutiérrez is interested in public health aspects of snakebite envenoming as well, and participates in extension programs aimed at improving the prevention and management of snakebites in Costa Rica and abroad. For his contributions, Dr. Gutiérrez has received several national and international awards, including the Redi Award (2015) of the International Society on Toxinology (Figure 1).

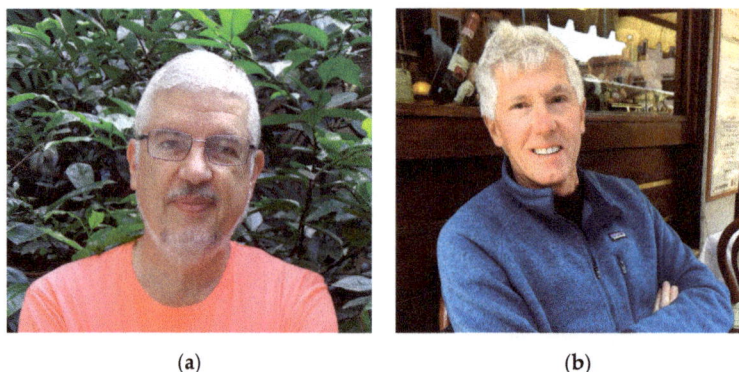

(a) (b)

Figure 1. (a) José María Gutiérrez; (b) Jay W. Fox.

Q. When did you first become interested in snake venoms and how did you get involved in research on this topic?

Jay Fox: I began my academic life as an undergraduate student at Monmouth College studying biology, chemistry, and philosophy. While I had no idea at the time what career I wanted to pursue, I knew I loved science. As I neared the end of college I took a deep interest in organic and biochemistry and my advisor Dr. John Kettering suggested I consider graduate school. None of my family had progressed beyond a bachelor's degree so I had no idea what this entailed but since I had no other plans ... why not? I was accepted at Colorado State University and matriculated without concern mainly based on total ignorance of just what earning a Ph.D. would entail or for that matter what it would ultimately prepare me for in terms of a career. As a first year graduate student studying biochemistry at Colorado State University, a more senior student, Jon Bjarnason, who was in Professor Anthony Tu's laboratory, befriended me. Jon was studying snake venoms and told me how interesting it was trying to isolate toxins and understand their mechanism of action. Dr. Tu's laboratory was very well equipped and I met with him and we discussed possible projects. One was to isolate sea snake neurotoxins and in order to collect the venom I needed, he would send me to Asia. That sounded very exciting given that I had never travelled much so I signed on with Dr. Tu. As it turned out it was an excellent decision in that I not only learned about venoms and toxins, but I also received an excellent education on protein chemistry and protein structure and function which has served me well throughout my career regardless of what biomedical subject I am investigating.

After graduating I did post-doctoral work first in the laboratory of Dr. Marshall Elzinga at Brookhaven National Laboratory, Upton, N.Y. and then with Professor John Stewart at the University of Colorado Medical School, Denver, Colorado. Dr. Elzinga was a superb protein chemist who was a leader in sequencing large muscle proteins. He had just moved to Brookhaven when I arrived and together we set up his spinning cup Edman sequencer. To identify the amino acids from the sequence we did a combination of thin layer chromatography and back hydrolysis of the phenylthiohydantoin (PTH)-amino acids using a home built amino acid analyzer Dr. Elzinga acquired from Dr. Stein at the Rockefeller University. While I was there we also began using the new technique of HPLC to analyze the PTH amino acids. Brookhaven at the time was a focal point of outstanding protein studies and it was a wonderful experience and I met many leaders in the field of protein characterization. At Colorado, working with Professor Stewart was also an honor and privilege for me. John had previously worked with Nobel Prize winner, Professor Bruce Merrifield, developing an automated peptide synthesizer. John knew all about peptide synthesis, synthesizers and peptide isolation. John's book on the subject was in every synthesis lab at the time and always well-worn with use. My project as an NIH fellow in John's lab was to synthesize novel ACTH analogs looking for novel activities. John was a very generous scientist giving me time to work on my own ideas as well. One project I conducted was to synthesize the active site loop of a sea snake neurotoxin to determine if the peptide could recapitulate some of the neurotoxin's activities. The project was successful as we made an active peptide, but I never published it as my formal project with John took precedent with my time. Hence, as I say later in this piece, if you do not publish your work it did not happen . . . at least as far as the rest of the world is concerned.

While finishing up my studies in Denver I had two job offers, one at the University of Virginia and one at the Coors Brewery in Golden, Colorado. I was tormented with this choice; to go into academics or stay in Colorado, a place I loved for its hiking and backpacking. I chose Coors, but on the day I was to show for work, I had a change of heart and called Virginia and told them I would be there in two weeks. Over the first few years I often wondered if I had made the correct decision. Now nearly four decades later, it was clearly the best decision of my life.

I arrived at the University of Virginia School of Medicine as an Assistant Professor of Microbiology with the charge to bring modern protein chemistry technology to the school. This I did by starting a protein-sequencing core, then a peptide synthesis core, and eventually a DNA sequencing core. These activities solidified in my mind the value of utilizing cutting edge technology applied to whatever area you may be studying. Simultaneously I pursued my interests in toxinology, beginning with isolating a number of snake venom metalloproteinases, characterizing them and ultimately determining their protein and cDNA sequences, which were some of the first ever published. Over the intervening years I have strived to always do something novel in the field following the admonition of my sabbatical host at the Max-Planck Institute for Biochemistry in Munchen, Dr. Rupert Timpl, who always said with regards to a project "You must do it first or do it a lot better; generally it is easier to do it first". Something I always tell my students as well when thinking about projects.

J.M. Gutiérrez: I started working as an undergraduate research assistant at the Instituto Clodomiro Picado in 1975, under the supervision of Róger Bolaños, the founder and first director of this institute. As a mentor, Dr. Bolaños instilled in me the vision that research can be done with passion and joy, and also the belief in the relevance of the social implications of scientific work, in this case in relation to snakebite envenomings and the human suffering they inflict. This went hand in hand with my own social and political beliefs. My first research projects had to do with the study of karyotypes of venomous snakes, but rapidly I became interested in venoms and antivenoms. At that time relatively little was known on the pathogenesis of the local tissue damage induced by viperid snake venoms, a very important aspect of snakebite envenoming since it may lead to permanent tissue damage and other sequelae in the victims. Initially, I studied the local pathology induced by Costa Rican snake venoms in mice by using light microscopic techniques. In 1980, I had the opportunity to perform my PhD studies under the supervision of Prof. Charlotte L. Ownby at Oklahoma State University, with the support of a scholarship provided by the University of Costa Rica. Charlotte made significant contributions to the study of venom-induced pathology. In her laboratory we were able to isolate and characterize a myotoxic phospholipase A2 from the venom of *Bothrops asper*, the most important snake in Central America. In addition, we studied the action of this toxin in muscle tissue, by using electron microscopy and other techniques, and proposed a mechanism of action for myotoxic phospholipases A2. In 1984 I returned to Costa Rica and continued my research at the Instituto Clodomiro Picado, in collaboration with a highly qualified group of Costa Rican colleagues and international collaborators with whom I have worked for over 40 years. The philosophy of cooperation and partnership that has characterized the work of Instituto Clodomiro Picado, and its relationships with groups in our own country, and in countries of Latin America, North America, Europe, Asia, Africa, and Oceania, has allowed our team to contribute to toxinological research and antivenom development.

Q. Can you describe your research group's current work? How has it changed over the past ten years, and where do you see it going in the future? Is there an area of the field that you are particularly excited about at the moment?

Jay Fox: Currently my laboratory focuses on the role of damage-associated molecular pattern molecules (DAMPs) on snake envenomation pathophysiology. We work closely with José's group in Costa Rica. This area, which we discovered, is going to play an important role in understanding the non-lethal aspects of snake envenomation that are associated with envenomation morbidity. My lab is also working on the role of stroma in carcinogenesis in dense breast tissue. Ironically, there are some features of these areas of research which intersect, such as the role of stroma in envenomation and in carcinogenesis and tumor invasion. While there is certainly a lot yet to discover regarding toxins and their activities, one must admit that much has been learned over the recent past by virtue of the explosion of proteomic, transcriptomic, and now genomic studies on snake venoms and snakes themselves. As we have written, in the end it is ultimately a systems biology issue in terms of how all the toxins in the venom collectively give rise to the effects observed in the host as well as how the biology of the snake and its environment also impinges on what effects the venom may cause. So, for me, the future will be in discovering what previously unknown activities some toxins may have and how all the toxins work together under the biological systems of the snake to give rise to the observed pathophysiology in envenomated hosts.

J.M. Gutiérrez: I participate with several research groups at the Instituto Clodomiro Picado, since we have a cooperative and integrative philosophy of doing research. Specifically on the subject of snake venom metalloproteinases (SVMPs), I work with Alexandra Rucavado, Teresa Escalante, Erika Camacho, and Cristina Herrera, in addition to several graduate and undergraduate students. For many years, we have also collaborated with a number of research groups from other countries; in this particular subject of SVMPs we have had fruitful collaborations with the groups of Jay W. Fox (University of Virginia, USA), Michael Ovadia (University of Tel Aviv, Israel), Catarina F.P. Teixeira and Ana M. Moura-da-Silva (Instituto Butantan, Brazil), and Juan J. Calvete (Instituto de Biomedicina de Valencia, Spain), among other groups. Since the early 1990s our main goal on the topic of SVMPs has been to understand how these toxins induce hemorrhage, one of the main manifestations of viperid snakebite envenomings. Initially we isolated and characterized a number of SVMPs, and studied their action using transmission electron microscopy and other microscopic approaches. Then, we investigated the action of hemorrhagic and non-hemorrhagic SVMPs on the basement membrane of capillary blood vessels, by combining histology, ultrastructure, immunohistochemistry, and immunoblotting. More recently, and in a close collaboration with J. W. Fox, we have introduced the proteomics analysis of exudates collected in the vicinity of SVMP-damaged tissue as a tool to have a deeper view of the pathological alterations occurring in the tissue. As an outcome of these investigations, a model for the mechanism of action of hemorrhagic SVMPs has been proposed, based on the cleavage of structurally-relevant basement membrane components, especially type IV collagen, followed by the mechanical disruption of vessels due to hemodynamic biophysical forces operating in the circulation. In the near future we are interested in the identification of the regions in the molecular structure of SVMPs that determine their ability to bind to microvessels, as well as of the cleavage sites of basement membrane proteins that determine the disruption of capillary blood vessels. Moreover, our more recent studies in collaboration with Jay Fox indicate that fragments of extracellular matrix proteins and other types of proteins released in the tissues as a consequence of SVMP action may contribute to tissue alterations and may play roles in the processes of repair and regeneration, a hitherto unknown subject which may bring novel clues for understanding the pathogenesis of venom-induced tissue damage. An additional challenge for our future studies is to understand the role of SVMPs in envenomings from an integrative perspective, i.e., in the light of the overall picture of envenoming, which involves studying the synergistic actions of SVMPs and other venom components, a poorly studied aspect of envenomings.

Q. It is well-known that snake venom metalloproteinases (SVMPs) are the primary factors responsible for hemorrhage. How does an improved understanding of actions of the SVMPs advance our understanding of snakebite envenoming? Do you think sufficient research has been done for the SVMPs? Are there any aspects that need further exploration?

Jay Fox: When Dr. Solange Serrano and I coined the name and classifications of the SVMPs, it an incredibly exciting time in the field. For a long period when only limited sequence data was known for the SVMPs (hemorrhagic and non-hemorrhagic alike), Dr. Jon Bjarnason, my close colleague, and I were leaning toward SVMPs being members of the matrix metalloproteinase family and we were pushing for recognition of this classification. However, Dr. Hideaki Nagase told us, based on his studies, these SVMPs were not matrix metalloproteinases (MMPs), and as it turns out, he was right! Also at this time I had the good fortune to hear a talk by Dr. Judith White on a new class of fertility related proteins found on sperm that seemed to have proteinase activity. When she shared some preliminary sequence data on these proteins with me, I had an epiphany; they had sequences similar to the SVMPs! We were no longer alone; there were normal orthologs to these toxins, something which is now very well known about all venom proteins. The class of proteins Judy was working on became known as A disintegrin like and metalloproteinase proteins (ADAMs) and the study of ADAMs and SVMPs exploded hand in hand. SVMPs were often referred to for insights into the structure and function of ADAMs and this is now a paradigm of toxinology in that the study of toxins often provides outstanding insights into the function of their normal orthologs and comparison of the toxins with the orthologs can also yield insights into how small changes in structure and/or location can cause a somewhat "normal" protein to become toxic.

There are still some lingering aspects of SVMPs that deserve study. Certainly their evolution and gene structure merit additional investigation. Also, higher resolution understanding of substrate specificity as associated with functional activity is still of some interest, and as mentioned above, how these data may inform on the ADAMs' functions is also of value.

J.M. Gutiérrez: Many years ago, through the pioneering work of A. Ohsaka, A.T. Tu, J.B. Bjarnason, J.W. Fox, and F.R. Mandelbaum, among other researchers, it was demonstrated that SVMPs were responsible for the hemorrhagic activity of snake venoms. Local and systemic hemorrhages are key aspects of viperid snakebite envenomings, since they are associated with local tissue damage and with systemic bleeding leading to cerebrovascular accident and hemodynamic perturbations. In addition, SVMPs also participate in the alterations of blood coagulation in these envenomings, and release pharmacologically-active components from proteins of the extracellular matrix, which are likely to contribute to tissue damage and repair. Thus, SVMPs are at the center of the pathophysiology of viperid snakebite envenomings. The study of the mechanisms of action of SVMPs has provided valuable clues to our understanding of the overall pathophysiology of these envenomings. Even though there has been a great deal of research on SVMPs, there are still pending issues for understanding their molecular evolution, regulation of expression, mechanisms for generating the great diversity of these enzymes in venoms, mechanisms of action, structural determinants of toxicity, and their overall role in the pathophysiology of envenomings, as well as their ecological role, i.e., their action in natural prey. Likewise, the search for novel and more potent SVMP inhibitors is an area of research that might offer novel therapeutic alternatives. These aspects require renewed investigation in the near future, hopefully through interdisciplinary approaches that will pave the way for more integrative and holistic perspectives of this fascinating group of venom components.

Q. Looking back at research on snake venoms, what do you think have been the most important milestones? What are the most important techniques that will contribute to research on snake venoms in the future?

Jay Fox: The key discoveries that have led us to our current understanding, in my opinion, are in this order:

(a)   Venoms are comprised of toxic and non-toxic components (many of which are proteins and peptides).

(b)   In many venoms, the activity of toxins is dependent on enzymatic activities directed at specific substrates.

(c)   In the case of the SVMPs, the hemorrhagic SVMPs were demonstrated to disrupt the basement membranes of capillaries allowing the extravasation of contents into the stroma. This finding was critical in that it focused our research on the proteolysis of matrix by SVMPs and how this is related to the observed pathology (a topic we continue to study by our current focus on the roles of DAMPs in snake envenomation).

(d)   Venoms are somewhat complex and this is demonstrated by proteomics and transcriptomics.

(e)   The toxic effects of venom are due to the collective action of toxins and non-toxic components in the venom and to understand envenomation one must ultimately take a systems approach.

(f)   Development of anti-venom agents must always consider the venom as a system; not a simple collection of unilaterally acting toxins.

(g)   By discovering the activities of toxins we will better understand the function and structure of normal orthologs, and by understanding the targets and the nature of the toxin mechanisms, the possibility of developing novel drugs becomes a reality.

Certainly, the techniques of omics will continue to be important in venom studies as well as advanced microscopy; however, I would also suggest that modern physiology and molecular physiology will play an increasingly important role if we are to fully understand the function of individual toxins as well as venoms.

J.M. Gutiérrez: Over the last decades there have been a number of relevant achievements in our understanding of snake venoms, their molecular evolution and composition, mechanism of action and identification of their targets in tissues and blood, and ecological and medical roles. Likewise, the understanding of the structure-function of many snake toxins has paved the way, as has the study of other animal toxins, to the discovery of promising leads which may derive into new drugs for the treatment of a variety of diseases. It is the combination and integration of molecular evolution and structural studies, in parallel with the characterization of venom proteomes and the understanding of the structure-function relationship, as well as the mechanism of action of many toxins, that I consider major achievements in the field of toxinology.

Many techniques will impact on the future of toxinology. The growing impact of the 'omics', with the power of bioinformatics, is generating and will continue to provide vast amounts of information on snake venoms, but this has to be linked to the study of the mechanisms of action of venom toxins from both ecological and medical perspectives. It is the integration of such vast volumes of information, from a Systems Biology perspective, that will bring a more complete understanding of SVMPs and venoms in general from the evolutionary, ecological, and medical perspectives. The ongoing work on the genomes of several snakes is an exciting development, together with the completion of novel venom proteomes and venom gland transcriptomes, the structural characterization of novel venom's toxins, and the understanding of tissue targets and mechanisms of action, with the aid of many techniques, including recent developments in imaging.

Q. Are there any challenges for future research on snake venoms?

Jay Fox: There are several challenges to snake venom research facing us. First, the easy work has been done. What is left requires very clever, insightful thinking and a judicious use of resources and application of instrumentation/technology. Simply isolating additional isoforms of venom components is of little impact unless one has very clear and clever ideas about novel activities for these isoforms and ways to test for them. Also, there is the risk that we lose sight of why we are doing this research. We must always remember it is to advance science and as such we should be improving the human condition. Whether we are doing basic, translational, or clinical research we must always be able to connect what we do, in a coherent fashion, to improving the human condition. And, we must be able to explain this to the lay public. If we cannot do that we will just become a group of disaffected hobbyists without impact or interest to the wider world.

J.M. Gutiérrez: The main challenge, in my view, is how to integrate a massive volume of data that is being generated on snake venoms in order to provide understanding in addition to information. A big challenge is to develop inter- and transdisciplinary research approaches aimed at generating deeper insights on the evolutionary, ecological, and medical aspects of snake venoms. The field of toxinology should move in the future from a predominantly 'reductionist' approach to a more 'holistic' perspective of snake venoms. Another big challenge for the future is to reduce the gaps between basic toxinological research and the clinical and antivenom manufacturing areas, i.e., to generate more translational research in toxinology. The cross-talk between basic toxinologists and clinical toxinologists and with antivenom developers is of paramount relevance in order to bring the scientific advances to the improvement of the treatment of envenomed patients.

Q. How do you anticipate research on snake venoms will progress over the coming years?

Jay Fox: I believe there will be more drugs developed based on research stemming from venoms by virtue of new target discovery as well as new activities found for some toxins. I also believe we will become much more efficient and effective in developing antivenoms (by virtue of the work of my colleagues in Jose's group and others) based on a scientific understanding of venoms and venom action. Also, with the closer association of a number of regional societies with interests in toxinology with the International Society on Toxinology, I believe all the science in the arena will be enhanced with an improved outcome for those who suffer from envenomation both in terms of health as well as financial burdens.

J.M. Gutiérrez: I foresee advances in the following areas:

(a) Vast volumes of new information on snake genomes, venom gland transcriptomes, and venom proteomes, all of which will enlighten our comprehension of snake venom evolution.

(b) A deeper understanding on the structure-function of venom components, particularly in those playing a key role in toxicity. Additionally, identification of the key toxic components in medically-relevant venoms, with the consequent impact in the design and evaluation of antivenoms.

(c) Significant progress in our understanding of the mechanisms of action of toxins and identification of their targets, with implications both for clinical management of envenomings and for drug design.

(d) Deeper understanding of the ecological role of venoms, i.e., promotion of stronger links between toxinology and natural history.

(e) Exploring poorly studied venoms, such as those of 'colubrid' snakes (*sensu lato*) and other taxa, with the aim of discovering new types of toxins and mechanisms of action. Such novel toxins might become useful tools to understand basic physiological processes and as sources of new lead compounds for drug design.

(f) Harnessing the vast volume of toxinological information for improving the design and development of antivenoms, either animal-derived products or novel antivenoms based on recombinant antibody technology. Additionally, the discovery and development of molecules with strong inhibitory action that could complement the action of antivenoms in the therapy of envenoming.

(g) Developing the field of Public Health within the area of snakebite envenoming in order to understand the role of snake venoms in disease from a broader perspective.

Q. For the Special Issue "Snake Venom Metalloproteinases" that you edited for *Toxins*, what new information do you think it has brought forward and what gaps in the literature has it filled? How will our current research on SVMPs shed light on the future snakebite envenoming therapy?

Jay Fox: This Special Issue represents an outstanding mix of comprehensive reviews, where many such gaps have been addressed, as well as new research opening novel areas of scientific pursuit in toxinology, such as the work on DAMPs by Rucavado and colleagues and concepts on serine proteinases by Kini and colleagues. I personally would recommend this Special Issue to all new students to the field and to those more senior who wish to understand what is the current state of thinking in the field.

I am certain a careful reading of this Special Issue will sharpen our understanding of snake envenomation and it is my sincere hope that this will be translated to new science and new modes of therapy for both snake envenomation as well as possible new drug leads for other diseases.

J.M. Gutiérrez: This Special Issue includes highly relevant contributions that present the state of the art in SVMP research. The principal aim of this Special Issue is to provide a synthesis of our current knowledge of this fascinating field of toxinology. In addition, novel findings are presented which bring new concepts to the field of SVMPs. The readers will find in these contributions the key issues behind our current knowledge on SVMPs, with highly suggestive views on several unsolved aspects in SVMP research. The papers cover a wide range of topics, from molecular evolution and structure to the mechanisms involved in the generation of SVMP diversity, from the mechanism of action of SVMPs to their role in the pathophysiology of envenomings, with implications for improving the therapy of envenomings. Any person interested in SVMP will find in this Special Issue a collection of papers that summarize the current knowledge on these types of enzymes, and at the same time presenting the most relevant open questions in this topic, as a stimulus for future research efforts.

The understanding of SVMP variability and structure, as well as mechanisms of action and targets, will undoubtedly provide highly useful information for improving the therapy of snakebite envenoming, particularly in the case of species, such as the majority of viperid snakes, whose pathophysiological activities largely depend on the action of SVMPs. Understanding the structural determinants of toxicity, the targets of these toxins in the tissues, and their toxicokinetic profiles will lead to renewed knowledge-based antivenom design, as well as to the development of novel recombinant antivenoms and new inhibitors of high efficacy that could be used as a first aid in the field as a complement to antivenom therapy.

Q. What advice do you have for a young scientist who wants to start a career in snake venom research?

Jay Fox: Frankly, I must give much of the same advice I was given by my mentor Professor Tu. Some things do not change. Regardless of what you are investigating, snake venoms or whatever, you must always be thinking not only about how your work will impact this field but how it can have impact beyond your field; and ultimately how it can affect the human condition. The critical thinking, the sophisticated tools, experimental design, and data analysis involved in high quality venom research is no different than that required in all fields. It is applicable in all domains. Whatever you may be researching in the field of toxinology, always do so looking over the horizon to see how what you are learning can be applied to other fields and what scientists in other fields are doing that you can co-opt for your use in your studies. Science in a vacuum is irrelevant. You must present your work; you must publish your work. Someone spent their hard-earned money trusting you to do good work for the good of the world. Do not waste their money and do not waste your time. Make it count. Also, collective science working with a variety of colleagues often gives rise to the best science and likely is a lot more enjoyable. Science is too complex now to do it all alone.

Finally, have fun and if should you lose interest in a field, then move to another; and should you lose interest in science, there are many other admirable things to do. Find your passion and follow it.

J.M. Gutiérrez: There are many points of advice I would give to young people interested in snake venom research. Some of them are:

(a) Have a broad knowledge and understanding of the field of snake venoms. In addition to specializing in a particular type of toxin or venom, understand snake venoms from a wide perspective, i.e., their biological relevance, evolution, variation, mechanisms of action, and medical significance. Such broad views will help you place your particular questions in a larger landscape and will allow you to generate more relevant questions for your research. For this you need to follow the scientific literature in the subject on a regular basis and with discipline.

(b) Find topics for your research which are, at the same time, exciting for you and novel. Avoid the 'me too' type of research which may bring abundant data but few new ideas, and try to find unexplored or poorly explored niches where your contributions will be more meaningful. For this, it is necessary to have an ample knowledge on snake venoms (see item (a) above).

(c) Independently of your specific research interests, cultivate your knowledge in general biological and biomedical topics such as evolutionary biology, biochemistry, bioinformatics, structural biology, cellular biology, and physiology, as well as pathophysiology, among others. Having a background in these general subjects will allow you to place your toxinological research interests in a broad scenario which, in turn, will more likely provide you with new ideas.

(d) As much as possible, try to think 'out of the box'. This will give you the opportunity to generate new ideas and concepts, and new experimental tools and models. The collection of experimental data should be guided and complemented by the generation of new ideas and hypotheses. When possible, try to get out of the predominant paradigms and take the risk of generating new ways of looking at things, regardless of the natural resistance that this may provoke. After all, scientific work is, to a large extent, about generating novel ways to view the topics that are being investigated.

(e) Develop a cooperative philosophy for doing scientific research. Be part of interdisciplinary teams and groups in order to approach questions of wide interest from your own area of expertise. Be humble and accept that you need the contribution of other scientists for developing your own research projects. Be in contact with colleagues and actively discuss subjects of common interest with them. Procure the people that can help you and, at the same time, be generous with your colleagues and students by sharing your knowledge. Keep in mind that scientific research is a collective undertaking.

(f) Finally, in addition to being a dedicated scientist, strive to become also a person interested in general issues related to society and culture as a whole. Scientists have a huge social responsibility, and toxinologists are no exception. Develop a compassionate and generous way of doing science and in shaping your relation with societal issues in general. Introduce and follow ethical considerations when doing scientific research and in your life in general.

Conflicts of Interest: J.M. Gutiérrez and Jay W. Fox declare that they have no conflicts of interest regarding the content of this interview.

Section 1:
Reviews

Review

A Disintegrin and Metalloprotease (ADAM): Historical Overview of Their Functions

Nives Giebeler and Paola Zigrino *

Department of Dermatology and Venerology, University of Cologne, Cologne 50937, Germany;
nives.giebeler@uk-koeln.de
* Correspondence: paola.zigrino@uni-koeln.de; Tel.: +49-221-478-97443

Academic Editors: Jay Fox and José María Gutiérrez
Received: 14 March 2016; Accepted: 19 April 2016; Published: 23 April 2016

Abstract: Since the discovery of the first disintegrin protein from snake venom and the following identification of a mammalian membrane-anchored metalloprotease-disintegrin implicated in fertilization, almost three decades of studies have identified additional members of these families and several biochemical mechanisms regulating their expression and activity in the cell. Most importantly, new *in vivo* functions have been recognized for these proteins including cell partitioning during development, modulation of inflammatory reactions, and development of cancers. In this review, we will overview the a disintegrin and metalloprotease (ADAM) family of proteases highlighting some of the major research achievements in the analysis of ADAMs' function that have underscored the importance of these proteins in physiological and pathological processes over the years.

Keywords: ADAM; disintegrin; SVMP

1. Some Generalities

ADAMs (a disintegrin and metalloproteinases), originally also known as MDC proteins (metalloproteinase/disintegrin/cysteine-rich), belong to the Metzincins superfamily of metalloproteases. The timeline of key events in ADAM research is shown in Figure 1.

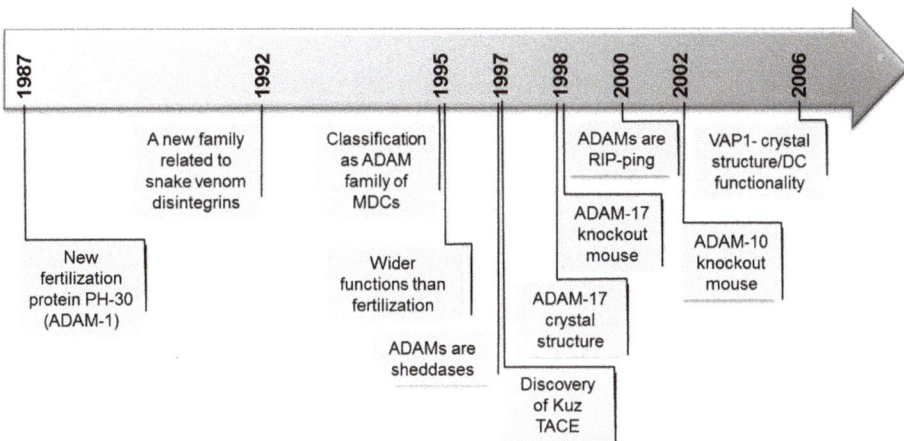

Figure 1. Timeline of key events in ADAMs research. The dates correspond to the publication year of the first article related to the event.

Together with snake venom metalloproteases (SVMPs), ADAM and ADAMTS (a disintegrin and metalloproteinases with thrombospondin motif) proteins form the M12 (MEROP database; https://merops.sanger.ac.uk/) Adamalysin subfamily of metallopeptidases. ADAMs have been detected in various species, from *Ciona intestinalis* to mice and humans [1]. Phylogenetic and molecular evolutions studies on these proteins have identified several gene duplications followed by pseudogene formation and/or positive selection of those genes mostly related to reproduction thus ensuring survival of the species [2]. Duplications and speciation have probably contributed to the divergence of SVMP and ADAM from the common ancestor gene [3]. Similarities in domain organization and sequences exist between the ADAMs and the P-III SVMPs [4]. Both protein families contain a pro-domain, a metalloproteinase and a disintegrin domain, and a cysteine domain. The latter in ADAMs has cell adhesive and fusogenic potential. ADAMs also contain an EGF-like repeat, a transmembrane domain, and a cytoplasmic tail (Figure 2). In addition, the tails of some ADAMs have intrinsic signaling activity and regulate proteolysis [5]. Alternative splicing of ADAMs produces proteins with different localization and activity [6].

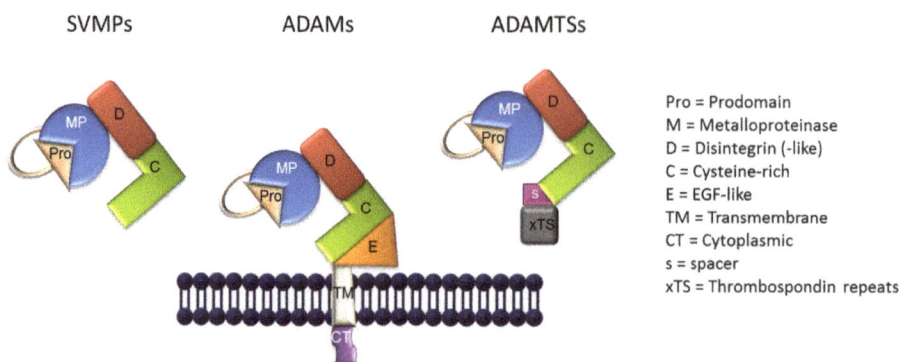

Figure 2. General structure of ADAMs (a disintegrin and metalloproteinases), ADAMTSs (a disintegrin and metalloproteinases with thrombospondin motif), and SVMPs (snake venom metalloproteases).

The label "disintegrin" was initially given to describe snake venom cysteine-rich, RGD-containing proteins able to adhere to integrins and inhibit platelet aggregation and cause hemorrhage in snake bite victims [7]. Similar to SVMPs, ADAMs adhere to integrins even though their binding sequence mostly contains an aspartic acid-containing sequence ECD (or xCD sequence) instead of the typical RGD amino acid sequence, except for human ADAM-15 (Figure 3). For this reason these domains are referred to as "disintegrin-like" domains [8]. Structural analysis of resolved structures of the ADAM and SVMP domains has been extensively reviewed elsewhere [9,10]. Only half of the known ADAMs contain a catalytic-Zn binding signature for metalloproteases (HExGHxxGxxHD) in their metalloprotease domain and can potentially be catalytically active. Those mammalian ADAMs that are catalytically active use a cysteine-switch mechanism to maintain enzyme latency [11]. Interestingly, the pro-domain not only is implicated in this process, but is also important for the correct protein folding and intracellular transport through the secretory pathway as shown for example for ADAM-9, -12, and -17 [12–14].

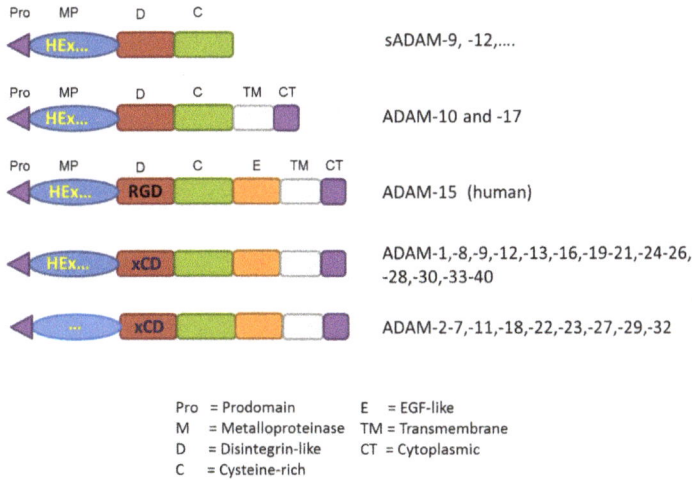

Figure 3. Protein domain structure comparison between ADAMs. Metalloproteinase domains with consensus sequence HEx (HExGHxxGxxHD, HEx ...) are predicted to be proteolytic active (" ... ", lack of a consensus sequence). Only human ADAM-15 contains a RGD amino acid sequence, all other ADAMs contain a conserved consensus binding motif xCD in their disintegrin-like domains. Soluble ADAMs, sADAM.

2. ADAMs Functions

2.1. ADAMs and Cell Adhesion

The initial studies on ADAMs focused on the function of the disintegrin domain in cell-cell and cell-matrix interactions. As for the SVMP, the disintegrin-like domain of ADAMs was also supposed to bind to integrins on the cell surface. Similarly to SVMP, human, but not mouse ADAM-15 contains an integrin binding motif RGD in the disintegrin domain. However, human ADAM-15 can also bind integrin in a RGD-independent manner as shown for the binding of $\alpha9\beta1$ integrin [15]. An extensive review on ADAM-15 structural and functional characteristics has been provided by Lu and colleagues [16]. Several *in vitro* studies using recombinant domains, mutation studies, or peptide sequences supported this role for a variety of ADAMs and suggested their function as cellular counter receptors [17,18]. This ADAMs-integrin interaction was not receptor-specific as each ADAM could interact with several integrins and may depend on the cell type.

Additional studies suggested that interactions with substrates and integrin receptors are not mediated solely by the disintegrin domain, but additional sequences outside this domain are involved in binding. In this respect, Takeda and colleagues analyzed the crystal structure of VAP1 (vascular apoptosis-inducing protein-1, a P-III SVMP), which has a conserved MDC structure, and identified the disintegrin-loop of this protein packed inside the C-shaped MDC architecture, which is therefore not available for binding [19]. Interestingly, these authors identified another sequence, the highly variable region (HVR) of the cysteine rich domain as a potential protein-protein adhesive interface [19]. It is therefore possible that the cysteine-rich domain drives adhesion or complements the binding capacity of the disintegrin domain possibly conferring specificity to the mediated interactions. The disintegrin and the cysteine-rich domains can also mediate cellular interactions via binding not only to integrins, but also to heparansulfate proteoglycans, such as syndecans [20]. Despite all these studies, the question remained whether the ADAM–integrin interactions identified *in vitro* could be relevant *in vivo*.

Few studies analyzed the crystal structure of ADAM proteases (reviewed in [9,10]). An example is the analysis of the crystal structure of the mature ADAM-22 ectodomain, one of the catalytically inactive ADAMs. In this analysis, ADAM-22 displays a four-leafed clover structure of four domains, the MDCE (metalloproteinase/disintegrin-like/cysteine-rich/EGF-like), without the pro-domain. According to this, the authors proposed that ADAMs function is modulated by dynamic structural changes in M domain and DCE domains that allow for the opening and closing of the protein configuration [21]. Electron microscopy studies on the full-length ADAM-12-s showed a similar four-leafed structure. In this structure, as for ADAM-22, the pro-domain is an integral domain of mature ADAM12 non-covalently associated to M domain [22].

Analysis of the crystal structure of ADAM-17 metalloproteinase domain bound to a hydroxamic acid inhibitor highlighted some unique features of the active site distinct from matrix metalloproteinases that may contribute to substrate specificity of TACE (TNF alpha converting enzyme) [23].

However, the substrate specificity depends on the entire MDC structure. Guan and colleagues [24] recently performed crystallographic analysis of two new SVMPs: atragin and kaouthiagin-like (K-like). This study pointed out that the MDC structure of atragin is C-shaped whereas that of K-like has an I-shaped structure depending on the disulfide bond patterns present in the D domain of both enzymes. Thus, the D domain would be important in orientating the M and C domains for their function and substrate specificity [24].

The physiological relevance of the interaction of ADAMs with integrins was clearly shown for the fertilization process. The first studies were done on ADAM-1, previously known as PH-30 (found on guinea pig sperm surface) and later as fertilin alpha (found in mouse and monkey), and on ADAM-2, or fertilin beta [25–27]. Interestingly, both proteins lose their metalloproteinase domain during the maturation process and are present on the cell surface exposing directly their disintegrin domain [28]. Initially, the disintegrin-like domain of ADAM-2 was shown to bind to the integrin $\alpha 6 \beta 1$ on the egg plasma membrane, and that this binding was required for membrane fusion [29]. Additional ADAMs were suggested to be involved in the egg-sperm membrane fusion fertilization process based on the fact that the first ADAMs discovered (ADAM 1–6) were found to be expressed in mammalian male reproductive organs such as testis and epididymis (reviewed by Cho [30]). However, although the first discovered members of the ADAM family, ADAM-1, ADAM-2, and -3 are critical in mediating fertilization processes, this mediation does not depend on their adhesive and fusogenic potential [31]. Indeed, deficiency of either ADAM-1a or -2 revealed an indirect involvement of these ADAMs in fertilization. Both ADAM-1 and -2 deficient sperms were defective in migration, but the process that leads to cell-cell adhesion and fusion was only minorly altered [32–34]. ADAM-2 is not dispensable for the sperm-egg fusion process, however, ~50% decreased fusion sperm to oocytes was detected in the deficient mice [32,35]. Nine years later Yamaguchi and colleagues demonstrated that the ADAM-2–ADAM-3 complex is critical for *in vivo* sperm migration from the uterus to the oviduct [36]. Interestingly, ADAM-1 and -3 are pseudogenes in humans leaving ADAM-2 as the key ADAM mediating human fertilization [37]. Apart from the proposed role in fertilization, ADAMs have been implicated in cell fusion during development as described for ADAM-12, promoting myoblast fusion during myogenesis [38,39].

In addition to mediating interactions with cell surface receptors for adhesive and fusogenic purposes, the disintegrin and cysteine domains were suggested to regulate the proteolytic function of ADAMs as shown for the shedding of interleukin-1 receptor-II by ADAM-17 [19]. Recently, Dusterhoft *et al.* [40] demonstrated that an MDC adjacent to the conserved region in ADAM-17, named CANDIS (Conserved ADAM seventeeN Dynamic Interaction Sequence), is able to bind membranes, thereby regulating shedding activity. Therefore, further studies are necessary to elucidate the precise function of the ADAM disintegrin-loop for physiological functions of ADAMs.

2.2. ADAMs Are Active Proteases

By analogy to the SVMPs, that also have a metalloprotease and disintegrin domain, it was suggested that some ADAMs cleave extracellular matrix components. Indeed, early in 1989

Chantry *et al.* found that brain myelin membrane preparations contain a metalloproteinase activity which degrades myelin basic protein (MBP) [41]. The responsible metalloprotease was cloned in 1996 and named MADM/ADAM-10 [42]. In 1994, a metalloprotease implicated in TNFα (tumor necrosis factor alpha) processing was described [43] and identified a few years later as a member of the Adamalysin family, ADAM-17. ADAM-17 was and described simultaneously by two research groups as the enzyme that releases membrane bound tumor necrosis factor (TNF)-α precursor to a soluble form earning a name TACE (TNF alpha converting enzyme) [44,45]. ADAM-17 had a unique structure with very similar sequence to ADAM-10 [44,46].

In the following years, new discoveries on the proteolytic functions of ADAM-10 and -17 opened a larger field of research activities. ADAM-10 was identified in Drosophila, called Kuzbanian (gene *Kuz*), and is required for the cleavage of Notch and its ligand Delta leading to lateral inhibition and to neuronal fate specification during neural development [47–50]. Given its highly conserved structure in Drosophila, Xenopus, and C. Elegans, ADAM-10 was proposed to be of great importance in vertebrate development. ADAM-10 has a wide variety of functions including ECM degradation, localized shedding of various cell surface proteins, and influence on cell signaling patterns (reviewed in [51]). Importantly, shedding activities mediated by ADAM proteases may affect both cell autocrine and paracrine signaling. Indeed, shedding may occur on the surface of one cell or two adjacent cells, and the outcome may not only affect the receptor but also the ligand-bearing cell, thus resulting in the generation of reciprocal signals. As a consequence of cell surface protein shedding, the released soluble ectodomain with signaling or decoying function may act on another distant cell (reviewed in [52]).

2.2.1. ADAMs in EGFR Transactivation

Over time ADAMs were shown to process a wide variety of substrates among which are the EGFR (epidermal growth factor receptor) ligands. Aberrant EGFR activation has been frequently found in hyper proliferative diseases such as cancer, and the discovery of the ligand-dependent EGFR signal transactivation pathway may explain how autocrine signaling loops involving GPCR (G-protein coupled receptor) ligands are likely to contribute to and drive autocrine EGFR stimulatory mechanisms [53]. Prenzel and colleagues [54] first showed that EGFR transactivation by ligand involved the activity of a metalloproteinase. The transactivation involved an upstream signal acting through a G-protein-coupled receptor that activated a metalloprotease to shed an EGF receptor ligand. The result of this activity ultimately led to the modulation of mitogen responses [54]. The first studies directly implicating ADAM-10 in EGFR transactivation were presented by two groups in 2002 [55,56]. Several other studies have followed and identified additional ADAMs that fulfil this function in various cell types and tissues including ADAM-12, -15, and -17 [57]. ADAMs have been implicated in the shedding of six out of the seven known EGFR ligands (transforming growth factor (TGF)α, EGF, HB-EGF, betacellulin, epiregulin, and amphiregulin) [52]. ADAM activation by GPCRs, and the mechanisms and pathophysiological role of ADAM-dependent EGFR transactivation have been reviewed elsewhere [57,58].

2.2.2. RIPping by ADAMs

Upon shedding of the extracellular domain of some cell surface proteins, the activity of a second protease cleaving within the membrane leads to the release of an intracellular fragment of the protein, generally the cytoplasmic tail, which exerts signaling activity. This process is known as Regulated Intramembrane Proteolysis (RIP). RIP starts with the release of the protein ectodomain by the activity of ADAMs that is followed by intramembrane-cleavage and final release of the intracellular stub within the cell. Aspartyl proteases, S2P-metalloproteases, and rhomboid serine proteases catalyze the intramembrane cleavage [59]. RIPping is an evolutionary highly conserved mechanism to release messenger peptides from transmembrane proteins [60]. The classical examples for RIP activity are the processing of amyloid-precursor-protein (APP) and Notch signaling [20]. Ligation of Notch to its ligand leads to a conformational change exposing the protease-sensitive sequence to ADAM-10 that sheds and releases Notch ectodomain. This primes truncated Notch for additional intramembrane cleavage by gamma-secretase, thereby releasing the intracellular domain (NICD, notch intracellular domain) that translocates to the nucleus and mediates transcriptional activities (reviewed by [61]). ADAM-10 and -17 can cleave the extracellular domain of Notch when it is bound to ligands like Delta or Jagged, and from this follows subsequent intramembrane processing of Notch by γ-secretase activity [62,63]. Recently, Notch was shown to enhance its own activity by transcriptionally inducing expression of furin, which in turns leads to the activation of MMP (matrix metalloproteinase)-14 and ADAM-10 and ultimately to the amplification of Notch signaling [64]. This would further provide a mechanism of signal amplification during tumor progression.

APP also undergoes RIPping, most likely in a ligand-independent fashion. Among ADAMs, ADAM-9, -10, and -17, were shown to cleave APP *in vitro* acting as α-secretases [65]. Processing of the APP extracellular domain by either α- or β-secretase is followed by intramembrane proteolysis by γ-secretases, but, in the case of the APP, the intracellular role of the intracellular released fragment remains unclear [59]. However, whereas fragments generated by α-secretases are physiological, those released by β-secretase are neurotoxic and key factors in Alzheimer's disease [66].

RIPping is not limited to Notch and APP, as this mechanism is active for several other proteins shed by ADAMs. ADAM-10 is able to mediate RIPping of a variety of additional proteins such as Notch2, Notch3, N-, and E-cadherin, and ADAM-17 may mediate RIPping of EpCAM and ErbB4 [67–71]. Interestingly, there is also evidence for a reciprocal RIPping of ADAMs themselves. For example, RIPping of ADAM-10 may be executed by ADAM-9 and -15 leading, upon presenilin intramembrane proteolysis, to the release of the ADAM-10 intracellular domain (ICD) that localizes to nuclear speckles [72]. ADAM cleavage sites are usually in close proximity to O-glycosylation sites. Most recently Goth and colleagues investigated the potential of site-specific O-glycosylation on peptides from known ADAM-17 substrates and proposed that O-glycosylation might co-regulate ectodomain shedding by ADAMs [73].

2.2.3. Inhibition of Proteolytic Activity

Apart from being active towards matrix metalloproteinase, TIMP (tissue inhibitors of metalloproteinases) were shown to be active towards ADAMs [74]. After the first discovery of ADAM-17 inhibition by TIMP-3 [75], additional ADAMs were shown to be selectively inhibited by TIMPs. For example, ADAM-10 can be inhibited by TIMP-1 and -3 but not by TIMP-2 and -4, and ADAM-8 and -9 are not sensitive to inhibition by TIMPs at all [20]. ADAM-9, -10, and -17 prodomains can inhibit protease activity once released from the activated enzyme [76–78]. A number of synthetic inhibitors binding to the catalytic zinc ion, but with a broad inhibitory spectrum towards ADAMs and MMPs, have been described. Although the majority of those inhibitors could inhibit both MMPs and ADAMs, some were relatively selective for specific ADAMs. An example is the INCB3619 which inhibits ADAM-8, -9, -10, -17, and -33, but with lower IC50 for ADAM-10 and -17 [79]. Other synthetic inhibitors include CGS 27023, GW280264, and GI254023 displaying a certain degree of specificity for ADAM-9, -10, and -17 [14,80]. An interesting recent report indicated that glycosylation of substrates may also play a role in modulating ADAM activity, thus glycosylation of TNF-α enhanced ADAM-8 and -17 activities and decreased ADAM-10 activity [81]. The importance of this study was the discovery of a novel class of ADAM-17-selective inhibitors that act via a non-zinc-binding mechanism. Thus, additional type of inhibitory molecules targeted to these unique exosites within ADAM structures could be used for specific protease targeting [81].

3. Lessons from *in Vivo* Models

An important milestone in ADAM research was the generation and analysis of *in vivo* models for ADAM deficiency. Particularly, the generation of ADAM-10 and -17 *in vivo* mutants helped by elucidating their substrate specificity in a physiologically relevant context. ADAM-10 knockout mice die at day 9.5 of embryogenesis due to multiple defects of the developing central nervous system, somites, cardiovascular system, and missing Notch signaling thus providing a key evidence for the functional role of ADAM-10 in Notch processing *in vivo* [82]. ADAM-17 knockout mice die between embryonic day 17.5 and the first day of birth, not due to a lack of TNF-α shedding or dysregulated TNF-α signaling, but because of a lack in processing of multiple EGFR-ligands including TGF-α, HB-EGF, and amphiregulin. These *in vivo* findings are extensively reviewed elsewhere [52,83,84]. Recently the two new proteins iRhoms 1 and 2 (inactive Rhomboid proteins, catalytically inactive members of the rhomboid family of intramembrane serine proteases) were identified as upstream ADAM-17 regulators, controlling the substrate selectivity of ADAM-17-dependent shedding [85,86]. Additionally, truncated iRhom 1 or iRhom 2 enhance ADAM-17 activity towards TNFR shedding, thereby triggering resistance against TNF-induced cell death [87]. In agreement with these functions, the knockout mouse of both iRhoms closely resembles ADAM-17-deficient mouse phenotype, with lack of functional ADAM-17 as well as lack of EGFR phosphorylation [88].

Although ADAM-10 and -17 have become the most prominent members of this protease family, there are a few other ADAMs with significant importance in mouse development *in vivo*. ADAM-22 deficiency leads to early death due to severe ataxia and hypomyelination of peripheral nerves indicating an important role of this protease in the development of the peripheral nervous system [89]. In addition, ADAM-19 is important for heart development as 80% of ADAM-19 deficient animals die postnatally from congenital heart defects [90]. Deletion of ADAM-12 results in 30% neonatal lethality and a 30% reduction in brown adipose tissue [91], and ADAM-11 knockout animals show deficits in spatial learning, motor coordination, and altered nociception responses [92,93]. Unexpectedly, ADAM-9-deficient mice show no obvious developmental phenotype [94], but in adulthood, 20 months after birth, mice display retinal degeneration [95]. However, depletion of ADAM-9 showed its implication in pathological induced retinal vascularization, in skin repair, and in melanoma growth [96–98]. Similarly, ADAM-15 depletion showed reduced pathological neovascularization and tumor metastasis [99,100].

Given the early lethality of ADAM-10 and -17, the generation of conditional tissue specific mutants has permitted the opportunity to learn more about the role of these proteases in tissue homeostasis. For instance, nestin-Cre driven ADAM-10 deletion, with inactivation of ADAM-10 in neuronal progenitor cells, has identified ADAM-10 as the main α-secretase for APP [101]. Further, ADAM-10 depletion in epidermis leads to disturbed epidermal homeostasis [102] whereas its depletion in the endothelial cells leads to various vascular defects in organs conducible to defects in endothelial cell fate determination [103]. Overall, ADAM-10 has emerged as one of most important members of the ADAM family with broad spectrum of protein shedding (for further studies see [104,105]).

Specific loss of ADAM-17 function in epidermis leads to pronounced defects in epidermal barrier integrity and development of chronic dermatitis, closely resembling mice with epidermal loss of the EGFR [106]. Furthermore, ADAM-17 ablation in the endothelial cell progeny does not appear to be required for normal developmental angiogenesis or vascular homeostasis, but for pathological neovascularization [107]. Lisi and colleagues have recently summarized the phenotypes of ADAM-17 genetic models [108].

Thus, more studies with tissue-specific genetic targeting of ADAMs or *in vivo* challenging will be necessary to clarify protease or adhesive function of these proteins in an *in vivo* physiological and pathological context.

4. Impact on Human Disease

In various human cancers there is an increased expression of ADAM-8, -9, -10, -15, -17, -19, and -28 [109,110].

ADAM-9 null mutations were found in patients from families with recessively inherited cone-rod dystrophy (CRD), an inherited progressive retinal dystrophy. Interestingly, retinal degeneration in adulthood was also detected in the ADAM-9-deficient mouse [95].

ADAM-17 is involved in human inflammatory diseases such as rheumatoid arthritis (a systemic inflammatory autoimmune disorder), the Guillain-Barré syndrome (an acute autoimmune polyneuropathy), multiple sclerosis (an inflammatory, autoimmune, demyelinating disease of the central nervous system), and systemic lupus erythematosus (an autoimmune disease affecting nearly all organs) [111]. ADAM-17 deficiency is very rare in humans, as it has been reported only in three patients, and causes severe multiorgan disorders [112–114].

ADAM-33 has been identified as the susceptibility gene for asthma and airway hyper responsiveness, but its biological function is yet unclear [115].

In Alzheimer's disease ADAM-10 could have beneficial properties, as APP processing by ADAM-10 reduces both APP cleavage by BACE1 and β-amyloid generation, a physiologically active APP fragment clumping into neurotoxic aggregates. Indeed, accumulation of β-amyloid peptide in the brain leads to development of Alzheimer's disease [116]. β-amyloid peptide is produced by the processing of APP by β-secretases, but a soluble form of APPα is produced by the action of ADAM-9, -10, and -17 which oppose the adverse effects of the β-amyloid peptide [117–119]. Thus, the upregulation of various zinc metalloproteinase activities may represent a possible alternative therapeutic strategy for the treatment of the disease [120].

5. Future Perspectives

ADAMs are implicated in a variety of cellular functions *in vitro* and *in vivo*, and many substrates for some of these proteases are already known. However, additional substrates are yet to be discovered, and further protein degradation analyses may help to identify the protease implicated. This analysis, together with the analysis of processing characteristics, such as shedding followed by intramembrane proteolysis and generation of an ICD signaling fragment, will further identify downstream effects of such proteolytic events.

Over the years it has become clear that ADAMs are implicated in several diseases, and because of the broad specificity of several synthetic inhibitors to MMPs or ADAMs targeting the metalloproteinase activity, it is now important to develop alternative strategies for *in vivo* targeting. Recombinant pro-domains or the generation of specific antibodies may help with increasing targeting specificity. Moreover, targeting exosites has recently gained importance in drug development. One of the outstanding questions that still need to be addressed is, what are the protein modifications on substrates or substrate structure that determine specificity of binding and enzyme efficacy? Protein modifications have just started to be analyzed, and will certainly lead to new clues for selective targeting strategies.

Acknowledgments: We thank Barbara Boggetti for her critical reading of this manuscript. We apologize to the authors whose work was not cited. This work was supported by the Deutsche Forschungsgemeinschaft through the SFB 829, B4 at the University of Cologne.

Conflicts of Interest: The authors declare no conflict of interest.

References

1. Huxley-Jones, J.; Clarke, T.K.; Beck, C.; Toubaris, G.; Robertson, D.L.; Boot-Handford, R.P. The evolution of the vertebrate metzincins; insights from Ciona intestinalis and Danio rerio. *BMC Evol. Biol.* **2007**, *7*. [CrossRef] [PubMed]
2. Long, J.; Li, M.; Ren, Q.; Zhang, C.; Fan, J.; Duan, Y.; Chen, J.; Li, B.; Deng, L. Phylogenetic and molecular evolution of the ADAM (a disintegrin and metalloprotease) gene family from Xenopus tropicalis, to Mus musculus, Rattus norvegicus, and Homo sapiens. *Gene* **2012**, *507*, 36–43. [CrossRef] [PubMed]
3. Glassey, B.; Civetta, A. Positive selection at reproductive adam genes with potential intercellular binding activity. *Mol. Biol. Evol.* **2004**, *21*, 851–859. [CrossRef] [PubMed]
4. Bjarnason, J.B.; Fox, J.W. Snake venom metalloendopeptidases: Reprolysins. *Methods Enzymol.* **1995**, *248*, 345–368. [PubMed]
5. Stone, A.L.; Kroeger, M.; Sang, Q.X. Structure-function analysis of the ADAM family of disintegrin-like and metalloproteinase-containing proteins (review). *J. Protein Chem.* **1999**, *18*, 447–465. [CrossRef] [PubMed]

6. Seals, D.F.; Courtneidge, S.A. The adams family of metalloproteases: Multidomain proteins with multiple functions. *Genes Dev.* **2003**, *17*, 7–30. [CrossRef] [PubMed]
7. Gould, R.J.; Polokoff, M.A.; Friedman, P.A.; Huang, T.F.; Holt, J.C.; Cook, J.J.; Niewiarowski, S. Disintegrins: A family of integrin inhibitory proteins from viper venoms. *Proc. Soc. Exp. Biol. Med.* **1990**, *195*, 168–171. [CrossRef] [PubMed]
8. Blobel, C.P. Metalloprotease-disintegrins: Links to cell adhesion and cleavage of tnf alpha and notch. *Cell* **1997**, *90*, 589–592. [CrossRef]
9. Takeda, S. Three-dimensional domain architecture of the adam family proteinases. *Semin. Cell Dev. Biol.* **2009**, *20*, 146–152. [CrossRef] [PubMed]
10. Takeda, S.; Takeya, H.; Iwanaga, S. Snake venom metalloproteinases: Structure, function and relevance to the mammalian adam/adamts family proteins. *Biochim. Biophys. Acta* **2012**, *1824*, 164–176. [CrossRef] [PubMed]
11. Bode, W.; Gomis-Ruth, F.X.; Stockler, W. Astacins, serralysins, snake venom and matrix metalloproteinases exhibit identical zinc-binding environments (HEXXHXXGXXH and Met-turn) and topologies and should be grouped into a common family, the 'metzincins'. *FEBS Lett.* **1993**, *331*, 134–140. [CrossRef]
12. Leonard, J.D.; Lin, F.; Milla, M.E. Chaperone-like properties of the prodomain of TNFalpha-converting enzyme (TACE) and the functional role of its cysteine switch. *Biochem. J.* **2005**, *387*, 797–805. [CrossRef] [PubMed]
13. Loechel, F.; Overgaard, M.T.; Oxvig, C.; Albrechtsen, R.; Wewer, U.M. Regulation of human ADAM 12 protease by the prodomain. Evidence for a functional cysteine switch. *J. Biol. Chem.* **1999**, *274*, 13427–13433. [CrossRef] [PubMed]
14. Roghani, M.; Becherer, J.D.; Moss, M.L.; Atherton, R.E.; Erdjument-Bromage, H.; Arribas, J.; Blackburn, R.K.; Weskamp, G.; Tempst, P.; Blobel, C.P. Metalloprotease-disintegrin MDC9: Intracellular maturation and catalytic activity. *J. Biol. Chem.* **1999**, *274*, 3531–3540. [CrossRef] [PubMed]
15. Eto, K.; Puzon-McLaughlin, W.; Sheppard, D.; Sehara-Fujisawa, A.; Zhang, X.P.; Takada, Y. RGD-independent binding of integrin $\alpha 9 \beta 1$ to the ADAM-12 and -15 disintegrin domains mediates cell-cell interaction. *J. Biol. Chem.* **2000**, *275*, 34922–34930. [CrossRef] [PubMed]
16. Lu, D.; Scully, M.; Kakkar, V.; Lu, X. ADAM-15 disintegrin-like domain structure and function. *Toxins* **2010**, *2*, 2411–2427. [CrossRef] [PubMed]
17. Bridges, L.C.; Bowditch, R.D. Adam-integrin interactions: Potential integrin regulated ectodomain shedding activity. *Curr. Pharm. Des.* **2005**, *11*, 837–847. [CrossRef] [PubMed]
18. White, J.M. ADAMs: Modulators of cell-cell and cell-matrix interactions. *Curr. Opin. Cell Biol.* **2003**, *15*, 598–606. [CrossRef] [PubMed]
19. Takeda, S.; Igarashi, T.; Mori, H.; Araki, S. Crystal structures of VAP1 reveal ADAMs' MDC domain architecture and its unique C-shaped scaffold. *EMBO J.* **2006**, *25*, 2388–2396. [CrossRef] [PubMed]
20. Edwards, D.R.; Handsley, M.M.; Pennington, C.J. The ADAM metalloproteinases. *Mol. Aspects Med.* **2008**, *29*, 258–289. [CrossRef] [PubMed]
21. Liu, H.; Shim, A.H.; He, X. Structural characterization of the ectodomain of a disintegrin and metalloproteinase-22 (ADAM22), a neural adhesion receptor instead of metalloproteinase: Insights on adam function. *J. Biol. Chem.* **2009**, *284*, 29077–29086. [CrossRef] [PubMed]
22. Wewer, U.M.; Morgelin, M.; Holck, P.; Jacobsen, J.; Lydolph, M.C.; Johnsen, A.H.; Kveiborg, M.; Albrechtsen, R. ADAM12 is a four-leafed clover: The excised prodomain remains bound to the mature enzyme. *J. Biol. Chem.* **2006**, *281*, 9418–9422. [CrossRef] [PubMed]
23. Maskos, K.; Fernandez-Catalan, C.; Huber, R.; Bourenkov, G.P.; Bartunik, H.; Ellestad, G.A.; Reddy, P.; Wolfson, M.F.; Rauch, C.T.; Castner, B.J.; *et al.* Crystal structure of the catalytic domain of human tumor necrosis factor-alpha-converting enzyme. *Proc. Natl. Acad. Sci. USA* **1998**, *95*, 3408–3412. [CrossRef] [PubMed]
24. Guan, H.H.; Goh, K.S.; Davamani, F.; Wu, P.L.; Huang, Y.W.; Jeyakanthan, J.; Wu, W.G.; Chen, C.J. Structures of two elapid snake venom metalloproteases with distinct activities highlight the disulfide patterns in the d domain of adamalysin family proteins. *J. Struct. Biol.* **2010**, *169*, 294–303. [CrossRef] [PubMed]
25. Blobel, C.P.; Wolfsberg, T.G.; Turck, C.W.; Myles, D.G.; Primakoff, P.; White, J.M. A potential fusion peptide and an integrin ligand domain in a protein active in sperm-egg fusion. *Nature* **1992**, *356*, 248–252. [CrossRef] [PubMed]

26. Evans, J.P.; Schultz, R.M.; Kopf, G.S. Mouse sperm-egg plasma membrane interactions: Analysis of roles of egg integrins and the mouse sperm homologue of PH-30 (fertilin) beta. *J. Cell Sci.* **1995**, *108*, 3267–3278. [PubMed]

27. Primakoff, P.; Hyatt, H.; Tredick-Kline, J. Identification and purification of a sperm surface protein with a potential role in sperm-egg membrane fusion. *J. Cell Biol.* **1987**, *104*, 141–149. [CrossRef] [PubMed]

28. Blobel, C.P. Functional processing of fertilin: Evidence for a critical role of proteolysis in sperm maturation and activation. *Rev. Reprod.* **2000**, *5*, 75–83. [CrossRef] [PubMed]

29. Chen, H.; Sampson, N.S. Mediation of sperm-egg fusion: Evidence that mouse egg alpha6beta1 integrin is the receptor for sperm fertilinbeta. *Chem. Biol.* **1999**, *6*, 1–10. [CrossRef]

30. Cho, C. Mammalian adams with testis-specific or -predominant expression. In *The Adam Family of Proteases*; Hooper, N.M., Lendeckel, U., Eds.; Springer US: Boston, MA, USA, 2005; pp. 239–259.

31. Kim, E.; Nishimura, H.; Iwase, S.; Yamagata, K.; Kashiwabara, S.; Baba, T. Synthesis, processing, and subcellular localization of mouse ADAM3 during spermatogenesis and epididymal sperm transport. *J. Reprod. Dev.* **2004**, *50*, 571–578. [CrossRef] [PubMed]

32. Cho, C.; Bunch, D.O.; Faure, J.E.; Goulding, E.H.; Eddy, E.M.; Primakoff, P.; Myles, D.G. Fertilization defects in sperm from mice lacking fertilin beta. *Science* **1998**, *281*, 1857–1859. [CrossRef] [PubMed]

33. Nishimura, H.; Kim, E.; Nakanishi, T.; Baba, T. Possible function of the ADAM1a/ADAM2 fertilin complex in the appearance of ADAM3 on the sperm surface. *J. Biol. Chem.* **2004**, *279*, 34957–34962. [CrossRef] [PubMed]

34. Stein, K.K.; Primakoff, P.; Myles, D. Sperm-egg fusion: Events at the plasma membrane. *J. Cell Sci.* **2004**, *117*, 6269–6274. [CrossRef] [PubMed]

35. Tomczuk, M.; Takahashi, Y.; Huang, J.; Murase, S.; Mistretta, M.; Klaffky, E.; Sutherland, A.; Bolling, L.; Coonrod, S.; Marcinkiewicz, C.; *et al.* Role of multiple beta1 integrins in cell adhesion to the disintegrin domains of ADAMs 2 and 3. *Exp. Cell Res.* **2003**, *290*, 68–81. [CrossRef]

36. Yamaguchi, R.; Muro, Y.; Isotani, A.; Tokuhiro, K.; Takumi, K.; Adham, I.; Ikawa, M.; Okabe, M. Disruption of ADAM3 impairs the migration of sperm into oviduct in mouse. *Biol. Reprod.* **2009**, *81*, 142–146. [CrossRef] [PubMed]

37. Takahashi, Y.; Bigler, D.; Ito, Y.; White, J.M. Sequence-specific interaction between the disintegrin domain of mouse ADAM 3 and murine eggs: Role of beta1 integrin-associated proteins CD9, CD81, and CD98. *Mol. Biol. Cell* **2001**, *12*, 809–820. [CrossRef] [PubMed]

38. Fritsche, J.; Moser, M.; Faust, S.; Peuker, A.; Buttner, R.; Andreesen, R.; Kreutz, M. Molecular cloning and characterization of a human metalloprotease disintegrin—A novel marker for dendritic cell differentiation. *Blood* **2000**, *96*, 732–739. [PubMed]

39. Yagami-Hiromasa, T.; Sato, T.; Kurisaki, T.; Kamijo, K.; Nabeshima, Y.; Fujisawa-Sehara, A. A metalloprotease-disintegrin participating in myoblast fusion. *Nature* **1995**, *377*, 652–656. [CrossRef] [PubMed]

40. Dusterhoft, S.; Michalek, M.; Kordowski, F.; Oldefest, M.; Sommer, A.; Roseler, J.; Reiss, K.; Grotzinger, J.; Lorenzen, I. Extracellular juxtamembrane segment of ADAM17 interacts with membranes and is essential for its shedding activity. *Biochemistry* **2015**, *54*, 5791–5801. [CrossRef] [PubMed]

41. Chantry, A.; Gregson, N.A.; Glynn, P. A novel metalloproteinase associated with brain myelin membranes. Isolation and characterization. *J. Biol. Chem.* **1989**, *264*, 21603–21607. [PubMed]

42. Howard, L.; Lu, X.; Mitchell, S.; Griffiths, S.; Glynn, P. Molecular cloning of MADM: A catalytically active mammalian disintegrin-metalloprotease expressed in various cell types. *Biochem. J.* **1996**, *317*, 45–50. [CrossRef] [PubMed]

43. McGeehan, G.M.; Becherer, J.D.; Bast, R.C., Jr.; Boyer, C.M.; Champion, B.; Connolly, K.M.; Conway, J.G.; Furdon, P.; Karp, S.; Kidao, S.; *et al.* Regulation of tumour necrosis factor-alpha processing by a metalloproteinase inhibitor. *Nature* **1994**, *370*, 558–561. [CrossRef] [PubMed]

44. Black, R.A.; Rauch, C.T.; Kozlosky, C.J.; Peschon, J.J.; Slack, J.L.; Wolfson, M.F.; Castner, B.J.; Stocking, K.L.; Reddy, P.; Srinivasan, S.; *et al.* A metalloproteinase disintegrin that releases tumour-necrosis factor-alpha from cells. *Nature* **1997**, *385*, 729–733. [CrossRef] [PubMed]

45. Moss, M.L.; Jin, S.L.; Becherer, J.D.; Bickett, D.M.; Burkhart, W.; Chen, W.J.; Hassler, D.; Leesnitzer, M.T.; McGeehan, G.; Milla, M.; *et al.* Structural features and biochemical properties of TNF-alpha converting enzyme (TACE). *J. Neuroimmunol.* **1997**, *72*, 127–129. [CrossRef]

46. Moss, M.L.; Jin, S.L.; Milla, M.E.; Bickett, D.M.; Burkhart, W.; Carter, H.L.; Chen, W.J.; Clay, W.C.; Didsbury, J.R.; Hassler, D.; *et al.* Cloning of a disintegrin metalloproteinase that processes precursor tumour-necrosis factor-alpha. *Nature* **1997**, *385*, 733–736. [CrossRef] [PubMed]

47. Pan, D.; Rubin, G.M. Kuzbanian controls proteolytic processing of notch and mediates lateral inhibition during drosophila and vertebrate neurogenesis. *Cell* **1997**, *90*, 271–280. [CrossRef]

48. Qi, H.; Rand, M.D.; Wu, X.; Sestan, N.; Wang, W.; Rakic, P.; Xu, T.; Artavanis-Tsakonas, S. Processing of the notch ligand delta by the metalloprotease kuzbanian. *Science* **1999**, *283*, 91–94. [CrossRef] [PubMed]

49. Rooke, J.; Pan, D.; Xu, T.; Rubin, G.M. Kuz, a conserved metalloprotease-disintegrin protein with two roles in drosophila neurogenesis. *Science* **1996**, *273*, 1227–1231. [CrossRef] [PubMed]

50. Sotillos, S.; Roch, F.; Campuzano, S. The metalloprotease-disintegrin kuzbanian participates in notch activation during growth and patterning of drosophila imaginal discs. *Development* **1997**, *124*, 4769–4779. [PubMed]

51. Saftig, P.; Hartmann, D. ADAM10. In *The ADAM Family of Proteases*; Hooper, N.M., Lendeckel, U., Eds.; Springer US: Boston, MA, USA, 2005; pp. 85–121.

52. Blobel, C.P. ADAMs: Key components in EGFR signalling and development. *Nat. Rev. Mol. Cell Biol.* **2005**, *6*, 32–43. [CrossRef] [PubMed]

53. Fischer, O.M.; Hart, S.; Gschwind, A.; Ullrich, A. EGFR signal transactivation in cancer cells. *Biochem. Soc. Trans.* **2003**, *31*, 1203–1208. [CrossRef] [PubMed]

54. Prenzel, N.; Zwick, E.; Daub, H.; Leserer, M.; Abraham, R.; Wallasch, C.; Ullrich, A. EGF receptor transactivation by G-protein-coupled receptors requires metalloproteinase cleavage of proHB-EGF. *Nature* **1999**, *402*, 884–888. [PubMed]

55. Yan, Y.; Shirakabe, K.; Werb, Z. The metalloprotease Kuzbanian (ADAM10) mediates the transactivation of EGF receptor by G protein-coupled receptors. *J. Cell Biol.* **2002**, *158*, 221–226. [CrossRef] [PubMed]

56. Lemjabbar, H.; Basbaum, C. Platelet-activating factor receptor and ADAM10 mediate responses to staphylococcus aureus in epithelial cells. *Nat. Med.* **2002**, *8*, 41–46. [CrossRef] [PubMed]

57. Ohtsu, H.; Dempsey, P.J.; Eguchi, S. Adams as mediators of EGF receptor transactivation by G protein-coupled receptors. *Am. J. Physiol. Cell Physiol.* **2006**, *291*, C1–C10. [CrossRef] [PubMed]

58. Kataoka, H. EGFR ligands and their signaling scissors, ADAMs, as new molecular targets for anticancer treatments. *J. Dermatol. Sci.* **2009**, *56*, 148–153. [CrossRef] [PubMed]

59. Lichtenthaler, S.F.; Haass, C.; Steiner, H. Regulated intramembrane proteolysis—Lessons from amyloid precursor protein processing. *J. Neurochem.* **2011**, *117*, 779–796. [CrossRef] [PubMed]

60. Brown, M.S.; Ye, J.; Rawson, R.B.; Goldstein, J.L. Regulated intramembrane proteolysis: A control mechanism conserved from bacteria to humans. *Cell* **2000**, *100*, 391–398. [CrossRef]

61. Groot, A.J.; Vooijs, M.A. The role of adams in notch signaling. *Adv. Exp. Med. Biol.* **2012**, *727*, 15–36. [PubMed]

62. Brou, C.; Logeat, F.; Gupta, N.; Bessia, C.; LeBail, O.; Doedens, J.R.; Cumano, A.; Roux, P.; Black, R.A.; Israel, A. A novel proteolytic cleavage involved in Notch signaling: The role of the disintegrin-metalloprotease TACE. *Mol. Cell* **2000**, *5*, 207–216. [CrossRef]

63. Van Tetering, G.; van Diest, P.; Verlaan, I.; van der Wall, E.; Kopan, R.; Vooijs, M. Metalloprotease ADAM10 is required for Notch1 site 2 cleavage. *J. Biol. Chem.* **2009**, *284*, 31018–31027. [CrossRef] [PubMed]

64. Qiu, H.; Tang, X.; Ma, J.; Shaverdashvili, K.; Zhang, K.; Bedogni, B. Notch1 autoactivation via transcriptional regulation of furin, which sustains Notch1 signaling by processing Notch1-activating proteases ADAM10 and membrane type 1 matrix metalloproteinase. *Mol. Cell. Biol.* **2015**, *35*, 3622–3632. [CrossRef] [PubMed]

65. Asai, M.; Hattori, C.; Szabo, B.; Sasagawa, N.; Maruyama, K.; Tanuma, S.; Ishiura, S. Putative function of ADAM9, ADAM10, and ADAM17 as app alpha-secretase. *Biochem. Biophys. Res. Commun.* **2003**, *301*, 231–235. [CrossRef]

66. O'Brien, R.J.; Wong, P.C. Amyloid precursor protein processing and alzheimer's disease. *Annu. Rev. Neurosci.* **2011**, *34*, 185–204. [CrossRef] [PubMed]

67. Groot, A.J.; Habets, R.; Yahyanejad, S.; Hodin, C.M.; Reiss, K.; Saftig, P.; Theys, J.; Vooijs, M. Regulated proteolysis of NOTCH2 and NOTCH3 receptors by ADAM10 and presenilins. *Mol. Cell. Biol.* **2014**, *34*, 2822–2832. [CrossRef] [PubMed]

68. Maetzel, D.; Denzel, S.; Mack, B.; Canis, M.; Went, P.; Benk, M.; Kieu, C.; Papior, P.; Baeuerle, P.A.; Munz, M.; *et al.* Nuclear signalling by tumour-associated antigen epcam. *Nat. Cell Biol.* **2009**, *11*, 162–171. [CrossRef] [PubMed]

69. Maretzky, T.; Reiss, K.; Ludwig, A.; Buchholz, J.; Scholz, F.; Proksch, E.; de Strooper, B.; Hartmann, D.; Saftig, P. ADAM10 mediates e-cadherin shedding and regulates epithelial cell-cell adhesion, migration, and beta-catenin translocation. *Proc. Natl. Acad. Sci. USA* **2005**, *102*, 9182–9187. [CrossRef] [PubMed]

70. Reiss, K.; Maretzky, T.; Ludwig, A.; Tousseyn, T.; de Strooper, B.; Hartmann, D.; Saftig, P. ADAM10 cleavage of n-cadherin and regulation of cell-cell adhesion and beta-catenin nuclear signalling. *EMBO J.* **2005**, *24*, 742–752. [CrossRef] [PubMed]

71. Rio, C.; Buxbaum, J.D.; Peschon, J.J.; Corfas, G. Tumor necrosis factor-alpha-converting enzyme is required for cleavage of erbB4/HER4. *J. Biol. Chem.* **2000**, *275*, 10379–10387. [CrossRef] [PubMed]

72. Tousseyn, T.; Thathiah, A.; Jorissen, E.; Raemaekers, T.; Konietzko, U.; Reiss, K.; Maes, E.; Snellinx, A.; Serneels, L.; Nyabi, O.; *et al.* ADAM10, the rate-limiting protease of regulated intramembrane proteolysis of Notch and other proteins, is processed by ADAMS-9, ADAMS-15, and the gamma-secretase. *J. Biol. Chem.* **2009**, *284*, 11738–11747. [CrossRef] [PubMed]

73. Goth, C.K.; Halim, A.; Khetarpal, S.A.; Rader, D.J.; Clausen, H.; Schjoldager, K.T. A systematic study of modulation of adam-mediated ectodomain shedding by site-specific o-glycosylation. *Proc. Natl. Acad. Sci. USA* **2015**, *112*, 14623–14628. [CrossRef] [PubMed]

74. Brew, K.; Nagase, H. The tissue inhibitors of metalloproteinases (TIMPs): An ancient family with structural and functional diversity. *Biochim. Biophys. Acta* **2010**, *1803*, 55–71. [CrossRef] [PubMed]

75. Amour, A.; Slocombe, P.M.; Webster, A.; Butler, M.; Knight, C.G.; Smith, B.J.; Stephens, P.E.; Shelley, C.; Hutton, M.; Knauper, V.; *et al.* TNF-alpha converting enzyme (TACE) is inhibited by TIMP-3. *FEBS Lett.* **1998**, *435*, 39–44. [CrossRef]

76. Gonzales, P.E.; Solomon, A.; Miller, A.B.; Leesnitzer, M.A.; Sagi, I.; Milla, M.E. Inhibition of the tumor necrosis factor-alpha-converting enzyme by its pro domain. *J. Biol. Chem.* **2004**, *279*, 31638–31645. [CrossRef] [PubMed]

77. Moss, M.L.; Bomar, M.; Liu, Q.; Sage, H.; Dempsey, P.; Lenhart, P.M.; Gillispie, P.A.; Stoeck, A.; Wildeboer, D.; Bartsch, J.W.; *et al.* The ADAM10 prodomain is a specific inhibitor of ADAM10 proteolytic activity and inhibits cellular shedding events. *J. Biol. Chem.* **2007**, *282*, 35712–35721. [CrossRef] [PubMed]

78. Moss, M.L.; Powell, G.; Miller, M.A.; Edwards, L.; Qi, B.; Sang, Q.X.; de Strooper, B.; Tesseur, I.; Lichtenthaler, S.F.; Taverna, M.; *et al.* ADAM9 inhibition increases membrane activity of ADAM10 and controls alpha-secretase processing of amyloid precursor protein. *J. Biol. Chem.* **2011**, *286*, 40443–40451. [CrossRef] [PubMed]

79. Fridman, J.S.; Caulder, E.; Hansbury, M.; Liu, X.; Yang, G.; Wang, Q.; Lo, Y.; Zhou, B.B.; Pan, M.; Thomas, S.M.; *et al.* Selective inhibition of adam metalloproteases as a novel approach for modulating erbb pathways in cancer. *Clin. Cancer Res.* **2007**, *13*, 1892–1902. [CrossRef] [PubMed]

80. Ludwig, A.; Hundhausen, C.; Lambert, M.H.; Broadway, N.; Andrews, R.C.; Bickett, D.M.; Leesnitzer, M.A.; Becherer, J.D. Metalloproteinase inhibitors for the disintegrin-like metalloproteinases ADAM10 and ADAM17 that differentially block constitutive and phorbol ester-inducible shedding of cell surface molecules. *Comb. Chem. High Throughput Screen.* **2005**, *8*, 161–171. [CrossRef] [PubMed]

81. Minond, D.; Cudic, M.; Bionda, N.; Giulianotti, M.; Maida, L.; Houghten, R.A.; Fields, G.B. Discovery of novel inhibitors of a disintegrin and metalloprotease 17 (ADAM17) using glycosylated and non-glycosylated substrates. *J. Biol. Chem.* **2012**, *287*, 36473–36487. [CrossRef] [PubMed]

82. Hartmann, D.; de Strooper, B.; Serneels, L.; Craessaerts, K.; Herreman, A.; Annaert, W.; Umans, L.; Lubke, T.; Lena Illert, A.; von Figura, K.; *et al.* The disintegrin/metalloprotease ADAM 10 is essential for Notch signalling but not for α-secretase activity in fibroblasts. *Hum. Mol. Genet.* **2002**, *11*, 2615–2624. [CrossRef] [PubMed]

83. Black, R.A. Tumor necrosis factor-alpha converting enzyme. *Int. J. Biochem. Cell Biol.* **2002**, *34*, 1–5. [CrossRef]

84. Weber, S.; Saftig, P. Ectodomain shedding and adams in development. *Development* **2012**, *139*, 3693–3709. [CrossRef] [PubMed]

85. Christova, Y.; Adrain, C.; Bambrough, P.; Ibrahim, A.; Freeman, M. Mammalian irhoms have distinct physiological functions including an essential role in tace regulation. *EMBO Rep.* **2013**, *14*, 884–890. [CrossRef] [PubMed]

86. Maretzky, T.; McIlwain, D.R.; Issuree, P.D.; Li, X.; Malapeira, J.; Amin, S.; Lang, P.A.; Mak, T.W.; Blobel, C.P. iRhom2 controls the substrate selectivity of stimulated ADAM17-dependent ectodomain shedding. *Proc. Natl. Acad. Sci. USA* **2013**, *110*, 11433–11438. [CrossRef] [PubMed]

87. Maney, S.K.; McIlwain, D.R.; Polz, R.; Pandyra, A.A.; Sundaram, B.; Wolff, D.; Ohishi, K.; Maretzky, T.; Brooke, M.A.; Evers, A.; *et al.* Deletions in the cytoplasmic domain of iRhom1 and iRhom2 promote shedding of the TNF receptor by the protease ADAM17. *Sci. Signal.* **2015**, *8*, ra109. [CrossRef] [PubMed]

88. Li, X.; Maretzky, T.; Weskamp, G.; Monette, S.; Qing, X.; Issuree, P.D.; Crawford, H.C.; McIlwain, D.R.; Mak, T.W.; Salmon, J.E.; *et al.* iRhoms 1 and 2 are essential upstream regulators of ADAM17-dependent EGFR signaling. *Proc. Natl. Acad. Sci. USA* **2015**, *112*, 6080–6085. [CrossRef] [PubMed]

89. Sagane, K.; Hayakawa, K.; Kai, J.; Hirohashi, T.; Takahashi, E.; Miyamoto, N.; Ino, M.; Oki, T.; Yamazaki, K.; Nagasu, T. Ataxia and peripheral nerve hypomyelination in ADAM22-deficient mice. *BMC Neurosci.* **2005**, *6*, 33. [CrossRef] [PubMed]

90. Zhou, H.M.; Weskamp, G.; Chesneau, V.; Sahin, U.; Vortkamp, A.; Horiuchi, K.; Chiusaroli, R.; Hahn, R.; Wilkes, D.; Fisher, P.; *et al.* Essential role for ADAM19 in cardiovascular morphogenesis. *Mol. Cell. Biol.* **2004**, *24*, 96–104. [CrossRef] [PubMed]

91. Kurisaki, T.; Masuda, A.; Sudo, K.; Sakagami, J.; Higashiyama, S.; Matsuda, Y.; Nagabukuro, A.; Tsuji, A.; Nabeshima, Y.; Asano, M.; *et al.* Phenotypic analysis of meltrin alpha (ADAM12)-deficient mice: Involvement of meltrin alpha in adipogenesis and myogenesis. *Mol. Cell. Biol.* **2003**, *23*, 55–61. [CrossRef] [PubMed]

92. Takahashi, E.; Sagane, K.; Nagasu, T.; Kuromitsu, J. Altered nociceptive response in ADAM11-deficient mice. *Brain Res.* **2006**, *1097*, 39–42. [CrossRef] [PubMed]

93. Takahashi, E.; Sagane, K.; Oki, T.; Yamazaki, K.; Nagasu, T.; Kuromitsu, J. Deficits in spatial learning and motor coordination in ADAM11-deficient mice. *BMC Neurosci.* **2006**, *7*, 19. [CrossRef] [PubMed]

94. Weskamp, G.; Cai, H.; Brodie, T.A.; Higashyama, S.; Manova, K.; Ludwig, T.; Blobel, C.P. Mice lacking the metalloprotease-disintegrin MDC9 (ADAM9) have no evident major abnormalities during development or adult life. *Mol. Cell. Biol.* **2002**, *22*, 1537–1544. [CrossRef] [PubMed]

95. Parry, D.A.; Toomes, C.; Bida, L.; Danciger, M.; Towns, K.V.; McKibbin, M.; Jacobson, S.G.; Logan, C.V.; Ali, M.; Bond, J.; *et al.* Loss of the metalloprotease ADAM9 leads to cone-rod dystrophy in humans and retinal degeneration in mice. *Am. J. Hum. Genet.* **2009**, *84*, 683–691. [CrossRef] [PubMed]

96. Abety, A.N.; Fox, J.W.; Schonefuss, A.; Zamek, J.; Landsberg, J.; Krieg, T.; Blobel, C.; Mauch, C.; Zigrino, P. Stromal fibroblast-specific expression of ADAM-9 modulates proliferation and apoptosis in melanoma cells *in vitro* and *in vivo*. *J. Investig. Dermatol.* **2012**, *132*, 2451–2458. [CrossRef] [PubMed]

97. Guaiquil, V.; Swendeman, S.; Yoshida, T.; Chavala, S.; Campochiaro, P.A.; Blobel, C.P. ADAM9 is involved in pathological retinal neovascularization. *Mol. Cell. Biol.* **2009**, *29*, 2694–2703. [CrossRef] [PubMed]

98. Mauch, C.; Zamek, J.; Abety, A.N.; Grimberg, G.; Fox, J.W.; Zigrino, P. Accelerated wound repair in ADAM-9 knockout animals. *J. Investig. Dermatol.* **2010**, *130*, 2120–2130. [CrossRef] [PubMed]

99. Horiuchi, K.; Weskamp, G.; Lum, L.; Hammes, H.P.; Cai, H.; Brodie, T.A.; Ludwig, T.; Chiusaroli, R.; Baron, R.; Preissner, K.T.; *et al.* Potential role for ADAM15 in pathological neovascularization in mice. *Mol. Cell. Biol.* **2003**, *23*, 5614–5624. [CrossRef] [PubMed]

100. Schonefuss, A.; Abety, A.N.; Zamek, J.; Mauch, C.; Zigrino, P. Role of ADAM-15 in wound healing and melanoma development. *Exp. Dermatol.* **2012**, *21*, 437–442. [CrossRef] [PubMed]

101. Jorissen, E.; Prox, J.; Bernreuther, C.; Weber, S.; Schwanbeck, R.; Serneels, L.; Snellinx, A.; Craessaerts, K.; Thathiah, A.; Tesseur, I.; *et al.* The disintegrin/metalloproteinase ADAM10 is essential for the establishment of the brain cortex. *J. Neurosci.* **2010**, *30*, 4833–4844. [CrossRef] [PubMed]

102. Weber, S.; Niessen, M.T.; Prox, J.; Lullmann-Rauch, R.; Schmitz, A.; Schwanbeck, R.; Blobel, C.P.; Jorissen, E.; de Strooper, B.; Niessen, C.M.; *et al.* The disintegrin/metalloproteinase ADAM10 is essential for epidermal integrity and notch-mediated signaling. *Development* **2011**, *138*, 495–505. [CrossRef] [PubMed]

103. Glomski, K.; Monette, S.; Manova, K.; de Strooper, B.; Saftig, P.; Blobel, C.P. Deletion of ADAM10 in endothelial cells leads to defects in organ-specific vascular structures. *Blood* **2011**, *118*, 1163–1174. [CrossRef] [PubMed]

104. Reiss, K.; Saftig, P. The "a disintegrin and metalloprotease" (ADAM) family of sheddases: Physiological and cellular functions. *Semin. Cell Dev. Biol.* **2009**, *20*, 126–137. [CrossRef] [PubMed]

105. Saftig, P.; Lichtenthaler, S.F. The alpha secretase ADAM10: A metalloprotease with multiple functions in the brain. *Prog. Neurobiol.* **2015**, *135*, 1–20. [CrossRef] [PubMed]

106. Franzke, C.W.; Cobzaru, C.; Triantafyllopoulou, A.; Loffek, S.; Horiuchi, K.; Threadgill, D.W.; Kurz, T.; van Rooijen, N.; Bruckner-Tuderman, L.; Blobel, C.P. Epidermal ADAM17 maintains the skin barrier by regulating EGFR ligand-dependent terminal keratinocyte differentiation. *J. Exp. Med.* **2012**, *209*, 1105–1119. [CrossRef] [PubMed]

107. Weskamp, G.; Mendelson, K.; Swendeman, S.; Le Gall, S.; Ma, Y.; Lyman, S.; Hinoki, A.; Eguchi, S.; Guaiquil, V.; Horiuchi, K.; *et al.* Pathological neovascularization is reduced by inactivation of ADAM17 in endothelial cells but not in pericytes. *Circ. Res.* **2010**, *106*, 932–940. [CrossRef] [PubMed]

108. Lisi, S.; D'Amore, M.; Sisto, M. ADAM17 at the interface between inflammation and autoimmunity. *Immunol. Lett.* **2014**, *162*, 159–169. [CrossRef] [PubMed]

109. Duffy, M.J.; Mullooly, M.; O'Donovan, N.; Sukor, S.; Crown, J.; Pierce, A.; McGowan, P.M. The ADAMs family of proteases: New biomarkers and therapeutic targets for cancer? *Clin. Proteomics* **2011**, *8*, 9. [CrossRef] [PubMed]

110. Saftig, P.; Reiss, K. The "a disintegrin and metalloproteases" ADAM10 and ADAM17: Novel drug targets with therapeutic potential? *Eur. J. Cell Biol.* **2011**, *90*, 527–535. [CrossRef] [PubMed]

111. Kurz, M.; Pischel, H.; Hartung, H.P.; Kieseier, B.C. Tumor necrosis factor-alpha-converting enzyme is expressed in the inflamed peripheral nervous system. *J. Peripher. Nerv. Syst.* **2005**, *10*, 311–318. [CrossRef] [PubMed]

112. Bandsma, R.H.; van Goor, H.; Yourshaw, M.; Horlings, R.K.; Jonkman, M.F.; Scholvinck, E.H.; Karrenbeld, A.; Scheenstra, R.; Komhoff, M.; Rump, P.; *et al.* Loss of ADAM17 is associated with severe multiorgan dysfunction. *Hum. Pathol.* **2015**, *46*, 923–928. [CrossRef] [PubMed]

113. Blaydon, D.C.; Biancheri, P.; Di, W.L.; Plagnol, V.; Cabral, R.M.; Brooke, M.A.; van Heel, D.A.; Ruschendorf, F.; Toynbee, M.; Walne, A.; *et al.* Inflammatory skin and bowel disease linked to ADAM17 deletion. *N. Engl. J. Med.* **2011**, *365*, 1502–1508. [CrossRef] [PubMed]

114. Tsukerman, P.; Eisenstein, E.M.; Chavkin, M.; Schmiedel, D.; Wong, E.; Werner, M.; Yaacov, B.; Averbuch, D.; Molho-Pessach, V.; Stepensky, P.; *et al.* Cytokine secretion and NK cell activity in human ADAM17 deficiency. *Oncotarget* **2015**, *6*, 44151–44160. [PubMed]

115. Tripathi, P.; Awasthi, S.; Gao, P. ADAM metallopeptidase domain 33 (ADAM33): A promising target for asthma. *Mediators Inflamm.* **2014**, *2014*, 572025. [CrossRef] [PubMed]

116. Gandy, S.; Petanceska, S. Neurohormonal signalling pathways and the regulation of alzheimer beta-amyloid metabolism. *Novartis Found. Symp.* **2000**, *230*, 239–251. [PubMed]

117. Amour, A.; Knight, C.G.; Webster, A.; Slocombe, P.M.; Stephens, P.E.; Knauper, V.; Docherty, A.J.; Murphy, G. The *in vitro* activity of ADAM-10 is inhibited by TIMP-1 and TIMP-3. *FEBS Lett.* **2000**, *473*, 275–279. [CrossRef]

118. Hotoda, N.; Koike, H.; Sasagawa, N.; Ishiura, S. A secreted form of human ADAM9 has an α-secretase activity for app. *Biochem. Biophys. Res. Commun.* **2002**, *293*, 800–805. [CrossRef]

119. Slack, B.E.; Ma, L.K.; Seah, C.C. Constitutive shedding of the amyloid precursor protein ectodomain is up-regulated by tumour necrosis factor-alpha converting enzyme. *Biochem. J.* **2001**, *357*, 787–794. [CrossRef] [PubMed]

120. Gough, M.; Parr-Sturgess, C.; Parkin, E. Zinc metalloproteinases and amyloid beta-peptide metabolism: The positive side of proteolysis in alzheimer's disease. *Biochem. Res. Int.* **2011**, *2011*, 721463. [CrossRef] [PubMed]

Review

ADAM and ADAMTS Family Proteins and Snake Venom Metalloproteinases: A Structural Overview

Soichi Takeda

Department of Cardiac Physiology, National Cerebral and Cardiovascular Center Research Institute, 5-7-1, Fujishirodai, Suita, Osaka 565-8565, Japan; stakeda@ri.ncvc.go.jp; Tel.: +81-6-6833-5012

Academic Editors: Jay Fox and José María Gutiérrez
Received: 8 April 2016; Accepted: 4 May 2016; Published: 17 May 2016

Abstract: A disintegrin and metalloproteinase (ADAM) family proteins constitute a major class of membrane-anchored multidomain proteinases that are responsible for the shedding of cell-surface protein ectodomains, including the latent forms of growth factors, cytokines, receptors and other molecules. Snake venom metalloproteinases (SVMPs) are major components in most viper venoms. SVMPs are primarily responsible for hemorrhagic activity and may also interfere with the hemostatic system in envenomed animals. SVMPs are phylogenetically most closely related to ADAMs and, together with ADAMs and related ADAM with thrombospondin motifs (ADAMTS) family proteinases, constitute adamalysins/reprolysins or the M12B clan (MEROPS database) of metalloproteinases. Although the catalytic domain structure is topologically similar to that of other metalloproteinases such as matrix metalloproteinases, the M12B proteinases have a modular structure with multiple non-catalytic ancillary domains that are not found in other proteinases. Notably, crystallographic studies revealed that, in addition to the conserved metalloproteinase domain, M12B members share a hallmark cysteine-rich domain designated as the "ADAM_CR" domain. Despite their name, ADAMTSs lack disintegrin-like structures and instead comprise two ADAM_CR domains. This review highlights the current state of our knowledge on the three-dimensional structures of M12B proteinases, focusing on their unique domains that may collaboratively participate in directing these proteinases to specific substrates.

Keywords: snake venom; metalloproteinase; disintegrin; ADAM; ADAMTS; MDC; reprolysin; adamalysin; shedding; crystal structure

---

1. Introduction

A disintegrin and metalloproteinase (ADAM) family proteins, also known as metalloproteinase-disintegrins or metalloproteinase/disintegrin-like/cysteine-rich (MDC) proteins, are type-I transmembrane and soluble glycoproteins that have diverse functions in cell adhesion, migration, proteolysis and signaling [1–3]. The best-characterized function of the membrane-anchored ADAMs is their involvement in ectodomain shedding of various cell-surface proteins, including the latent forms of growth factors, cytokines and their receptors and cell-adhesion molecules. For example, ADAM17 (TACE, TNF-α converting enzyme) is a sheddase involved in the processing of tumor necrosis factor-α [4,5] and a broad range of other cell-surface molecules [1]. Identification of a patient lacking ADAM17 revealed that ADAM17 is involved in the protection of the skin and intestinal barrier [6]. Another major family member, ADAM10, is a principal player in signaling via the Notch and Eph/ephrin pathways [7]. ADAMs play key roles in normal development and morphogenesis. Dysregulation of shedding activity is a crucial factor in a number of pathologies, such as inflammation, neurodegenerative disease, cardiovascular disease, asthma, cancer and others [1,3,8–11]. So far, 40 family members have been identified in the mammalian genome, of which 37 are expressed in mice (most of them in a testis-specific

27

manner) and 20, excluding presumed pseudogenes, are expressed in humans [3]. However, only 12 of the human ADAM members (ADAM8, 9, 10, 12, 15, 17, 19, 20, 21, 28, 30 and 33) contain a functional catalytic consensus sequence (HEXGEHXXGXXH, see below). The physiological function of the proteinase-inactive ADAMs (ADAM2, 7, 11, 18, 22, 23, 29 and 32) remains largely unknown, although some members of this group play important roles in development and function as adhesion molecules rather than proteinases [12,13]. ADAMs are widely expressed in mammalian tissues, and the observed phenotypes of ADAM knockout mice are subsequently diverse, although only ADAM10, 17 and 19 are essential for mouse development [3]. An increasing number of transmembrane proteins have been identified as the targets of ADAM-mediated proteolysis [1]. Some of these substrates can be cleaved by different ADAMs while others appear to be specific to an individual ADAM. In addition, no clear consensus sequences have so far been identified around the scissile bonds of the ADAM substrates. These observations highlight the need for a better understanding of how the substrate specificity and proteolytic activity of ADAMs are determined.

The ADAM with thrombospondin motifs (ADAMTS) family is a close relative of the ADAM family. ADAMTS members contain a varying number of C-terminal thrombospondin type-1 motifs in place of the ADAM transmembrane and cytoplasmic domains and thus function as secreted proteinases [14,15]. Unlike ADAMs, all ADAMTS share the catalytic consensus sequence mentioned above and thus encode active metalloproteinases. The human ADAMTS family includes 19 members that can be sub-grouped on the basis of their known substrates, namely aggrecanases or proteoglycanases (ADAMTS1, 4, 5, 8, 9, 15 and 20), procollagen N-propeptidases (ADAMTS2, 3 and 14), cartilage oligomeric matrix protein (also known as thrombospondin-5) cleaving proteinases (ADAMTS7 and 12), von Willebrand factor (VWF) cleaving proteinase (ADAMTS13) and a group of orphan enzymes (ADAMTS6, 10, 16, 17, 18 and 19) [14,15]. The gene name ADAMTS11 was assigned in error to a gene already designated as ADAMTS5, and thus the term ADAMTS11 is no longer used. Mendelian disorders resulting from mutations in ADAMTS2, 10, 13 and 17 identified essential roles for each gene [16]. ADAMTS13 is one of the most studied ADAMTSs because of its critical involvement in a thrombotic disorder [17,18]. ADAMTS13 is the sole VWF-cleaving enzyme in blood plasma and regulates the multimerization state of VWF for proper blood coagulation, and has no other known substrates. A deficiency in plasma ADAMTS13 activity causes thrombotic thrombocytopenic purpura (TTP), a hereditary or acquired (idiopathic) life-threatening disease [17–19]. Various lines of evidence indicate that ADAMT4 and 5 are the principal enzymes involved in the degradation of aggrecan, the major proteoglycan in articular cartilage, resulting in the development of osteoarthritis and have thus become targets for therapeutic inhibition [20]. Replacement of the C-terminal ancillary domains of ADAMTS5 with those of ADAMTS13 confers the ability to cleave VWF, suggesting that the non-catalytic C-terminal domains strongly determine the specificity of ADAMTS5 and ADAMTS13 [21]. The importance of the non-catalytic domains is also supported by the observation that autoantibodies against the ancillary domain of ADAMTS13 can inhibit proteinase activity sufficiently to cause TTP [22]. Different ADAMTS recognize very distinct substrates but the non-catalytic domains that characterize each ADAMTS family member may perform similar functions in other ADAMTS.

Snake venom is a complex mixture of bioactive enzymes and non-enzymatic proteins [23,24]. These toxic compounds appear to have resulted from the convergent or divergent evolution of physiological molecules to have a role in killing and paralyzing prey [25,26]. Snake venom metalloproteinases (SVMPs) have been inferred to be derived through recruitment, duplication and neofunctionalization of ancestral gene encoding closely related ADAM7, 28 and ADAMDEC-1 [27]; therefore, SVMPs are also referred to as snake ADAMs. Actually, large SVMPs, categorized into the P-III class SVMPs, have a modular structure that is homologous to the ectodomain of membrane-anchored ADAMs [28]. SVMPs identified so far share the catalytic consensus sequence and thus are soluble proteinases. Proteomic analyses of snake venoms show that SVMPs constitute more than 30% of the total proteins in many Viperidae venoms and are also present, but are less significant, in the venoms of Elapidae, Atractaspididae and some species of Colubridae [29,30]. These observations suggest that SVMPs play potentially sigc-nificant roles in envenomation-related pathogeneses, such as bleeding, intravascular clotting, edema, inflammation and necrosis [31,32]. SVMPs are the primary factors responsible for local and systemic hemorrhage and may also interfere with the hemostatic system through fibrinogenolytic or fibrinolytic activities, activation of prothrombin or factor *X*, and inhibition of platelet aggregation [33–35]. SVMPs are grouped into several classes according to their domain organization (see below). The high molecular weight P-III class SVMPs are characterized by higher hemorrhagic activity than the P-I class of SVMPs, which only have a catalytic metalloproteinase domain. Although SVMP-induced hemorrhages are primarily dependent on SVMP proteolytic activity, the proteolytic activities themselves do not parallel the potency of these activities. The stronger hemorrhagic activity of P-III SVMPs is, at least in part, likely caused by the resistance to inhibition by the plasma proteinase inhibitor α2-macroglobulin (α2M) probably because of the large molecular size of P-III SVMPs: P-I SVMPs are readily inhibited by α2M [36]. P-III SVMPs are capable of inducing not only local but also systemic bleeding, whereas P-I SVMPs mainly induce local hemorrhage [37–39]. Therefore, it is more likely that the *C*-terminal non-catalytic domains may contribute to the targeting of P-III SVMPs to relevant molecules in the extracellular matrix of capillaries. P-III SVMPs represent not only higher hemorrhagic activities, but also more diverse and specific biological activities than P-I SVMPs. These observations strongly suggest the importance of the non-catalytic ancillary domains of P-III SVMPs for their functions.

ADAMs, ADAMTSs and SVMPs share a topological similarity with matrix metalloproteinases (MMPs) in the structure of their catalytic domain [40]. However, their non-catalytic ancillary domains are clearly distinct from those of MMPs and other metalloproteinases, thus comprising the M12B clan of zinc metalloendoproteinases (MEROPS classification, https://merops.sanger.ac.uk/). The M12B proteinases are also referred to as adamalysins or repropysins, nomenclatures chosen to reflect the two distinct origins of proteins in this class: the first family member to be structurally characterized was adamalysin II from *rept*ile venom, whereas others belong to a group of proteinases initially described in male *reprod*uctive tissues [41–43]. Structure–function studies of the M12B proteinases were reviewed several years ago [44–47]. This review will update our knowledge of the three-dimensional structures of M12B proteinases and describe details of the structural features of their unique domains that may collaboratively participate in directing these proteinases to specific substrates.

2. Modular Architecture of ADAMs, ADAMTSs and SVMPs

Figure 1 depicts the modular domain architectures of M12B clan members. The mature ADAMs generally possess, from *N* to *C* terminus, metalloproteinase (M), disintegrin-like (D), cysteine-rich (C) and epidermal growth factor (EGF) domains, a short connecting linker, a hydrophobic transmembrane (TM) segment and a cytoplasmic tail. ADAM10 and 17 lack an EGF domain and thus, the TM segment follows the MDC domains [28,48]. The D and C domains can be structurally further divided into two subdomains, D_a and D_s, and C_w and C_h, respectively (see below) [28]. The *C*-terminal cytoplasmic tails of ADAMs are very diverse in terms of length (40–250 amino acids) and sequence, and probably do not adopt stable three-dimensional structures. Some ADAMs (ADAM9, 12 and 28) have splicing variants that are expressed as soluble active proteinases without the transmembrane and cytoplasmic regions [49–51]. The ADAMDEC-1 (decysin-1) is a unique protein comprising an M domain and a short disintegrin-like domain and is predicted to be secreted as a soluble proteinase [52]. ADAMDEC-1 harbors a putative zinc-binding sequence (HEXXHXXGXXD). However, the third zinc-coordinating residue in ADAMDEC-1 is an Asp instead of the His residues found in all other proteolytically active ADAMs, and thus ADAMDEC-1 is regarded as a member only of a novel subgroup of ADAMs [52].

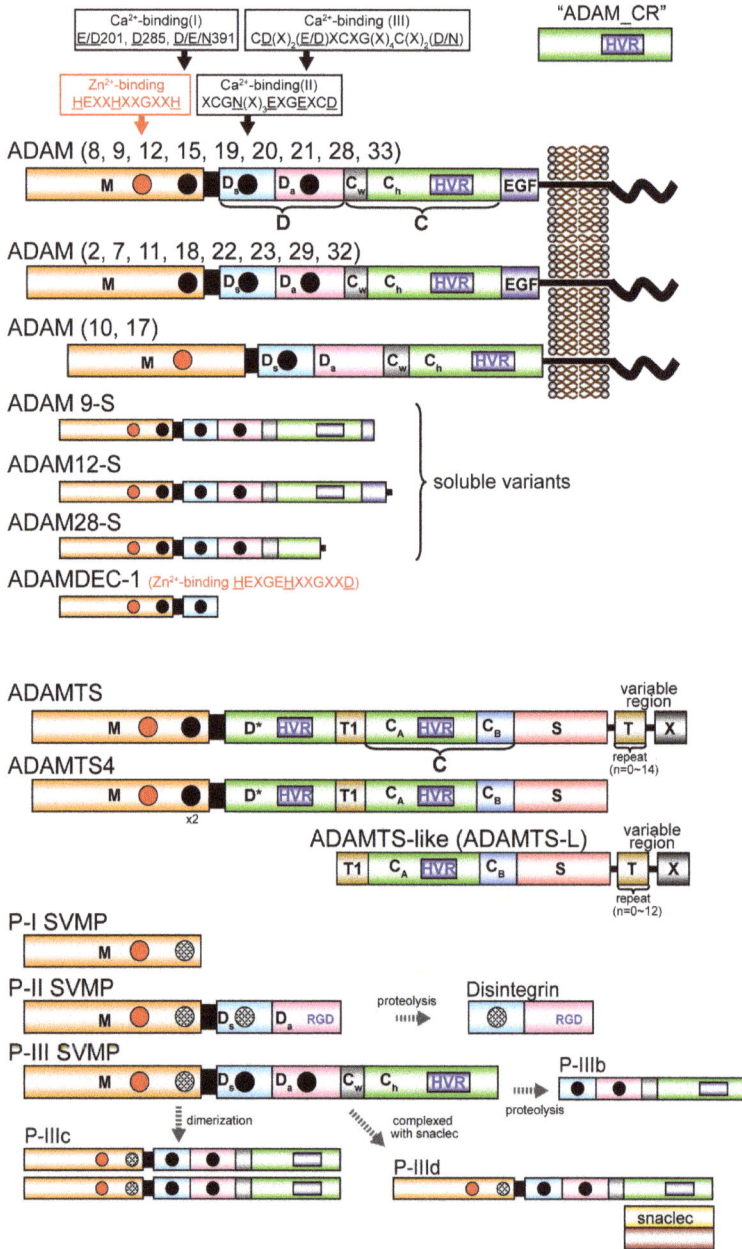

Figure 1. Schematic diagram of the domain structure of M12B proteinases. Each domain or subdomain is represented by a distinct color. The C_h subdomain of ADAMs and P-III SVMPs, D* domain of ADAMTSs, and C_A subdomain of ADAMTSs and ADAMTS-Ls, adopt the ADAM_CR domain fold and are thus shown in the same color. The region that is variable among ADAMTSs and ADAMTS-Ls is shown as X.

All mature ADAMTS members commonly possess, from the *N*- to *C*-terminus, metalloproteinase (M), disintegrin-like (D), central thrombospondin type-1 repeat (TSR) motif (T1), cysteine-rich (C) and spacer (S) domains. Despite its name, the D domain of ADAMTSs actually does not adopt a classic "disintegrin-like" tertiary structure, but has an ADAM_CR domain fold (see below) and is thus indicated as "D*" hereafter. The C domain in ADAMTS can be structurally further divided into two distinct subdomains, C_A and C_B (see below) [53]. ADAMTS4 has this basic core MD*TCS domain organization and other family members have a variety of more distal *C*-terminal domains, including one or more additional TSRs and additional domains denoted as "X" in Figure 1, which are characteristic of particular subgroups [14,15]. In the *C*-terminal region, ADAMTS9 and 20 have a GON-1 domain whereas ADAMTS13 has two CUB (complement components C1rC1s/urinary epidermal growth factor/bone morphogenic protein-1) domains. Several ADAMTSs (ADAMTS2, 3, 6, 7, 10, 12, 14, 16, 17 18 and 19) have a PLAC (protease and lacumin) domain, and ADAMTS7 and 12 have a mucin/proteoglycan domain interposed between TSR4 and TSR5 [14,15]. There are six ADAMTS-like (ADAMTS-L) proteins, which include ADAMT-L1 to 5 and papilin, resemble ADAMTS ancillary domains but lack the M and D domains. ADAMTS-Ls are products of distinct genes, not alternatively spliced variants of ADAMTS genes. ADAMTS-Ls appear to have architectural or regulatory roles in the extracellular matrix instead of a catalytic activity [15]. ADAMTS-L2 is implicated in an inherited connective tissue disorder named geleophysic dysplasia [54]. A homozygous *ADAMTS-L4* mutation was identified in isolated ectopia lentis [55].

SVMPs are classified into three major classes, P-I, P-II and P-III, according to their domain organization [34,56]. P-I SVMPs are composed of a single catalytic M domain. P-II SVMPs are synthesized as an M domain and a D domain. P-III SVMPs have a modular structure homologous to the MDC domains of the membrane-anchored ADAMs. In venoms, P-I and P-III SVMPs are abundant, but P-II SVMPs are frequently found in processed forms containing only their disintegrin domain, *i.e.*, classic disintegrins. P-III SVMPs can be divided further into subclasses depending on their post-translational modifications, such as proteolytic processing between the M and D domains (P-IIIb) or dimerization (P-IIIc), complexation (P-IIId) with additional snake venom C-type lectin-like proteins (snaclecs) [57], in addition to the canonical P-IIIa SVMPs. SVMPs of different classes are often present in the same viper venom. P-III SVMPs are present in the venoms of species of the families Viperidae, Elapidae, Atractaspididae and Colubridae, whereas P-I and P-II SVMPs have been described only in venoms of viperid species [58]. The evolutionary history of viperid SVMPs is characterized by repeated domain loss; the loss of the C domain precedes the formation of the P-II SVMPs, which in turn precedes the evolution of the P-I SVMPs through loss of the D domain [58–60].

All M12B proteinase members possess an *N*-terminal signal sequence that directs the proteinase into the secretory pathway. Adjacent to this signal sequence is the pro domain (typically approximately 200 amino acid residues) that has been suggested to assist with the correct folding of the protein and to maintain the proteinase in a latent state via a cysteine-switch [61] or other mechanism [62] until its cleavage either by a pro-protein convertase or by autocatalysis during its transit through the Golgi apparatus. Unlike other M12B members, the pro domain of ADAMTS13 is relatively short (only 41 residues) and is not required for its secretion and function [63].

3. Three-Dimensional Structures

The three-dimensional structures currently available for the M12B members are summarized in Table 1. Adamalysin II is a P-I SVMP isolated from *Crotalus adamanteus* and is the first M12B proteinase for which a crystal structure was solved in 1993 [42]. The first mammalian member, the M domain of human ADAM17 (TACE) structure was reported in 1998 [64]. To date, the isolated M domains or M-domain-containing structures of ten P-I SVMPs, seven P-III SVMPs, four ADAMs and three ADAMTSs are available in the Protein Data Bank (PDB). A significant advance in the field was the characterization of the crystal structure of the first P-III SVMP, vascular apoptosis-inducing protein-1 (VAP-1) in 2006 [28]. The structural determination of six P-III SVMPs, including almost all

P-III subclasses, followed that of VAP-1. The entire ectodomain structure of mammalian ADAMs is currently only available for ADAM22, which was reported in 2009 [65]. The ADAM22 structure was also the only non-catalytic ADAM for which a crystal structure was solved [65]. Other significant advances are the structural determination of the MD* domains of ADAMTS1 in 2007 [66] and the D*TCS domains of ADAMTS13 in 2009 [53]. The MD*-domain-containing structures of ADAMTS4 and 5 are also available in the PDB. Although no three-dimensional structure of the intact ADAMTS has been determined, a structural model of the core MD*TCS domain of ADAMTS13 has been proposed [53]. No pro domain-containing structures are currently available for M12B proteinases although several zymogen structures of MMPs have been deposited in the PDB [67].

Table 1. Selection of the 3D structures of the M12B proteinases deposited in the PDB.

Protein	Source	Domains	PDB ID	Year	Reference
ADAMs					
ADAM8	human	M	4DD8	2012	[68]
ADAM10	bovine	DC	2AO7	2005	[48]
ADAM17	human	M	1BKC	1998	[64]
ADAM17	human	C	2M2F (NMR)	2013	[69]
ADAM22	human	MDCE	3G5C	2009	[65]
ADAM33	human	M	1R54, 1R55	2004	[70]
ADAMTSs					
ADAMTS1	human	MD*	2JIH, 2V4B	2007	[66]
ADAMTS4	human	MD*	2RJP, 3B2Z	2008	[71]
ADAMTS5	human	M	3B8Z	2008	[72]
ADAMTS5	human	MD*	2RJQ	2008	[71]
ADAMTS13	human	D*TCS	3GHM, 3GHN, 3VN4	2009	[53,73]
P-I SVMPs					
acutolysin A	*A. acutus*	M	1BSW, 1BUD	1998	[74]
acutolysin C	*A. acutus*	M	1QUA	1999	[75]
adamalysin II	*C. adamantus*	M	1IAG	1993	[42,76]
atrolysin C	*C. atrox*	M	1ATL, 1HTD	1994	[77]
BaP1	*B. asper*	M	1ND1	2003	[78]
BmooMPα-I	*B. moojeni*	M	3GBO	2010	[79]
F II	*A. acutus*	M	1YP1	2005	[80]
H2 proteinase	*T. flavoviridis*	M	1WNI	1996	[81]
TM-1	*T. mucrosquamatus*	M	4J4M	2013	[82]
TM-3	*T. mucrosquamatus*	M	1KUF, 1KUG, 1KUI, 1KUK	2002	[83]
P-IIIa/b SVMPs					
AaHIV	*A. acutus*	MDC	3HDB	2009	[84]
atragin	*N. atra*	MDC	3K7L	2010	[85]
bothropasin	*B. jararaca*	MDC	3DSL	2008	[86]
catrocollastain/ VAP2B	*C. atrox*	MDC	2DW0, 2DQ1, 2DW2	2007	[87]
K-like	*N. atra*	MDC	3K7N	2010	[85]
P-IIIc SVMPs					
VAP1	*C. atrox*	2x(MDC)	2ERO, 2ERP, 2ERQ	2006	[28]
P-IIId SVMPs					
RVV-X	*D. russelli*	MDC+snaclec	2E3X	2007	[88]
multactivase	*E. multisquamatus*	DC+snaclec		unpublished	

* Despite its name, the D domain of ADAMTSs actually does not adopt a classic "disintegrin-like" tertiary structure, but has an ADAM_CR domain fold and is thus indicated as "D*".

3.1. M Domain

The M domains of M12B proteinases range from 180 to 260 (typically 200–210) residues in length [33,87]. The currently available M domain structures of ADAMs, ADAMTSs and all classes of SVMPs are very similar to each other, although comparison of the amino acid sequences of various members shows high variability (typically 20%–50% identity). Interestingly, although the human ADAM8 M domain is most similar in sequence to the human ADAM33 M domain (44% identity), its crystal structure is most similar to that of P-I SVMP adamalysin II [68]. The M domain of the non-catalytic ADAM22 also adopts a very similar backbone structure to those of other catalytic ADAMs, ADAMTs and SVMPs [65]. The M domain of M12B proteinases has a core structure with a conserved molecular topology consisting of a five-stranded β-sheet, four long α-helices, and one short *N*-terminal α-helix. Figure 2A depicts the M domain structure of catrocollastatin/VAP2B, a representative of P-III SVMPs, in complex with the hydroxamic inhibitor GM6001 as viewed from the so-called standard orientation, a frontal view of the horizontally-aligned active site-cleft proposed for the general description of structural features of metalloproteinses [89]. The M domain has an oblate ellipsoidal shape with a notch in its flat side that separates the upper subdomain (about 150 *N*-terminal residues, colored in olive) from an irregularly folded lower subdomain (about 50 *C*-terminal residues, colored in magenta). The active site cleft extends horizontally across the flat surface of the M domain to accommodate the peptidic inhibitor (Figure 2B). The amino acid sequence of the irregular lower domain region is highly divergent among M12B members and is therefore important for substrate recognition because it forms part of the wall of the substrate-binding pocket. Crystal structures of inhibitor-bound M domain complexes suggest that the hydrogen-bond network formed between the extended substrate and the adjacent pocket-flanking regions of the enzyme resembles that of an antiparallel β-sheet, in essence extending the central β-sheet by two strands [46]. The catalytic site is characterized by a consensus HEXXHXXGXXH sequence (residues 333–343 in catrocollastatin/VAP2B sequence), which is conserved not only in M12B members but also across the metzincin superfamily of metalloproteinases, which also contains MMPs, astacins, and serralysins [40,90]. The three conserved histidine residues (His333, His337 and His 343) coordinate the catalytic zinc ion and Glu334 functions as a catalytic base at the bottom of the active site cleft. The conserved Met357, located 12–24 residues downstream of the catalytic consensus sequence, folds into a so-called Met-turn and forms a hydrophobic base beneath the three zinc-binding imidazole rings, a hallmark of the metzincin superfamily of proteinases.

The secondary structure arrangement of the M domain is similar to that of other metzincins, such as astacin and MMPs, except for the large insertion of the H3 helix and the loop between strand S1 and helix H3 [40]. This insertion contributes to the creation of a Ca^{2+}-binding site(S), which is unique to M12B proteinases. Most M12B members have one or two (the case for some ADAMTSs) structural calcium ions (Ca^{2+}-binding site I) in close proximity to the crossover point of the *N*- and *C*-termini of the M domain opposite the catalytic site (Figure 2C). In catrocollastatin/VAP2B, the Ca^{2+} ion is coordinated by the side-chains of Asp285, Asn391 and Glu201, the main-chain carbonyl oxygen atom of Cys388, and two water molecules in a pentagonal bipyramidal arrangement. Some ADAMs (e.g., human ADAM10 and 17) and SVMPs have substitutions in these Ca^{2+}-coordinating residues and thus lack Ca^{2+}-binding at this site. For example, Glu201 and Asn391 are replaced by Lys202 and Lys392, respectively, and the distal Nε atom of Lys202 substitutes for the Ca^{2+} ion in VAP1 (Figure 2C) [28]. Replacement of the Ca^{2+}-coordinating Glu residue with Lys is also observed in other SVMPs and ADAMs. The high degree of conservation of residues involved in Ca^{2+}-binding or in mimicking Ca^{2+}-binding might reflect the importance of this region for the structural link between the M and D domains. In addition, Ca^{2+} protects against autoproteolysis at this M/D domain junction [76,91]. In ADAMTS1, 4 and 5, a second bound Ca^{2+} ion is found with a metal-metal distance to the first conserved site of around 4Å. The residues coordinating the second Ca^{2+} ion are not conserved in all ADAMTS sequences, hence the second Ca^{2+} ion at this site may not necessarily be a feature of the ADAMTS family. A distinctive feature of the M domain of the M12B proteinases, when compared to that of MMPs, is the presence of two to four disulfide bonds that stabilize the structure (whereas MMPs have none).

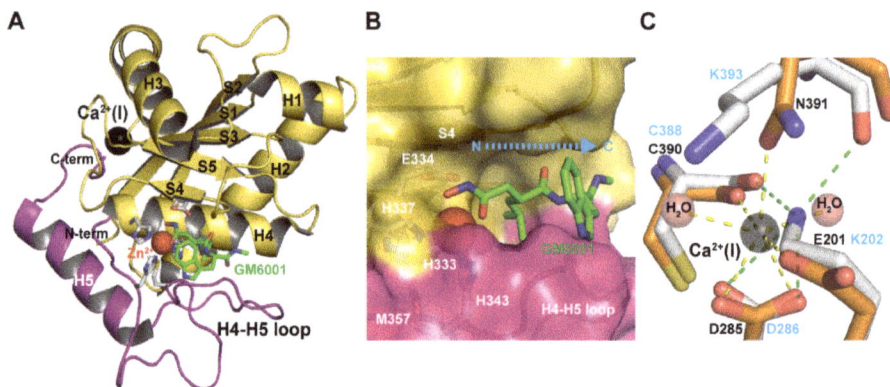

Figure 2. Catalytic M domain structure. (**A**) Structure of the M domain of catrocollastatin/VAP2B in complex with GM6001 (2DW0). The upper and lower subdomains are colored in gold and magenta respectively. (**B**) Close up view of the catalytic site. (**C**) Close up view of the Ca^{2+}-binding site of catrocollastatin/VAP2B (shown in orange) overlaid on the corresponding region of VAP1 (shown in gray). Residues in catrocollastatin/VAP2B and VAP1 are indicated in black and cyan, respectively.

In ADAMTS4 and 5, in addition to the two Ca^{2+} ions at site I, another bound Ca^{2+} ion has been observed in the M domain in close proximity to the active site (Figure 3A). Crystal structures of the MD* domain-containing fragment of ADAMTS4 in the presence or absence of the inhibitor revealed that the active site of ADAMTS4 adopts two alternative conformations that may exist in equilibrium: an inhibitor-bound "open" structure with an additional Ca^{2+} ion bound (Figure 3B) and an apo "closed" inaccessible structure without a bound Ca^{2+} ion (Figure 3C) [71]. In the open form, the Ca^{2+} ion is coordinated by the side-chain oxygen atoms of Asp320 and Glu349 and the main-chain carbonyl oxygens of Leu321, Cys327 and Thr329. The major difference between these two states is found in the position and conformation of the short disulfide-containing "S2'-loop" encompassing residues 322–330. In the apo state, the S2'-loop moves from its "open" position toward the catalytic Zn^{2+} ion by ~8Å and folds into the active site in a "closed" autoinhibited state in which the side-chain carboxylate of Asp289 chelates the Zn^{2+} ion, resulting in the removal of bound Ca^{2+} ion (Figure 3D). Owing to the strong sequence similarity among ADAMTS4 and other ADAMTSs (ADAMTS1, 5, 8 and 15) in the S2'-loop, which has the consensus CGXXXCDTL sequence, and the Ca^{2+}-coordinating Asp320 and Glu349 (Figure 3E), it seems likely that these ADAMTSs may also bind Ca^{2+} and adopt two alternative conformations. ADAMTS13 does not share the S2'-loop sequence with ADAMTS5, but a site-directed mutagenesis study suggested that Ca^{2+}-binding to the residues constituting this loop strongly affects the catalytic activity of ADAMTS13 [92]. The crystal structure of the M domain of ADAMTS13 remains to be elucidated. The above consensus sequence and the existence of two distinct conformational states in the M domain have not been observed in either ADAMs or SVMPs.

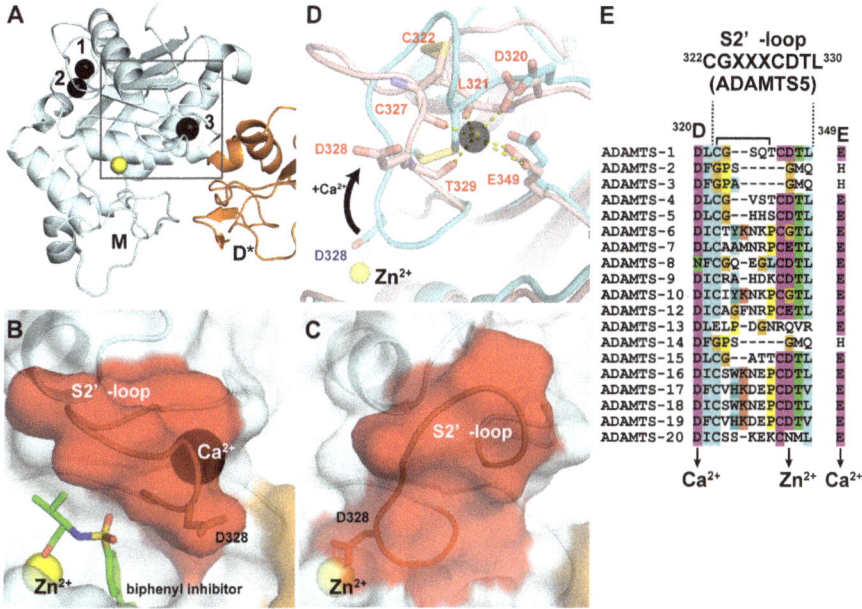

Figure 3. Crystal structures of ADAMTS5-MD*. (**A**) Overall structure of ADAMTS-MD*. Zn^{2+} and Ca^{2+} ions are shown in yellow and black spheres, respectively. Molecular surface of the active site in the inhibitor-bound "open" (**B**) and the apo "closed" form (**C**). (**D**) Superimposition of the two loop configurations: The closed conformation is depicted in cyan, the open conformation is shown in pink. (**E**) Amino acid sequence alignments of ADAMTSs around the "S2'-loop". The Genbank IDs for each ADAMTS sequence are, ADAMTS1 (Genbank ID (GI): 50845384), ADAMTS2 (GI: 3928000), ADAMTS3 (GI: 21265037), ADAMTS4 (GI: 12643637), ADAMTS5 (GI: 12643903), ADAMTS6 (GI: 64276808), ADAMTS7 (GI: 38197242), ADAMTS8 (GI: 153792351), ADAMTS9 (GI: 33624896), ADAMTS10 (GI: 56121815), ADAMTS12 (GI: 51558724), ADAMTS13 (GI: 21265034), ADAMTS14 (GI: 21265052), ADAMTS15 (GI: 21265058), ADAMTS16 (GI: 32363141), ADAMTS17 (GI: 37999850), ADAMTS18 (GI: 76800647), ADAMTS19 (GI: 29336810) and ADAMTS20 (GI: 28316229).

3.2. C-Shaped MDC Domains of ADAMs and P-III SVMPs

Figure 4A depicts the crystal structure of catrocollastatin/VAP2B, the first monomeric P-III SVMP structure to be solved [87], representing a structural prototype of P-III SVMPs. The crystal structures of P-III SVMPs reveal that the MDC domains fold into a C-shaped configuration in which the distal HVR portion (see below) of the C domain is situated near to, and faces towards, the catalytic site in the M domain. The complete ectodomain (M/D/C/EGF domains) structure of ADAM22 (Figure 4B) shows that four domains assemble together like a four-leaf clover, each leaf representing one of the four domains [65]. ADAM22 structure reveals that the C-shaped configuration of the MDC domains found in SVMPs are conserved in mammalian ADAMs, and the additional EGF domain is tightly associated with both the D and C domains forming a continuous D/C/E module. In catalytically active ADAMs, the EGF domain may form a rigid spacer that correctly positions the MDC domains against the membrane for the subsequent shedding of membrane-anchored molecules. The D domain is linked to the M domain by a short linker (7–12 amino acid residues) that allows variable orientation and positioning between the M and D domains [28,46,65,87]. Consistent with this, comparison of the available P-III SVMP and ADAM structures reveals substantial diversity in the relative position of the M and D domains [87]. For example, catrocollastatin/VAP2B shows an open C-shaped molecule with no direct interaction between the M and D domains except at the domain junction, whereas the two domains directly interact with each other in ADAM22 and thus adopt a closed C-shaped structure (Figure 4C). The flexibility of the molecule is reflected in the ability of the same proteins to crystallize in different crystal forms, and *vice versa* [93]. The structures of ADAMs and P-III SVMPs are most likely dynamic, allowing for a varying distance between the M domain and the rest of the molecule. This intrinsic flexibility may be important for fine-tuning substrate recognition, by adjusting the spatial alignment between the catalytic region and the exosite (see below) during the catalytic cycle.

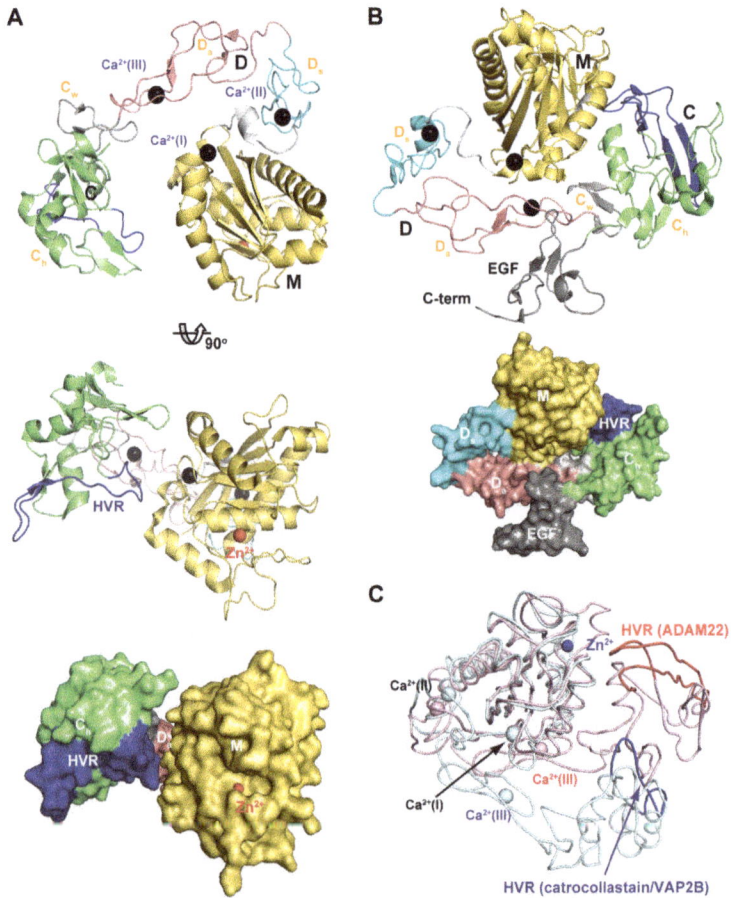

Figure 4. C-shaped MDC-domain configuration of ADAMs and P-III SVMPs. Ribbon and molecular surface representations of the crystal structure of catrocollastatin/VAP2B (**A**) and ADAM22 (**B**). (**C**) Superimposition of the M domains of catrocollastatin/VAP2B (shown in cyan) and ADAM22 (shown in pink).

In some instances, substantial amounts of processed DC fragments of P-IIIb SVMPs have been identified in venoms alongside their unprocessed counterparts [94,95]. Although lacking proteolytic activity, such isolated DC fragments display diverse biological activities, such as inhibition of collagen-stimulated platelet aggregation and the modulation of cell adhesion, migration, and proliferation, implying that the DC fragments derived from P-IIIb SVMPs are also important in the toxicity of the venoms [33,56]. Some membrane-anchored ADAMs, such as ADAM2 (fertilin-β) and

ADAM1 (fertilin-α), undergo proteolytic processing within the M/D-linker and the Ca^{2+}-binding site III (see below), respectively, at different stages of sperm maturation [12,96]. A flexible modular structure, in addition to Ca^{2+}-binding, may also play a role in differential proteolytic processing of precursor proteins, giving rise to the functional complexity of snake venoms, as well as in the post-translational regulation of ADAMs' functions, probably by modifying the capabilities of protein–protein interactions.

3.3. Arm Structure in ADAMs and P-III SVMPs

The D domain that follows the M domain of ADAMs and P-III SVMPs can be further subdivided into two structural subdomains, the "shoulder" (D_s, residues 403–436 in catrocollastatin/VAP2B sequence) and the "arm" (D_a, residues 437–486) [28] (Figure 5). Both subdomains consist largely of a series of turns and constitute an elongated curved arm structure together with the immediately subsequent region of the primary sequence, the N-terminal region of the C domain designated as the "wrist" (C_w, residues 437–503) subdomain (Figure 5A). The structure of the entire C-shaped arm ($D_s/D_a/C_w$) itself seems to be rigid because it is stabilized by a number of disulfide bonds and structural Ca^{2+} ions. There are three disulfide bonds in each D_s and D_a, and one in C_w, with the subdomains (e.g., D_s/D_a and D_a/C_w) connected by single additional disulfide bonds. The numbers and spacing of the cysteine residues involved in these disulfide bonds are strictly conserved among ADAMs and P-III SVMPs [28,87] (Figure 5F), with few exceptions, one of which is the kaouthiagin-like (K-like) SVMP from *Naja atra*. The K-like proteinase lacks the 17-amino acid segment at the junction of the D_s and D_a subdomains, resulting in a different disulfide-bond pattern in the D domain. Consequently, the K-like proteinase has a different orientation between the D_s and D_a subdomains when compared to that of catrocollastatin/VAP2B (Figure 5B), and thus the MDC domains of K-like proteinase adopt a more elongated, I-shaped configuration [85]. However, how this I-shaped structure correlates with the proteinase function remains to be elucidated.

Both the D_s and D_a subdomains contain structural Ca^{2+}-binding sites that were not predicted from the amino acid sequences [28,87]. In the D_s subdomain, the side-chain oxygen atoms of the highly conserved Asn408, Glu412, Glu415 and Asp418 (represented by the consensus sequence XCGN(X)$_3$EXGEXCD, in which the side-chains of underlined residues are involved in Ca^{2+}-binding) and the main-chain carbonyl oxygen atoms of Val405 and Phe410 are involved in pentagonal bipyramid coordination of the Ca^{2+}-binding site II (Figure 5C). On the other hand, the side-chain oxygen atoms of Asp469, Asp472 and Asp483 and the main-chain carbonyl oxygen atoms Met470 and Arg484, as well as a water molecule, coordinate the Ca^{2+} ion at the corner of a pentagonal bipyramid and constitute the Ca^{2+}-binding site III in the D_a subdomain (Figure 5D). These residues are also highly conserved among all known ADAMs and P-III SVMPs, with the exception of ADAM10 and 17, and are represented by the consensus sequence CD(X)$_2$(E/D)XCXG(X)$_4$C(X)$_2$(D/N) [28,87]. Both bound Ca^{2+} ions in sites II and III are deeply buried and tightly coordinated and cannot be stripped from ADAM22, even using EDTA [65]. Therefore, these Ca^{2+} ions are likely to remain permanently in place once the D domain is folded.

The overall structures of the D domain of P-III SVMPs and ADAM22 are similar to that of trimestatin, an RGD (Arg-Gly-Asp sequence)-containing classic disintegrin [97] (Figure 5E). The integrin-binding ability of disintegrins has been attributed to a highly mobile hairpin loop (disintegrin loop) that contains the cell-adhesion sequence RGD at its tip. In ADAMs and P-III SVMPs, the RGD sequence is usually replaced by an (D/S)XCD sequence (residues 466–469 in the catrocollastatin/VAP2B sequence). The disintegrin-like loops of P-III SVMPs and ADAMs are packed against the subsequent C_w subdomain, and a disulfide bond (Cys468/Cys499) and bound Ca^{2+} ion at site III further stabilize the continuous rigid D_a/C_w structure. Therefore, in ADAMs and P-III SVMPs, the disintegrin-like loop is inaccessible for protein–protein interactions due to steric hindrance. Disintegrins (40–100 amino acids) are typically generated by proteolytic processing of larger precursor P-II SVMPs [98–100], albeit with some exceptions [101]. Most P-II SVMPs have two to four fewer cysteine residues in the D_s subdomain than P-III SVMPs, and thus one or two fewer disulfide bonds.

In addition, there are substitutions of the key residues constituting the Ca^{2+}-binding site II and III in most P-II SVMPs [87]. Although a number of disintegrin structures have been determined by NMR and X-ray crystallography [100], no structural Ca^{2+}-binding has been identified in these structures and the D_s subdomain region of disintegrins is generally shorter and less ordered than the corresponding regions of ADAMs and P-III SVMPs. Because of the lack of structural Ca^{2+} ions, disintegrin structures are more flexible throughout the molecule, than the corresponding region of ADAMs and P-III SVMPs. The flexibility of RGD-containing disintegrin loops is probably important for the binding of integrins. As previously mentioned, P-II SVMPs may have evolved from ancestral P-III SVMP genes after losing the genetic information encoding the protein regions downstream of the D domain [58–60]. Removal of structural constraints (disulfide bonds and structural Ca^{2+}-binding sites), imposed both on the disintegrin loop and the D_s subdomain in the ancestral P-II SVMPs, has been postulated as the key event that permitted the subsequent evolution of both integrin-binding activity and the proteolytic release mechanism.

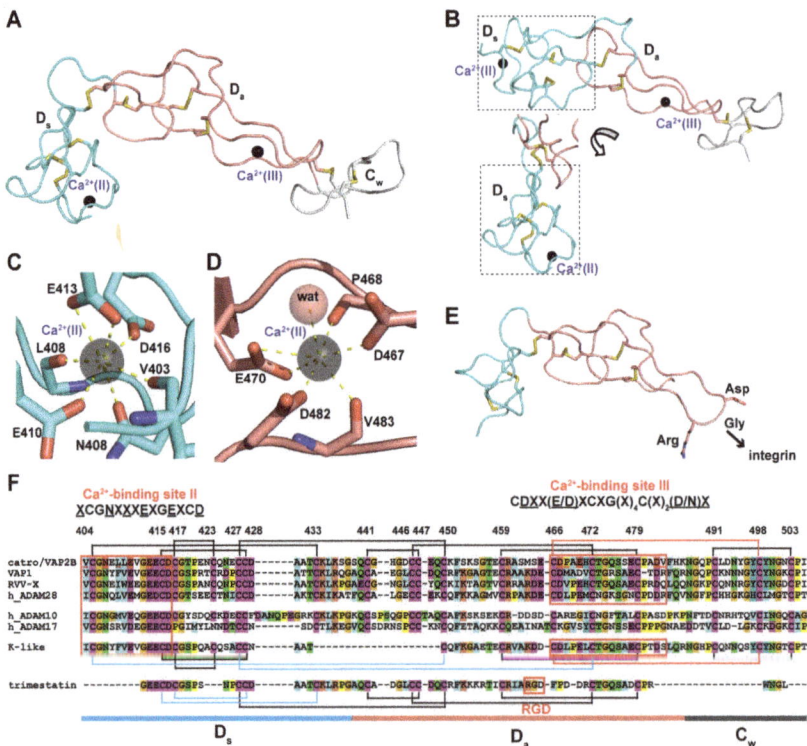

Figure 5. Arm structure. (**A**) The D_s, D_a, and C_w subdomains of catrocollastatin/VAP2B (2DW0) are shown in cyan, pink, and gray, respectively. (**B**) The D domain of K-like proteinase (3K7N) with two different views of the D_s subdomain (in dotted line boxes). Close up views of the Ca^{2+}-binding sites, II (**C**) and III (**D**) in catrocollastatin/VAP2B. (**E**) Structure of an RGD-containing disintegrin, trimestatin (1J2L). Suggested integrin-binding residues are indicated. (**F**) Amino acid sequence alignment of catrocollastatin/VAP2B (PDB: 2DW0_A), VAP1 (PDB: 2ERO_A), RVV-X (PDB: 2E3X_A), human ADAM28 (Genbank ID (GI): 98985828), human ADAM10 (GI: 29337031), human ADAM17 (GI: 14423632), K-like (PDB: 3K7Y_A) and trimestatin (1J2L_A) generated using Clustal X2 (http://www.clustal.org/clustal2/). Disulfide bonds and the boundaries of the subdomains are schematically indicated. Ca^{2+}-binding sites II and III are boxed in red.

While the pattern of disulfide-bond pairing in the D domain determined thus far is strictly conserved among ADAMs and P-III SVMPs, with the exception of K-like proteinase, it may be possible that multiple structural isoforms of the same SVMPs exist in the venom, perhaps as the result of alternative disulfide-bond pairing [102]. For example, the disintegrin bitistatin, which is derived from the precursor P-II SVMP, adopts at least two distinct conformations, the result of different disulfide-bonding patterns [103]. Recently, protein-disulfide isomerase (PDI) was implicated in the regulation of shedding activity of ADAM17 [104], and an NMR structural analysis of the C_h subdomain of ADAM17 revealed that PDI can act on this subdomain and convert it from the inactive to the active conformation by disulfide-bond isomerization [69].

3.4. ADAM_CR Domain, Another Hallmark of M12B Proteinases

The C domain of ADAMs and P-III SVMPs, typically about 80–150 amino acid residues, can be structurally subdivided into the "wrist" (C_w, residues 437–503) and the "hand" (C_h, residues 504–609 in catrocollastatin/VAP2B sequence) subdomains [28,87]. As mentioned, the C_w subdomain tightly associates with the D domain, and the two are integrated into one continuous structure. On the other hand, the C_h subdomain constitutes a separate unit and has a unique structure consisting of irregularly folded loops with a core α/β-fold and four to five disulfide bonds. The C_h subdomain has a novel fold with no structural similarity to any currently known proteins, with the exception of the corresponding segments of M12B proteinases. The whole C domain of P-III SVMPs and ADAMs has been deposited in the Conserved Domain Database (CDD, http://www.ncbi.nlm.nih.gov/cdd) and the Pfam database (http://pfam.xfam.org/) as the ADAM_CR domain (cl15456 and PF08516, respectively). Here, we define the C_h subdomain of ADAMs and SVMPs and corresponding regions of ADAMTSs (D* domain and C_A subdomain, see below) as the ADAM_CR domain in a more restricted sense.

Crystallographic studies on the D* domain-containing fragments of ADAMTS1, 4, 5 and 13 revealed that the D* domain of ADAMTSs has no structural similarity to classic snake disintegrins, but is very similar in structure to the C_h subdomain of ADAMs and P-III SVMPs [44,53,66,71]. The N-terminal portion of the C domain of ADAMTSs (the C_A subdomain) also possesses essentially the same fold as the C_h subdomain, even though the two share no apparent sequence similarity [53]. Thus while the "disintegrin" nomenclature has been used to describe ADAMTS family proteinases, ADAMTSs actually contain no disintegrin-like structures, but instead have two homologous domains that belongs to the ADAM_CR. Therefore, it is now obvious that the presence of the evolutionarily-conserved ADAM_CR domain, not the disintegrin domain, is another hallmark of the M12B members in addition to the catalytic M domain architecture.

Figure 6A,B depict ribbon representations of the C_h subdomain of catrocollastatin/VAP2B and the D* domain of ADAMTS5, respectively, two typical ADAM_CR domain structures. Although there is negligible sequence identity between these two protein portions (~16%), they clearly show similar topologies. The topology diagram of these two protein portions is shown in Figure 6C. The conserved regions are a core α-helix (shown in red), two sets of short β-sheets (shown in yellow), and four disulfide bonds (shown in orange). Major differences between the two molecules are observed in the segment between the two N-terminal strands, S1 and S4, shown in gray. A short connecting loop of six amino acids in ADAMTS5 is replaced by a 27 amino acid residue insertion forming a central α-helix and two consecutive hairpin loops protruding out the top of the molecule in the case of catrocollastatin/VAP2B. This segment is named variable loop (V-loop) [28,53]. Current ADAM_CR domain structures can be classified into two groups according to the length of their V-loop. All of the C_h subdomains of SVMPs determined thus far and ADAM22 show a catrocollastatin/VAP2B type long V-loop structure (classified as group-A, Figure 6C), whereas ADAM10 and 17, and the D* domains and C_A subdomains of ADAMTSs have a short ADAMTS5-D* type V-loop (classified as group-B, Figure 6D). Inspection of the amino acid sequence alignments of other M12B members suggests that the C_h subdomains of all known P-III SVMPs and ADAMs, except for ADAM10 and 17, are classified into group-A, whereas the D* and C_A domains of ADAMTSs are classified into group-B. The V-loop exhibits a high level of variability among the group-B ADAM_CR structures (Figure 6D), comparable to that of the HVR (see below), while the structure of the V-loop in group-B molecules in general is quite mobile and potentially functions as a protein-protein interaction site in addition to the HVR (see below).

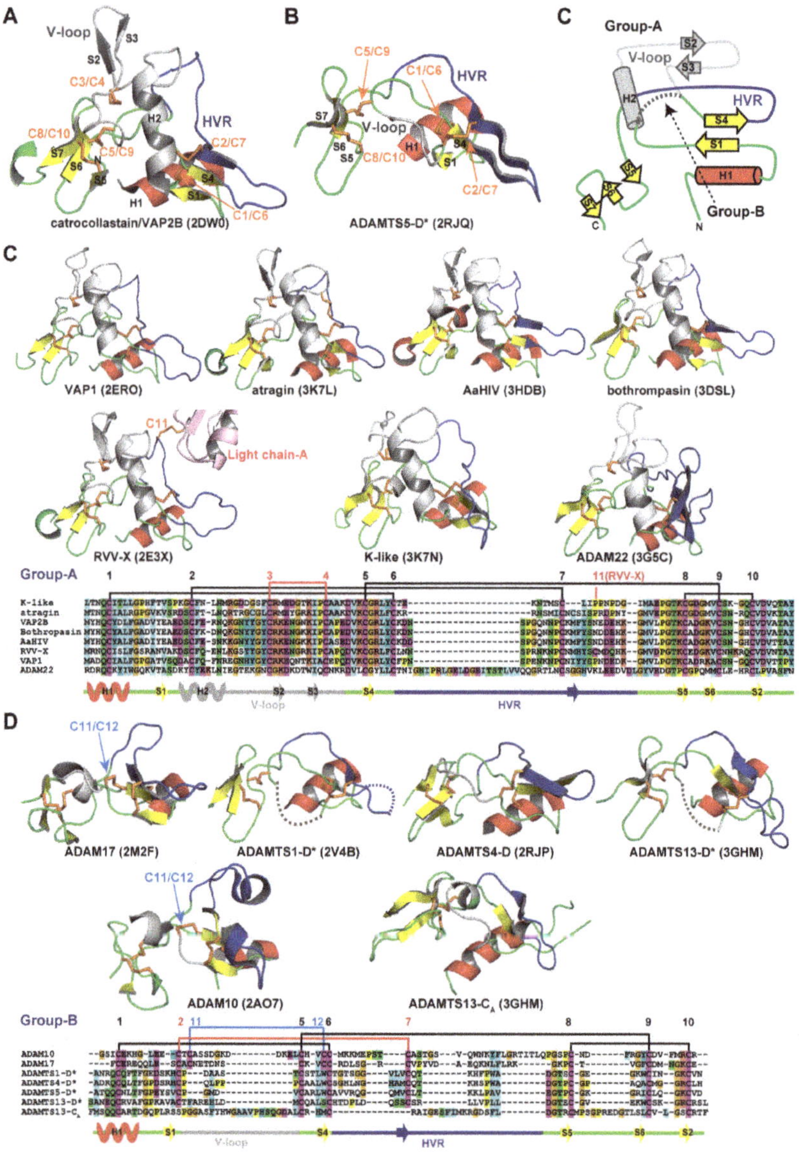

Figure 6. ADAM_CR domain. Ribbon representation of the C_h subdomains of catrocollastain/VAP2B (**A**) and the D* domain of ADAMTS5 (**B**). (**C**) Topology diagram of the ADAM_CR domain. Gallery of the group-A (**C**) and group-B (**D**) ADAM_CR domain structures. Conserved α-helix and β-strands are shown in red and yellow, respectively. Disulfide bonds, residues in HVR, and the residues in the V-loop are shown in orange, blue, and gray, respectively. The PDB ID for each protein structure is indicated in parentheses.

The overall structure of the C_h subdomain of catrocollastatin/VAP2B is very similar to that of six other SVMPs and that of ADAM22, with variability occurring mostly in loop regions. Of note, aside from the V-loop, the loop encompassing residues 561–582 (catrocollastatin/VAP2B sequence, shown in blue in Figure 6A,C) and extending across the central region of the C_h subdomain is the most variable both in length (16–22 amino acids in SVMPs and 27–55 amino acids in human ADAMs) and in amino acid composition. Therefore, this region has been designated as the hypervariable region (HVR) [28,44]. The HVRs in ADAMTSs are relatively short (13–17), but also show variability in their amino acid sequences when compared with different ADAMTSs and ADAMTS-Ls [53]. In ADAM22 and SVMP structures, the HVR is present at the distal end of the C-shaped MDC domains, and points toward and is situated close to the catalytic site of the M domain (Figure 4). This raises the intriguing possibility that the HVR creates an exosite for substrate binding [28,44]. Different ADAMs and SVMPs have distinct HVR sequences, resulting in distinct molecular surface features. Therefore, in addition to the V-loop, the HVR might have a role in specific protein-protein interactions for the cleavage by the M domain, providing a structural correlate for the diversity of biological activities characteristic of ADAMs and P-III SVMPs. The D domain is located opposite to and apart from the M domain active site and thus plays a primary role as a scaffold that spatially allocates two functional units, the catalytic site and exosite, to both ends of the C-shaped molecule.

Several reports suggest that the HVR region directly contributes to the substrate recognition of ADAMs and SVMPs. Most of these studies, however, used synthetic peptides derived from the HVR region or the isolated domains expressed in *E. coli* for functional assays. It should be noted that short peptides or *E. coli* expressed cysteine-rich proteins do not always mimic their counterparts in the intact molecule. The whole C domain or DC domains of ADAMs are suggested to be involved in protein–protein interactions [105–108]. The acidic surface pocket, which is located apart from both HVR and the V-loop within the C domain of ADAM10, defines cleavage specificity in Eph/ephrin signaling [48]. Recently, the membrane proximal domain (MPD, corresponding to the C_h subdomain in this text) of ADAM17, was shown to be responsible for recognition of two type-I transmembrane substrates, the IL-6R and the IL-1RII, but not for the interaction with the type-II transmembrane molecule TNF-α [109]. Further studies identified that the membrane proximal amphipathic 17 amino acid segment, which has the ability to bind lipid bilayers *in vitro*, is also involved both in substrate recognition and in regulating the shedding activity of ADAM17 [110,111], as well as MPD, which functions as a PDI-dependent molecular switch [69]. Most of these studies, however, do not identify specific regions of the C domain involved in the interactions, and the molecular mechanisms underlying substrate recognition remain to be elucidated. There are no systematic structure-based mutagenesis studies of the HVR region or the V-loop of particular ADAMs or SVMPs, and thus there is still no clear evidence establishing that these regions actually form an exosite. In contrast to the situation for ADAMs and SVMPs, the HVR and the V-loop in the D* and the C_A domains of ADAMTS13 have actually been shown to constitute VWF-binding exosites (see below).

3.5. Structures of Subclasses of P-III SVMP

Proteins with multimer and/or heterogeneous complex structures are frequently observed in snake venoms. Such multimers or protein complexes generally exhibit markedly enhanced pharmacological activities compared to the individual components and thus may play significant roles in snake venom toxicity [112]. Some SVMPs exist as a homo- or hetero dimer (P-IIIc) or as a hetero trimer (P-IIId). The formation of dimers or higher-order oligomers is not uncommon within M12B members. ADAMTS5 can form oligomers and this oligomerization is required for full aggrecanase activity [113]. Early purifications of ADAMTS2 and 13 indicated that these enzymes formed oligomers [114,115], however, there has been no further characterization of these oligomers. Membrane-bound ADAMs, ADAM17 [116] and the sperm-specific ADAMs, such as ADAM2 and 3 [12], exist as multimers in the cell membrane. However, how the multimeric state of these ADAMs and ADAMTSs relates to their functions is largely unknown.

Figure 7A depicts the crystal structure of VAP1, a homodimeric P-IIIc SVMP. The structure revealed an inter-chain disulfide bond formed between symmetry-related Cys365 residues and some features that characterize P-IIIc SVMPs [28]. The top of the dimer interface is capped by hydrophobic interactions involving Tyr209, Ile210, Leu213, and Tyr215 and the aliphatic portion of Lys214 (Figure 7B). At the middle, there are specific interactions that are best characterized by the QDHSK sequence (residues 320–324 in VAP1) (Figure 7C). The C-terminal region of this segment (residues 322–324) forms an antiparallel β-sheet with its counterpart. In addition, water molecules are bound to the side-chain oxygen atoms of His322 and Ser323 and form a hydrogen-bond network that further stabilizes the interface between the monomers. Lys324 plays a pivotal role in the key-to-keyhole recognition between the monomers. The Nε amino group of Lys324 is coordinated by six oxygen atoms, which belong to the opposite chain and are located at the corners of a pentagonal pyramid. The six atoms include the side-chain oxygen atoms of Asn295 and Gln320, the carbonyl oxygen atoms of Phe296, Gly298 and Thr300, and a water molecule (Figure 7C). The intermolecular disulfide bond, located at the bottom of the dimer interface, and the residues in the QDHSK sequence constitute the wall of the substrate-binding S3' pocket which merges with its counterpart inside the molecule (Figure 7D). Therefore, the two catalytic sites in the dimer are located back-to-back and share their S3' pockets, suggesting that the two catalytic sites in P-IIIc SVMPs may work in a cooperative manner. VAP1 induces cell death in vascular endothelial cells in culture with all the characteristic features of apoptosis [117]. However, the physiological target(s) of VAP1, the underlying mechanism of VAP1-induced apoptosis, and how dimerization relates to the substrate preference and/or activity of VAP1 remain totally unknown. In addition to VAP-1, HV1 (Genbank ID (GI): 14325767), halysase (GI: 60729695), VLAIP (GI: 82228618), TSV-DM (UniProt ID: Q2LD49.1) and VaH3 (GI: 496537199) are reported to exist in their native states as homo- or heterodimers. In addition to these SVMPs, agkihagin (Uniprot ID: Q1PS45) and halysetin (Uniprot ID: Q90Y44) also share Cys365 and the QDH(S/N)K sequence and thus, these SVMPs can be considered to be P-IIIc SVMPs. Bilitoxin-1 (GI: 172044534) [118], a unique homodimeric P-II SVMP, has neither a cysteine residue at position 365 nor the QDH(S/N)K sequence, suggesting that its dimer interface is different from that of VAP1. Cys365 and the QDH(S/N)K sequence are not found in either ADAMs or ADAMTSs.

A few P-III SVMP members exist as heterocomplexes due to the existence of an extra subunit that interacts through covalent or non-covalent interactions. The venom of Russell's viper (*Daboia russelli*) has been recognized for its potent coagulation activity. Two major components, RVV-X and RVV-V, of this venom can collaboratively accelerate formation of the prothrombinase complex (Factor Xa (FXa)/Factor Va (FVa) complex) that converts prothrombin to thrombin, resulting in a disseminated intravascular coagulation in the body of the prey [119]. RVV-X is a unique high molecular weight metalloproteinase, a representative of P-IIId SVMPs. RVV-X activates factor X (FX) by cleaving the Arg194-Ile195 bond in FX, which is also cleaved by factors IXa and VIIa during physiological coagulation [120,121]. Because of its extremely high specificity for FX, RVV-X is widely used in coagulation research and in diagnostic applications. A similar FX-activating P-IIId SVMP, VLFXA, has also been isolated from *Vipera lebetina* venom [121,122]. On the other hand, another component RVV-V is a thrombin-like serine proteinase that specifically activates factor V (FV) [123,124].

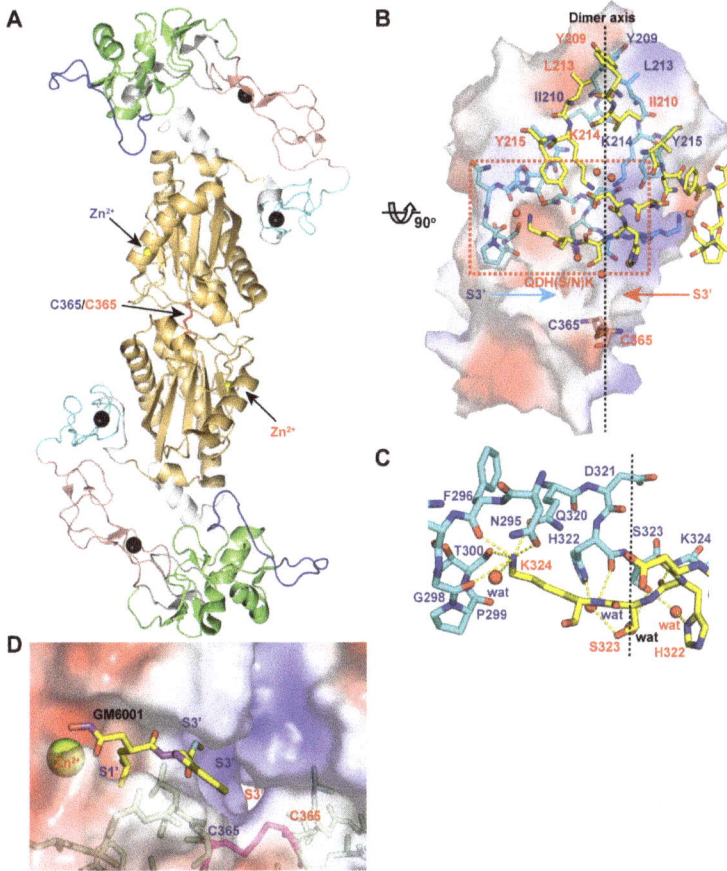

Figure 7. Structure of VAP1, representative of P-IIIc SVMPs. (**A**) Crystal structure of VAP1 (2ERO) viewed from the dimer axis. (**B**) Dimer interface viewed from a direction nearly perpendicular to the dimer axis. The molecular surface of the cyan molecule in the back is colored according to electrochemical potential (red to blue). (**C**) Close up view of the dimer interface. The residues involved in the inter-chain interactions are indicated with blue and red letters for cyan and yellow molecules. (**D**) Close up view of the catalytic cleft of the VAP1 (shown with the molecular surface)/GM6001 (shown in yellow) complex structure.

RVV-X is a heterotrimeric complex consisting of an MDC-containing heavy and two light chains [120,125]. Two light chains form a domain-swapped dimer [126] with features characteristic of snake venom C-type lectins (snaclecs [57]). Instead of binding to carbohydrate moieties, snaclecs bind to membrane receptors, coagulation factors and other proteins essential for hemostasis. The crystal structure of RVV-X revealed its unique hook-spanner-wrench configuration (Figure 8A), in which the MD domains constitute the hook, and the remainder of the molecule forms the handle [88,127]. The backbone structure of the heavy chain is essentially the same as those of other P-III SVMPs. RVV-X has a unique cysteine residue (Cys389), not found in other classes of SVMPs, in the middle of the HVR in the C_h subdomain. Cys389 forms a disulfide bond with the C-terminal cysteine residue (Cys133) of the light chain-A (LA). In addition, the residues in the HVR and the surrounding regions in the heavy chain form multiple aromatic and hydrophobic interactions and hydrogen bonds with the N- and

C-terminal residues in LA, further stabilizing the continuous C/LA structure. The RVV-X structure provides the first direct observation of a protein–protein interaction mediated by HVR.

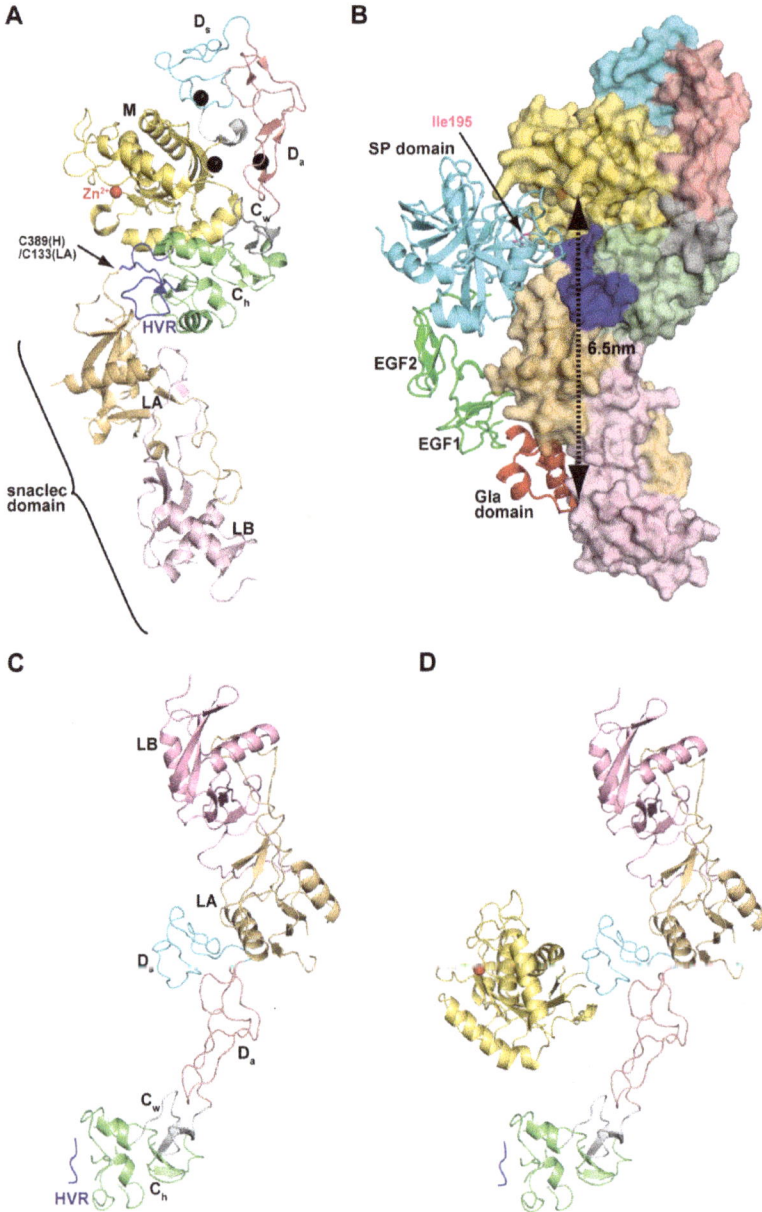

Figure 8. Structures of RVV-X and multactivase, two representatives of P-IIId SVMPs. (**A**) Ribbon representation of the crystal structure of RVV-X. (**B**) Factor Xa docking model. (**C**) Ribbon representation of the crystal structure of multactivase-ΔM. (**D**) A model of the whole multactivase molecule. The model was constructed by a superimposition of the crystal structures of multactivase-ΔM and of the M/D$_s$ domains of catrocollastatin/VAP2B (2DW0). Each subdomain is in a different color.

The structure of the snaclec domain of RVV-X is quite similar to that of the FX-binding protein (X-Bp) whose crystal structure was solved in complex with the γ-carboxyglutamic acid (Gla) domain of FX [128]. This structural similarity, along with the surface chemical properties and previous biochemical observations, suggests a docking model for FX (Figure 8B) [88,127]. The snaclec domain forms a Gla-domain-binding exosite that may serve as the Ca^{2+}-dependent primary capture site for circulating FX. The docking model indicates that the C_h/snaclec domains act as a scaffold to accommodate the elongated FX model. The relatively large separation (~65 Å) between the catalytic site and the exosite explains the high specificity of RVV-X for FX. This is in sharp contrast to thrombin-like RVV-V which cleaves the Arg1545-Ser1546 bond specifically by recognizing the side-chains of Ile1539 (P7)-Arg1545 (P1) located in close proximity to the scissile bond of FV [129]. The RVV-X structure represents a good example of the evolutionary acquisition of ligand-binding specificity by ADAMs and SVMPs.

Carinactivase-1 and multactivase are potent prothrombin activators isolated from the venom of *Echis carinatus* and *Echis multisquamatus*, respectively [130,131]. They have a snaclec domain in addition to MDC domains, and also use their snaclec domain for prothrombin recognition. Therefore, they are considered to be another example of P-IIId SVMPs. Unlike RVV-X, these two P-IIId SVMPs do not possess a disulfide bond between the heavy chain and snaclec domains, and thus how the catalytic and the regulatory domains interact and are oriented with respect to each other remains unclear. A crystal structure of the proteolytic fragment of multactivase, named multactivase-ΔM because it lacks the M domain from the intact molecule, was recently determined at 2.6Å resolution (Figure 8C) and a structural model of the entire multactivase molecule (Figure 8D) was constructed (*S. Takeda* and *T. Morita*, unpublished work). Each subdomain in multactivase is similar in structure to the corresponding one in RVV-X. However, the interactions between the heavy chain and the snaclec domain are remarkably different. The snaclec domain interacts with the D_s subdomain in multactivase but the C_h subdomain in RVV-X, resulting in a different overall shape and configuration of the catalytic site and the exosite between these two P-IIId SVMPs. The multactivase structure represents the first crystallographic observation of the interaction between an ADAM D domain and another polypeptide chain, providing additional insights into protein–protein interactions by the M12B clan of proteinases.

3.6. Core Structure of ADAMTSs

Figure 9A depicts a structural model of the MD*TCS domains of ADAMTS13 constructed based on the crystal structures of the MD* domains of ADAMTS5 [71] and the D*TCS domains of ADAMTS13 [53]. This model represents the basic architecture of the core portion commonly found in ADAMTS family proteinases. The structure of the core MD*TCS domains consists of three globular knobs, corresponding to the MD*, C_A and S domains, which are connected by two elongated structural modules, T1 and C_B. Unlike ADAMs, ADAMTSs lack the D_s/D_a/C_w arm structure, and the D* domain with an ADAM_CR domain fold is directly connected to the M domain by a connector loop (16–20 residues) that wraps around the opposite surface of the catalytic site [66,71]. The D* domain stacks against the M domain active site cleft, forming a continuous MD* unit, and potentially provides an auxiliary substrate-binding surface (see below). The side-chain of Phe216 in the M domain points toward, and makes a number of van der Waals contacts with, the small hydrophobic pocket formed in the D domain, thus playing a pivotal role in the interaction between the M and D domains. The F216E mutant, designed to impair the interactions between the M and D* domains, completely lost catalytic activity for the synthetic ADAMTS13 substrate FRET-VWF73 [132] although the secretion level was not greatly reduced [53]. On the other hand, the mutant that increased the stability of the association between the M and D* domains due to the introduction of an extra disulfide bond between the two domains, retained a catalytic activity indistinguishable from that of wild-type ADAMTS13 [53]. These results indicate that the M and D* domains may form a stable association that is not altered during the catalytic cycle and constitute a functional part of the proteinase domain. This is supported by absence of the D* domain in all ADAMTS-L proteins (Figure 1).

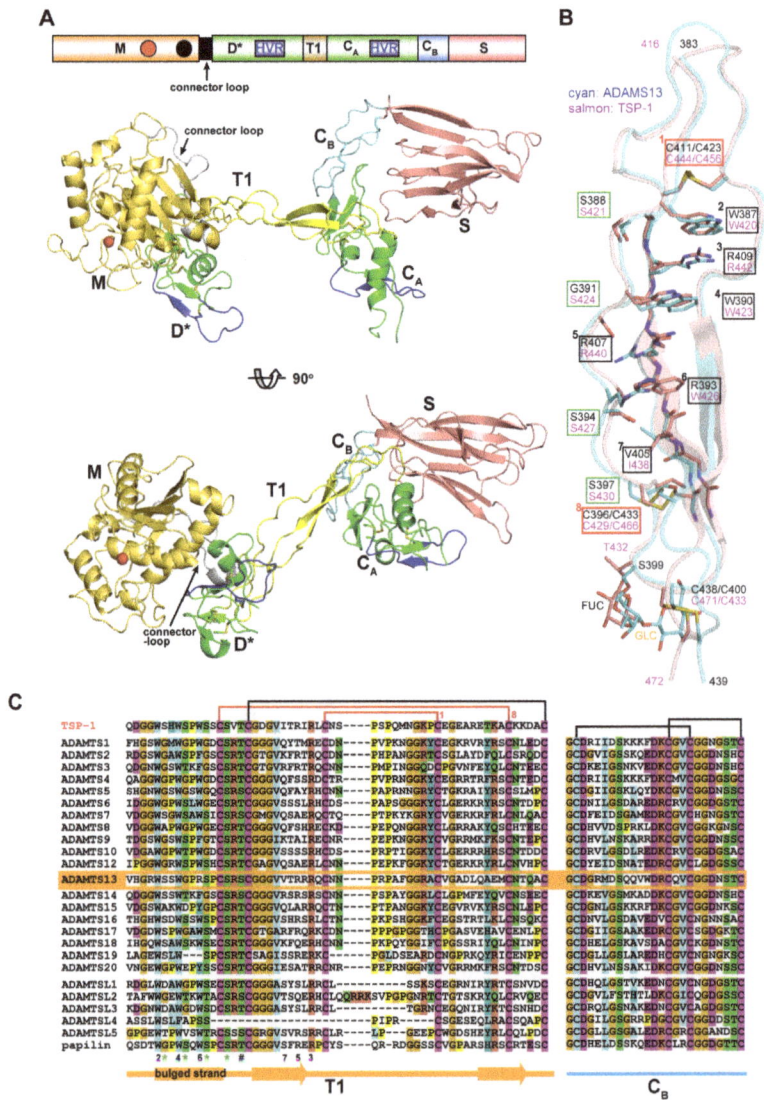

Figure 9. Structure of the core MD*TCS domains of ADAMTS proteinases. (**A**) A structural model of the MD*TCS domains of ADAMTS13 in two different views. (**B**) T1 structure of ADAMTS13 (shown in cyan) superposed onto the TSR2 in TSP-1 (PDB ID: 1LSL, shown in salmon). The residues that form the CWR-layered core (boxed in black and red), the serine residues in the bulged strand that form the hydrogen bond network (boxed in green) and the O-linked carbohydrate are indicated. (**C**) Sequence alignment of the T1 and the C_B subdomain regions of human TSP-1, ADAMTSs, ADAMTS-L and papilin. The residues involved in the CWR-layered core are indicated by layer number. Conserved serine residues and O–linked glycosylation sites are marked with * and #, respectively. The GI numbers for each ADAMTS-L sequence are, ADAMTS-L1 (GI: 37181773), ADAMTS-L2 (GI: 1232266328), ADAMTS-L3 (GI: 145275198), ADAMTS-L4 (GI: 187954849), ADAMTS-L5 (GI: 115311311) and papilin (GI: 145309328).

The homologous ADAM_CR domains, D* and C_A, are separated by about 45Å along T1. T1 has a very similar structure to the prototypical TSR, TSR2 in TSP-1 [133] adopting a long, twisted and antiparallel three-stranded fold (Figure 9B). The core of the T1 structure is stabilized by stacked layers of tryptophan, arginine, and hydrophobic residues, and is capped by disulfide bonds at both ends (Cys411/Cys423 and Cys396/Cys433), which has been referred to as the "CWR-layered core" [133]. In addition to the CWR-layered core, the second and third strands in T1 form a regular antiparallel β-sheet, whereas the bulged third strand is stabilized by hydrogen bonds between the side chains of three serine residues (Ser388, Ser394 and Ser397) and backbone nitrogen atoms from the neighboring strand. The residues involved in the CWR-layered core and the serine residues in the bulged strand are highly conserved among the T1 portions of ADAMTS and ADAMTS-L members [53] (Figure 9C). The β-sheet in T1 stacks against the C-terminal β-sheet in the C_A subdomain, forming a mini β-sandwich structure with a hydrophobic core that strengthens the interactions between T1 and C_A, thus fixing the C_A domain position relative to T1. On the other hand, there are few specific interactions between the D* and T1 domains in the crystal structure of ADAMTS13-DTCS, suggesting that the relative orientation between the D* and T1 domains may be fixed by crystal packing and would be variable in solution. The flexibility of the molecule between the D* and T1 domains is reflected by the low isomorphism of the ADAMTS13-DTCS crystals [53,134]. The C_B subdomain has no apparent secondary structure but comprises a series of turns stabilized by a pair of disulfide bonds and forms a rod shape with its N and C termini about 25Å apart (Figure 9A). The C_A and S domains are bridged by the C_B subdomain whose amino-acid sequence is highly conserved among ADAMTSs and ADAMTS-Ls [53] (Figure 9C). In the crystal structure of ADAMTS13-DTCS, direct contact exists between the C_A domain and the extended loop in the S domain. The mutants with an extra disulfide bond formed between the C_A and S domains affected nether secretion nor enzymatic activity, suggesting that the C_A and S domains form a stable association and that functional detachment between the domains does not occur during ADAMTS13 function [53]. The residues involved in the interaction between the C_A and S domains are conserved among ADAMTS13s from different species, but not among other ADAMTS members. Therefore, whether the stable association between the C_A and S domains is conserved in other ADAMTS members remains to be elucidated.

The structure of the S domain of ADAMTSs is currently only available for ADAMTS13 [53]. The nomenclature of the "spacer" domain of ADAMTSs comes from the fact that this region is a long cysteine-less segment and its primary structure shows no apparent homology to known structural motifs. However, the crystal structure of the ADAMTS13 S domain and the structure-based sequence alignments revealed that all ADAMTS and ADAMTS-L members share the single globular S domain structure with 10 β-strands in a jelly-roll topology, forming two antiparallel β-sheets that lie almost parallel to each other [53] (Figure 10A). Conserved hydrophobic residues form the core of the β-sandwich (Figure 10B,C), while loops located at the distal end of the molecule are highly variable in both in length and amino acid sequences among ADAMTSs and ADAMTS-Ls (Figure 10C), suggesting these loops could form protein–protein interaction sites. The N and C termini of the S domain lie in close proximity to one another, and thus the T2 domain that follows the S domain should be protruding out from near the C_B/S-domain junction but not from the distal side of the S domain.

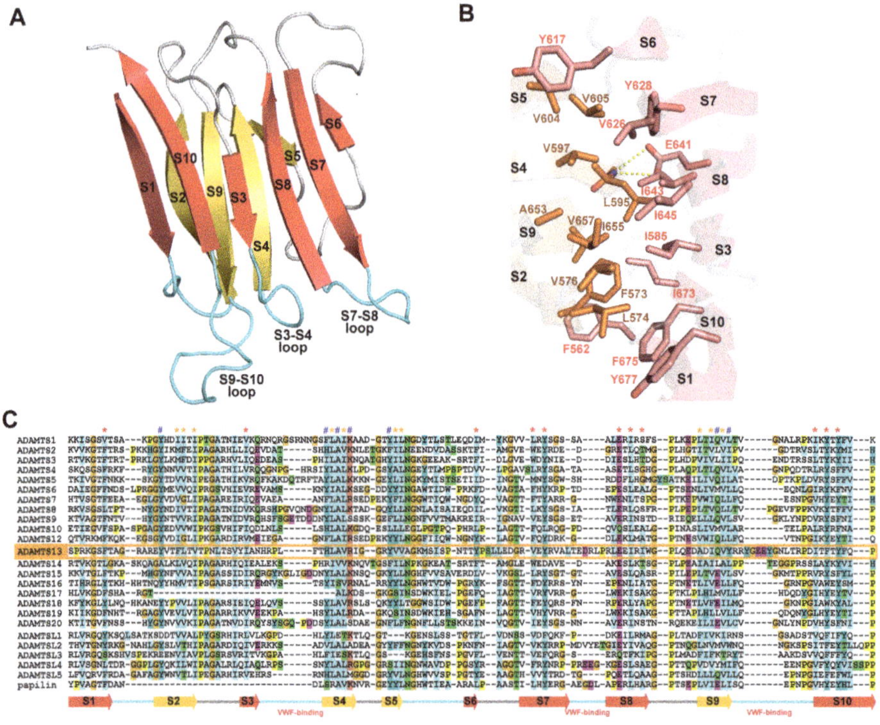

Figure 10. S domain structure. (A) Ribbon representation of the crystal structure of the S domain of
ADAMTS13. The strands in the two β-sheets are shown in red and orange. (B) Close-up view of the
hydrophobic core between the β-sheets. Side chains forming the hydrophobic core and the conserved
Glu641, whose side-chain oxygen atoms make hydrogen bonds with the backbone nitrogen atom of
Leu595 in the opposing strand, are indicated. (C) Sequence alignment of the S domain of human
ADAMTSs, ADAMTS-L and papilin. The residues in the hydrophobic core and the conserved aromatic
surface cluster [53] are marked with * and #, respectively.

4. ADAMTS13 and VWF Interaction

Significant progress in our knowledge of the structure-function relationship of the M12B clan
proteinases has been made by studies on ADAMTS13 [135,136], including the demonstration of
the actual involvement of the ADAM_CR and S domains in substrate recognition by intensive
mutagenesis experiments.

Von Willebrand Factor (VWF) is a plasma glycoprotein that plays an essential role in platelet
dependent hemostasis [137,138]. VWF (2050 amino acid residues) circulates in blood in multimeric
forms of highly variable size, ranging from dimers to species that may exceed 60-mers (UL-VWF
multimers) [139]. In healthy individuals, UL-VWF multimers undergo limited proteolytic processing
by ADAMTS13 [18]. Deficiency in ADAMTS13 activity either by genetic mutations in the ADAMTS13
gene or by acquired inhibitory autoantibodies directed against the ADAMTS13 protein, result in the
accumulation of UL-VWF in the plasma. UL-VWF accumulation leads to the formation of disseminated
platelet-rich micro thrombi in the micro-vasculature, which results in the life-threatening disease
TTP [17–19,140,141]. ADAMTS13 specifically cleaves the Tyr1605-Met1606 peptidyl bond within the
A2 domain of VWF [142] in a fluid shear-stress-dependent manner [143,144]. The MD*TCS domains
of ADAMTS13 (ADAMTS13-MD*TCS) are necessary and sufficient for specific proteolytic cleavage
of VWF *in vitro* [145–148]. VWF73 (residues 1595–1668 in the VWF A2 domain) was identified as

a minimum specific substrate for ADAMTS13 and suggested that a segment (residues 1607–1668) of VWF73 contains essential residues for recognition by ADAMTS13 [149]. Recent studies have added to our understanding of this recognition, revealing that specific regions in ADAMTS13, namely exosites-1, -2 and -3, in the D*, C$_A$ and S domains respectively, are all required for its interaction with VWF. These exosites of ADAMTS13 directly interact in a linear fashion with various segments in the central VWF-A2 domain between residues Ala1612 and Arg1668. In addition, fine mapping of epitopes of anti-ADAMTS13 antibodies derived from TTP patients, has provided further insight into the structural elements in ADAMTS13 that are essential for VWF binding. Figure 11A represents a summary of our understanding of the VWF-interacting sites in ADAMTS13 mapped on the molecular surface. Corresponding ADAMTS13-binding sites within VWF (residues 1596–1668) are schematically indicated in Figure 11B.

The segment corresponding to the HVR runs across the middle of the D* domain of ADAMTSs in close proximity to the active site, suggesting that the HVR might be ideally positioned to directly influence cleavage of the substrate [44]. In the D* domain in ADAMTS13, the HVR, together with the V-loop located beside it, was shown to form part of exosite I. ADAMTS13 variants carrying a point mutation, R349A or L350G [150], or R349D [53] in the HVR or a deletion of the V-loop (residues 324–330) [53] displayed a dramatically reduced proteolytic activity. Further studies demonstrated that residues Arg349 and Leu350 of the D domain of ADAMTS13 may interact with residues Asp1614 and Ala1612, respectively, in the central A2 domain of VWF [150]. These interactions, in addition to the direct active site cleft interactions in the M domain, may help orientate the scissile bond toward the active site center of ADAMTS13 [151].

The C$_A$ subdomain adopts an ADAM_CR domain fold and thus potentially functions as a protein–protein interaction site. As expected, ΔV-loop, a triple alanine substitution (H476A/S477A/Q478A) in the V-loop and R488E in the HVR mutants had significantly reduced proteolytic activity, suggesting that these hydrophilic or charged residues play a pivotal role in VWF recognition and constitute exosite-2 [53]. Recently, de Groot *et al.*, reported the results of a comprehensive analysis of the C domain in ADAMTS13 that identified its functional importance for interacting with VWF [152]. They found that mutagenesis of the 11 predominantly-charged residues in the C domain (actually in the C$_A$ subdomain) had no major effect on ADAMTS13 function, and five out of six engineered glycans on the C domain also had no effect on ADAMTS13 function. However, glycans attached at position 476 appreciably reduced both VWF binding and proteolysis.

By substituting the segments of the C domain with the corresponding regions in ADAMTS1, they identified that residues Gly471-Val474 at the base of the V-loop within the C$_A$ subdomain form a hydrophobic pocket that appears to be involved in binding hydrophobic residues Ile1642, Trp1644, Ile1649, Leu1650 and Ile1651 in VWF. The east Asian-specific P475S polymorphism in the ADAMTS13 gene causes approximately 16% reduction in plasma ADAMTS13 activity [154]. The crystal structure of ADAMTS13-DTCS (P475S) revealed that the conformation of the V-loop in the C$_A$ subdomain of this mutant was significantly different from that of the wild type [73].

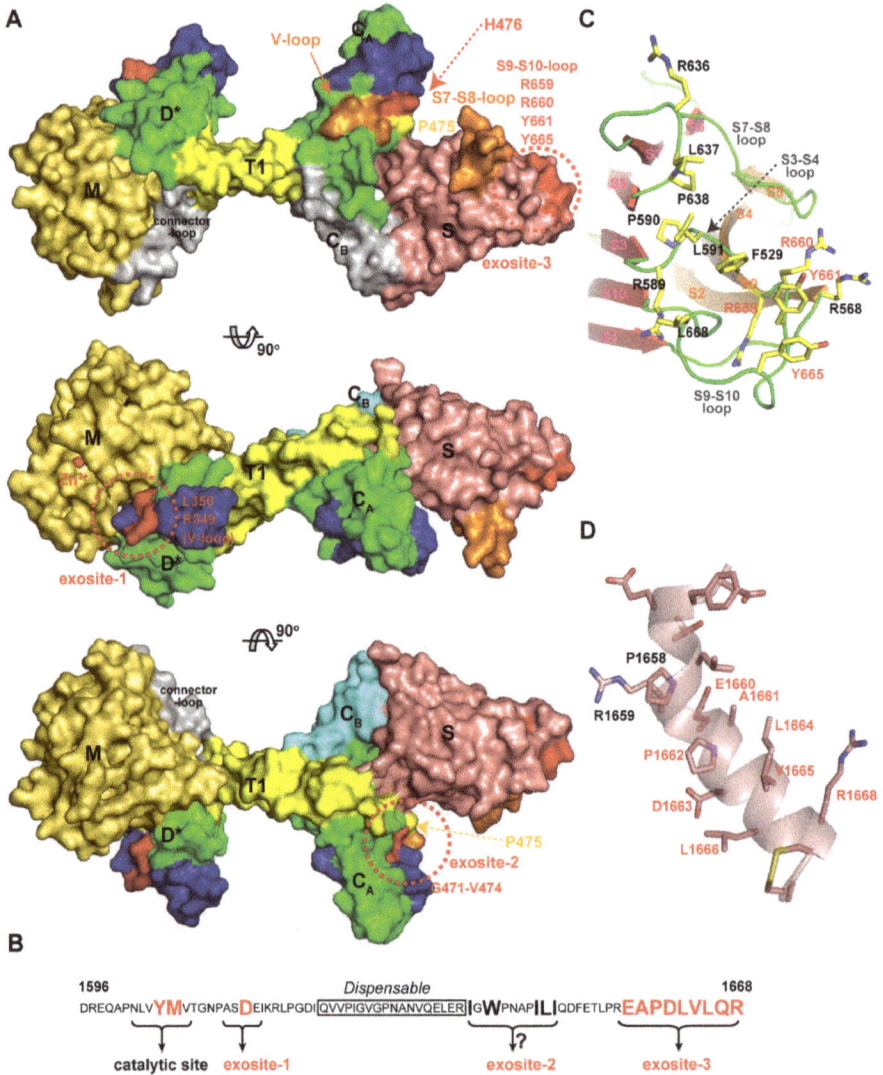

Figure 11. ADAMTS13 and VWF interaction. (**A**) Models of the ADAMTS13-MDTCS shown at 90 degree rotations. The residues important for the interaction with VWF are shown in red and other potential VWF-binding sites are shown in orange. Domains are colored as in Figure 7A. (**B**) Amino acid sequence of VWF (D1596-R1668) and the ADAMTS13-interacting sites are schematically represented. (**C**) Close-up view of the exosite-3 in the S domain, the hydrophobic cluster surrounded by arginine residues. The residues important for VWF-binding are indicated with red letters. (**D**) Close-up view of the VWF segment (D1653-C1670) in an α-helical conformation observed in the crystal structure of the VWF A2 domain [153]. The figure was created in reference to the original drawing by de Groot *et al.* [152].

The S domain in ADAMTS13 has the highest binding affinity for the A2 site of VWF. C-terminal deletion mutants of the VWF115 (VWF residues 1554–1668) and VWF73 fragments demonstrated that VWF A2 domain residues Glu1660-Arg1668 appreciably contribute to the cleavage of the Try1605-Met1606 scissile bond [149] and that the S domain of ADAMTS13 binds to this sequence [155]. As previously mentioned, the distal loops in the S domain are highly variable among ADAMTS/ADAMTS-L members (Figure 10C) and are thus suggested to create a substrate-binding exosite [53]. Mutants in which the S7-S8-loop (residues 606–611) and S9-S10-loop were replaced by short linkers, showed greatly reduced enzymatic activity for FRET-VWF73 [53]. In ADAMTS13, these variable loops create a hydrophobic cluster that is surrounded by arginine residues (Figure 11C). Systematic site-directed mutagenesis identified that this hydrophobic cluster rimmed with arginine residues actually constitutes another VWF-binding exosite (exosite-3) [53], and further identified Arg659, Arg660 and Tyr661 as critical residues for VWF cleavage [156]. It was also demonstrated that Arg660, Tyr661, and Tyr665 in the S domain of ADAMTS13 represent a core binding site for autoantibodies isolated from patients with acquired TTP [157]. The ADAMTS13 variants, R600K/F592Y/R568K/Y661F and R660/F592Y/R568K/Y661/Y665F, exhibit increased specific activity for both peptide substrates and multimeric VWF [158]. These gain-of-function ADAMTS13 variants were more resistant to inhibition by autoantibodies from idiopathic TTP patients because of reduced binding by anti-ADAMTS13 IgGs [158]. Both the surface properties and the size of exosite-3 imply that it binds to VWF, such that the VWF segment (residues 1653–1668) forms an amphiphilic α-helix (Figure 11D) and makes contact with ADAMTS13 by facing its hydrophobic surface toward exosite-3 [53]. Similar to ADAMTS13, the removal of the S domain dramatically reduces the aggrecanaolytic activity of ADAMTS5 and further removal of the C domain essentially abolished the activity [159]. An antibody reacting with the S domain of ADAMTS5 was shown to block the cleavage of aggrecan by the enzyme [160]; however, the exact site of ADAMTS5 that reacts with the antibody has not been identified.

5. Concluding Remarks

Tremendous progress has been made in the past decade towards our understanding of the structure–function relationship of the M12B clan of proteinases. Crystallographic studies have revealed the structures and spatial relationships of the functionally important domains of both ADAM and ADAMTS family proteinases. Most of the structural information of the overall MDC domains of ADAMs has come from SVMPs. The higher abundance, stability and resistance to proteolysis of SVMPs compared to mammalian ADAMs have made them attractive models for structural studies. The key message from these findings is that the MDC domains adopt a C-shaped configuration, whereby the HVR in the ADAM_CR domain faces toward the catalytic site. This raises the intriguing possibility that the HVR creates an exosite for capturing substrates directly or via binding to an associated protein. The RVV-X structure is consistent with this hypothesis. The multactivase structure suggests a potential function of the D domain in protein–protein interactions. These structural studies on SVMPs have provided radical new insights into the structure–function relationship of ADAMs. Some molecules have been shown to work as cofactors in the process of ectodomain shedding by membrane-bound ADAMs [161–163]; however, how these molecules function with ADAMs at a molecular level remains to be elucidated. Moreover, fundamental aspects of the functions of ADAMs, such as how membrane-bound ADAMs select their substrate and how their activity is regulated, are still largely unknown. A crystal structure of ADAM in complex with a substrate and/or such a cofactor would greatly improve our understanding of ADAMs' functions. Recent advances in ADAMTS13 research have provided invaluable information not only for our understanding of the mechanisms underlying TTP but also for designing structure-function studies for other family members. Notably, ADAMTSs contain no disintegrin-like structures but instead have two ADAM_CR domains that actually constitute VWF-binding exosites in ADAMTS13. This finding strongly supports the idea that the ADAM_CR domain functions as a novel protein-protein interaction module. The S domain, uniquely found in ADAMTSs and ADAMTS-Ls among the 12B members, may also provide another protein-protein interaction site for these members. The functions of the distal domains, which are variable among ADAMTS members, are still largely unknown. Recently, the distal T2-CUB2 domains were shown to directly interact with the proximal MD*TCS domains and inhibit substrate cleavage, while binding of VWF to the distal ADAMTS13 domains relieves this autoinhibition. Thus, ADAMTS13 is regulated by substrate-induced allosteric activation [164,165]. Whether the distal domains of other ADAMTSs have similar allosteric properties or not remains to be determined. The growing number of links with human diseases makes ADAMs and ADAMTSs attractive targets for novel therapies. To date, no successful treatment exists involving specific ADAM/ADAMTS inhibitors targeting the catalytic site. MMPs were also considered valuable therapeutic targets; however, early trials of small-molecule inhibitors (SMIs) toward their catalytic site failed due to poor inhibitor specificity profiles [166]. Because of the structural similarity of the catalytic sites of MMPs and ADAMSs/ADAMTSs, there is a limitation in generating active-site-targeted SMIs that are selective to one metalloproteinase species. Although we still have limited knowledge of how the prodomain controls enzymatic activity because of a lack of crystal structures, recombinant prodomains of ADAMs can act as inhibitors and might be used as alternatives to SMIs [167,168]. A unique cross-domain inhibitory antibody against ADAM17 has also been proposed [169]. Exosite or allosteric inhibitors may have more advantages in increasing the selectivity against specific ADAMs/ADAMTSs. Recently, three groups reported antibody-based exosite inhibitors of ADAMTS5, which were generated for therapeutic purposes to protect the destruction of articular cartilage in osteoarthritis [160,170,171]. Further structural knowledge of the exosite interactions of ADAM/ADAMTS family proteinases and their substrates will facilitate the development of novel inhibitors that may block cleavage of specific substrates, while leaving other catalytic functions of the targeted enzyme unaltered.

Acknowledgments: This work was supported in part by a Grant-in-Aid for Scientific Research from the Ministry of Education, Culture, Sports, Science and Technology.

Conflicts of Interest: The author declares no conflict of interest.

Abbreviations

The following abbreviations are used in this manuscript:

ADAM	a disintegrin and metalloproteinase
ADAMTS	a disintegrin-like and metalloproteinase with thrombospondin type-1 motif
ADAMTS-L	ADAMTS-like proteins
EGF	epidermal growth factor
Gla	γ-carboxyglutamic acid
MMP	matrix metalloproteinase
PDI	protein-disulfide isomerase
RVV-X	Russell's viper venom factor X activator
SMI	small-molecule inhibitor
Snaclec	snake venom C-type lectin like protein
SVMP	snake venom metalloproteinase
TSP	thrombospondin
TSR	thrombospondin type-1 repeat
TTP	thrombotic thrombocytopenic purpura
VAP1	vascular apoptosis inducing protein-1
VWF	von Willebrand Factor

References

1. Edwards, D.R.; Handsley, M.M.; Pennington, C.J. The ADAM metalloproteinases. *Mol. Asp. Med.* **2009**, *29*, 258–289. [CrossRef] [PubMed]
2. Reiss, K.; Saftig, P. The "A Disintegrin and Metalloprotease" (ADAM) family of sheddases: Physiological and cellular functions. *Semin. Cell Dev. Biol.* **2009**, *20*, 126–137. [CrossRef] [PubMed]
3. Weber, S.; Saftig, P. Ectodomain shedding and ADAMs in development. *Development* **2012**, *139*, 3693–3709. [CrossRef] [PubMed]
4. Black, R.A.; Rauch, C.T.; Kozlosky, C.J.; Peschon, J.J.; Slack, J.L.; Wolfson, M.F.; Castner, B.J.; Stocking, K.L.; Reddy, P.; Srinivasan, S.; *et al.* A metalloproteinase disintegrin that releases tumour-necrosis factor-alpha from cells. *Nature* **1997**, *385*, 729–733. [CrossRef] [PubMed]
5. Moss, M.L.; Jin, S.L.; Milla, M.E.; Bickett, D.M.; Burkhart, W.; Carter, H.L.; Chen, W.J.; Clay, W.C.; Didsbury, J.R.; Hassler, D.; *et al.* Cloning of a disintegrin metalloproteinase that processes precursor tumour-necrosis factor-alpha. *Nature* **1997**, *385*, 733–736. [CrossRef] [PubMed]
6. Blaydon, D.C.; Biancheri, P.; Di, W.L.; Plagnol, V.; Cabral, R.M.; Brooke, M.A.; van Heel, D.A.; Ruschendorf, F.; Toynbee, M.; Walne, A.; *et al.* Inflammatory skin and bowel disease linked to ADAM17 deletion. *N. Engl. J. Med.* **2011**, *365*, 1502–1508. [CrossRef] [PubMed]
7. Atapattu, L.; Lackmann, M.; Janes, P.W. The role of proteases in regulating Eph/ephrin signaling. *Cell Adh. Migr.* **2014**, *8*, 294–307. [CrossRef] [PubMed]
8. Murphy, G. The ADAMs: Signalling scissors in the tumour microenvironment. *Nat. Rev. Cancer* **2008**, *8*, 929–941. [CrossRef] [PubMed]
9. Mochizuki, S.; Okada, Y. ADAMs in cancer cell proliferation and progression. *Cancer Sci.* **2007**, *98*, 621–628. [CrossRef] [PubMed]
10. Zhang, P.; Shen, M.; Fernandez-Patron, C.; Kassiri, Z. ADAMs family and relatives in cardiovascular physiology and pathology. *J. Mol. Cell Cardiol.* **2015**, *93*, 186–199. [CrossRef] [PubMed]
11. Van Eerdewegh, P.; Little, R.D.; Dupuis, J.; Del Mastro, R.G.; Falls, K.; Simon, J.; Torrey, D.; Pandit, S.; McKenny, J.; Braunschweiger, K.; *et al.* Association of the ADAM33 gene with asthma and bronchial hyperresponsiveness. *Nature* **2002**, *418*, 426–430. [CrossRef] [PubMed]
12. Blobel, C.P.; Wolfsberg, T.G.; Turck, C.W.; Myles, D.G.; Primakoff, P.; White, J.M. A potential fusion peptide and an integrin ligand domain in a protein active in sperm-egg fusion. *Nature* **1992**, *356*, 248–252. [CrossRef] [PubMed]
13. Fukata, Y.; Adesnik, H.; Iwanaga, T.; Bredt, D.S.; Nicoll, R.A.; Fukata, M. Epilepsy-related ligand/receptor complex LGI1 and ADAM22 regulate synaptic transmission. *Science* **2006**, *313*, 1792–1795. [CrossRef] [PubMed]
14. Kelwick, R.; Desanlis, I.; Wheeler, G.N.; Edwards, D.R. The ADAMTS (A Disintegrin and Metalloproteinase with Thrombospondin motifs) family. *Genome Biol.* **2015**, *16*, 113. [CrossRef] [PubMed]
15. Apte, S.S. A disintegrin-like and metalloprotease (reprolysin-type) with thrombospondin type 1 motif (ADAMTS) superfamily-functions and mechanisms. *J. Biol. Chem.* **2009**, *284*, 31493–31497. [CrossRef] [PubMed]
16. Dubail, J.; Apte, S.S. Insights on ADAMTS proteases and ADAMTS-like proteins from mammalian genetics. *Matrix Biol.* **2015**, *44–46*, 24–37. [CrossRef] [PubMed]
17. Levy, G.G.; Nichols, W.C.; Lian, E.C.; Foroud, T.; McClintick, J.N.; McGee, B.M.; Yang, A.Y.; Siemieniak, D.R.; Stark, K.R.; Gruppo, R.; *et al.* Mutations in a member of the ADAMTS gene family cause thrombotic thrombocytopenic purpura. *Nature* **2001**, *413*, 488–494. [CrossRef] [PubMed]
18. Sadler, J.E.; Moake, J.L.; Miyata, T.; George, J.N. Recent advances in thrombotic thrombocytopenic purpura. *Hematol. Am. Soc. Hematol. Educ. Progr.* **2004**, *2004*, 407–423. [CrossRef] [PubMed]
19. Tsai, H.M.; Lian, E.C. Antibodies to von Willebrand factor-cleaving protease in acute thrombotic thrombocytopenic purpura. *N. Engl. J. Med.* **1998**, *339*, 1585–1594. [CrossRef] [PubMed]
20. Troeberg, L.; Nagase, H. Proteases involved in cartilage matrix degradation in osteoarthritis. *Biochim. Biophys. Acta* **2012**, *1824*, 133–145. [CrossRef] [PubMed]
21. Gao, W.; Zhu, J.; Westfield, L.A.; Tuley, E.A.; Anderson, P.J.; Sadler, J.E. Rearranging exosites in noncatalytic domains can redirect the substrate specificity of ADAMTS proteases. *J. Biol. Chem.* **2012**, *287*, 26944–26952. [CrossRef] [PubMed]
22. Zheng, X.L. ADAMTS13 and von Willebrand Factor in Thrombotic Thrombocytopenic Purpura. *Annu. Rev. Med.* **2015**, *66*, 211–225. [CrossRef] [PubMed]

23. Kang, T.S.; Georgieva, D.; Genov, N.; Murakami, M.T.; Sinha, M.; Kumar, R.P.; Kaur, P.; Kumar, S.; Dey, S.; Sharma, S.; *et al.* Enzymatic toxins from snake venom: Structural characterization and mechanism of catalysis. *FEBS J.* **2011**, *278*, 4544–4576. [CrossRef] [PubMed]

24. McCleary, R.J.R.; Kini, R.M. Non-enzymatic proteins from snake venoms: A gold mine of pharmacological tools and drug leads. *Toxicon* **2013**, *62*, 56–74. [CrossRef] [PubMed]

25. Ivanov, C.P.; Ivanov, O.C. The evolution and ancestors of toxic proteins. *Toxicon* **1979**, *17*, 205–220. [CrossRef]

26. Fry, B.G.; Vidal, N.; Norman, J.A.; Vonk, F.J.; Scheib, H.; Ramjan, S.F.R.; Kuruppu, S.; Fung, K.; Hedges, S.B.; Richardson, M.K.; *et al.* Early evolution of the venom system in lizards and snakes. *Nature* **2006**, *439*, 584–588. [CrossRef] [PubMed]

27. Casewell, N.R. On the ancestral recruitment of metalloproteinases into the venom of snakes. *Toxicon* **2012**, *60*, 449–454. [CrossRef] [PubMed]

28. Takeda, S.; Igarashi, T.; Mori, H.; Araki, S. Crystal structures of VAP1 reveal ADAMs' MDC domain architecture and its unique C-shaped scaffold. *EMBO J.* **2006**, *25*, 2388–2396. [CrossRef] [PubMed]

29. Calvete, J.J.; Juarez, P.; Sanz, L. Snake venomics. Strategy and applications. *J. Mass Spectrom.* **2007**, *42*, 1405–1414. [CrossRef] [PubMed]

30. Fox, J.W.; Serrano, S.M. Exploring snake venom proteomes: Multifaceted analyses for complex toxin mixtures. *Proteomics* **2008**, *8*, 909–920. [CrossRef] [PubMed]

31. Gutierrez, J.M.; Rucavado, A.; Gutiérrez, J.M.; Rucavado, A. Snake venom metalloproteinases: Their role in the pathogenesis of local tissue damage. *Biochimie* **2000**, *82*, 841–850. [CrossRef]

32. Moura-da-Silva, A.M.; Butera, D.; Tanjoni, I. Importance of snake venom metalloproteinases in cell biology: Effects on platelets, inflammatory and endothelial cells. *Curr. Pharm. Des.* **2007**, *13*, 2893–2905. [CrossRef] [PubMed]

33. Fox, J.W.; Serrano, S.M. Structural considerations of the snake venom metalloproteinases, key members of the M12 reprolysin family of metalloproteinases. *Toxicon* **2005**, *45*, 969–985. [CrossRef] [PubMed]

34. Fox, J.W.; Serrano, S.M. Timeline of key events in snake venom metalloproteinase research. *J. Proteom.* **2009**, *72*, 200–209. [CrossRef] [PubMed]

35. Markland, F.S.; Swenson, S. Snake venom metalloproteinases. *Toxicon* **2013**, *62*, 3–18. [CrossRef] [PubMed]

36. Baramova, E.N.; Shannon, J.D.; Bjarnason, J.B.; Gonias, S.L.; Fox, J.W. Interaction of hemorrhagic metalloproteinases with human alpha 2-macroglobulin. *Biochemistry* **1990**, *29*, 1069–1074. [CrossRef] [PubMed]

37. Gutierrez, J.M.; Rucavado, A.; Escalante, T.; Diaz, C. Hemorrhage induced by snake venom metalloproteinases: Biochemical and biophysical mechanisms involved in microvessel damage. *Toxicon* **2005**, *45*, 997–1011. [CrossRef] [PubMed]

38. Escalante, T.; Rucavado, A.; Fox, J.W.; Gutiérrez, J.M. Key events in microvascular damage induced by snake venom hemorrhagic metalloproteinases. *J. Proteom.* **2011**, *74*, 1781–1794. [CrossRef] [PubMed]

39. Bjarnason, J.B.; Fox, J.W. Hemorrhagic metalloproteinases from snake venoms. *Pharmacol. Ther.* **1994**, *62*, 325–372. [CrossRef]

40. Gomis-Ruth, F.X. Structural aspects of the metzincin clan of metalloendopeptidases. *Mol. Biotechnol.* **2003**, *24*, 157–202. [CrossRef]

41. Bjarnason, J.B.; Fox, J.W. Snake venom metalloendopeptidases: Reprolysins. *Methods Enzym.* **1995**, *248*, 345–368.

42. Gomis-Ruth, F.X.; Kress, L.F.; Bode, W. First structure of a snake venom metalloproteinase: A prototype for matrix metalloproteinases/collagenases. *EMBO J.* **1993**, *12*, 4151–4157. [PubMed]

43. Wolfsberg, T.G.; Straight, P.D.; Gerena, R.L.; Huovila, A.P.; Primakoff, P.; Myles, D.G.; White, J.M. ADAM, a widely distributed and developmentally regulated gene family encoding membrane proteins with a disintegrin and metalloprotease domain. *Dev. Biol.* **1995**, *169*, 378–383. [CrossRef] [PubMed]

44. Takeda, S. Three-dimensional domain architecture of the ADAM family proteinases. *Semin. Cell. Dev. Biol.* **2009**, *20*, 146–152. [CrossRef] [PubMed]

45. Takeda, S. VAP1: Snake venom homolog of mammalian ADAMs. In *Handbook of Metalloproteins*; Messerschmidt, A., Ed.; John Wiley & Sons Inc.: Chichester, UK, 2010; Volume 5, pp. 699–713.

46. Takeda, S.; Takeya, H.; Iwanaga, S. Snake venom metalloproteinases: Structure, function and relevance to the mammalian ADAM/ADAMTS family proteins. *Biochim. Biophys. Acta* **2012**, *1824*, 164–176. [CrossRef] [PubMed]

47. Takeda, S. Structure-Function Relationship of Modular Domains of P-III Class Snake Venom Metalloproteinases. In *Springer Reference (Toxinology)*; Gopalakrishnakone, P., Calvete, J.J., Eds.; Springer Science Business & Media: Dordrecht, The Netherlands, 2014; pp. 1–22.

48. Janes, P.W.; Saha, N.; Barton, W.A.; Kolev, M.V.; Wimmer-Kleikamp, S.H.; Nievergall, E.; Blobel, C.P.; Himanen, J.P.; Lackmann, M.; Nikolov, D.B. Adam meets Eph: An ADAM substrate recognition module acts as a molecular switch for ephrin cleavagein trans. *Cell* **2005**, *123*, 291–304. [CrossRef] [PubMed]

49. Wewer, U.M.; Morgelin, M.; Holck, P.; Jacobsen, J.; Lydolph, M.C.; Johnsen, A.H.; Kveiborg, M.; Albrechtsen, R. ADAM12 is a four-leafed clover: The excised prodomain remains bound to the mature enzyme. *J. Biol. Chem.* **2006**, *281*, 9418–9422. [CrossRef] [PubMed]

50. Mazzocca, A.; Coppari, R.; De Franco, R.; Cho, J.Y.; Libermann, T.A.; Pinzani, M.; Toker, A. A secreted form of ADAM9 promotes carcinoma invasion through tumor-stromal interactions. *Cancer Res.* **2005**, *65*, 4728–4738. [CrossRef] [PubMed]

51. Roberts, C.M.; Tani, P.H.; Bridges, L.C.; Laszik, Z.; Bowditch, R.D. MDC-L, a novel metalloprotease disintegrin cysteine-rich protein family member expressed by human lymphocytes. *J. Biol. Chem.* **1999**, *274*, 29251–29259. [CrossRef] [PubMed]

52. Mueller, C.G.; Rissoan, M.C.; Salinas, B.; Ait-Yahia, S.; Ravel, O.; Bridon, J.M.; Briere, F.; Lebecque, S.; Liu, Y.J. Polymerase chain reaction selects a novel disintegrin proteinase from CD40-activated germinal center dendritic cells. *J. Exp. Med.* **1997**, *186*, 655–663. [CrossRef] [PubMed]

53. Akiyama, M.; Takeda, S.; Kokame, K.; Takagi, J.; Miyata, T. Crystal structures of the noncatalytic domains of ADAMTS13 reveal multiple discontinuous exosites for von Willebrand factor. *Proc. Natl. Acad. Sci. USA* **2009**, *106*, 19274–19279. [CrossRef] [PubMed]

54. Le G, C.; Morice-Picard, F.; Dagoneau, N.; Wang, L.W.; Perrot, C.; Crow, Y.J.; Bauer, F.; Flori, E.; Prost-Squarcioni, C.; Krakow, D.; *et al.* ADAMTSL2 mutations in geleophysic dysplasia demonstrate a role for ADAMTS-like proteins in TGF-beta bioavailability regulation. *Nat. Genet.* **2008**, *40*, 1119–1123.

55. Ahram, D.; Sato, T.S.; Kohilan, A.; Tayeh, M.; Chen, S.; Leal, S.; Al-Salem, M.; El-Shanti, H. A homozygous mutation in ADAMTSL4 causes autosomal-recessive isolated ectopia lentis. *Am. J. Hum. Genet.* **2008**, *84*, 274–278. [CrossRef] [PubMed]

56. Fox, J.W.; Serrano, S.M. Insights into and speculations about snake venom metalloproteinase (SVMP) synthesis, folding and disulfide bond formation and their contribution to venom complexity. *FEBS J.* **2008**, *275*, 3016–3030. [CrossRef] [PubMed]

57. Clemetson, K.J. Snaclecs (snake C-type lectins) that inhibit or activate platelets by binding to receptors. *Toxicon* **2010**, *56*, 1236–1246. [CrossRef] [PubMed]

58. Casewell, N.R.; Wagstaff, S.C.; Harrison, R.A.; Renjifo, C.; Wüster, W. Domain loss facilitates accelerated evolution and neofunctionalization of duplicate snake venom metalloproteinase toxin genes. *Mol. Biol. Evol.* **2011**, *28*, 2637–2649. [CrossRef] [PubMed]

59. Moura-da-Silva, A.M.; Theakston, R.D.G.; Crampton, J.M. Evolution of disintegrin cysteine-rich and mammalian matrix-degrading metalloproteinases: Gene duplication and divergence of a common ancestor rather than convergent evolution. *J. Mol. Evol.* **1996**, *43*, 263–269. [CrossRef] [PubMed]

60. Juarez, P.; Comas, I.; Gonzalez-Candelas, F.; Calvete, J.J. Evolution of snake venom disintegrins by positive darwinian selection. *Mol. Biol. Evol.* **2008**, *25*, 2391–2407. [CrossRef] [PubMed]

61. Grams, F.; Huber, R.; Kress, L.F.; Moroder, L.; Bode, W. Activation of snake venom metalloproteinases by a cysteine switch-like mechanism. *FEBS Lett.* **1993**, *335*, 76–80. [CrossRef]

62. Wong, E.; Maretzky, T.; Peleg, Y.; Blobel, C.; Sagi, I. The Functional Maturation of A Disintegrin and Metalloproteinase (ADAM) 9, 10 and 17 Requires Processing at a Newly Identified Proprotein Convertase (PC) Cleavage Site. *J. Biol. Chem.* **2015**, *290*, 1–21. [CrossRef] [PubMed]

63. Majerus, E.M.; Zheng, X.; Tuley, E.A.; Sadler, J.E. Cleavage of the ADAMTS13 propeptide is not required for protease activity. *J. Biol. Chem.* **2003**, *278*, 46643–46648. [CrossRef] [PubMed]

64. Maskos, K.; Fernandez-Catalan, C.; Huber, R.; Bourenkov, G.P.; Bartunik, H.; Ellestad, G.A.; Reddy, P.; Wolfson, M.F.; Rauch, C.T.; Castner, B.J.; *et al.* Crystal structure of the catalytic domain of human tumor necrosis factor-alpha-converting enzyme. *Proc. Natl. Acad. Sci. USA* **1998**, *95*, 3408–3412. [CrossRef] [PubMed]

65. Liu, H.; Shim, A.H.; He, X. Structural characterization of the ectodomain of a disintegrin and metalloproteinase-22 (ADAM22), a neural adhesion receptor instead of metalloproteinase: Insights on ADAM function. *J. Biol. Chem.* **2009**, *284*, 29077–29086. [CrossRef] [PubMed]

66. Gerhardt, S.; Hassall, G.; Hawtin, P.; McCall, E.; Flavell, L.; Minshull, C.; Hargreaves, D.; Ting, A.; Pauptit, R.A.; Parker, A.E.; *et al.* Crystal structures of human ADAMTS-1 reveal a conserved catalytic domain and a disintegrin-like domain with a fold homologous to cysteine-rich domains. *J. Mol. Biol.* **2007**, *373*, 891–902. [CrossRef] [PubMed]

67. Tallant, C.; Marrero, A.; Gomis-Rüth, F.X. Matrix metalloproteinases: Fold and function of their catalytic domains. *Biochim. Biophys. Acta Mol. Cell Res.* **2010**, *1803*, 20–28. [CrossRef] [PubMed]

68. Hall, T.; Shieh, H.S.; Day, J.E.; Caspers, N.; Chrencik, J.E.; Williams, J.M.; Pegg, L.E.; Pauley, A.M.; Moon, A.F.; Krahn, J.M.; *et al.* Structure of human ADAM-8 catalytic domain complexed with batimastat. *Acta Crystallogr. Sect. F* **2012**, *68*, 616–621. [CrossRef] [PubMed]

69. Düsterhöft, S.; Jung, S.; Hung, C.W.; Tholey, A.; Sönnichsen, F.D.; Grötzinger, J.; Lorenzen, I. Membrane-proximal domain of a disintegrin and metalloprotease-17 represents the putative molecular switch of its shedding activity operated by protein-disulfide isomerase. *J. Am. Chem. Soc.* **2013**, *135*, 5776–5781. [CrossRef] [PubMed]

70. Orth, P.; Reichert, P.; Wang, W.; Prosise, W.W.; Yarosh-Tomaine, T.; Hammond, G.; Ingram, R.N.; Xiao, L.; Mirza, U.A.; Zou, J.; *et al.* Crystal structure of the catalytic domain of human ADAM33. *J. Mol. Biol.* **2004**, *335*, 129–137. [CrossRef] [PubMed]

71. Mosyak, L.; Georgiadis, K.; Shane, T.; Svenson, K.; Hebert, T.; McDonagh, T.; Mackie, S.; Olland, S.; Lin, L.; Zhong, X.; *et al.* Crystal structures of the two major aggrecan degrading enzymes, ADAMTS4 and ADAMTS5. *Protein Sci.* **2008**, *17*, 16–21. [CrossRef] [PubMed]

72. Shieh, H.S.; Mathis, K.J.; Williams, J.M.; Hills, R.L.; Wiese, J.F.; Benson, T.E.; Kiefer, J.R.; Marino, M.H.; Carroll, J.N.; Leone, J.W.; *et al.* A. High resolution crystal structure of the catalytic domain of ADAMTS-5 (aggrecanase-2). *J. Biol. Chem.* **2008**, *283*, 1501–1507. [CrossRef] [PubMed]

73. Akiyama, M.; Nakayama, D.; Takeda, S.; Kokame, K.; Takagi, J.; Miyata, T. Crystal structure and enzymatic activity of an ADAMTS-13 mutant with the East Asian-specific P475S polymorphism. *J. Thromb. Haemost.* **2013**, *11*, 1399–1406. [CrossRef] [PubMed]

74. Gong, W.; Zhu, X.; Liu, S.; Teng, M.; Niu, L. Crystal structures of acutolysin A, a three-disulfide hemorrhagic zinc metalloproteinase from the snake venom of *Agkistrodon acutus*. *J. Mol. Biol.* **1998**, *283*, 657–668. [CrossRef] [PubMed]

75. Zhu, X.Y.; Teng, M.K.; Niu, L.W. Structure of acutolysin-C, a haemorrhagic toxin from the venom of *Agkistrodon acutus*, providing further evidence for the mechanism of the pH-dependent proteolytic reaction of zinc metalloproteinases. *Acta Crystallogr. Sect. D* **1999**, *55*, 1834–1841. [CrossRef]

76. Gomis-Rüth, F.X.; Kress, L.F.; Kellermann, J.; Mayr, I.; Lee, X.; Huber, R.; Bode, W. Refined 2.0 A X-ray crystal structure of the snake venom zinc-endopeptidase adamalysin II. Primary and tertiary structure determination, refinement, molecular structure and comparison with astacin, collagenase and thermolysin. *J. Mol. Biol.* **1994**, *239*, 513–544. [CrossRef] [PubMed]

77. Zhang, D.; Botos, I.; Gomis-Ruth, F.X.; Doll, R.; Blood, C.; Njoroge, F.G.; Fox, J.W.; Bode, W.; Meyer, E.F. Structural interaction of natural and synthetic inhibitors with the venom metalloproteinase, atrolysin C (form d). *Proc. Natl. Acad. Sci. USA* **1994**, *91*, 8447–8451. [CrossRef] [PubMed]

78. Watanabe, L.; Shannon, J.D.; Valente, R.H.; Rucavado, A.; Alape-Giron, A.; Kamiguti, A.S.; Theakston, R.D.; Fox, J.W.; Gutierrez, J.M.; Arni, R.K. Amino acid sequence and crystal structure of BaP1, a metalloproteinase from *Bothrops asper* snake venom that exerts multiple tissue-damaging activities. *Protein Sci.* **2003**, *12*, 2273–2281. [CrossRef] [PubMed]

79. Akao, P.K.; Tonoli, C.C.C.; Navarro, M.S.; Cintra, A.C.O.; Neto, J.R.; Arni, R.K.; Murakami, M.T. Structural studies of BmooMPalpha-I, a non-hemorrhagic metalloproteinase from Bothrops moojeni venom. *Toxicon* **2010**, *55*, 361–368. [CrossRef] [PubMed]

80. Lou, Z.; Hou, J.; Liang, X.; Chen, J.; Qiu, P.; Liu, Y.; Li, M.; Rao, Z.; Yan, G. Crystal structure of a non-hemorrhagic fibrin(ogen)olytic metalloproteinase complexed with a novel natural tri-peptide inhibitor from venom of *Agkistrodon acutus*. *J. Struct. Biol.* **2005**, *152*, 195–203. [CrossRef] [PubMed]

81. Kumasaka, T.; Yamamoto, M.; Moriyama, H.; Tanaka, N.; Sato, M.; Katsube, Y.; Yamakawa, Y.; Omori-Satoh, T.; Iwanaga, S.; Ueki, T. Crystal structure of H2-proteinase from the venom of *Trimeresurus flavoviridis*. *J. Biochem.* **1996**, *119*, 49–57. [CrossRef] [PubMed]

82. Chou, T.L.; Wu, C.H.; Huang, K.F.; Wang, A.H.J. Crystal structure of a Trimeresurus mucrosquamatus venom metalloproteinase providing new insights into the inhibition by endogenous tripeptide inhibitors. *Toxicon* **2013**, *71*, 140–146. [CrossRef] [PubMed]

83. Huang, K.F.; Chiou, S.H.; Ko, T.P.; Yuann, J.M.; Wang, A.H. The 1.35 A structure of cadmium-substituted TM-3, a snake-venom metalloproteinase from Taiwan habu: Elucidation of a TNFalpha-converting enzyme-like active-site structure with a distorted octahedral geometry of cadmium. *Acta Crystallogr. D* **2002**, *58*, 1118–1128. [CrossRef] [PubMed]

84. Zhu, Z.; Gao, Y.; Yu, Y.; Zhang, X.; Zang, J.; Teng, M.; Niu, L. Structural basis of the autolysis of AaHIV suggests a novel target recognizing model for ADAM/reprolysin family proteins. *Biochem. Biophys. Res. Commun.* **2009**, *386*, 159–164. [CrossRef] [PubMed]

85. Guan, H.H.; Goh, K.S.; Davamani, F.; Wu, P.L.; Huang, Y.W.; Jeyakanthan, J.; Wu, W.G.; Chen, C.J. Structures of two elapid snake venom metalloproteases with distinct activities highlight the disulfide patterns in the D domain of ADAMalysin family proteins. *J. Struct. Biol.* **2009**, *169*, 294–303. [CrossRef] [PubMed]

86. Muniz, J.R.; Ambrosio, A.L.; Selistre-de-Araujo, H.S.; Cominetti, M.R.; Moura-da-Silva, A.M.; Oliva, G.; Garratt, R.C.; Souza, D.H. The three-dimensional structure of bothropasin, the main hemorrhagic factor from Bothrops jararaca venom: Insights for a new classification of snake venom metalloprotease subgroups. *Toxicon* **2008**, *52*, 807–816. [CrossRef] [PubMed]

87. Igarashi, T.; Araki, S.; Mori, H.; Takeda, S. Crystal structures of catrocollastatin/VAP2B reveal a dynamic, modular architecture of ADAM/adamalysin/reprolysin family proteins. *FEBS Lett.* **2007**, *581*, 2416–2422. [CrossRef] [PubMed]

88. Takeda, S.; Igarashi, T.; Mori, H. Crystal structure of RVV-X: An example of evolutionary gain of specificity by ADAM proteinases. *FEBS Lett.* **2007**, *581*, 5859–5864. [CrossRef] [PubMed]

89. Gomis-Ruth, F.X.; Botelho, T.O.; Bode, W. A standard orientation for metallopeptidases. *Biochim. Biophys. Acta* **2012**, *1824*, 157–163. [CrossRef] [PubMed]

90. Gomis-Ruth, F.X. Catalytic domain architecture of metzincin metalloproteases. *J. Biol. Chem.* **2009**, *284*, 15353–15357. [CrossRef] [PubMed]

91. Takeya, H.; Nishida, S.; Nishino, N.; Makinose, Y.; Omori-Satoh, T.; Nikai, T.; Sugihara, H.; Iwanaga, S. Primary structures of platelet aggregation inhibitors (disintegrins) autoproteolytically released from snake venom hemorrhagic metalloproteinases and new fluorogenic peptide substrates for these enzymes. *J. Biochem.* **1993**, *113*, 473–483. [PubMed]

92. Gardner, M.D.; Chion, C.K.; De Groot, R.; Shah, A.; Crawley, J.T.; Lane, D.A. A functional calcium-binding site in the metalloprotease domain of ADAMTS13. *Blood* **2009**, *113*, 1149–1157. [CrossRef] [PubMed]

93. Igarashi, T.; Oishi, Y.; Araki, S.; Mori, H.; Takeda, S. Crystallization and preliminary X-ray crystallographic analysis of two vascular apoptosis-inducing proteins (VAPs) from Crotalus atrox venom. *Acta Crystallogr. Sect. F Struct. Biol. Cryst. Commun.* **2006**, *62*, 688–691. [CrossRef] [PubMed]

94. Shimokawa, K.; Shannon, J.D.; Jia, L.G.; Fox, J.W. Sequence and biological activity of catrocollastatin-C: A disintegrin-like/cysteine-rich two-domain protein from Crotalus atrox venom. *Arch. Biochem. Biophys.* **1997**, *343*, 35–43. [CrossRef] [PubMed]

95. Usami, Y.; Fujimura, Y.; Miura, S.; Shima, H.; Yoshida, E.; Yoshioka, A.; Hirano, K.; Suzuki, M.; Titani, K. A 28 kDa-protein with disintegrin-like structure (jararhagin-C) purified from Bothrops jararaca venom inhibits collagen- and ADP-induced platelet aggregation. *Biochem. Biophys. Res. Commun.* **1994**, *201*, 331–339. [CrossRef] [PubMed]

96. Blobel, C.P.; Myles, D.G.; Primakoff, P.; White, J.M. Proteolytic processing of a protein involved in sperm-egg fusion correlates with acquisition of fertilization competence. *J. Cell Biol.* **1990**, *111*, 69–78. [CrossRef] [PubMed]

97. Fujii, Y.; Okuda, D.; Fujimoto, Z.; Horii, K.; Morita, T.; Mizuno, H. Crystal structure of trimestatin, a disintegrin containing a cell adhesion recognition motif RGD. *J. Mol. Biol.* **2003**, *332*, 1115–1122. [CrossRef]

98. Kini, R.M.; Evans, H.J. Structural domains in venom proteins: Evidence that metalloproteinases and nonenzymatic platelet aggregation inhibitors (disintegrins) from snake venoms are derived by proteolysis from a common precursor. *Toxicon* **1992**, *30*, 265–293. [CrossRef]

99. Calvete, J.J.; Marcinkiewicz, C.; Monleon, D.; Esteve, V.; Celda, B.; Juarez, P.; Sanz, L. Snake venom disintegrins: Evolution of structure and function. *Toxicon* **2005**, *45*, 1063–1074. [CrossRef] [PubMed]

100. Calvete, J.J. The continuing saga of snake venom disintegrins. *Toxicon* **2013**, *62*, 40–49. [CrossRef] [PubMed]

101. Okuda, D.; Koike, H.; Morita, T. A new gene structure of the disintegrin family: A subunit of dimeric disintegrin has a short coding region. *Biochemistry* **2002**, *41*, 14248–14254. [CrossRef] [PubMed]
102. Moura-da-Silva, A.M.; Della-Casa, M.S.; David, A.S.; Assakura, M.T.; Butera, D.; Lebrun, I.; Shannon, J.D.; Serrano, S.M.; Fox, J.W. Evidence for heterogeneous forms of the snake venom metalloproteinase jararhagin: A factor contributing to snake venom variability. *Arch. Biochem. Biophys.* **2003**, *409*, 395–401. [CrossRef]
103. Carbajo, R.J.; Sanz, L.; Perez, A.; Calvete, J.J. NMR structure of bitistatin—A missing piece in the evolutionary pathway of snake venom disintegrins. *FEBS J.* **2015**, *282*, 341–360. [CrossRef] [PubMed]
104. Willems, S.H.; Tape, C.J.; Stanley, P.L.; Taylor, N.A.; Mills, I.G.; Neal, D.E.; McCafferty, J.; Murphy, G. Thiol isomerases negatively regulate the cellular shedding activity of ADAM17. *Biochem. J.* **2010**, *428*, 439–450. [CrossRef] [PubMed]
105. Smith, K.M.; Gaultier, A.; Cousin, H.; Alfandari, D.; White, J.M.; DeSimone, D.W. The cysteine-rich domain regulates ADAM protease function *in vivo*. *J. Cell Biol.* **2002**, *159*, 893–902. [CrossRef] [PubMed]
106. Reddy, P.; Slack, J.L.; Davis, R.; Cerretti, D.P.; Kozlosky, C.J.; Blanton, R.A.; Shows, D.; Peschon, J.J.; Black, R.A. Functional analysis of the domain structure of tumor necrosis factor-alpha converting enzyme. *J. Biol. Chem.* **2000**, *275*, 14608–14614. [CrossRef] [PubMed]
107. Gaultier, A.; Cousin, H.; Darribere, T.; Alfandari, D.; Darribe, T. ADAM13 disintegrin and cysteine-rich domains bind to the second heparin-binding domain of fibronectin. *J. Biol. Chem.* **2002**, *277*, 23336–23344. [CrossRef] [PubMed]
108. Iba, K.; Albrechtsen, R.; Gilpin, B.; Frohlich, C.; Loechel, F.; Zolkiewska, A.; Ishiguro, K.; Kojima, T.; Liu, W.; Langford, J.K.; *et al.* The cysteine-rich domain of human ADAM 12 supports cell adhesion through syndecans and triggers signaling events that lead to beta1 integrin-dependent cell spreading. *J. Cell. Biol.* **2000**, *149*, 1143–1156. [CrossRef] [PubMed]
109. Lorenzen, I.; Lokau, J.; Dusterhoft, S.; Trad, A.; Garbers, C.; Scheller, J.; Rose-John, S.; Grotzinger, J. The membrane-proximal domain of A Disintegrin and Metalloprotease 17 (ADAM17) is responsible for recognition of the interleukin-6 receptor and interleukin-1 receptor II. *FEBS Lett.* **2012**, *586*, 1093–1100. [CrossRef] [PubMed]
110. Düsterhöft, S.; Höbel, K.; Oldefest, M.; Lokau, J.; Waetzig, G.H.; Chalaris, A.; Garbers, C.; Scheller, J.; Rose-John, S.; Lorenzen, I.; *et al.* A disintegrin and metalloprotease 17 dynamic interaction sequence, the sweet tooth for the human interleukin 6 receptor. *J. Biol. Chem.* **2014**, *289*, 16336–16348. [CrossRef] [PubMed]
111. Düsterhöft, S.; Michalek, M.; Kordowski, F.; Oldefest, M.; Sommer, A.; Röseler, J.; Reiss, K.; Grötzinger, J.; Lorenzen, I. Extracellular Juxtamembrane Segment of ADAM17 Interacts with Membranes and is Essential for Its Shedding Activity. *Biochemistry* **2015**, *54*, 5791–5801. [CrossRef] [PubMed]
112. Doley, R.; Kini, R.M. Protein complexes in snake venom. *Cell Mol. Life Sci.* **2009**, *66*, 2851–2871. [CrossRef] [PubMed]
113. Kosasih, H.J.; Last, K.; Rogerson, F.M.; Golub, S.B.; Gauci, S.J.; Russo, V.C.; Stanton, H.; Wilson, R.; Lamande, S.R.; Holden, P.; *et al.* A Disintegrin and Metalloproteinase with Thrombospondin Motifs-5 (ADAMTS-5) Forms Catalytically Active Oligomers. *J. Biol. Chem.* **2016**, *291*, 3197–3208. [CrossRef] [PubMed]
114. Hojima, Y.; Mckenzie, J.; van der Rest, M.; Prockop, D.J. Type I Procollagen N-proteinase from Chick Embryo Tendons. *Enzyme* **1989**, *264*, 11336–11345.
115. Furlan, M.; Robles, R.; Lamie, B. Partial purification and characterization of a protease from human plasma cleaving von Willebrand factor to fragments produced by *in vivo* proteolysis. *Blood* **1996**, *87*, 4223–4234. [PubMed]
116. Lorenzen, I.; Trad, A.; Grötzinger, J. Multimerisation of A disintegrin and metalloprotease protein-17 (ADAM17) is mediated by its EGF-like domain. *Biochem. Biophys. Res. Commun.* **2011**, *415*, 330–336. [CrossRef] [PubMed]
117. Masuda, S.; Araki, S.; Yamamoto, T.; Kaji, K.; Hayashi, H. Purification of a vascular apoptosis-inducing factor from hemorrhagic snake venom. *Biochem. Biophys. Res. Commun.* **1997**, *235*, 59–63. [CrossRef] [PubMed]
118. Nikai, T.; Taniguchi, K.; Komori, Y.; Masuda, K.; Fox, J.W.; Sugihara, H. Primary structure and functional characterization of bilitoxin-1, a novel dimeric P-II snake venom metalloproteinase from *Agkistrodon bilineatus* venom. *Arch. Biochem. Biophys.* **2000**, *378*, 6–15. [CrossRef] [PubMed]
119. Schiffman, S.; Theodor, I.; Rapaport, S.I. Separation from Russell's viper venom of one fraction reacting with factor X and another reacting with factor V. *Biochemistry* **1969**, *8*, 1397–1405. [CrossRef] [PubMed]
120. Morita, T. Proteases which activate factor X. In *Enzymes from Snake Venom*; Bailey, G.S., Ed.; InterScience Publishers: Alaken, CO, USA, 1998; pp. 179–208.

121. Siigur, J.; Siigur, E. Activation of factor X by snake venom proteases. In *Toxins and Hemostasis: From Bench to Bedside*; Kini, R., Clemetson, K.J., Markland, F.S., McLane, M.A., Morita, T., Eds.; Springer Science & Business Media: Dordrecht, The Netherlands, 2010; pp. 447–464.

122. Siigur, E.; Tonismagi, K.; Trummal, K.; Samel, M.; Vija, H.; Subbi, J.; Siigur, J. Factor X activator from *Vipera lebetina* snake venom, molecular characterization and substrate specificity. *Biochim. Biophys. Acta* **2001**, *1568*, 90–98. [CrossRef]

123. Hjort, P.F. Intermediate reactions in the coagulation of blood with tissue thromboplastin; convertin, accelerin, prothrombinase. *Scand. J. Clin. Lab. Investig.* **1957**, *9*, 1–183. [PubMed]

124. Tokunaga, F.; Nagasawa, K.; Tamura, S.; Miyata, T.; Iwanaga, S.; Kisiel, W. The factor V-activating enzyme (RVV-V) from Russell's viper venom. Identification of isoproteins RVV-V alpha, -V beta, and -V gamma and their complete amino acid sequences. *J. Biol. Chem.* **1988**, *263*, 17471–17481. [PubMed]

125. Takeya, H.; Nishida, S.; Miyata, T.; Kawada, S.; Saisaka, Y.; Morita, T.; Iwanaga, S. Coagulation factor X activating enzyme from Russell's viper venom (RVV-X). A novel metalloproteinase with disintegrin (platelet aggregation inhibitor)-like and C-type lectin-like domains. *J. Biol. Chem.* **1992**, *267*, 14109–14117. [PubMed]

126. Mizuno, H.; Fujimoto, Z.; Koizumi, M.; Kano, H.; Atoda, H.; Morita, T. Structure of coagulation factors IX/X-binding protein, a heterodimer of C-type lectin domains. *Nat. Struct. Biol.* **1997**, *4*, 438–441. [CrossRef] [PubMed]

127. Takeda, S. Structural aspects of the factor X activator RVV-X from Russell's viper venom. In *Toxins and Hemostasis: From Bench to Bedside*; Kini, R., Clemetson, K.J., Markland, F.S., McLane, M.A., Morita, T., Eds.; Springer Science Business & Media: Dordrecht, The Netherlands, 2010; pp. 465–484.

128. Mizuno, H.; Fujimoto, Z.; Atoda, H.; Morita, T. Crystal structure of an anticoagulant protein in complex with the Gla domain of factor X. *Proc. Natl. Acad. Sci. USA* **2001**, *98*, 7230–7234. [CrossRef] [PubMed]

129. Nakayama, D.; Ben Ammar, Y.; Miyata, T.; Takeda, S. Structural basis of coagulation factor V recognition for cleavage by RVV-V. *FEBS Lett.* **2011**, *585*, 3020–3025. [CrossRef] [PubMed]

130. Yamada, D.; Sekiya, F.; Morita, T. Isolation and characterization of carinactivase, a novel prothrombin activator in *Echis carinatus* venom with a unique catalytic mechanism. *J. Biol. Chem.* **1996**, *271*, 5200–5207. [PubMed]

131. Yamada, D.; Morita, T. Purification and characterization of a Ca^{2+}-dependent prothrombin activator, multactivase, from the venom of Echis multisquamatus. *J. Biochem.* **1997**, *122*, 991–997. [CrossRef] [PubMed]

132. Kokame, K.; Nobe, Y.; Kokubo, Y.; Okayama, A.; Miyata, T. FRETS-VWF73, a first fluorogenic substrate for ADAMTS13 assay. *Br. J. Haematol.* **2005**, *129*, 93–100. [CrossRef] [PubMed]

133. Tan, K.; Duquette, M.; Liu, J.H.; Dong, Y.; Zhang, R.; Joachimiak, A.; Lawler, J.; Wang, J.H. Crystal structure of the TSP-1 type 1 repeats: A novel layered fold and its biological implication. *J. Cell Biol.* **2002**, *159*, 373–382. [CrossRef] [PubMed]

134. Akiyama, M.; Takeda, S.; Kokame, K.; Takagi, J.; Miyata, T. Production, crystallization and preliminary crystallographic analysis of an exosite-containing fragment of human von Willebrand factor-cleaving proteinase ADAMTS13. *Acta Crystallogr. Sect. F* **2009**, *65*, 739–742. [CrossRef] [PubMed]

135. Zander, C.B.; Cao, W.; Zheng, X.L. ADAMTS13 and von Willebrand factor interaction. *Curr. Opin. Hematol.* **2015**, *22*, 452–459. [CrossRef] [PubMed]

136. Crawley, J.T.B.; De Groot, R.; Xiang, Y.; Luken, B.M.; Lane, D.A. Untravelling the scissile bond: How ADAMTS13 recognises and cleaves von Willebrand factor. *Blood* **2011**, *118*, 3212–3221. [CrossRef] [PubMed]

137. Sadler, J.E. Biochemistry and genetics of von Willebrand factor. *Annu. Rev. Biochem.* **1998**, *67*, 395–424. [CrossRef] [PubMed]

138. Ruggeri, Z.M. Von Willebrand factor, platelets and endothelial cell interactions. *J. Thromb. Haemost.* **2003**, *1*, 1335–1342. [CrossRef] [PubMed]

139. Springer, T. a. Biology and physics of von Willebrand factor concatamers. *J. Thromb. Haemost.* **2011**. [CrossRef] [PubMed]

140. Moake, J.L.; Rudy, C.K.; Troll, J.H.; Weinstein, M.J.; Colannino, N.M.; Azocar, J.; Seder, R.H.; Hong, S.L.; Deykin, D. Unusually large plasma factor VIII: Von Willebrand factor multimers in chronic relapsing thrombotic thrombocytopenic purpura. *N. Engl. J. Med.* **1982**, *307*, 1432–1435. [CrossRef] [PubMed]

141. Furlan, M.; Robles, R.; Galbusera, M.; Remuzzi, G.; Kyrle, P.A.; Brenner, B.; Krause, M.; Scharrer, I.; Aumann, V.; Mittler, U.; *et al.* Von Willebrand factor-cleaving protease in thrombotic thrombocytopenic purpura and the hemolytic-uremic syndrome. *N. Engl. J. Med.* **1998**, *339*, 1578–1584. [CrossRef] [PubMed]

142. Dent, J.A.; Berkowitz, S.D.; Ware, J.; Kasper, C.K.; Ruggeri, Z.M. Identification of a cleavage site directing the immunochemical detection of molecular abnormalities in type IIA von Willebrand factor. *Proc. Natl. Acad. Sci. USA* **1990**, *87*, 6306–6310. [CrossRef] [PubMed]

143. Tsai, H.M.; Sussman, I.I.; Nagel, R.L. Shear stress enhances the proteolysis of von Willebrand factor in normal plasma. *Blood* **1994**, *83*, 2171–2179. [PubMed]

144. Zhang, X.; Halvorsen, K.; Zhang, C.Z.; Wong, W.P.; Springer, T.A. Mechanoenzymatic cleavage of the ultralarge vascular protein von Willebrand factor. *Science* **2009**, *324*, 1330–1334. [CrossRef] [PubMed]

145. Soejima, K.; Matsumoto, M.; Kokame, K.; Yagi, H.; Ishizashi, H.; Maeda, H.; Nozaki, C.; Miyata, T.; Fujimura, Y.; Nakagaki, T. ADAMTS-13 cysteine-rich/spacer domains are functionally essential for von Willebrand factor cleavage. *Blood* **2003**, *102*, 3232–3237. [CrossRef] [PubMed]

146. Zheng, X.; Nishio, K.; Majerus, E.M.; Sadler, J.E. Cleavage of von Willebrand factor requires the spacer domain of the metalloprotease ADAMTS13. *J. Biol. Chem.* **2003**, *278*, 30136–30141. [CrossRef] [PubMed]

147. Ai, J.; Smith, P.; Wang, S.; Zhang, P.; Zheng, X.L. The Proximal Carboxyl-terminal Domains of ADAMTS13 Determine Substrate Specificity and Are All Required for Cleavage of von Willebrand Factor. *J. Biol. Chem.* **2005**, *280*, 29428–29434. [CrossRef] [PubMed]

148. Gao, W.; Anderson, P.J.; Sadler, J.E. Extensive contacts between ADAMTS13 exosites and von Willebrand factor domain A2 contribute to substrate specificity. *Blood* **2008**, *112*, 1713–1719. [CrossRef] [PubMed]

149. Kokame, K.; Matsumoto, M.; Fujimura, Y.; Miyata, T. VWF73, a region from D1596 to R1668 of von Willebrand factor, provides a minimal substrate for ADAMTS-13. *Blood* **2004**, *103*, 607–612. [CrossRef] [PubMed]

150. De Groot, R.; Bardhan, A.; Ramroop, N.; Lane, D.A.; Crawley, J.T.B. Essential role of the disintegrin-like domain in ADAMTS13 function. *Blood* **2009**, *113*, 5609–5616. [CrossRef] [PubMed]

151. Xiang, Y.; De Groot, R.; Crawley, J.T.B.; Lane, D.A. Mechanism of von Willebrand factor scissile bond cleavage by a disintegrin and metalloproteinase with a thrombospondin type 1 motif, member 13 (ADAMTS13). *Proc. Natl. Acad. Sci. USA* **2011**, *108*, 11602–11607. [CrossRef] [PubMed]

152. De Groot, R.; Lane, D.A.; Crawley, J.T.B. The role of the ADAMTS13 cysteine-rich domain in VWF binding and proteolysis. *Blood* **2015**, *125*, 1968–1975. [CrossRef] [PubMed]

153. Zhang, Q.; Zhou, Y.F.; Zhang, C.Z.; Zhang, X.; Lu, C.; Springer, T.A. Structural specializations of A2, a force-sensing domain in the ultralarge vascular protein von Willebrand factor. *Proc. Natl. Acad. Sci. USA* **2009**, *106*, 9226–9231. [CrossRef] [PubMed]

154. Kokame, K.; Matsumoto, M.; Soejima, K.; Yagi, H.; Ishizashi, H.; Funato, M.; Tamai, H.; Konno, M.; Kamide, K.; Kawano, Y.; *et al.* Mutations and common polymorphisms in ADAMTS13 gene responsible for von Willebrand factor-cleaving protease activity. *Proc. Natl. Acad. Sci. USA* **2002**, *99*, 11902–11907. [CrossRef] [PubMed]

155. Gao, W.; Anderson, P.J.; Majerus, E.M.; Tuley, E.A.; Sadler, J.E. Exosite interactions contribute to tension-induced cleavage of von Willebrand factor by the antithrombotic ADAMTS13 metalloprotease. *Proc. Natl. Acad. Sci. USA* **2006**, *103*, 19099–19104. [CrossRef] [PubMed]

156. Jin, S.; Skipwith, C.G.; Zheng, X.L.; Dc, W. Amino acid residues Arg 659, Arg 660, and Tyr 661 in the spacer domain of ADAMTS13 are critical for cleavage of von Willebrand factor. *Blood* **2010**, *115*, 2300–2310. [CrossRef] [PubMed]

157. Pos, W.; Crawley, J.T. B.; Fijnheer, R.; Voorberg, J.; Lane, D.A.; Luken, B.M. An autoantibody epitope comprising residues R660, Y661, and Y665 in the ADAMTS13 spacer domain identifies a binding site for the A2 domain of VWF. *Blood* **2010**, *115*, 1640–1649. [CrossRef] [PubMed]

158. Jian, C.; Xiao, J.; Gong, L.; Skipwith, C.G.; Jin, S.Y.; Kwaan, H.C.; Zheng, X.L. Gain-of-function ADAMTS13 variants that are resistant to autoantibodies against ADAMTS13 in patients with acquired thrombotic thrombocytopenic purpura. *Blood* **2012**, *119*, 3836–3843. [CrossRef] [PubMed]

159. Gendron, C.; Kashiwagi, M.; Lim, N.H.; Enghild, J.J.; Thøgersen, I.B.; Hughes, C.; Caterson, B.; Nagase, H. Proteolytic Activities of Human ADAMTS-5. *J. Biol. Chem.* **2007**, *282*, 18294–18306. [CrossRef] [PubMed]

160. Santamaria, S.; Yamamoto, K.; Botkjaer, K.; Tape, C.; Dyson, M.R.; McCafferty, J.; Murphy, G.; Nagase, H. Antibody-based exosite inhibitors of ADAMTS-5 (aggrecanase-2). *Biochem. J.* **2015**, *471*, 391–401. [CrossRef] [PubMed]

161. Nishi, E.; Hiraoka, Y.; Yoshida, K.; Okawa, K.; Kita, T. Nardilysin Enhances Ectodomain Shedding of Heparin-binding Epidermal Growth Factor-like Growth Factor through Activation of Tumor Necrosis Factor-alpha-converting Enzyme. *J. Biol. Chem.* **2006**, *281*, 31164–31172. [CrossRef] [PubMed]

162. Nakayama, H.; Fukuda, S.; Inoue, H.; Nishida-Fukuda, H.; Shirakata, Y.; Hashimoto, K.; Higashiyama, S. Cell surface annexins regulate ADAM-mediated ectodomain shedding of proamphiregulin. *Mol. Biol. Cell.* **2012**, *23*, 1964–1975. [CrossRef] [PubMed]

163. Maney, S.K.; McIlwain, D.R.; Polz, R.; Pandyra, A.A.; Sundaram, B.; Wolff, D.; Ohishi, K.; Maretzky, T.; Brooke, M.A.; Evers, A.; *et al.* Deletions in the cytoplasmic domain of iRhom1 and iRhom2 promote shedding of the TNF receptor by the protease ADAM17. *Sci. Signal.* **2015**, *8*, ra109. [CrossRef] [PubMed]

164. Muia, J.; Zhu, J.; Gupta, G.; Haberichter, S.L.; Friedman, K.D.; Feys, H.B.; Deforche, L.; Vanhoorelbeke, K.; Westfield, L.A.; Roth, R.; *et al.* Allosteric activation of ADAMTS13 by von Willebrand factor. *Proc. Natl. Acad. Sci. USA* **2014**, *111*, 18584–18589. [CrossRef] [PubMed]

165. South, K.; Luken, B.M.; Crawley, J.T.B.; Phillips, R.; Thomas, M.; Collins, R.F.; Deforche, L.; Vanhoorelbeke, K.; Lane, D.A. Conformational activation of ADAMTS13. *Proc. Natl. Acad. Sci. USA* **2014**, *111*, 18578–18583. [CrossRef] [PubMed]

166. Coussens, L.M. Matrix Metalloproteinase Inhibitors and Cancer: Trials and Tribulations. *Siecnce* **2009**, *295*, 2387–2393. [CrossRef] [PubMed]

167. Moss, M.L.; Bomar, M.; Liu, Q.; Sage, H.; Dempsey, P.; Lenhart, P.M.; Gillispie, P.A.; Stoeck, A.; Wildeboer, D.; Bartsch, J.W.; *et al.* The ADAM10 prodomain is a specific inhibitor of ADAM10 proteolytic activity and inhibits cellular shedding events. *J. Biol. Chem.* **2007**, *282*, 35712–35721. [CrossRef] [PubMed]

168. Miller, M.A.; Moss, M.L.; Powell, G.; Petrovich, R.; Edwards, L.; Meyer, A.S.; Griffith, L.G.; Lauffenburger, D.A. Targeting autocrine HB-EGF signaling with specific ADAM12 inhibition using recombinant ADAM12 prodomain. *Sci. Rep.* **2015**, *5*. [CrossRef] [PubMed]

169. Tape, C.J.; Willems, S.H.; Dombernowsky, S.L.; Stanley, P.L.; Fogarasi, M.; Ouwehand, W.; McCafferty, J.; Murphy, G. Cross-domain inhibition of TACE ectodomain. *Proc. Natl. Acad. Sci. USA* **2011**, *108*, 5578–5583. [CrossRef] [PubMed]

170. Larkin, J.; Lohr, T.A.; Elefante, L.; Shearin, J.; Matico, R.; Su, J.; Xue, Y.; Liu, F.; Genell, C.; Miller, R.E.; *et al.* Translational development of an ADAMTS-5 antibody for osteoarthritis disease modification. *Osteoarthr. Cartil.* **2015**, *23*, 1254–1266. [CrossRef] [PubMed]

171. Shiraishi, A.; Mochizuki, S.; Miyakoshi, A.; Kojoh, K.; Okada, Y. Development of human neutralizing antibody to ADAMTS4 (aggrecanase-1) and ADAMTS5 (aggrecanase-2). *Biochem. Biophys. Res. Commun.* **2015**, *4*, 1–8. [CrossRef] [PubMed]

toxins

MDPI

Review

Processing of Snake Venom Metalloproteinases: Generation of Toxin Diversity and Enzyme Inactivation

Ana M. Moura-da-Silva [1,*], Michelle T. Almeida [1], José A. Portes-Junior [1], Carolina A. Nicolau [2], Francisco Gomes-Neto [2] and Richard H. Valente [2]

[1]   Laboratório de Imunopatologia, Instituto Butantan, São Paulo CEP 05503-900, Brazil; michelle.almeida@butantan.gov.br (M.T.A.); portes.junior@butantan.gov.br (J.A.P.-J.)
[2]   Laboratório de Toxinologia, Instituto Oswaldo Cruz, Rio de Janeiro CEP 21040-360, Brazil; carolnicolau.bio@gmail.com (C.A.N.); gomes.netof@gmail.com (F.G.-N.); richardhemmi@gmail.com (R.H.V.)
*   Correspondence: ana.moura@butantan.gov.br; Tel.: +55-11-2627-9779

Academic Editors: Jay Fox and José María Gutiérrez
Received: 4 May 2016; Accepted: 3 June 2016; Published: 9 June 2016

Abstract: Snake venom metalloproteinases (SVMPs) are abundant in the venoms of vipers and rattlesnakes, playing important roles for the snake adaptation to different environments, and are related to most of the pathological effects of these venoms in human victims. The effectiveness of SVMPs is greatly due to their functional diversity, targeting important physiological proteins or receptors in different tissues and in the coagulation system. Functional diversity is often related to the genetic diversification of the snake venom. In this review, we discuss some published evidence that posit that processing and post-translational modifications are great contributors for the generation of functional diversity and for maintaining latency or inactivation of enzymes belonging to this relevant family of venom toxins.

Keywords: snake venom; metalloproteinase; post-translational processing; enzyme inhibitor; hemorrhage

1. Introduction

Generation of diversity is a very important feature in the evolution of different species of animals, especially in systems in which fast adaptation to the environment is required. The most relevant system in which the generation of diversity plays a key role is the immune system. A large repertoire of antibody molecules, T cell receptors, and MHC antigens are mostly generated by intrinsic mechanisms of genetic recombination, together with post-transcriptional and post-translational processing [1–7], generating molecules responsible for host protection against aggressors and for self-maintenance. Although such a large repertoire is not necessary for many other systems, generation of diversity is also used as mechanism for fitness enhancement in most venomous animals, from cone snails [8] to advanced snakes [9], generating a toxin array that interacts with functionally-relevant receptors of different species [10], enabling capture of a greater diversity of prey, or evasion from different predators. In advanced snakes, generation of diversity of venom components was a great adaptive advantage that allowed the radiation of several taxa after the appearance of the venom glands, recruitment and neofunctionalization of toxin genes, and development of the venom injection system [9,10]. A few gene families have been recruited for snake venom production [11]. However, these genes are under accelerated evolution and undergo a number of duplications followed by distinct genetic modification mechanisms as accumulation of substitutive mutations, domain loss, recombination, and neofunctionalization that result in the large diversity within venom toxin gene families [12–17].

Snake venom metalloproteinases (SVMPs) are particularly important for the adaptation of snakes to different environments. In the venoms of most species of viper snakes, SVMPs are the most abundant component [18,19] and, as discussed above, the evolutionary mechanisms of this gene family allowed the structural and functional diversity of SVMPs in viper venoms. SVMPs are able to interact with different targets that control hemostasis or relevant tissues related to essential physiological functions in prey and predators [20,21]. The most evident effect of SVMPs is hemorrhage, as a result of a combined disruption of capillary vessels integrity and impairment of the blood coagulation system, resulting in consumption of coagulation plasma factors [20]. The mechanisms of action of distinct SVMPs involve different targets as, for example, activation of coagulation Factor X [22], activation of Factor II [23], fibrino(gen)olytic activity [24], binding and damage of capillary vessels [25–27], among others. SVMPs interacting with distinct hemostatic targets may be found in the same pool of venom from a single species [21] and, together, these different enzymes interfere with the whole hemostatic system, subduing prey usually by shock [28].

The structural diversity of SVMPs is well known [29–31] and three classes (P-I, P-II, and P-III), further subdivided into at least 11 subclasses, have been described based on their domain structure [31]. This classification is based on the presence of different domains in the zymogens predicted by the mRNA sequences and the mature form of the enzymes. They are synthesized as pro-enzymes with pre- and prodomains responsible for directing the nascent proteins to the endoplasmic reticulum and for maintaining the latency of the enzyme before secretion, respectively [32,33]. The mechanism involved in the activation of SVMPs (step of biosynthesis involving the removal of the prodomain) is still understudied. Furthermore, an eventual role played by the free prodomain (or its fragments) in enzyme activity after secretion is elusive. In addition to pre- and prodomains, a catalytic domain is present in P-I, P-II, and P-III classes at the C-terminus of the prodomain and is the only domain present in mature class P-I SVMPs. P-II and P-III SVMP classes differ from the former by the presence of non-catalytic domains included at the C-terminus of the catalytic domain: the disintegrin domain in P-II class and disintegrin-like plus cysteine-rich domains in the P-III class [31].

Genes coding for P-III class SVMPs appear to have been the first recruited to the snake venom while P-II and P-I SVMP genes appeared in viperids later, mostly by domain loss [16]. However, genetic mechanisms are not the only ones responsible for generating diversity in SVMPs. Due to the recent increase of data generated by venom proteomes and transcriptomes, it has also become evident that post-transcriptional [34] and post-translational [35] modifications represent additional sources of diversity generation in venom composition, increasing the possibilities of mechanisms of predation and resulting in an adaptive advantage for snakes. In this review, we will focus on the role of processing of nascent SVMPs in the generation of diversity and in the inactivation of these enzymes during and after their biosynthesis.

2. Biosynthesis and Post-Translational Processing of SVMPs

SVMP transcripts predict proteins with multi-domain structure that undergo different post-translational processing generating distinct mature proteins (Figure 1). As other secreted proteins, SVMPs include a signal-peptide/predomain (p) responsible for driving the nascent SVMP to the endoplasmic reticulum where most of the protein modifications take place. In the endoplasmic reticulum, the pre-domain is removed by signal peptidases (Figure 1-①) resulting in zymogens that are subjected to further modifications [36]. Activation of the enzymes occurs by hydrolysis of the prodomain (Figure 1-②) [37] and, after this step, disintegrin or disintegrin-like/cysteine-rich domains can also be released by proteolysis (Figure 1-③) [38]. Mature forms may present other modifications such as cyclization of amino-terminal glutamyl residues to pyro-glutamate (Figure 1-④), glycosylation (Figure 1-⑤), addition of new domains (Figure 1-⑥), or dimerization of protein chains (Figure 1-⑦) [36]. These steps occur to different extents depending on the primary structure of the precursors predicted by the paralogue genes that originate the transcripts. As a result, different biological activities are associated to each particular isoform. Therefore, post-translational modifications are essential for activity and stability of the proteins, and also for diversifying their specific targets. Some of the issues related to each of these processing steps will be discussed below.

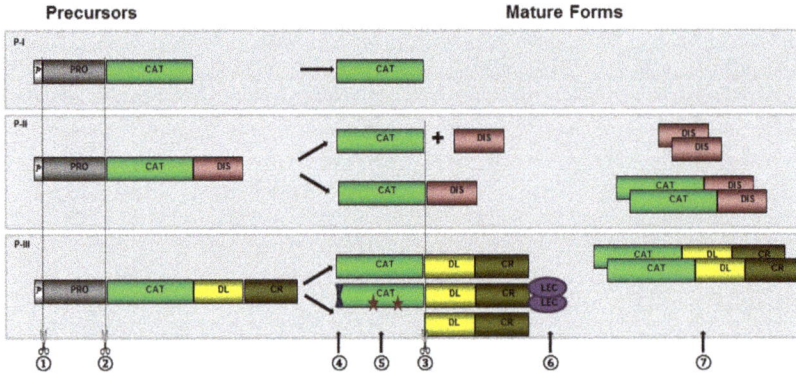

Figure 1. Schematic representation of the most typical post-translational modification steps occurring during the maturation of nascent SVMPs: SVMP precursors are composed of signal-peptide/pre-(p), pro- (PRO), catalytic or metalloproteinase (CAT), disintegrin (DIS), disintegrin-like (DL), cysteine-rich (CR), and lectin-like (LEC) domains. Processing of nascent SVMP involves removal of signal-peptide/pre-domain (①) hydrolysis of the prodomain (②) and disintegrin or disintegrin-like/cysteine-rich domains (③), cyclization of amino-terminal glutamyl residues to pyro-glutamate (④), glycosylation (represented by stars—⑤), addition of new domains (⑥) or dimerization of protein chains (⑦).

2.1. Hydrolysis of the Prodomain

Activation of SVMPs is regulated by hydrolysis of their prodomains, as happens with matrix metalloproteinases (MMPs) and disintegrin and metalloproteinase (ADAM) proteins. Prodomains of SVMPs include a conserved motif (PKMCGVT), also found in ADAM and MMP precursors [33]. In this motif, a free cysteine residue is a key factor for maintaining enzyme latency via a cysteine-switch mechanism. This process controls the activation state of enzymes by blocking the catalytic site (inactivated state) before the proteolytic processing of the prodomain (active state) [33,39].

In MMPs, activation generally occurs at the extracellular space catalyzed by members of the plasminogen/plasmin cascade, by other MMPs or by chemical modification of the conserved cysteine residue in the cysteine switch motif [40–42]. A different mechanism of activation is verified in ADAMs, as the prodomain is generally removed intracellularly by pro-protein convertases [43], or by autocatalytic mechanisms [31,44,45]. In SVMPs, only a few studies attempted to explain enzyme activation and/or hydrolysis of their prodomains. However, studying the activation of recombinant pro-atrolysin-E, Shimokawa and collaborators [38] suggested that chemical modifications are not efficient for activation, which probably occurs by proteolysis by metalloproteinase present on the crude venom. In a recent study from our group [46], we used antibodies specific to jararhagin prodomain to search for the presence of prodomains in different compartments of snake venom glands, either as zymogens or in the processed form. Using gland extracts obtained at different times of the venom production cycle, we immunodetected electrophoretic bands matching to the SVMP zymogen molecular mass (in high abundance), and only faint bands with molecular masses corresponding to different forms of cleaved prodomain. However, the presence of zymogens in the venom was rare, detected only as faint bands in samples collected from the lumen of venom gland at the peak of venom production. In milked venom, only weak bands corresponding to free prodomain were detected in samples collected at the peak of venom production, suggesting that most of the prodomain molecules promptly undergo further hydrolysis, generating diverse peptides that are not immunoreactive with anti-PD-Jar (Figure 2). In agreement to this suggestion, SVMP prodomain peptides are very rare in proteomes of viper venoms, with a few exceptions [47], but they were recently found in the proteopeptidome of *B. jararaca* venom [48], and in proteomes of *B. jararaca* gland extracts [49], which is consistent with our hypothesis.

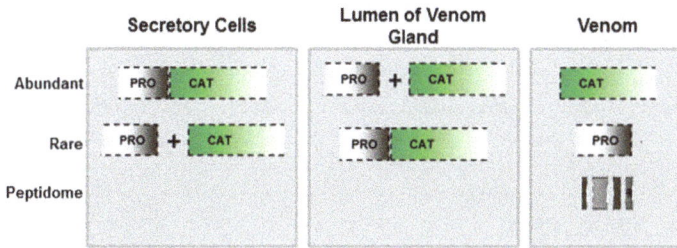

Figure 2. Schematic representation of prodomain processing: Antibodies against jararhagin prodomain detected predominantly bands of zymogen molecular mass in secretory cells and processed form in the lumen of the venom gland. Prodomain was poorly detected in the venom, suggesting that SVMPs are mostly in the active form.

Using immunohistochemistry and immunogold electromicroscopy, prodomain detection was concentrated in secretory vesicles of secretory cells (Figure 3). According to these images, we suggested that SVMPs are stored at secretory cell vesicles mostly as zymogens; the processing of prodomains starts within the secretory vesicles but reaches its maximal level during secretion or as soon as it reaches the lumen of the venom gland.

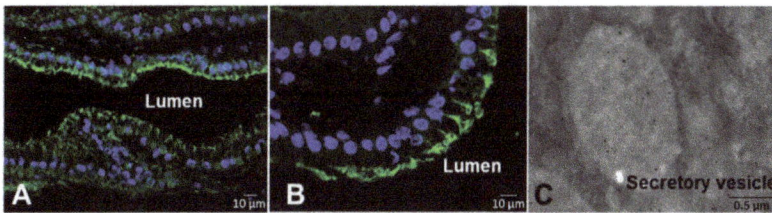

Figure 3. Cellular localization of prodomains. Venom glands collected before (**A**) or seven days after (**B**, **C**) venom extraction were sectioned and subjected to immunofluorescence (**A**, **B**) stained with DAPI (blue) and mouse anti-PD-Jar serum (green), which concentrated in the apical region of secretory cells, or electron microscopy (**C**) after staining with anti-PD-Jar serum, which highlighted spots in the secretory vesicles [46].

According to these data, processing and activation of SVMPs undergo distinct routes than MMPs or ADAMs. MMPs are critical enzymes for remodeling the extracellular matrix in a series of physiological and pathological processes as angiogenesis, wound healing, inflammation, cancer, and infections [40,50]. The regulatory role of MMPs in such processes requires a well-controlled mechanism of activation for which the secretion of latent enzymes is of great advantage. On the other hand, most ADAMs are transmembrane proteins that regulate mostly cell migration, adhesion, signaling and, eventually, proteolysis. In this case, processing of ADAMs through the secretion pathway by furins and other processing enzymes is the most common processing route [43,51]. SVMPs apparently undergo different processing routes since the release of the prodomain is very likely to occur during the secretion of vesicle contents. The enzymes responsible for the processing have not yet been identified, but it can be speculated that venom serine proteinases or even metalloproteinases could be involved. Moreover, a series of convertases have been detected in proteomic and transcriptomic studies [49]. One issue that remains unsolved and will be discussed below is whether SVMPs are maintained in the lumen of the venom gland in the active form or are kept inactivated by peptides liberated by prodomain hydrolysis or by other inhibitory factors present in the venom as the acidic pH environment, high citrate concentrations and tripeptides containing pyroglutamate.

2.2. Generation of Disintegrin and Disintegrin-Like Domains

Disintegrins are generated by proteolysis of SVMPs originated from class P-II transcripts. These small molecules are abundant in venoms of viper snakes that usually contain the RGD or a related (XGD) motif in a surface exposed loop that binds to RGD-dependent integrins, such as $\alpha_{IIb}\beta_3$, $\alpha_5\beta_1$, and $\alpha_v\beta_3$ or, in a few cases, they may display a MLD motif, targeting $\alpha_4\beta_1$, $\alpha_4\beta_7$, and $\alpha_9\beta_1$, or a K/RTS motif that is very selective for binding to $\alpha_1\beta_1$ integrin [52]. These are important receptors of different cell types, particularly platelets, inflammatory, and vascular endothelial cells, in which they are responsible for inhibition of platelet aggregation or endothelial cell adhesion, migration, and angiogenesis [53,54]. For these reasons, the inclusion of disintegrins in the venom of viper snakes conferred a great adaptive advantage for using hemostatic targets to surrender prey.

Genes coding for class P-II SVMPs have evolved from P-III ancestor genes by a single loss of the cysteine-rich domain followed by convergent losses of the disintegrin domain at different phylogenetic branches that were further responsible for the generation of distinct P-I SVMP structures [16]. After the cysteine-rich domain loss, evolution of P-II genes was continued by gene duplication and neofunctionalization of the disintegrin domain in some of the duplicated copies [16,55]. Other genetic mechanisms of recombination as exon shuffling or pre- or post-transcriptional recombination could also play a role in the diversification of class P-II SVMP structures. The first draft of the genomic organization of a PIII-SVMP gene revealed a series of nuclear retroelements and transposons within introns that could provide genomic explanations for the emergence of distinct class P-II messengers [56]. Evidence for post-transcriptional modification arose when *B. neuwiedi* SVMP cDNA sequences were analyzed: three distinct types of P-II sequences were noted including a typical transcript of class P-II SVMP and other transcripts that presented clear indications of recombination between P-II disintegrin domain coding regions with either P-I or P-III catalytic domain coding regions [34]. The data suggest that recombination between genes encoding SVMPs might have occurred after the emergence of the primary gene copies coding for each scaffold. Moreover, it has also been reported by different authors the occurrence of SVMP structures that might have been assembled by the P-III catalytic domain with the P-II disintegrin domain [57] or even by PII catalytic domains with the P-III disintegrin-like domains lacking the cysteine-rich domain [58]. Unfortunately, these mechanisms of recombination are still speculative since, up to now, genomic sequences coding for SVMPs were not completely disclosed and the exon/intron distribution at catalytic domain is still unknown.

In viper venoms, the products of P-II genes are diverse and the precursors undergo proteolytic steps depending on the structure predicted by the paralogue gene coding for each different toxin. Most P-II precursors are hydrolyzed at the spacer region, located between catalytic and disintegrin domains [31] generating free disintegrins and catalytic domains that are frequently found in venoms, and also recognized as classical disintegrins and P-I class SVMPs, respectively. However, some P-II precursors are not hydrolyzed and are expressed in the venom as single chained molecules, containing catalytic and disintegrin domains. The enzymes involved in the cleavage of P-II precursors to generate free disintegrins are still unrecognized and the mechanisms by which different P-II precursors are processed (or not) are still speculative. A mainstream hypothesis, postulated by Serrano and Fox [31], suggests that the presence of cysteinyl residues, particularly at the spacer region and at the N-terminus of the disintegrin domain, would confer more resistance to hydrolysis, acting in favor of maintenance of P-II SVMPs in the catalytic form. A few of these enzymes have been characterized [53,59–61] and recent reports indicate potent hemorrhagic activity in catalytic P-II SVMPs that may be achieved by their capability to cleave ECM proteins combined to their potential to inhibit platelet aggregation and/or to bind to basal lamina [62].

Some P-III class SVMPs are also cleaved generating fragments which correspond to the disintegrin-like/cysteine-rich domains [31]. However, cleavage mechanisms and the fate of the domains after cleavage are apparently different than the ones observed in P-II SVMPs. Most P-III SVMPs are found in venoms in their multi-domain form, containing catalytic, disintegrin-like and cysteine-rich domains, although examples of autolysis at the spacer region of isolated P-III SVMPs have been reported for jararhagin, from *Bothrops jararaca* [63], HR1A and HR1B from *Trimeresurus flavoviridis* [64], HT-1 from *Crotalus ruber ruber* [64], brevilysin H6, from *Gloydius halys brevicaudus* [65], alternagin, from *Bothrops alternatus* [66], batroxhagin, from *Bothrops atrox* [67,68], and catrocollastatin, from *Crotalus atrox* [69]. Autolysis of these proteins usually results in combined disintegrin-like/cysteine-rich domain fragments, known as "C" proteins, which are characterized as inhibitors of collagen-induced platelet-aggregation [63], but may also display pro-inflammatory activity [70] or stimulate endothelial cells to release pro-angiogenic mediators [71]. On the other hand, the free catalytic domain that results from this autolytic process has never been found in venoms, suggesting that a fast hydrolysis of the catalytic domain to small peptides occurs after autolysis. In *B. jararaca* venom, the presence of both intact jararhagin and processed jararhagin-C is currently detected [63,72]. Moreover, a third form of processed protein was detected that comprised a processed form of jararhagin-C linked to the catalytic domain by disulfide bonds [73]. This evidence suggests that at least three different proteoforms of jararhagin may exist and they probably display distinct pairing of cysteinyl residues that will drive to three different autolytic pathways. The three possibilities of autolysis appear to occur in venoms of other snakes. We recently reported the same three forms of processed batroxrhagin, a P-III SVMP isolated from the venom of *B. atrox* [67]. The presence of these different forms of P-III SVMPs in venoms contributes to greater structural and functional complexity of the venom, and may be a common feature among other class P-III SVMPs.

2.3. Dimerization and Inclusion of Other Domains

The position and pairing of cysteinyl residues certainly play a role in the liberation of disintegrins, but they are also very important in generating multimeric structures of some nascent SVMPs, increasing their structural and functional diversity. One example is the linkage of lectin-like domains to the cysteine-rich domain of class P-III SVMPs generating very active pro-coagulant toxins as RVV-X, from *Vipera russelli* [32–34], classified as a PIIId, or previously as a P-IV SVMP [30]. However, homodimers or heterodimers of homologous domains are most commonly found in the venoms and dimerization apparently contributes to the enhancement of the toxin activity. Class P-III SVMPs VAP1 and VAP2 from *Crotalus atrox* have been crystallized in their dimeric form [74,75] exerting potent pro-apoptotic activity in endothelial cell cultures [76]. Bilitoxin and BlatH1 are also examples of non-processed P-II SVMP homodimers [77,78]. Interestingly, in both cases the RGD sequence displayed in the disintegrin domain is replaced by MGD and TDN, respectively, resulting in toxins that are unable to block the platelet fibrinogen receptor; however, both dimeric P-II SVMPs are potent hemorrhagins with activity levels comparable to those of class P-III SVMPs [62,77].

The most common dimers of SVMPs present in venoms are undoubtedly homo- and heterodimeric disintegrins. For example, contortrostatin, a homodimeric disintegrin that displays RGD motifs in both chains, has been produced in a recombinant form [79] and presented substantial anti-angiogenic and anti-cancer effects [79] with some efficacy as an adjuvant in chemotherapy for melanoma [80] and for viral infections [81]. Heterodimers composed of distinct disintegrin domains have also been isolated. Usually, one chain contains the conserved RGD motif and the other chain displays alternative motifs that modulate specificity and selectivity. The first heterodimeric disintegrin group identified included EMF-10 and CC-8 disintegrins, with a RGD motif in one chain and a WGD in the other [82]. As a result, the heterodimer is the strongest blocker of the fibrinogen receptor and effective to modulate megakaryocyte activity [83]. Other interesting group of heterodimeric disintegrins includes EC3, VLO5, and EO5 with V/RGD motif in one chain and the MLD motif in the other [84]. These toxins target mainly $\alpha_4\beta_1$, $\alpha_4\beta_7$, and $\alpha_9\beta_1$ integrins, mostly related to inflammatory cell receptors [85]. Heterodimeric disintegrins are common in venoms of *Viperinae* subfamily of vipers, and the substitution of the RGD motif at least in one chain may decrease the effect of these toxins on platelets related to impairment of hemostasis. Although this is an apparent disadvantage for the snakes, the non-RGD disintegrins are undoubtedly good leading molecules for drug development or for producing biotechnological tools to address the mechanisms of action of integrin receptors.

2.4. Other Post-Translational Modifications

Most of snake venom proteins undergo glycosylation during their biosynthesis pathway. In eukaryotic cells, glycosylation influences important biochemical properties of the proteins, such as folding, stability, solubility, and ligand binding. In spite of the importance of glycosylation for eukaryotic-secreted proteins, very little is known about the carbohydrate structures present in venom glycoproteins.

In SVMPs, primary structures predict several putative *N*-glycosylation sites [31], and important functions have already been correlated to glycan moieties. However, glycosylation sites identified in the available crystal structures of SVMPs indicate a significant variability, and suggest that the presence of glycan moieties is not predictable based on primary structure information only [86]. Studying the cobra venom glycome, Huang and co-authors [86] identified four major *N*-glycan moieties on the biantennary glycan core, and a high variability of *N*-glycan composition in SVMPs from individual snake specimens. In the same study, the authors reported that these glycoproteins elicit much higher antibody response in antiserum when compared to other high-abundance cobra venom toxins, such as small molecular mass CTXs. The higher immunogenicity of SVMPs compared to other venom components has been also shown by our group [18], and it can be at least partially attributed to the glycan moieties present on these molecules. Moreover, *Bothrops jararaca* SVMPs bothropasin, BJ-PI, and HF3 were subjected to *N*-deglycosylation that induced loss of structural stability of bothropasin and BJ-PI. Although HF3 remained apparently intact, its hemorrhagic and fibrinogenolytic activities were partially impaired, suggesting the importance of glycans for stability, and also for the interaction with substrates [87].

The role of glycosylation in the generation of toxin diversity has been recently addressed in a few studies approaching ontogenetic or gender-related venom variability. The *N*-glycan composition of newborn and adult venoms did not vary significantly [88], but gender-based variations contributed to different glycosylation levels in toxins [35]. The studies demonstrated a complexity of carbohydrate moieties found in glycoproteins, indicating another level of complexity in snake venoms that could be related to the diversification of biological activities.

Another form of post-translational modification observed in many SVMPs is the cyclization of amino-terminal glutaminyl residues to pyro-glutamate. The cyclization of glutaminyl residues by the acyltransferase glutaminyl cyclase is a common occurrence in many organisms. For many bioactive peptides, cyclization of amino-terminal glutaminyl residues renders the peptide resistant to proteolytic processing by exopeptidases, thus protecting their biological activities [89]. Glutaminyl cyclase has been identified in the venom of viperid snakes [90] particularly in venoms collected at the seventh and tenth days after venom extraction [46]. Most class P-III SVMPs possess, in their mature form, a pyroglutamic acid as the *N*-terminus [31], and this modification provides protection to these enzymes from further digestion by aminopeptidases, or even further processing steps resulting in the release of disintegrin domains.

3. The Role of Prodomains for Enzyme Inactivation

One still-unsolved issue is the maintenance of mature enzymes in the lumen of the venom gland. Since SVMPs degrade extracellular matrix components [91,92] and can lead to loss of viability of different cell types [20] (including epithelial cells [93]), they can be considered a potential risk for the maintenance of venom gland integrity. It is currently accepted that the acidic pH environment in the lumen of the venom gland could limit proteolytic activity of SVMPs [94]. Additionally, high citrate concentrations and tripeptides containing pyroglutamate found in venoms could inhibit SVMPs that would be activated after venom injection due to dilution factors [95,96]. Previously, secretion of SVMPs into the lumen of the gland as zymogens was also considered as a mechanism for maintaining the latency of these enzymes. However, we have recently shown that the activation of SVMPs mainly occurs during the secretion to the lumen of the venom gland, and cleaved prodomains undergo further hydrolysis to small peptides [46].

The potential of peptides containing the cysteine-switch motif for the inactivation of SVMPs has already been shown [37]. This work led us to test if processed prodomain or prodomain degradation peptides could play a role in the inhibition of activated SVMPs, within the gland environment. To test this hypothesis, we produced jararhagin recombinant prodomain (PD-Jar) as described [46], and synthesized a 14-mer C-terminally amidated peptide (SynPep), based on a naturally-occurring prodomain peptide fragment that was abundantly detected in *Bothrops jararaca* peptidome [48]. Interestingly, our preliminary data show that both recombinant PD-Jar and SynPep inhibited jararhagin catalytic activity, and also toxic activities such as induction of fibrinolysis and hemorrhage (Table 1).

Table 1. Inhibition of jararhagin activities by its recombinant prodomain (PD-Jar) or a prodomain degradation peptide (SynPep).

Activity	PD-Jar [1]		SynPep [1]	
	Molar Ratio	% Inhibition	Molar Ratio	% Inhibition
Catalytic [2]	1:10	98	1:5000	90
Fibrinolytic [3]	1:14	100	1:200	100
Hemorrhagic [4]	1:9	100	1:500	100

[1] Values correspond to enzyme to PD-Jar/SynPep molar ratios that resulted in inhibition of jararhagin activity; [2] Inhibition of enzymatic activity was tested by incubation with Abz-A-G-L-A-EDDnp as fluorescence quenched metalloproteinase substrate and compared according to the relative fluorescence units (RFU/min/μg) of each reaction [97]; [3] Inhibition of jararhagin fibrinolytic activity was calculated by measuring the hydrolysis halo in fibrin-containing agarose plates [98]; [4] Hemorrhage levels were calculated by measuring the hemorrhagic area 30 min after intradermal injection in the dorsal region of four mice [98].

Inhibition of metalloproteinase activity by isolated prodomains has been recently addressed in the search for specific therapeutic tools for pathologies involving these enzymes. The cleaved prodomain of certain ADAMs can act as a selective inhibitor of the catalytic activity of the enzyme. Moss and coworkers [99] and Gonzales *et al.* [100] produced the recombinant prodomains of ADAM-9 and TACE, respectively, for the purpose of understanding their mechanism of inhibition and selectivity against these proteinases. In these studies, ADAM-9 prodomain was highly specific and the inhibition of ADAM-9, by its recombinant prodomain, regulated ADAM-10 activity controlling the release of soluble α-secretase enzyme, which is an important task in the therapy Alzheimer's disease [99]. Additionally, TACE prodomain was also specific for this enzyme and could be used as a potential inhibitor of TNF-α release in inflammatory diseases [100]. Considering the high selectivity of prodomains as metalloproteinase inhibitors, we are currently testing more accurately the selectivity and kinetics parameters of SVMPs' inhibition by PD-Jar and SynPep, and their effects on neutralization of local effects induced by viper venoms. These experiments may increase our understanding about the maintenance of inactivated SVMPs inside the venom glands, and could also lead to therapeutic alternatives to minimize local damage induced by snake venoms.

The animal protocols used in this work were evaluated and approved by the Animal Use and Ethic Committee (CEUAIB) of the Institute Butantan (Protocol 1271/14). They are in accordance with COBEA guidelines and the National law for Laboratory Animal Experimentation (Law no. 11.794, 8 October 2008).

4. Conclusions

The contribution of post-translational processing to the generation of venom diversity has been a recent issue offering important insights for understanding the complexity of animal venom arsenals. In this review, we present some current data that support the participation of post-translational processing for generation of diversity of snake venom metalloproteinases. Furthermore, there are other snake venom toxin families (serine proteinases, phospholipases, and C-type lectin-like proteins) represented by several proteoforms, and we predict that similar features discussed here for SVMPs could also be applicable to account, at least in part, for their diversity, and that the same would hold true for venom components from different animal species. Indeed, in a very elegant study, Dutertre and collaborators [101] explained the expanded peptide diversity in the cone snail *Conus marmoreus* revealing how a limited set of approximately 100 transcripts could generate thousands of conopeptides in the venom of a single species. More recently, Zhang and collaborators [102] working with peptide toxins from the tarantula *Haplopelma hainanum* went further into this aspect by showing the role of post-translational modifications in the generation of venom diversity and also in diversifying the functional venom arsenal. In this review, we addressed this issue for snake venom metalloproteinases. Although genetic mechanisms are essential for generating a great number of SVMP paralogue genes, post-translational processing appears as an important contributor for diversifying the SVMP arsenal able to interact with a greater number of physiological targets present in different prey. The articles revised in this paper may represent only the tip of the iceberg explaining SVMPs diversity. With analytical methods improvements, such as top-down proteomic approaches, characterization of proteoforms of complex molecules may present a larger number of possibilities for post-translational modification in SVMPs, supporting the role of processing for the stability, maintenance and functional diversification of this important toxin family present in snake venoms.

Acknowledgments: AMMS is a CNPq fellow (CNPq 304025/2014-3) and is supported by CAPES (AuxPE 1209/2011 and 1519/2011) and FAPESP (2014/26058-8). MTA was sponsored by a CNPq fellowship (131820/2014-1) and JAPR by a CAPES fellowship (AuxPE 1206/2011). RHV is supported by a CAPES grant, which also supports CAN as a post-doctoral fellow (AUXPE 1224/2011).

Author Contributions: A.M.M.-S., J.A.P.-J., and R.H.V. wrote the main body of this review, with contributions from C.A.N and F.G.-N. Regarding the original data displayed in Table 1, the experiments were conceived by A.M.M.-S. and J.A.P.-J., while performed by M.T.A. and J.A.P.-J.

Conflicts of Interest: The authors declare no conflict of interest.

Abbreviations

The following abbreviations are used in this manuscript:

SVMP	Snake Venom Metalloproteinase
ADAM	A Disintegrin and Metalloproteinase
MMP	Matrix Metalloproteinase
PD-Jar	Recombinant prodomain of Jararhagin
MHC	Major Histocompatibility Complex
ECM	Extracellular Matrix
TACE	Tumor Necrosis Factor-alpha Converting Enzyme
DAPI	4′,6-diamidino-2-phenylindole
SynPep	Synthetic Peptide of a prodomain hydrolysis product found in *B. jararaca* venom peptidome

References

1. Elhanati, Y.; Sethna, Z.; Marcou, Q.; Callan, C.G.; Mora, T.; Walczak, A.M. Inferring processes underlying b-cell repertoire diversity. *Philos. Trans. R. Soc. Lond. B Biol. Sci.* **2015**, *370*. [CrossRef] [PubMed]
2. Hoehn, K.B.; Fowler, A.; Lunter, G.; Pybus, O.G. The diversity and molecular evolution of b cell receptors during infection. *Mol. Biol. Evol.* **2016**, *33*, 1147–1157. [CrossRef] [PubMed]

3. Warren, R.L.; Freeman, J.D.; Zeng, T.; Choe, G.; Munro, S.; Moore, R.; Webb, J.R.; Holt, R.A. Exhaustive T-cell repertoire sequencing of human peripheral blood samples reveals signatures of antigen selection and a directly measured repertoire size of at least 1 million clonotypes. *Genome Res.* **2011**, *21*, 790–797. [CrossRef] [PubMed]
4. Qi, Q.; Liu, Y.; Cheng, Y.; Glanville, J.; Zhang, D.; Lee, J.Y.; Olshen, R.A.; Weyand, C.M.; Boyd, S.D.; Goronzy, J.J. Diversity and clonal selection in the human T-cell repertoire. *Proc. Natl. Acad. Sci. USA* **2014**, *111*, 13139–13144. [CrossRef] [PubMed]
5. Subedi, G.P.; Barb, A.W. The structural role of antibody *N*-glycosylation in receptor interactions. *Structure* **2015**, *23*, 1573–1583. [CrossRef] [PubMed]
6. Liu, L. Antibody glycosylation and its impact on the pharmacokinetics and pharmacodynamics of monoclonal antibodies and fc-fusion proteins. *J. Pharm. Sci.* **2015**, *104*, 1866–1884. [CrossRef] [PubMed]
7. Nikolich-Zugich, J.; Slifka, M.K.; Messaoudi, I. The many important facets of T-cell repertoire diversity. *Nat. Rev. Immunol.* **2004**, *4*, 123–132. [CrossRef]
8. Olivera, B.M.; Teichert, R.W. Diversity of the neurotoxic conus peptides: A model for concerted pharmacological discovery. *Mol. Interv.* **2007**, *7*, 251–260. [CrossRef] [PubMed]
9. Fry, B.G.; Vidal, N.; van der Weerd, L.; Kochva, E.; Renjifo, C. Evolution and diversification of the toxicofera reptile venom system. *J. Proteom.* **2009**, *72*, 127–136. [CrossRef]
10. Casewell, N.R.; Wüster, W.; Vonk, F.J.; Harrison, R.A.; Fry, B.G. Complex cocktails: The evolutionary novelty of venoms. *Trends Ecol. Evol.* **2013**, *28*, 219–229. [CrossRef] [PubMed]
11. Calvete, J.J.; Juárez, P.; Sanz, L. Snake venomics. Strategy and applications. *J. Mass Spectrom.* **2007**, *42*, 1405–1414. [CrossRef] [PubMed]
12. Ogawa, T.; Nakashima, K.; Nobuhisa, I.; Deshimaru, M.; Shimohigashi, Y.; Fukumaki, Y.; Sakaki, Y.; Hattori, S.; Ohno, M. Accelerated evolution of snake venom phospholipase A2 isozymes for acquisition of diverse physiological functions. *Toxicon* **1996**, *34*, 1229–1236. [CrossRef]
13. Ohno, M.; Ménez, R.; Ogawa, T.; Danse, J.M.; Shimohigashi, Y.; Fromen, C.; Ducancel, F.; Zinn-Justin, S.; Le Du, M.H.; Boulain, J.C.; *et al.* Molecular evolution of snake toxins: Is the functional diversity of snake toxins associated with a mechanism of accelerated evolution? *Prog. Nucleic Acid Res. Mol. Biol.* **1998**, *59*, 307–364. [PubMed]
14. Moura-da-Silva, A.M.; Theakston, R.D.G.; Crampton, J.M. Evolution of disintegrin cysteine-rich and mammalian matrix-degrading metalloproteinases: Gene duplication and divergence of a common ancestor rather than convergent evolution. *J. Mol. Evol.* **1996**, *43*, 263–269. [CrossRef] [PubMed]
15. Brust, A.; Sunagar, K.; Undheim, E.A.; Vetter, I.; Yang, D.C.; Casewell, N.R.; Jackson, T.N.; Koludarov, I.; Alewood, P.F.; Hodgson, W.C.; *et al.* Differential evolution and neofunctionalization of snake venom metalloprotease domains. *Mol. Cell. Proteom.* **2013**, *12*, 651–663. [CrossRef] [PubMed]
16. Casewell, N.R.; Wagstaff, S.C.; Harrison, R.A.; Renjifo, C.; Wüster, W. Domain loss facilitates accelerated evolution and neofunctionalization of duplicate snake venom metalloproteinase toxin genes. *Mol. Biol. Evol.* **2011**, *28*, 2637–2649. [CrossRef] [PubMed]
17. Casewell, N.R.; Wagstaff, S.C.; Wüster, W.; Cook, D.A.; Bolton, F.M.; King, S.I.; Pla, D.; Sanz, L.; Calvete, J.J.; Harrison, R.A. Medically important differences in snake venom composition are dictated by distinct postgenomic mechanisms. *Proc. Natl. Acad. Sci. USA* **2014**, *111*, 9205–9210. [CrossRef] [PubMed]
18. Sousa, L.F.; Nicolau, C.A.; Peixoto, P.S.; Bernardoni, J.L.; Oliveira, S.S.; Portes-Junior, J.A.; Mourão, R.H.; Lima-dos-Santos, I.; Sano-Martins, I.S.; Chalkidis, H.M.; *et al.* Comparison of phylogeny, venom composition and neutralization by antivenom in diverse species of bothrops complex. *PLoS Negl. Trop. Dis.* **2013**, *7*. [CrossRef] [PubMed]
19. Calvete, J.J.; Sanz, L.; Angulo, Y.; Lomonte, B.; Gutiérrez, J.M. Venoms, venomics, antivenomics. *FEBS Lett.* **2009**, *583*, 1736–1743. [CrossRef] [PubMed]
20. Moura-da-Silva, A.M.; Butera, D.; Tanjoni, I. Importance of snake venom metalloproteinases in cell biology: Effects on platelets, inflammatory and endothelial cells. *Curr. Pharm. Des.* **2007**, *13*, 2893–2905. [CrossRef] [PubMed]
21. Bernardoni, J.L.; Sousa, L.F.; Wermelinger, L.S.; Lopes, A.S.; Prezoto, B.C.; Serrano, S.M.T.; Zingali, R.B.; Moura-da-Silva, A.M. Functional variability of snake venom metalloproteinases: Adaptive advantages in targeting different prey and implications for human envenomation. *PLoS ONE* **2014**, *9*. [CrossRef]

22. Siigur, E.; Tõnismägi, K.; Trummal, K.; Samel, M.; Vija, H.; Subbi, J.; Siigur, J. Factor X activator from vipera lebetina snake venom, molecular characterization and substrate specificity. *Biochim. Biophys. Acta* **2001**, *1568*, 90–98. [CrossRef]

23. Modesto, J.C.; Junqueira-de-Azevedo, I.L.; Neves-Ferreira, A.G.; Fritzen, M.; Oliva, M.L.; Ho, P.L.; Perales, J.; Chudzinski-Tavassi, A.M. Insularinase A, a prothrombin activator from *Bothrops insularis* venom, is a metalloprotease derived from a gene encoding protease and disintegrin domains. *Biol. Chem.* **2005**, *386*, 589–600. [PubMed]

24. Kamiguti, A.S.; Slupsky, J.R.; Zuzel, M.; Hay, C.R. Properties of fibrinogen cleaved by jararhagin, a metalloproteinase from the venom of *Bothrops jararaca*. *Thromb. Haemost.* **1994**, *72*, 244–249. [PubMed]

25. Baldo, C.; Jamora, C.; Yamanouye, N.; Zorn, T.M.; Moura-da-Silva, A.M. Mechanisms of vascular damage by hemorrhagic snake venom metalloproteinases: Tissue distribution and *in situ* hydrolysis. *PLoS Negl. Trop. Dis.* **2010**, *4*. [CrossRef] [PubMed]

26. Escalante, T.; Shannon, J.; Moura-da-Silva, A.M.; Gutiérrez, J.M.; Fox, J.W. Novel insights into capillary vessel basement membrane damage by snake venom hemorrhagic metalloproteinases: A biochemical and immunohistochemical study. *Arch. Biochem. Biophys.* **2006**, *455*, 144–153. [CrossRef] [PubMed]

27. Escalante, T.; Rucavado, A.; Fox, J.W.; Gutierrez, J.M. Key events in microvascular damage induced by snake venom hemorrhagic metalloproteinases. *J. Proteom.* **2011**, *74*, 1781–1794. [CrossRef] [PubMed]

28. Kamiguti, A.S.; Cardoso, J.L.; Theakston, R.D.; Sano-Martins, I.S.; Hutton, R.A.; Rugman, F.P.; Warrell, D.A.; Hay, C.R. Coagulopathy and haemorrhage in human victims of *Bothrops jararaca* envenoming in brazil. *Toxicon* **1991**, *29*, 961–972. [CrossRef]

29. Bjarnason, J.B.; Fox, J.W. Snake venom metalloendopeptidases: Reprolysins. *Methods Enzymol.* **1995**, *248*, 345–368. [PubMed]

30. Fox, J.W.; Serrano, S.M.T. Structural considerations of the snake venom metalloproteinases, key members of the M12 reprolysin family of metalloproteinases. *Toxicon* **2005**, *45*, 969–985. [CrossRef] [PubMed]

31. Fox, J.W.; Serrano, S.M.T. Insights into and speculations about snake venom metalloproteinase (SVMP) synthesis, folding and disulfide bond formation and their contribution to venom complexity. *FEBS J.* **2008**, *275*, 3016–3030. [CrossRef] [PubMed]

32. Gomis-Rüth, F.X. Structural aspects of the metzincin clan of metalloendopeptidases. *Mol. Biotechnol.* **2003**, *24*, 157–202. [CrossRef]

33. Stöcker, W.; Grams, F.; Baumann, U.; Reinemer, P.; Gomis-Rüth, F.X.; McKay, D.B.; Bode, W. The metzincins—Topological and sequential relations between the astacins, adamalysins, serralysins, and matrixins (collagenases) define a superfamily of zinc-peptidases. *Protein Sci.* **1995**, *4*, 823–840. [CrossRef] [PubMed]

34. Moura-da-Silva, A.M.; Furlan, M.S.; Caporrino, M.C.; Grego, K.F.; Portes-Junior, J.A.; Clissa, P.B.; Valente, R.H.; Magalhães, G.S. Diversity of metalloproteinases in *Bothrops neuwiedi* snake venom transcripts: Evidences for recombination between different classes of SVMPs. *BMC Genet.* **2011**, *12*. [CrossRef] [PubMed]

35. Zelanis, A.; Menezes, M.C.; Kitano, E.S.; Liberato, T.; Tashima, A.K.; Pinto, A.F.; Sherman, N.E.; Ho, P.L.; Fox, J.W.; Serrano, S.M. Proteomic identification of gender molecular markers in *Bothrops jararaca* venom. *J. Proteom.* **2016**, *139*, 26–37. [CrossRef] [PubMed]

36. Moura-da-Silva, A.M.; Serrano, S.M.T.; Fox, J.W.; Gutierrez, J.M. Snake venom metalloproteinases: Structure, function and effects on snake bite pathology. In *Animal Toxins: State of the Art*; Lima, M.H., Pimenta, A.M.C., Martin-Euclaire, M.F., Zingali, R.B., Eds.; UFMG: Belo Horizonte, Brazil, 2009; pp. 525–546.

37. Grams, F.; Huber, R.; Kress, L.F.; Moroder, L.; Bode, W. Activation of snake venom metalloproteinases by a cysteine switch-like mechanism. *FEBS Lett.* **1993**, *335*, 76–80. [CrossRef]

38. Shimokawa, K.; Jia, L.G.; Wang, X.M.; Fox, J.W. Expression, activation, and processing of the recombinant snake venom metalloproteinase, pro-atrolysin e. *Arch. Biochem. Biophys.* **1996**, *335*, 283–294. [CrossRef] [PubMed]

39. Bode, W.; Gomis-Rüth, F.X.; Stöckler, W. Astacins, serralysins, snake venom and matrix metalloproteinases exhibit identical zinc-binding environments (HEXXHXXGXXH and Met-turn) and topologies and should be grouped into a common family, 'the metzincins'. *FEBS Lett.* **1993**, *331*, 134–140. [CrossRef]

40. Okamoto, T.; Akuta, T.; Tamura, F.; van der Vliet, A.; Akaike, T. Molecular mechanism for activation and regulation of matrix metalloproteinases during bacterial infections and respiratory inflammation. *Biol. Chem.* **2004**, *385*, 997–1006. [CrossRef] [PubMed]

41. Van Wart, H.E.; Birkedal-Hansen, H. The cysteine switch: A principle of regulation of metalloproteinase activity with potential applicability to the entire matrix metalloproteinase gene family. *Proc. Natl. Acad. Sci. USA* **1990**, *87*, 5578–5582. [CrossRef] [PubMed]

42. Tallant, C.; Marrero, A.; Gomis-Rüth, F.X. Matrix metalloproteinases: Fold and function of their catalytic domains. *Biochim. Biophys. Acta* **2010**, *1803*, 20–28. [CrossRef] [PubMed]

43. Lum, L.; Reid, M.S.; Blobel, C.P. Intracellular maturation of the mouse metalloprotease disintegrin MDC15. *J. Biol. Chem.* **1998**, *273*, 26236–26247. [CrossRef] [PubMed]

44. Howard, L.; Maciewicz, R.A.; Blobel, C.P. Cloning and characterization of ADAM28: Evidence for autocatalytic pro-domain removal and for cell surface localization of mature ADAM28. *Biochem. J.* **2000**, *348*, 21–27. [CrossRef] [PubMed]

45. Schlomann, U.; Wildeboer, D.; Webster, A.; Antropova, O.; Zeuschner, D.; Knight, C.G.; Docherty, A.J.; Lambert, M.; Skelton, L.; Jockusch, H.; *et al.* The metalloprotease disintegrin ADAM8. Processing by autocatalysis is required for proteolytic activity and cell adhesion. *J. Biol. Chem.* **2002**, *277*, 48210–48219. [CrossRef] [PubMed]

46. Portes-Junior, J.A.; Yamanouye, N.; Carneiro, S.M.; Knittel, P.S.; Sant'Anna, S.S.; Nogueira, F.C.; Junqueira, M.; Magalhães, G.S.; Domont, G.B.; Moura-da-Silva, A.M. Unraveling the processing and activation of snake venom metalloproteinases. *J. Proteom. Res.* **2014**, *13*, 3338–3348. [CrossRef] [PubMed]

47. Valente, R.H.; Guimarães, P.R.; Junqueira, M.; Neves-Ferreira, A.G.; Soares, M.R.; Chapeaurouge, A.; Trugilho, M.R.; León, I.R.; Rocha, S.L.; Oliveira-Carvalho, A.L.; *et al. Bothrops insularis* venomics: A proteomic analysis supported by transcriptomic-generated sequence data. *J. Proteom.* **2009**, *72*, 241–255. [CrossRef] [PubMed]

48. Nicolau, C.A.; Carvalho, P.C.; Junqueira-de-Azevedo, I.L.M.; Teixeira-Ferreira, A.; Junqueira, M.; Perales, J.; Neves-Ferreira, A.G.C.; Valente, R.H. An in-depth snake venom proteopeptidome characterization: Benchmarking *Bothrops jararaca*. *J. Proteom.* submitted for publication. 2016.

49. Luna, M.S.; Valente, R.H.; Perales, J.; Vieira, M.L.; Yamanouye, N. Activation of *Bothrops jararaca* snake venom gland and venom production: A proteomic approach. *J. Proteom.* **2013**, *94*, 460–472. [CrossRef] [PubMed]

50. Murphy, G.; Nagase, H. Progress in matrix metalloproteinase research. *Mol. Asp. Med.* **2008**, *29*, 290–308. [CrossRef] [PubMed]

51. Wewer, U.M.; Mörgelin, M.; Holck, P.; Jacobsen, J.; Lydolph, M.C.; Johnsen, A.H.; Kveiborg, M.; Albrechtsen, R. ADAM12 is a four-leafed clover: The excised prodomain remains bound to the mature enzyme. *J. Biol. Chem.* **2006**, *281*, 9418–9422. [CrossRef] [PubMed]

52. Marcinkiewicz, C. Applications of snake venom components to modulate integrin activities in cell-matrix interactions. *Int. J. Biochem. Cell Biol.* **2013**, *45*, 1974–1986. [CrossRef] [PubMed]

53. Calvete, J.J. The continuing saga of snake venom disintegrins. *Toxicon* **2013**, *62*, 40–49. [CrossRef] [PubMed]

54. Huang, T.F.; Yeh, C.H.; Wu, W.B. Viper venom components affecting angiogenesis. *Haemostasis* **2001**, *31*, 192–206. [CrossRef] [PubMed]

55. Juárez, P.; Comas, I.; González-Candelas, F.; Calvete, J.J. Evolution of snake venom disintegrins by positive Darwinian selection. *Mol. Biol. Evol.* **2008**, *25*, 2391–2407. [CrossRef] [PubMed]

56. Sanz, L.; Harrison, R.A.; Calvete, J.J. First draft of the genomic organization of a PIII-SVMP gene. *Toxicon* **2012**, *60*, 455–469. [CrossRef] [PubMed]

57. Mazzi, M.V.; Magro, A.J.; Amui, S.F.; Oliveira, C.Z.; Ticli, F.K.; Stábeli, R.G.; Fuly, A.L.; Rosa, J.C.; Braz, A.S.; Fontes, M.R.; *et al.* Molecular characterization and phylogenetic analysis of JussuMP-I: A RGD-P-III class hemorrhagic metalloprotease from *Bothrops jararacussu* snake venom. *J. Mol. Graph. Model.* **2007**, *26*, 69–85. [CrossRef] [PubMed]

58. Cidade, D.A.P.; Wermelinger, L.S.; Lobo-Hajdu, G.; Davila, A.M.R.; Bon, C.; Zingali, R.B.; Albano, R.M. Molecular diversity of disintegrin-like domains within metalloproteinase precursors of *Bothrops jararaca*. *Toxicon* **2006**, *48*, 590–599. [CrossRef] [PubMed]

59. Singhamatr, P.; Rojnuckarin, P. Molecular cloning of albolatin, a novel snake venom metalloprotease from green pit viper (*Trimeresurus albolabris*), and expression of its disintegrin domain. *Toxicon* **2007**, *50*, 1192–1200. [CrossRef] [PubMed]

60. Suntravat, M.; Jia, Y.; Lucena, S.E.; Sánchez, E.E.; Pérez, J.C. cDNA cloning of a snake venom metalloproteinase from the eastern diamondback rattlesnake (*Crotalus adamanteus*), and the expression of its disintegrin domain with anti-platelet effects. *Toxicon* **2013**, *64*, 43–54. [CrossRef] [PubMed]

61. Zhu, L.; Yuan, C.; Chen, Z.; Wang, W.; Huang, M. Expression, purification and characterization of recombinant jerdonitin, a P-II class snake venom metalloproteinase comprising metalloproteinase and disintegrin domains. *Toxicon* **2010**, *55*, 375–380. [CrossRef] [PubMed]

62. Herrera, C.; Escalante, T.; Voisin, M.B.; Rucavado, A.; Morazán, D.; Macêdo, J.K.; Calvete, J.J.; Sanz, L.; Nourshargh, S.; Gutiérrez, J.M.; *et al.* Tissue localization and extracellular matrix degradation by PI, PII and PIII snake venom metalloproteinases: Clues on the mechanisms of venom-induced hemorrhage. *PLoS Negl. Trop. Dis.* **2015**, *9*. [CrossRef] [PubMed]

63. Usami, Y.; Fujimura, Y.; Miura, S.; Shima, H.; Yoshida, E.; Yoshioka, A.; Hirano, K.; Suzuki, M.; Titani, K. A 28 kda-protein with disintegrin-like structure (jararhagin-C) purified from *Bothrops jararaca* venom inhibits collagen- and adp-induced platelet aggregation. *Biochem. Biophys. Res. Commun.* **1994**, *201*, 331–339. [CrossRef] [PubMed]

64. Takeya, H.; Nishida, S.; Nishino, N.; Makinose, Y.; Omori-Satoh, T.; Nikai, T.; Sugihara, H.; Iwanaga, S. Primary structures of platelet aggregation inhibitors (disintegrins) autoproteolytically released from snake venom hemorrhagic metalloproteinases and new fluorogenic peptide substrates for these enzymes. *J. Biochem.* **1993**, *113*, 473–483. [PubMed]

65. Fujimura, S.; Oshikawa, K.; Terada, S.; Kimoto, E. Primary structure and autoproteolysis of brevilysin h6 from the venom of *Gloydius halys* brevicaudus. *J. Biochem.* **2000**, *128*, 167–173. [CrossRef] [PubMed]

66. Souza, D.H.F.; Iemma, M.R.C.; Ferreira, L.L.; Faria, J.P.; Oliva, M.L.V.; Zingali, R.B.; Niewiarowski, S.; Selistre-de-Araujo, H.S. The disintegrin-like domain of the snake venom metalloprotease alternagin inhibits α2β1 integrin-mediated cell adhesion. *Arch. Biochem. Biophys.* **2000**, *384*, 341–350. [CrossRef] [PubMed]

67. Freitas-de-Sousa, L.A.; Amazonas, D.R.; Sousa, L.F.; Sant'Anna, S.S.; Nishiyama, M.Y., Jr.; Serrano, S.M.T.; Junqueira-de-Azevedo, I.L.M.; Chalkidis, H.M.; Moura-da-Silva, A.M.; Mourao, R.H.V. Comparison of venoms from wild and long-term captive bothrops atrox snakes and characterization of batroxrhagin, the predominant class piii metalloproteinase from the venom of this species. *Biochimie* **2015**, *118*, 60–70. [CrossRef] [PubMed]

68. Petretski, J.H.; Kanashiro, M.M.; Rodrigues, F.R.; Alves, E.W.; Machado, O.L.T.; Kipnis, T.L. Edema induction by the disintegrin-like/cysteine-rich domains from a Bothrops atrox hemorrhagin. *Biochem. Biophys. Res. Commun.* **2000**, *276*, 29–34. [CrossRef] [PubMed]

69. Shimokawa, K.; Shannon, J.D.; Jia, L.G.; Fox, J.W. Sequence and biological activity of catrocollastatin-c: A disintegrin-like/cysteine-rich two-domain protein from *Crotalus atrox* venom. *Arch. Biochem. Biophys.* **1997**, *343*, 35–43. [CrossRef] [PubMed]

70. Clissa, P.B.; Lopes-Ferreira, M.; Della-Casa, M.S.; Farsky, S.; Moura-da-Silva, A.M. Importance of jararhagin disintegrin-like and cysteine-rich domains in the early events of local inflammatory response. *Toxicon* **2006**, *47*, 591–596. [CrossRef] [PubMed]

71. Cominetti, M.; Terruggi, C.H.B.; Ramos, O.H.P.; Fox, J.W.; Mariano-Oliveira, A.; de Freitas, M.S.; Figueiredo, C.C.; Morandi, V.; Selistre-de-Araujo, H.S. Alternagin-C, a disintegrin-like protein, induces vascular endothelial cell growth factor (VEGF) expression and endothelial cell proliferation *in vitro*. *J. Biol. Chem.* **2004**, *279*, 18247–18255. [CrossRef] [PubMed]

72. Paine, M.J.; Desmond, H.P.; Theakston, R.D.; Crampton, J.M. Purification, cloning, and molecular characterization of a high molecular weight hemorrhagic metalloprotease, jararhagin, from *Bothrops jararaca* venom. Insights into the disintegrin gene family. *J. Biol. Chem.* **1992**, *267*, 22869–22876. [PubMed]

73. Moura-da-Silva, A.M.; Della-Casa, M.S.; David, A.S.; Assakura, M.T.; Butera, D.; Lebrun, I.; Shannon, J.D.; Serrano, S.M.T.; Fox, J.W. Evidence for heterogeneous forms of the snake venom metalloproteinase jararhagin: A factor contributing to snake venom variability. *Arch. Biochem. Biophys.* **2003**, *409*, 395–401. [CrossRef]

74. Igarashi, T.; Araki, S.; Mori, H.; Takeda, S. Crystal structures of catrocollastatin/VAP2B reveal a dynamic, modular architecture of adam/adamalysin/reprolysin family proteins. *FEBS Lett.* **2007**, *581*, 2416–2422. [CrossRef] [PubMed]

75. Takeda, S.; Igarashi, T.; Mori, H.; Araki, S. Crystal structures of VAP1 reveal ADAMS' MDC domain architecture and its unique c-shaped scaffold. *EMBO J.* **2006**, *25*, 2388–2396. [CrossRef] [PubMed]

76. Kikushima, E.; Nakamura, S.; Oshima, Y.; Shibuya, T.; Miao, J.Y.; Hayashi, H.; Nikai, T.; Araki, S. Hemorrhagic activity of the vascular apoptosis-inducing proteins VAP1 and VAP2 from *Crotalus atrox*. *Toxicon* **2008**, *52*, 589–593. [CrossRef] [PubMed]

77. Nikai, T.; Taniguchi, K.; Komori, Y.; Masuda, K.; Fox, J.W.; Sugihara, H. Primary structure and functional characterization of bilitoxin-1, a novel dimeric p-ii snake venom metalloproteinase from *Agkistrodon bilineatus* venom. *Arch. Biochem. Biophys.* **2000**, *378*, 6–15. [CrossRef] [PubMed]

78. Camacho, E.; Villalobos, E.; Sanz, L.; Pérez, A.; Escalante, T.; Lomonte, B.; Calvete, J.J.; Gutiérrez, J.M.; Rucavado, A. Understanding structural and functional aspects of PII snake venom metalloproteinases: Characterization of BlatH1, a hemorrhagic dimeric enzyme from the venom of *Bothriechis lateralis*. *Biochimie* **2014**, *101*, 145–155. [CrossRef] [PubMed]

79. Minea, R.; Swenson, S.; Costa, F.; Chen, T.C.; Markland, F.S. Development of a novel recombinant disintegrin, contortrostatin, as an effective anti-tumor and anti-angiogenic agent. *Pathophysiol. Haemost. Thromb.* **2005**, *34*, 177–183. [CrossRef] [PubMed]

80. Schwartz, M.A.; McRoberts, K.; Coyner, M.; Andarawewa, K.L.; Frierson, H.F.; Sanders, J.M.; Swenson, S.; Markland, F.; Conaway, M.R.; Theodorescu, D. Integrin agonists as adjuvants in chemotherapy for melanoma. *Clin. Cancer Res.* **2008**, *14*, 6193–6197. [CrossRef] [PubMed]

81. Hubbard, S.; Choudhary, S.; Maus, E.; Shukla, D.; Swenson, S.; Markland, F.S.; Tiwari, V. Contortrostatin, a homodimeric disintegrin isolated from snake venom inhibits herpes simplex virus entry and cell fusion. *Antivir. Ther.* **2012**, *17*, 1319–1326. [CrossRef] [PubMed]

82. Marcinkiewicz, C.; Calvete, J.J.; Vijay-Kumar, S.; Marcinkiewicz, M.M.; Raida, M.; Schick, P.; Lobb, R.R.; Niewiarowski, S. Structural and functional characterization of EMF10, a heterodimeric disintegrin from *Eristocophis macmahoni* venom that selectively inhibits α5β1 integrin. *Biochemistry* **1999**, *38*, 13302–13309. [CrossRef] [PubMed]

83. Calvete, J.J.; Fox, J.W.; Agelan, A.; Niewiarowski, S.; Marcinkiewicz, C. The presence of the WGD motif in CC8 heterodimeric disintegrin increases its inhibitory effect on αIIbβ3, αvβ3, and α5β1 integrins. *Biochemistry* **2002**, *41*, 2014–2021. [CrossRef] [PubMed]

84. Marcinkiewicz, C.; Calvete, J.J.; Marcinkiewicz, M.M.; Raida, M.; Vijay-Kumar, S.; Huang, Z.; Lobb, R.R.; Niewiarowski, S. EC3, a novel heterodimeric disintegrin from *Echis carinatus* venom, inhibits α4 and α5 integrins in an RGD-independent manner. *J. Biol. Chem.* **1999**, *274*, 12468–12473. [CrossRef] [PubMed]

85. Marcinkiewicz, C.; Taooka, Y.; Yokosaki, Y.; Calvete, J.J.; Marcinkiewicz, M.M.; Lobb, R.R.; Niewiarowski, S.; Sheppard, D. Inhibitory effects of MLDG-containing heterodimeric disintegrins reveal distinct structural requirements for interaction of the integrin α9β1 with vcam-1, tenascin-c, and osteopontin. *J. Biol. Chem.* **2000**, *275*, 31930–31937. [CrossRef] [PubMed]

86. Huang, H.W.; Liu, B.S.; Chien, K.Y.; Chiang, L.C.; Huang, S.Y.; Sung, W.C.; Wu, W.G. Cobra venom proteome and glycome determined from individual snakes of *Naja atra* reveal medically important dynamic range and systematic geographic variation. *J. Proteom.* **2015**, *128*, 92–104. [CrossRef] [PubMed]

87. Oliveira, A.K.; Leme, A.F.P.; Asega, A.F.; Camargo, A.C.M.; Fox, J.W.; Serrano, S.M.T. New insights into the structural elements involved in the skin haemorrhage induced by snake venom metalloproteinases. *Thromb. Haemost.* **2010**, *104*, 485–497. [CrossRef] [PubMed]

88. Zelanis, A.; Serrano, S.M.; Reinhold, V.N. N-glycome profiling of *Bothrops jararaca* newborn and adult venoms. *J. Proteom.* **2012**, *75*, 774–782. [CrossRef] [PubMed]

89. Fischer, W.H.; Spiess, J. Identification of a mammalian glutaminyl cyclase converting glutaminyl into pyroglutamyl peptides. *Proc. Natl. Acad. Sci. USA* **1987**, *84*, 3628–3632. [CrossRef] [PubMed]

90. Calvete, J.J.; Fasoli, E.; Sanz, L.; Boschetti, E.; Righetti, P.G. Exploring the venom proteome of the western diamondback rattlesnake, *Crotalus atrox*, via snake venomics and combinatorial peptide ligand library approaches. *J. Proteom. Res.* **2009**, *8*, 3055–3067. [CrossRef] [PubMed]

91. Baramova, E.N.; Shannon, J.D.; Bjarnason, J.B.; Fox, J.W. Degradation of extracellular matrix proteins by hemorrhagic metalloproteinases. *Arch. Biochem. Biophys.* **1989**, *275*, 63–71. [CrossRef]

92. Baramova, E.N.; Shannon, J.D.; Fox, J.W.; Bjarnason, J.B. Proteolytic digestion of non-collagenous basement membrane proteins by the hemorrhagic metalloproteinase Ht-e from *Crotalus atrox* venom. *Biomed. Biochim. Acta* **1991**, *50*, 763–768. [PubMed]

93. Costa, E.P.; Santos, M.F. Jararhagin, a snake venom metalloproteinase-disintegrin, stimulates epithelial cell migration in an *in vitro* restitution model. *Toxicon* **2004**, *44*, 861–870. [CrossRef] [PubMed]

94. Mackessy, S.P. Venom composition in rattlesnakes: Trends and biological significance. In *The Biology of Rattlesnakes*; Hayes, W.K., Beaman, K.R., Cardwell, M.D., Bush, S.P., Eds.; Loma Linda University Press: Loma Linda, CA, USA, 2008; pp. 495–510.

95. Marques-Porto, R.; Lebrun, I.; Pimenta, D.C. Self-proteolysis regulation in the *Bothrops jararaca* venom: The metallopeptidases and their intrinsic peptidic inhibitor. *Comp. Biochem. Physiol. C Toxicol. Pharmacol.* **2008**, *147*, 424–433. [CrossRef] [PubMed]

96. Munekiyo, S.M.; Mackessy, S.P. Presence of peptide inhibitors in rattlesnake venoms and their effects on endogenous metalloproteases. *Toxicon* **2005**, *45*, 255–263. [CrossRef] [PubMed]

97. Kuniyoshi, A.K.; Rocha, M.; Cajado Carvalho, D.; Juliano, M.A.; Juliano Neto, L.; Tambourgi, D.V.; Portaro, F.C. Angiotensin-degrading serine peptidase: A new chymotrypsin-like activity in the venom of *Bothrops jararaca* partially blocked by the commercial antivenom. *Toxicon* **2012**, *59*, 124–131. [CrossRef] [PubMed]

98. Baldo, C.; Tanjoni, I.; León, I.R.; Batista, I.F.C.; Della-Casa, M.S.; Clissa, P.B.; Weinlich, R.; Lopes-Ferreira, M.; Lebrun, I.; Amarante-Mendes, G.; *et al.* BnP1, a novel P-I metalloproteinase from *Bothrops neuwiedi* venom: Biological effects benchmarking relatively to jararhagin, a P-III SVMP. *Toxicon* **2008**, *51*, 54–65. [CrossRef] [PubMed]

99. Moss, M.L.; Powell, G.; Miller, M.A.; Edwards, L.; Qi, B.; Sang, Q.X.; de Strooper, B.; Tesseur, I.; Lichtenthaler, S.F.; Taverna, M.; *et al.* ADAM9 inhibition increases membrane activity of ADAM10 and controls α-secretase processing of amyloid precursor protein. *J. Biol. Chem.* **2011**, *286*, 40443–40451. [CrossRef] [PubMed]

100. Gonzales, P.E.; Solomon, A.; Miller, A.B.; Leesnitzer, M.A.; Sagi, I.; Milla, M.E. Inhibition of the tumor necrosis factor-alpha-converting enzyme by its pro domain. *J. Biol. Chem.* **2004**, *279*, 31638–31645. [CrossRef] [PubMed]

101. Dutertre, S.; Jin, A.H.; Kaas, Q.; Jones, A.; Alewood, P.F.; Lewis, R.J. Deep venomics reveals the mechanism for expanded peptide diversity in cone snail venom. *Mol. Cell. Proteom.* **2013**, *12*, 312–329. [CrossRef] [PubMed]

102. Zhang, Y.Y.; Huang, Y.; He, Q.Z.; Luo, J.; Zhu, L.; Lu, S.S.; Liu, J.Y.; Huang, P.F.; Zeng, X.Z.; Liang, S.P. Structural and functional diversity of peptide toxins from tarantula *Haplopelma hainanum* (*Ornithoctonus hainana*) venom revealed by transcriptomic, peptidomic, and patch clamp approaches. *J. Biol. Chem.* **2015**, *290*, 26471–26472. [CrossRef] [PubMed]

toxins

MDPI

Article

A Comprehensive View of the Structural and Functional Alterations of Extracellular Matrix by Snake Venom Metalloproteinases (SVMPs): Novel Perspectives on the Pathophysiology of Envenoming

José María Gutiérrez [1,*], Teresa Escalante [1], Alexandra Rucavado [1], Cristina Herrera [1,2] and Jay W. Fox [3,*]

1   Instituto Clodomiro Picado, Facultad de Microbiología, Universidad de Costa Rica, San José 11501-2060, Costa Rica; teresa.escalante@ucr.ac.cr (T.E.); alexandra.rucavado@ucr.ac.cr (A.R.); cristina.herreraarias@gmail.com (C.H.)
2   Facultad de Farmacia, Universidad de Costa Rica, San José 11501-2060, Costa Rica
3   School of Medicine, University of Virginia, Charlottesville, VA 22959, USA
*   Correspondence: jose.gutierrez@ucr.ac.cr (J.M.G.); jwf8x@virginia.edu (J.W.F.); Tel.: +506-2511-7865 (J.M.G.); +1-434-4924-0050 (J.W.F.)

Academic Editors: R. Manjunatha Kini and Stephen P. Mackessy
Received: 16 September 2016; Accepted: 14 October 2016; Published: 22 October 2016

Abstract: Snake venom metalloproteinases (SVMPs) affect the extracellular matrix (ECM) in multiple and complex ways. Previously, the combination of various methodological platforms, including electron microscopy, histochemistry, immunohistochemistry, and Western blot, has allowed a partial understanding of such complex pathology. In recent years, the proteomics analysis of exudates collected in the vicinity of tissues affected by SVMPs has provided novel and exciting information on SVMP-induced ECM alterations. The presence of fragments of an array of ECM proteins, including those of the basement membrane, has revealed a complex pathological scenario caused by the direct action of SVMPs. In addition, the time-course analysis of these changes has underscored that degradation of some fibrillar collagens is likely to depend on the action of endogenous proteinases, such as matrix metalloproteinases (MMPs), synthesized as a consequence of the inflammatory process. The action of SVMPs on the ECM also results in the release of ECM-derived biologically-active peptides that exert diverse actions in the tissue, some of which might be associated with reparative events or with further tissue damage. The study of the effects of SVMP on the ECM is an open field of research which may bring a renewed understanding of snake venom-induced pathology.

Keywords: proteomics; exudate; extracellular matrix; basement membrane; hemorrhage; snake venom metalloproteinases; FACITs

1. Extracellular Matrix Pathology: An Elusive Aspect in the Understanding of Snakebite Envenoming

Snakebite envenoming is a public health problem of high impact on a global basis, especially in tropical and subtropical regions of Africa, Asia, Latin America, and parts of Oceania, causing morbidity, mortality, and a wave of social suffering [1–4]. The spectrum of pathological and pathophysiological effects inflicted by snake venoms is very wide, and encompasses both local tissue damage and systemic, life-threatening alterations [5]. Venoms of snakes of the family Viperidae are rich in hydrolytic enzymes, having a high content of zinc-dependent metalloproteinases (SVMPs), phospholipases A_2 (PLA_2s), and serine proteinases (SVSPs), although the relative proportions of these enzymes varies among venoms [6,7]. SVMPs are known to play multiple roles in the local and systemic effects induced by viperid venoms in natural prey and humans [8–10]. SVMPs induce local and systemic hemorrhage, myonecrosis, blistering, dermonecrosis, edema, and coagulopathies, in addition to being algogenic and strongly pro-inflammatory [9–12]. Furthermore, the pathological alterations induced by SVMPs in skeletal muscle tissue contribute to the poor muscle regeneration characteristic of these envenomings [13,14].

Several aspects of the local pathological effects induced by SVMPs have been studied, such as the microvascular damage leading to hemorrhage [15,16], skeletal muscle necrosis and poor muscle regeneration [13,14,17], blistering, and dermonecrosis [18,19], as well as the identification of inflammatory mediators responsible for pain, edema, and leukocyte infiltration [20–22]. To a great extent, these alterations are considered to be associated in some manner with the action of SVMPs on the extracellular matrix (ECM). However, the specific effects induced by SVMPs as a result of their action on ECM have been investigated only to a limited extent, being mostly focused on the structural damage to basement membrane (BM) components of capillary blood vessels [15,16,23,24]. The reasons behind the paucity of information in this aspect of envenoming have to do mostly with methodological limitations and to the complexity of ECM structure and function. In his celebrated book The Logic of Life, François Jacob stated "The alternative approach to the history of Biology involves the attempt to discover how objects become accessible to investigation thus permitting new fields of science to be developed" [25]. It is then relevant to discuss how the alterations in ECM by snake venoms and SVMPs have become accessible to investigation.

The ability of SVMPs to degrade diverse ECM proteins has been assessed in vitro, mostly through SDS-PAGE, by observing the degradation patterns of ECM components incubated for various time intervals with the enzymes (Figure 1A). This has led to a wealth of information showing that SVMPs have a relatively wide spectrum of activity over substrates such as laminin, nidogen/entactin, type IV collagen, and fibronectin [23,26–33]. Likewise, SVMPs have been shown to hydrolyze proteoglycans in vitro, such as heparan sulphate proteoglycan and aggrecan [15,34]. However, this experimental approach has limitations, as hydrolysis has been studied in isolated ECM components and, therefore, the experimental conditions do not reproduce the complex landscape of these proteins in the tissues, which may determine their susceptibility to these enzymes. The ability of snake venoms to degrade hyaluronic acid, a glycosaminoglycan of the ECM, by the action of hyaluronidases, has been assessed using various in vitro methods [35].

Studies using classical ultrastructural methods, i.e., transmission electron microscopy (TEM), have focused on the alterations in BM structure by SVMPs (Figure 1B), as well as on the disorganization of fibrillar collagen bundles [36,37]. This approach, nevertheless, is not able to detect alterations in components of the ECM that are not observed at the ultrastructural level. Moreover, tissue processing for TEM reduces the actual thickness of BM, as demonstrated by atomic force microscopy [38]. Histochemistry and immunohistochemistry techniques, on the other hand, provide a more specific assessment of ECM components (Figure 1C,D), but the number of studies with SVMPs and hyaluronidases is limited (examples are [14,16,31,39,40]). Moreover, these procedures allow the detection of specific components, but do not provide a broad view of ECM alterations. More recently, the application of Western blot techniques to study the degradation of ECM components in vivo by SVMPs has provided novel clues to understand these phenomena, in particular regarding hydrolysis of BM components [15,24,41] (Figure 1E). Nevertheless, this method has the limitation that only the proteins to which antibodies are directed can be detected, thus precluding a comprehensive analysis of ECM alterations. Overall, the methodologies described have provided valuable, albeit limited, information of the action of SVMPs on the ECM.

To this end, the introduction of proteomic analysis to the field of pathology has represented a significant step forward in the study of disease at clinical and experimental levels, and in the search of biomarkers (Figure 1F) (see for example [42–47]). This methodological platform offers complementary and often advantageous outcomes as compared to other methods mentioned above. Of particular relevance is that it is not focused on the detection of particular tissue components, as occurs with the immunological-based methods, but instead provides unbiased information on many tissue components at a time. This approach thereby opens a greater aperture through which overt and subtle tissue alterations can be detected. In 2009, our group first utilized a proteomics-based approached to study the tissue damage induced by snake venoms and by specific venom components, such as SVMPs and myotoxic PLA$_2$s [40]. This initial watershed contribution demonstrated the great potential of this methodology to understand the pathological effects of snake venoms and toxins, and was followed by a series of studies on this topic [15,24,40,41,48,49]. The present review summarizes the key findings that have emerged from these investigations in relation to the alterations induced by SVMPs in the ECM and how these inform our understanding of the role of SVMPs and envenoming.

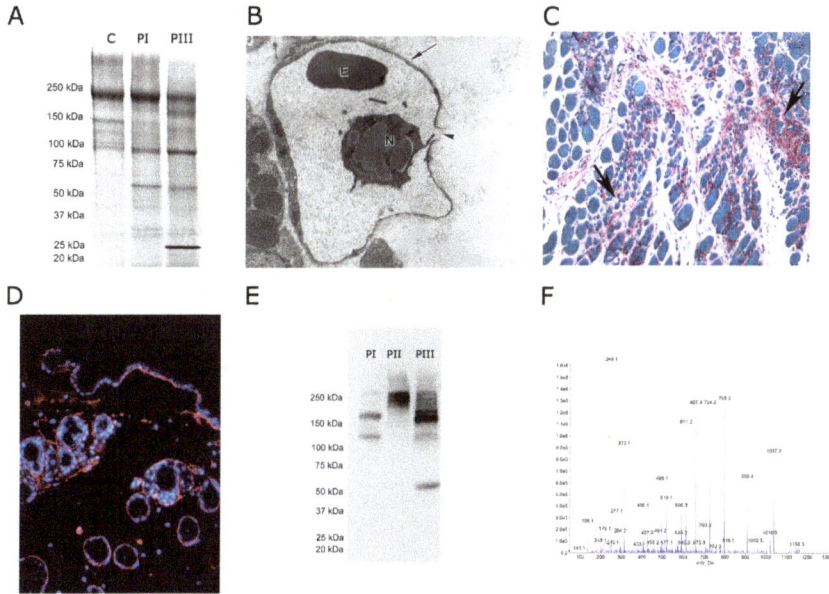

Figure 1. Experimental approaches to study the action of snake venom metalloproteinases (SVMPs) on the extracellular matrix (ECM). (**A**) In vitro analysis of hydrolysis of ECM proteins by SVMPs. Basement membrane (BM) preparations, such as Matrigel in this figure, or isolated ECM proteins, are incubated with SVMPs and the mixture is then analyzed by SDS-PAGE to assess the cleavage products; C: control Matrigel; degradation induced by PI and PIII SVMPs is shown. Molecular mass markers are shown to the left (reproduced from [31], copyright 2006 Elsevier); (**B**) Transmission electron microscopy assessment of ECM damage by analyzing the alterations in tissues from animals injected with SVMPs. A disrupted capillary vessel with damage to BM is shown after the injection of a hemorrhagic SVMP; 10,000 × (reproduced [50], copyright 2006 Elsevier); (**C**) Histochemical assessment of collagen degradation. A histology section of muscle tissue stained with Sirius Red, which stains collagen, and Fast Green, which stains proteins, is shown; 200 × (reproduced from [14], copyright 2011 PLOS); (**D**) Immunohistochemistry staining of a sample of skin injected with a SVMP. The blue staining corresponds to Hoechst 33258, which stains nuclei, whereas the red staining corresponds to immunostaining with a monoclonal antibody against type IV collagen; 400 ×; (**E**) Western blotting analysis of type VI collagen in samples of exudates collected from tissue injected with PI, PII and PIII SVMPs. Different patterns of hydrolyzed fragments are observed. Molecular mass markers are depicted to the left (reproduced from [24], copyright 2015 PLOS); (**F**) Mass spectrometry analysis of proteins in exudates collected in the vicinity of tissue injected with SVMPs allows the identification of degradation products of many types of ECM proteins. A mass spectrum is shown for illustrative purposes.

2. Methodological Aspects of Proteomics Studies

Proteomic analysis of tissue samples in pathological settings can be performed by studying tissue homogenates. This approach has been followed in the analysis of alterations induced by SVMPs of the venom of *Bothrops jararaca* in the skin [51]. One problem for analyzing proteomics of ECM in tissue homogenates is that extraction of ECM proteins is difficult and, therefore, the "matrisome", i.e., the ECM proteome, is often underrepresented in tissue homogenate samples [47]. As with most experimental approaches to identify markers of particular biological or pathological processes, proteomic assessment of compartments nearest to the site of interest is likely to give best results. Thus, our group has developed a strategy based on the proteomic analysis of exudates collected in the vicinity of tissues injected with snake venoms or isolated toxins, such as SVMPs. In these studies we employed a mouse model extensively used for the investigation of histological and ultrastructural alterations after injection of venoms or purified toxins. Specifically we inject SVMPs intramuscularly in the gastrocnemius muscle of mice and then, at various time intervals, animals are sacrificed and an incision made in the skin overlying the affected muscle. A heparinized glass capillary vessel is then introduced under the skin, and the exudate fluid is collected by capillarity (Figure 2). In this experimental setting, the effect of SVMP inhibitors or of antivenom antibodies can be assessed either by preincubating SVMPs with inhibitors/antibodies or by injecting these molecules after envenoming [48,49]. In parallel, the affected muscle tissue can be collected and either fixed and processed for histological, ultrastructural or immunohistochemical observation, or homogenized for immunological analyses, i.e., Western blots or ELISA. One limitation of this approach is the generation of appropriate controls. Unfortunately, exudates cannot be collected from control animals, i.e., mice injected with saline solution, because edema and exudate do not develop in these conditions. Therefore, these studies have to be performed using other types of controls, such as other toxins, and then comparing the differences in the outcomes of proteomics analysis between different treatments.

Once exudate samples are collected, they are rapidly freeze-dried in order to ensure the stability of the sample. Aliquots of exudates are separated by SDS-PAGE and stained with Coomassie Brilliant Blue. Then, the gel lanes corresponding to each sample are cut into ten equal size slices, corresponding to regions of varying ranges of molecular masses. After reduction and alkylation, gel slices are submitted to trypsinization, and tryptic peptides are analyzed by LC/MS/MS mass spectrometry analysis. Lists of peaks are generated from the raw data against the Uniprot Mouse database. The results from the searches are exported to Scaffold (version 4.3.2, Proteome Software Inc., Portland, OR, USA). Scaffold is used to validate MS/MS based peptide and protein identifications, and also to visualize multiple datasets in a comprehensive manner. Relative quantification of proteins is accomplished by combining all data from the 10 gel slices for a particular sample in Scaffold and then displaying the Quantitative Value from the program. This format of presentation allows for a comparison of the relative abundance of a specific protein presenting different samples. A detailed account on the methodology used in these studies can be found in Escalante et al. [40] (Figure 2).

The separation of protein bands in the gels into ten slices allows the determination of whether proteins in the samples are degraded or not. The amount of a given protein in a particular gel slice is determined as the percentage of that protein in all slices. Knowing the molecular mass of the native protein, the percentage of the protein migrating in regions of molecular mass lower than its native mass corresponds to the percentage of degradation of that protein in the sample [40]. The presence of a protein, or a protein fragment, in an exudate is likely to be due to one of the following reasons: (a) The protein has been degraded by proteinases present in the venom; (b) the protein has been degraded by endogenous proteinases derived from the inflammatory reaction to envenoming; (c) the protein has been released, without degradation, from a storage site in the ECM; (d) the protein has been synthesized during the process of envenoming and the ensuing inflammatory reaction; (e) the protein is present in the blood plasma and reaches the exudate as a consequence of the increment in vascular permeability; and (f) the protein has been released from cells due to the cytotoxic action of venom components (Figure 3). The first two possibilities can be detected by demonstrating the presence of fragments of the proteins in regions in the gel corresponding to molecular masses lower than those of the native proteins. Noteworthy, in addition to ECM-derived protein fragments, exudates collected from the site of SVMP injection may also include plasma proteins, intracellular proteins, and proteins of membrane origin, among others.

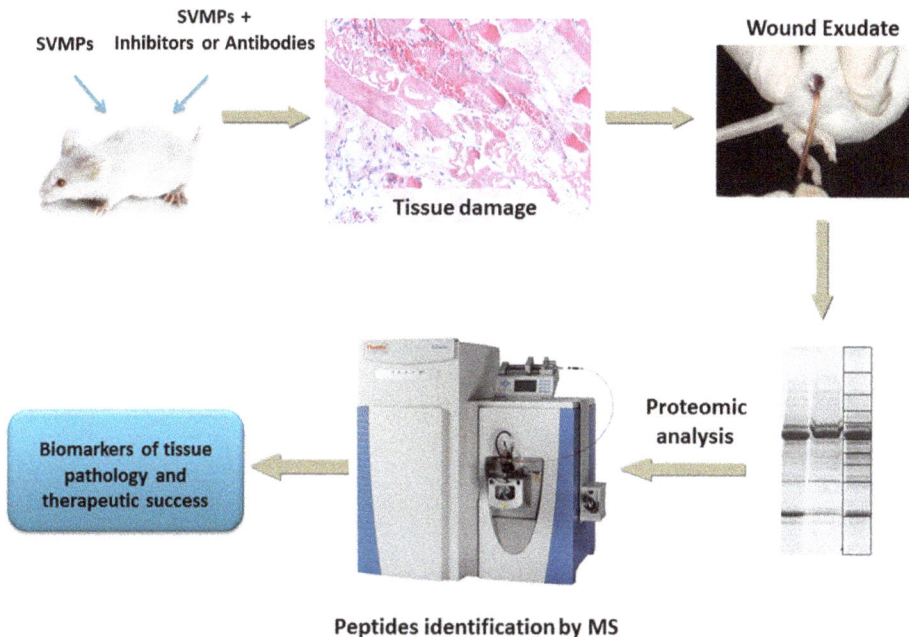

Figure 2. Basic experimental protocol for the proteomics analysis of exudates collected from tissues injected with SVMPs. Mice are injected intramuscularly in the gastrocnemius with SVMPs, or with mixtures of SVMPs and antibodies or inhibitors. At various time intervals animals are sacrificed and a sample of exudate is collected with a heparinized capillary vessel after sectioning the skin underlying the affected region. Upon separation of exudate proteins on SDS-PAGE and staining, sections of the gel are cut, reduced, carboxymethylated, and trypsin-digested, and then submitted to proteomic analysis (see text for more details). The identity of ECM proteins in the exudate and the extent of degradation are then assessed. Magnification of the histology section: 200 ×.

Table 1 shows the most abundant ECM proteins that have been detected in the proteomics analysis of exudates collected from tissues injected with SVMPs.

Proteomics analyses need to be validated by complementary experimental approaches. In our studies, Western blot analysis of proteins of particular interest has been utilized for validation. These analyses have been performed either in the same exudate samples on which proteomic analyses were performed or in homogenates of tissues injected with the SVMPs, such as skeletal muscle or skin [15,24]. Another complementary approach is the use of immunohistochemistry, which allows the identification of the areas of the tissue where ECM components are being altered [15,31,40]. Taken together, proteomic analysis and these complementary approaches constitute a robust experimental platform to assess the pathological alterations occurring in the ECM as a consequence of the action of SVMPs.

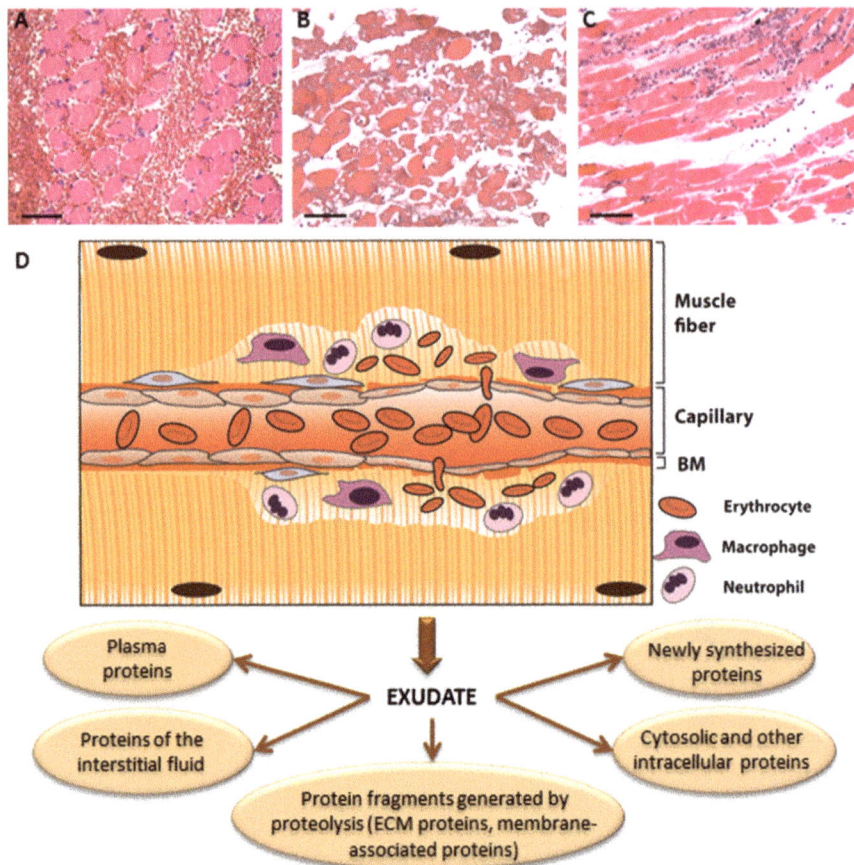

Figure 3. Scheme indicating the different sources of proteins that appear in exudates collected from tissues injected with snake venoms of with isolated SVMPs. After injection in skeletal muscle, viperid venoms or SVMPs induce direct pathological effects, such as degradation of BM components leading to hemorrhage (**A**); cytotoxicity on various cell types, such as skeletal muscle fibers (**B**); and degradation of other ECM components. As a consequence of direct tissue damage, resident tissue cells (mast cells, macrophages, fibroblasts) synthesize and release a number of mediators, favoring increments in vascular permeability leading to edema. An inflammatory infiltrate (**C**), composed mainly of neutrophils and macrophages, also contributes to the release of proteinases and other mediators. (**D**) Summary of SVMP-induced damage to muscle fibers and the microvasculature. As a consequence, the exudate that forms in the tissue is composed of proteins originating from different sources, as indicated in the bottom of the figure. Magnification in A, B and C: 200 ×.

Table 1. Extracellular matrix proteins detected in exudates collected from mice injected in the gastrocnemius muscle with snake venom metalloproteinases (SVMPs) [15,40,41].

Collagens
Collagen α-1 (I) chain (Isoform 1)
Collagen α-2 (I) chain
Collagen α-1 (II) chain (Isoform 2)

Table 1. *Cont.*

Collagens
Collagen α-1 (III) chain
Collagen α-1 (V) chain
Collagen α-3 (VI) chain
Collagen α-1 (VII) chain
Collagen α-2 (XI) chain (Isoform 7)
Collagen α-1 (XII) chain (Isoform 1)
Collagen α-1 (XIV) chain (Isoform 1)
Collagen α-1 (XV) chain
Collagen α-1 (XVI) chain (Isoform 1)
Collagen α-1 (XVIII) chain (Isoform 2)
Collagen α-1 (XIX) chain
Collagen α-1 (XXII) chain (Isoform 2)
Collagen α-1 (XXVII) chain
Collagen α-1 (XXVIII) chain (Isoform 1)

Laminins
Laminin subunit α-1
Laminin subunit α-3 (Isoform B)
Laminin subunit β-1
Laminin γ-2

Nidogens
Nidogen-1
Nidogen-2

Proteoglycans
Decorin
Lumican
Perlecan
Basement membrane—specific heparan sulfate proteoglycan core protein
Biglycan

Other extracellular matrix (ECM) proteins
Fibulin-1 (Isoform C)
Dystroglycan
Tenascin X
Thrombospondin-1
Thrombospondin-4
Tetranectin
Vitronectin
Fibronectin

3. Effects of SVMPs on the BM: Identifying Key Protein Targets of Hemorrhagic Toxins

Disruption of the integrity of microvessels leading to hemorrhage is one of the most important effects induced by viperid SVMPs [11,52,53]. The pioneering ultrastructural studies of McKay et al. [54] and Ownby et al. [36] described drastic alterations in endothelial cells and BM of capillary vessels in tissues injected with hemorrhagic SVMPs. Similar findings were then extended to other SVMPs from different venoms [37,55], and this mechanism of microvessel damage was named "hemorrhage *per rhexis*" [36]. The ability of SVMPs to hydrolyze components of the BM in vitro was demonstrated in several studies [26–32,56]. It was thus hypothesized that hydrolysis of BM components is a key event in the mechanism of hemorrhage by SVMPs.

Proteomic analysis of exudates collected at early time intervals (15 min and 1 h) after injection of crude venom of *Bothrops asper* and several hemorrhagic SVMPs purified from this and other viperid venoms revealed the presence of various BM components, such as laminin, nidogen, type IV collagen, and BM-specific heparan sulfate proteoglycan [15,24,40,41,48], which are the main components of BMs [38,57–61]. To a large extent, these proteins were degraded, as judged by the molecular mass of the fragments detected in the analyses. The fact that fragments of these BM components were present in exudates collected at early time periods after venom or SVMP injection strongly suggests that hydrolysis of these proteins is due to the direct action of SVMPs in the tissue, in agreement with in vitro observations. Such rapid degradation of BM proteins was corroborated by immunohistochemistry [15,16,31] and Western blotting [15,24,41]. Inhibition of *B. asper* venom with the peptidomimetic hydroxamante metalloproteinase inhibitor Batimastat, prior to injection in mice, resulted in the abrogation of the degradation of BM-specific heparan sulfate proteoglycan core protein (HSPG) [48]. This finding underscores the role of SVMPs in the proteolysis of this proteoglycan as well as a role for HSPG in the stabilization of microvessels.

For years, a puzzling finding regarding the mechanism of action of hemorrhagic SVMPs was that non-hemorrhagic SVMPs were also able to hydrolyze BM-associated proteins in vitro [15,62,63]. This in itself is not particularly surprising as most BM components are susceptible to proteolysis. A comparative analysis of BM degradation by a hemorrhagic and a non-hemorrhagic SVMP from *Bothrops* sp. venoms contributed to the clarification of this issue. No differences were observed between these enzymes regarding degradation of nidogen and laminin, as judged by proteomic analyses of exudates and by Western blotting of skeletal muscle homogenates [15]. However, a clear distinction occurred when comparing degradation of type IV collagen (by Western blot and immunohistochemistry) and HSPG (by proteomics and Western blot) [15]. In particular type IV collagen is known to play a key role in the mechanical stability of BM owing to the formation of interchain covalent bonds of various types and supramolecular networks between collagen chains [64,65]; these results strongly suggest that the ability of SVMPs to induce hemorrhage is related to their capacity to hydrolyze these BM components. This hypothesis was supported by a study comparing BM degradation by SVMPs of classes I, II, and III, which have a variable domain composition and different intrinsic hemorrhagic activity [24]. The doses of these enzymes injected were adjusted so as to induce the same extent of hemorrhage. In these conditions, there was a similar extent of degradation of type IV collagen and HSPG [24], thus reinforcing the concept that hydrolysis of these components seems to be critical for the onset of microvascular damage and hemorrhage.

The cleavage sites of type IV collagen by a hemorrhagic SVMP from the venom of the rattlesnake *Crotalus atrox* have been determined [23]. The relevance of type IV collagen hydrolysis in the pathogenesis of hemorrhage has been also shown in the case of a PIII SVMP from the venom of *Bothrops jararaca* [16,66,67] and of a hemorrhagic metalloproteinase from the prokaryote *Vibrio vulnificus* [68]. Moreover, genetic disorders affecting type IV collagen are associated with vascular alterations and hemorrhagic stroke [69–71]. HSPG is also known to contribute to the mechanical stabilization of BM in capillary vessels, and embryos having mutations in this proteoglycan show dilated microvessels in the brain and skin, associated with vessel disruption and severe bleedings [72–74]. This agrees with these proteins having a key role in the mechanical stabilization of BMs and therefore on the action of hemorrhagic SVMPs.

The ability of hemorrhagic SVMPs to hydrolyze components that contribute to the mechanical stability of capillaries has been integrated into a 'two-step' hypothesis to explain the mechanism of SVMP-induced hemorrhage [10,52,53]. The first step is the enzymatic hydrolysis of BM components, especially type IV collagen, and also HSPG, with the consequent weakening of the mechanical stability of the BM. Such hydrolysis may also affect cell-cell and cell matrix interactions. Then, the hemodynamic biophysical forces normally operating in the circulation, especially hydrostatic pressure-mediated wall tension and shear stress, cause a distention of the capillary wall, which ends up with the disruption in the integrity of endothelial cells and the vessel wall, with the consequent extravasation.

In addition to capillary BM, SVMPs also affect the BM of other tissue components. The presence of laminin subunit α3 in exudates collected after injection of a PI SVMP [40] suggests that this enzyme degrades laminin at the BM of the dermal-epidermal junction, since this laminin isoform is characteristic of the skin [75,76]. This SVMP induces skin blistering, suggesting that hydrolysis of laminin, and probably other components of the dermal-epidermal interface, is the basis for blister formation. Immunohistochemical observations revealed the presence of laminin in the two sides of the blister, thus supporting the contention of hydrolysis of the BM structure in the skin [40]. Moreover, BM hydrolysis by SVMPs is likely to affect tissue structure, since BM components, especially type IV collagen, are known to play a central role in the organization of tissue architecture [77]. Thus, alterations induced by SVMPs as a consequence of hydrolysis of BM components go beyond the acute effects associated with hemorrhage, blistering, and myonecrosis, since they also affect tissue organization and, probably, cell proliferation and regeneration occurring after tissue damage (Figure 4).

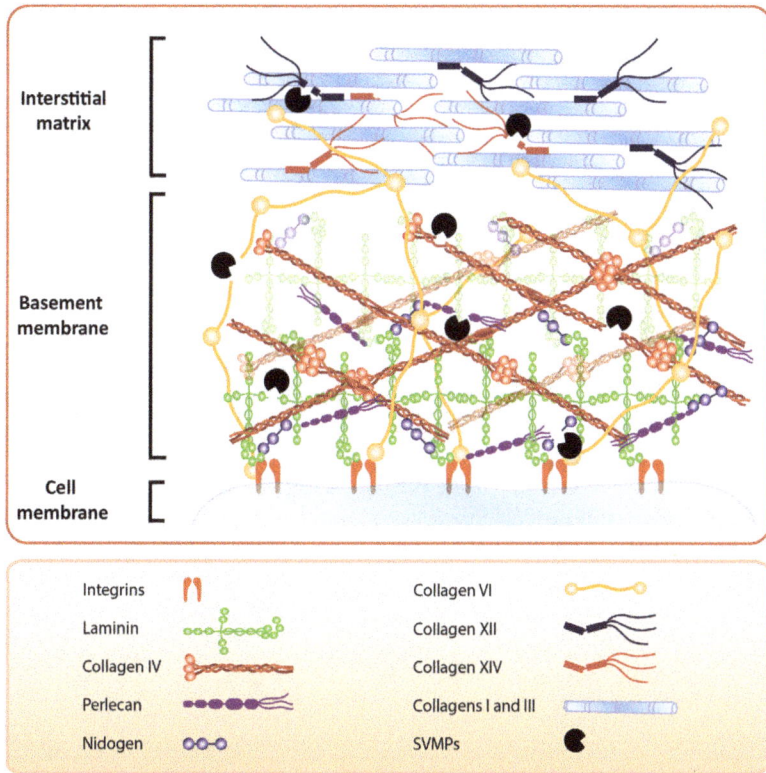

Figure 4. Hydrolysis of ECM components by SVMPs. Some SVMPs hydrolyze components at the BM of capillary vessels, skeletal muscle fibers, and dermal-epidermal junction. In the case of hemorrhagic SVMPs, it has been postulated that hydrolysis of type IV collagen is a key step in the destabilization of BM, which leads to extravasation. SVMPs also hydrolyze additional ECM proteins, such as FACITs, type VI collagen, and other components that connect the BM with the surrounding matrix stromal proteins. Moreover, SVMP degrade proteins that bind to and organize fibrillar collagens, leading to a disorganization of the ECM supramolecular structure. SVMPs may also hydrolyze plasma membrane components, such as integrins, that interact with BM components. All these hydrolytic actions result in a profound alteration of ECM, with consequences for the processes of venom-induced tissue damage, repair, and regeneration.

4. The Action of SVMPs on Proteins that Connect the BM with the Stromal Components of ECM

Proteomic analyses of exudates collected from tissues affected by snake venoms and SVMPs have allowed the detection of degradation products of proteins that play a role in the integration of the

BMs with the surrounding ECM (Figure 4). This was a hitherto unknown aspect of venom-induced ECM degradation, since the traditional experimental tools did not allow for the in vivo assessment of hydrolysis of these components. These ECM proteins are essential for the stability and mechanical integration of BMs with other ECM proteins, and for the assembly of fibrillar components of the matrix. For example, type VI collagen is a beaded-filament-forming collagen which integrates BM with fibrillar collagens and other components of the ECM. It interacts with types IV, XIV, I, and II collagens, and with perlecan, decorin, and lumican [78–80], and plays a key role in the mechanical stability of skeletal muscle cells. Deficiencies in type VI collagen have been associated with Ulrich syndrome, a muscle dystrophic condition [81,82] and with other myopathies [83,84]. Degradation products of type VI collagen have been found in exudates collected from tissues after injection of *B. asper* venom and SVMPs of the classes PI, PII, and PIII [15,24,40,41].

The potential implications of hydrolysis of this particular collagen in the action of SVMPs deserve additional consideration as this might affect the mechanical stability of skeletal muscle fibers. The resulting decreased stability of the fibers could contribute to the skeletal muscle pathology initially caused by myotoxic PLA$_2$s, which affect the integrity of muscle cell plasma membrane [13]. SVMP-induced hydrolysis of type VI collagen, and the consequent weakening of the mechanical stability of muscle BM, together with PLA$_2$-induced plasma membrane perturbation, may be an example of toxin-toxin synergism to give rise to lesions to the periphery of muscle fibers, leading to myonecrosis. Likewise, by affecting the stability of muscle cell BM, type VI collagen hydrolysis might hamper the process of skeletal muscle regeneration, which depends on the integrity of muscle BM [85,86]. The observation that fragments of type VI collagen are more abundant at early time intervals in exudates from mice injected with *B. asper* venom suggests that such degradation is due to the action of SVMPs [41]. Like laminin, type VI collagen is also important in the dermal-epidermal interface [87,88] and, therefore, its hydrolysis by SVMPs may be also involved in the pathogenesis of blistering in snakebite envenomings.

The possibility that hydrolysis of type VI collagen plays a role in the pathogenesis of hemorrhage also deserves discussion. Although the most likely mechanism by which SVMPs induce capillary damage and hemorrhage is their ability to hydrolyze type IV collagen and possibly HSPG at the BM, the degradation of ECM components that link the BM with fibrillar collagens needs to be considered as a possible mechanism of capillary damage as well. The observation that exudate collected from tissue injected with a PI hemorrhagic SVMP contains higher amounts of degradation products of type VI collagen than samples collected from mice injected with a non-hemorrhagic PI SVMP lends support to this hypothesis [15].

Proteomic analyses also identified degradation products of types XII, XIV, and XV collagens in exudates from tissues affected by *B. asper* venom and SVMPs [15,24,40,41]. Exosites in the Cys-rich domain of SVMPs mediate their interaction with type XII and XIV collagens [89], thus targeting PIII SVMPs to interact and hydrolyze these ECM proteins. These collagens are fibril-associated collagens with interrupted triple helices (FACITs) and play a role in the supramolecular organization of fibrillar collagens [90,91], as well as in the integration of BM with the ECM fibrillar components [92,93]. In skeletal muscle, these FACITs are important for connecting the muscle cell BM with the epimysium and the perimysium [94,95]. The fact that no differences were observed in the amounts of degradation products of types XII and XIV collagens in exudates from tissue injected with hemorrhagic and non-hemorrhagic PI SVMPs [15] argues against a role of hydrolysis of these proteins in the mechanism of hemorrhage. However, the possible involvement of such hydrolysis in the stability of muscle fibers, and on the integration of muscle cells with muscle connective tissue at epimysium and perimysium, has to be considered. Type XV collagen, on the other hand, is a proteoglycan often expressed in BM zones, in regions adjacent to BM where several proteins anchor BM to the subjacent ECM, where it has been proposed to act as a BM organizer [96–98]. Moreover, type XV collagen has a restricted and uniform presence in many tissues, including vascular and muscle BM zones [96,99]. Genetic mutations of this protein in mice have been associated with abnormal capillary morphology, extravasated erythrocytes, and cell degeneration in heart and skeletal muscle [61,100,101]. The relevance of this protein in the organization of the microvasculature has been also demonstrated [101]. Interestingly, as in the case of Type VI collagen, a hemorrhagic SVMP induces higher amounts of Type XV collagen in exudates than a non-hemorrhagic SVMP [15].

The ability of SVMPs to hydrolyze FACITs and proteoglycans having a role in the assembly of fibrillar collagens and in the integration of fibrillar collagens with BMs and other components of the connective tissue has implications for the ability of snake venoms to digest skeletal muscle tissue. The disruption of the connective tissue matrix resulting from the hydrolysis of these integrative components would facilitate the diffusion of snake venom components through the tissues and into the circulation [102]. This action could work in concert with the action of other venom hydrolases, such as hyaluronidase, a well-known spreading factor present in many venoms [35]. This would, in turn, favor the digestive role of SVMPs and venom serine proteinases, as a consequence of the disruption in the organization of muscle tissue. Since viperid venoms are often injected intramuscularly, and since muscle tissue comprises a significant mass of prey, the action of SVMPs on these ECM components is likely to represent a significant contribution to the digestion of muscle mass. Likewise, the 'softening' and disorganization of interstitial connective tissue described above may promote the digestion of the muscle mass of prey by proteinases of the gastric and pancreatic secretions of snakes after ingestion.

From the human pathology standpoint, such disruption of the components of connective tissue may play a role in venom dispersion in the tissues, thus facilitating the systemic action of venom toxins, and also may contribute to the extent of local tissue damage by making the connective tissue more amenable to digestion by endogenous proteinases, such as matrix metalloproteinases (MMPs), which are synthesized as part of the inflammatory response [103,104]. These effects on FACITs and related integrative ECM components may also affect the process of skeletal muscle repair and regeneration, an issue that deserves more investigation.

5. Action of SVMPs on Fibrillar Collagens: A Secondary Outcome of SVMP-Induced Local Tissue Damage

Proteomic analysis of exudates collected from mice injected with venom of *B. asper* and with hemorrhagic and non-hemorrhagic SVMPs has revealed the presence of degradation fragments of fibrillar collagens, i.e., types I, II, III collagens [15,24,40,41,48]. Since SVMPs are not able to hydrolyze fibrillar collagens lacking triple helical interruptions [105], the basis for this degradation is intriguing. A number of observations strongly suggest that it is due to the action of endogenous proteinases, especially MMPs, which are synthesized and secreted by resident and infiltrating cells in the course of the inflammatory response that follows the acute tissue damage induced by the venom. When comparing exudate proteomics from mice injected with a PI hemorrhagic SVMP to that from tissue injected with a myotoxic PLA_2, higher amounts of types I and III collagens were found with the latter. Myotoxic PLA_2s induce muscle necrosis of rapid onset by damaging the integrity of muscle fiber plasma membrane [106] and induce inflammation characterized by pain, edema, synthesis of cytokines and MMPs, and a prominent cellular infiltrate [103,107,108]. Thus, it is suggested that the hydrolysis of fibrillar collagens is a consequence of the action of endogenous MMPs and perhaps other proteinases derived from resident and inflammatory cells. Degradation products of fibrillar collagens were also detected in exudates collected from SVMP-injected muscle [15,24,40,48]. Interestingly, no differences were observed in the amounts of these fragments after injection of hemorrhagic and non-hemorrhagic SVMPs [15], both of which induce an inflammatory response in the tissue.

Additional evidence in support of the concept that hydrolysis of fibrillar collagens is due to endogenous proteinases synthesized during inflammation is that degradation products of types I and III collagens in exudates from mice injected with *B. asper* venom reach their highest amounts in samples collected 24 h after envenoming, whereas type IV collagen products are most abundant in samples collected at 1 h [41]. This suggests that type IV and VI collagens, as well as other BM components and FACITs are hydrolyzed by SVMPs during the early phase of envenoming, whereas fibrillar collagens are hydrolyzed by endogenous proteinases at later time intervals. In the biological context, such as the case of natural envenomings in prey, this second stage in ECM degradation of interstitial fibrillar collagen is likely to be accomplished by digestive proteinases from gastric and pancreatic secretions of the snakes.

6. Hydrolysis of ECM Proteins Alters Cell-Matrix Interactions and Generates Fragments with Diverse Physiological and Pathological Actions

In addition to the direct pathological consequences of degradation of ECM by SVMPs, another important consequence of this hydrolysis is the alteration of the interaction between ECM and cells. For instance, fibronectin interacts with cells through the integrin α5β1, and is involved with other extracellular signals to regulate morphogenesis and cellular differentiation [109]. SVMPs hydrolyze fibronectin in vitro [26,27,30] and fibronectin degradation products are detected in the proteomics analysis of exudates of tissues injected with SVMPs [24,40]. Although plasma fibronectin is probably present in exudates as a consequence of increments in vascular permeability, it is very likely that ECM fibronectin is also hydrolyzed and contributes to fragments in exudates.

SVMP-induced ECM degradation may also release proteins or protein fragments that exert a variety of physiological effects. For instance, hydrolysis of types XV and XVIII collagens results in the generation of endostatin, an inhibitor of angiogenesis [110–112], and the cleavage of the α3 chain of type IV collagen by MMPs releases the fragment tumstatin, which is a potent anti-angiogenic molecule [113,114]. Another BM-derived fragment is endorepelin, the C-terminal fragment of perlecan, which also exerts anti-angiogenic activity [115]. Since SVMPs release fragments of all of these proteins in the exudates [15,24,40,41,48], it is suggested that some of them exert anti-angiogenic activity and influence the tissue repair process. SVMPs may also release matrikines from ECM proteins which regulate a number of cellular activities [116]. Our own studies indicate that snake venom and SVMPs release a number of DAMPs in the affected tissues, which are likely to play diverse roles in the processes of tissue damage and repair (unpublished results). Another interesting protein that has been found elevated in exudates collected from mice injected with *B. asper* venom and SVMPs is thrombospondin-1 [15,41,48], a counter-adhesive protein that influences endothelial cell behavior by modulating cell-matrix and cell-cell interactions and by regulating growth factors [117]. This protein has been shown to play roles in hemostasis, inflammation, tissue regeneration, and angiogenesis [118–121]. Therefore, the release of this protein into the exudate by SVMPs could modulate the inflammatory process and the consequent tissue repair response.

BM and other ECM components act as storage sites for a variety of growth factors and other physiologically-active components, such as insulin growth factor (IGF), vascular endothelial growth factor (VEGF), fibroblast growth factor (FGF), transforming growth factor-β (TGF-β), hepatocyte growth factor (HGF), and platelet-derived growth factor (PDGF) [122]. Proteolytic processing of ECM components in inflammation releases growth factors thus influencing cell activation, differentiation, and proliferation [123]. Regarding angiogenesis, the action of SVMPs might result in the release of both pro-angiogenic, e.g., VEGF, and anti-angiogenic, e.g., endostatin and endorepelin, components. Hence, hydrolysis of ECM by SVMPs is likely to result in the release of diverse mediators, which in turn may expand tissue alterations, dysregulate cell-matrix interaction, promote and inhibit cell proliferation, and play reparative and regenerative roles in the complex tissue interactive landscape. This is an aspect of SVMP-induced ECM alterations that needs to be explored in detail.

Hydrolysis of ECM components might also result in changes in the mechanical properties of the matrix and on the interaction of ECM and cells. It is known that the stiffness of ECM varies depending on many factors, and that changes in such stiffness bring consequences for cellular behavior in many ways, including cell differentiation [124]. Likewise, the release of growth factors stored in the matrix may occur not only by direct proteolysis, but also by mechanical forces generated in the ECM as a consequence of hydrolysis by proteinases [125–127]. The biomechanical consequences of ECM degradation by SVMPs constitute an area of research that needs to be developed, since phenomena associated with changes in cellular behavior secondary to mechanical alterations in the matrix may have consequences in the processes of tissue inflammation, repair, and regeneration.

7. Concluding Remarks

The proteomic analysis of exudates collected from mice injected with snake venoms and isolated toxins has opened an avenue to study hitherto unknown aspects of the action of venom enzymes on the ECM, an issue that has been largely elusive in toxinological research. This methodological platform, when combined with histological, ultrastructural, immunohistochemical, and immunological methods, has provided new and valuable information on the pathogenesis of tissue damage induced by viperid snake venoms.

The studies reviewed in this communication uncovered a pathophysiological scenario characterized by various levels of ECM degradation by SVMPs. BM components are rapidly hydrolyzed upon venom and SVMP injection in the tissues. Of special significance for the pathogenesis of microvascular damage leading to hemorrhage is the hydrolysis of structurally-relevant BM components, especially type IV collagen and, possibly, HSPG. In addition to microvessels, hydrolysis of specific BM components detected in exudate could also affect the stability of skeletal muscle fibers as well as the muscle regenerative process. It can also be postulated that BM damage affects the spatial organization of cells in the tissue owing to the role of this ECM structure in the compartmentalization of cells, in addition to favoring the spread of venom components in the tissue and to the circulation. Figure 5 summarizes the main effects of SVMPs on the ECM.

Concomitantly, SVMPs also hydrolyze ECM proteins that play a role in the integration of BM with the surrounding matrix, and also in the assembly of fibrillar collagens in the matrix, such as types VI, XII, XV, and XIV collagens. The hydrolysis of these FACITs and related proteins may contribute to the collapse of BM, but also may result in the disorganization of the interstitial fibrillar matrix, favoring tissue disorganization, venom spreading, and digestion. On the other hand, the observed hydrolysis of fibrillar collagens I and III is likely to be a consequence of the action of endogenous proteinases, especially MMPs, and not to the direct action of SVMPs. Hence, as part of the inflammatory process that ensues as a consequence of venom induced acute tissue damage, MMPs and other endogenous proteinases, derived from resident tissue cells or invading leukocytes, hydrolyze fibrillar collagens, resulting in widespread ECM degradation, which further complicates venom-induced tissue damage. Moreover, hydrolysis of ECM provides protein fragments of diverse physiological actions that are likely to participate in tissue alterations, as well as in inflammation, repair, and regeneration. Likewise, biomechanical changes in the tissue occurring as a result of changes in the stiffness of ECM after SVMP action may affect the behavior of cells.

Understanding the mechanisms involved in ECM degradation in snakebite envenoming may pave the way for the search of novel therapeutic agents, aimed at the inhibition of SVMPs and at the modulation of the inflammatory response. A rapid administration of SVMP inhibitors in the field, in combination with the use of anti-inflammatory agents and the antivenom, may contribute to ameliorate the extent of local tissue damage and hence the magnitude of the sequelae in people envenomed by viperid snakebites. In addition, a deeper understanding of venom-induced ECM damage may provide information for designing interventions aimed at reducing snakebite envenoming morbidity by improving the processes of tissue repair and regeneration.

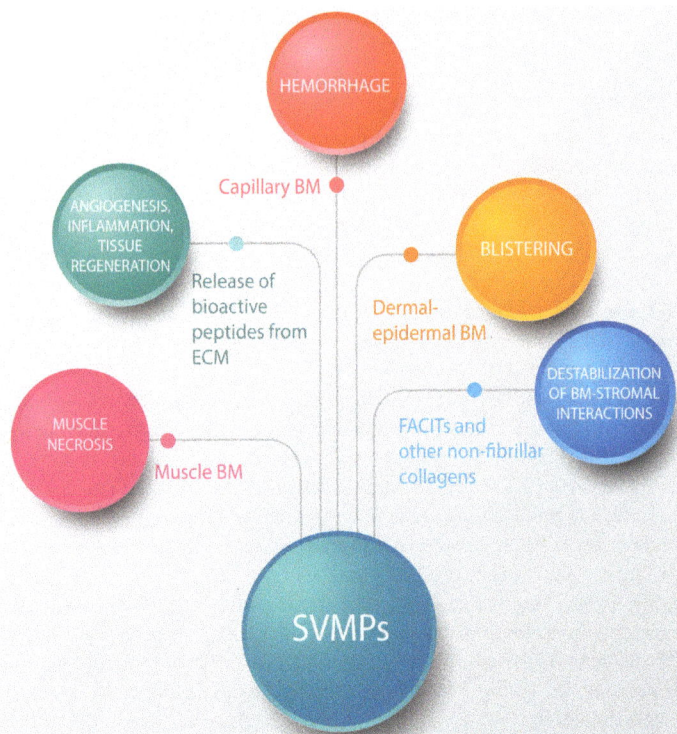

Figure 5. Summary of the effects induced by SVMPs on the ECM components. SVMPs hydrolyze components of the BM (type IV collagen, laminin, nidogen, heparan sulfate proteoglycan core protein (HSPG)) in capillary blood vessels, skeletal muscle fibers and dermal-epidermal junctions. As a consequence, hemorrhage and blistering occurs, and it is hypothesized that acute skeletal muscle damage also ensues. On the other hand, hydrolysis of FACITs, type VI collagen and other components results in alterations in the interactions between BM and the surrounding stromal components. In addition, hydrolysis of ECM proteins results in exposition of cryptic sites, release of growth factors stored in the matrix, and generation of a variety of protein fragments with potent biological activities, which are involved in pathological, reparative, and regenerative events.

Acknowledgments: The collaboration of colleagues and students at Instituto Clodomiro Picado (University of Costa Rica) and at the University of Virginia in some of the studies reviewed in this paper is greatly acknowledged. This study was supported by Vicerrectoría de Investigación, Universidad de Costa Rica (projects 741-A7-502, 741-A7-604, 741-B4-660 and 741-B6-125) and the University of Virgina School of Medicine.

Conflicts of Interest: The authors declare no conflict of interest. The founding sponsors had no role in the design of the study, in the writing of the manuscript, and in the decision to publish the review.

References

1. Chippaux, J.P. Snake-bites: Appraisal of the global situation. *Bull. World Health Organ.* **1998**, *76*, 515–524. [PubMed]
2. Gutiérrez, J.M.; Theakston, R.D.G.; Warrell, D.A. Confronting the neglected problem of snake bite envenoming: The need for a global partnership. *PLoS Med.* **2006**, *3*. [CrossRef] [PubMed]
3. Williams, D.; Gutiérrez, J.M.; Harrison, R.; Warrell, D.A.; White, J.; Winkel, K.D.; Gopalakrishnakone, P. Global Snake Bite Initiative Working Group. The Global Snake Bite Initiative: An antidote for snake bite. *Lancet* **2010**, *375*, 89–91. [CrossRef]

4. Kasturiratne, A.; Wickremasinghe, A.R.; de Silva, N.; Gunawardena, N.K.; Pathmeswaran, A.; Premaratna, R.; Savioli, L.; Lalloo, D.G.; de Silva, H.J. The global burden of snakebite: A literature analysis and modelling based on regional estimates of envenoming and deaths. *PLoS Med.* **2008**, *5*. [CrossRef] [PubMed]

5. Warrell, D.A. Snake bite. *Lancet* **2010**, *375*, 77–88. [CrossRef]

6. Calvete, J.J. Proteomic tools against the neglected pathology of snake bite envenoming. *Expert Rev. Proteomics* **2011**, *8*, 739–758. [CrossRef] [PubMed]

7. Lomonte, B.; Fernández, J.; Sanz, L.; Angulo, Y.; Sasa, M.; Gutiérrez, J.M.; Calvete, J.J. Venomous snakes of Costa Rica: Biological and medical implications of their venom proteomic profiles analyzed through the strategy of snake venomics. *J. Proteomics* **2014**, *105*, 323–339. [CrossRef] [PubMed]

8. Gutiérrez, J.M.; Rucavado, A. Snake venom metalloproteinases: Their role in the pathogenesis of local tissue damage. *Biochimie* **2000**, *82*, 841–850. [CrossRef]

9. Moura-da-Silva, A.; Serrano, S.M.T.; Fox, J.W.; Gutiérrez, J.M. Snake venom metalloproteinases. Structure, function and effects on snake bite pathology. In *Animal Toxins: State of the Art: Perspectives in Health and Biotechnology*; de Lima, M.E., de Castro, A.M., Martin-Eauclaire, M.F., Rochat, H., Eds.; Editora UFMG: Belo Horizonte, Brasil, 2009; pp. 525–546.

10. Gutiérrez, J.M.; Rucavado, A.; Escalante, T. Snake venom metalloproteinases. Biological roles and participation in the pathophysiology of envenomation. In *Handbook of Venoms and Toxins of Reptiles*; Mackessy, S.P., Ed.; CRC Press: Boca Raton, FL, USA, 2010; pp. 115–138.

11. Fox, J.W.; Serrano, S.M.T. Structural considerations of the snake venom metalloproteinases, key members of the M12 reprolysin family of metalloproteinases. *Toxicon* **2005**, *45*, 969–985. [CrossRef] [PubMed]

12. Teixeira, C.F.P.; Fernandes, C.M.; Zuliani, J.P.; Zamuner, S.F. Inflammatory effects of snake venom metalloproteinases. *Memórias Inst. Oswaldo Cruz* **2005**, *100*, 181–184. [CrossRef]

13. Gutiérrez, J.M.; Rucavado, A.; Chaves, F.; Díaz, C.; Escalante, T. Experimental pathology of local tissue damage induced by *Bothrops asper* snake venom. *Toxicon* **2009**, *54*, 958–975. [CrossRef] [PubMed]

14. Hernández, R.; Cabalceta, C.; Saravia-Otten, P.; Chaves, A.; Gutiérrez, J.M.; Rucavado, A. Poor regenerative outcome after skeletal muscle necrosis induced by *Bothrops asper* venom: Alterations in microvasculature and nerves. *PLoS ONE* **2011**, *6*. [CrossRef]

15. Escalante, T.; Ortiz, N.; Rucavado, A.; Sanchez, E.F.; Richardson, M.; Fox, J.W.; Gutiérrez, J.M. Role of collagens and perlecan in microvascular stability: Exploring the mechanism of capillary vessel damage by snake venom metalloproteinases. *PLoS ONE* **2011**, *6*. [CrossRef] [PubMed]

16. Baldo, C.; Jamora, C.; Yamanouye, N.; Zorn, T.M.; Moura-da-Silva, A.M. Mechanisms of vascular damage by hemorrhagic snake venom metalloproteinases: Tissue distribution and in situ hydrolysis. *PLoS Negl. Trop. Dis.* **2010**, *4*. [CrossRef] [PubMed]

17. Gutiérrez, J.M.; Romero, M.; Núñez, J.; Chaves, F.; Borkow, G.; Ovadia, M. Skeletal muscle necrosis and regeneration after injection of BaH1, a hemorrhagic metalloproteinase isolated from the venom of the snake *Bothrops asper* (Terciopelo). *Exp. Mol. Pathol.* **1995**, *62*, 28–41. [CrossRef] [PubMed]

18. Rucavado, A.; Núñez, J.; Gutiérrez, J.M. Blister formation and skin damage induced by BaP1, a haemorrhagic metalloproteinase from the venom of the snake *Bothrops asper*. *Int. J. Exp. Pathol.* **1998**, *79*, 245–254. [PubMed]

19. Jiménez, N.; Escalante, T.; Gutiérrez, J.M.; Rucavado, A. Skin pathology induced by snake venom metalloproteinase: Acute damage, revascularization, and re-epithelization in a mouse ear model. *J. Investig. Dermatol.* **2008**, *128*, 2421–2428. [CrossRef] [PubMed]

20. Fernandes, C.M.; Teixeira, C.F.P.; Leite, A.C.R.M.; Gutiérrez, J.M.; Rocha, F.A.C. The snake venom metalloproteinase BaP1 induces joint hypernociception through TNF-alpha and PGE2-dependent mechanisms. *Br. J. Pharmacol.* **2007**, *151*, 1254–1261. [CrossRef] [PubMed]

21. Fernandes, C.M.; Zamuner, S.R.; Zuliani, J.P.; Rucavado, A.; Gutiérrez, J.M.; Teixeira, C.F.P. Inflammatory effects of BaP1 a metalloproteinase isolated from *Bothrops asper* snake venom: Leukocyte recruitment and release of cytokines. *Toxicon* **2006**, *47*, 549–559. [CrossRef] [PubMed]

22. Moura-da-Silva, A.M.; Baldo, C. Jararhagin, a hemorrhagic snake venom metalloproteinase from *Bothrops jararaca*. *Toxicon* **2012**, *60*, 280–289. [CrossRef] [PubMed]

23. Baramova, E.N.; Shannon, J.D.; Bjarnason, J.B.; Fox, J.W. Identification of the cleavage sites by a hemorrhagic metalloproteinase in type IV collagen. *Matrix* **1990**, *10*, 91–97. [CrossRef]

24. Herrera, C.; Escalante, T.; Voisin, M.-B.; Rucavado, A.; Morazán, D.; Macêdo, J.K.A.; Calvete, J.J.; Sanz, L.; Nourshargh, S.; Gutiérrez, J.M.; et al. Tissue localization and extracellular matrix degradation by PI, PII and PIII snake venom metalloproteinases: Clues on the mechanisms of venom-induced hemorrhage. *PLoS Negl. Trop. Dis.* **2015**, *9*. [CrossRef] [PubMed]

25. Jacob, F. *The Logic of Life: A History of Heredity*; Vintage Books: New York, NY, USA, 1976.

26. Bjarnason, J.B.; Hamilton, D.; Fox, J.W. Studies on the mechanism of hemorrhage production by five proteolytic hemorrhagic toxins from *Crotalus atrox* venom. *Biol. Chem. Hoppe-Seyler* **1988**, *369*, 121–129. [PubMed]

27. Baramova, E.N.; Shannon, J.D.; Bjarnason, J.B.; Fox, J.W. Degradation of extracellular matrix proteins by hemorrhagic metalloproteinases. *Arch. Biochem. Biophys.* **1989**, *275*, 63–71. [CrossRef]

28. Baramova, E.N.; Shannon, J.D.; Fox, J.W.; Bjarnason, J.B. Proteolytic digestion of non-collagenous basement membrane proteins by the hemorrhagic metalloproteinase Ht-e from *Crotalus atrox* venom. *Biomed. Biochim. Acta* **1991**, *50*, 763–768. [PubMed]

29. Maruyama, M.; Sugiki, M.; Yoshida, E.; Shimaya, K.; Mihara, H. Broad substrate specificity of snake venom fibrinolytic enzymes: Possible role in haemorrhage. *Toxicon* **1992**, *30*, 1387–1397. [CrossRef]

30. Rucavado, A.; Lomonte, B.; Ovadia, M.; Gutiérrez, J.M. Local tissue damage induced by BaP1, a metalloproteinase isolated from *Bothrops asper* (Terciopelo) snake venom. *Exp. Mol. Pathol.* **1995**, *63*, 186–199. [CrossRef] [PubMed]

31. Escalante, T.; Shannon, J.; Moura-da-Silva, A.M.; Gutiérrez, J.M.; Fox, J.W. Novel insights into capillary vessel basement membrane damage by snake venom hemorrhagic metalloproteinases: A biochemical and immunohistochemical study. *Arch. Biochem. Biophys.* **2006**, *455*, 144–153. [CrossRef] [PubMed]

32. Oliveira, A.K.; Paes Leme, A.F.; Asega, A.F.; Camargo, A.C.M.; Fox, J.W.; Serrano, S.M.T. New insights into the structural elements involved in the skin haemorrhage induced by snake venom metalloproteinases. *Thromb. Haemost.* **2010**, *104*, 485–497. [CrossRef] [PubMed]

33. Macêdo, J.K.A.; Fox, J.W. Biological activities and assays of the snake venom metalloproteinases (SVMPs). In *Venom Genomics and Proteomics*; Springer: Dordrecht, The Netherlands, 2014; pp. 1–24.

34. Tortorella, M.D.; Pratta, M.A.; Fox, J.W.; Arner, E.C. The interglobular domain of cartilage aggrecan is cleaved by hemorrhagic metalloproteinase HT-d (atrolysin C) at the matrix metalloproteinase and aggrecanase sites. *J. Biol. Chem.* **1998**, *273*, 5846–5850. [CrossRef] [PubMed]

35. Kemparaju, K.; Girish, K.S.; Nagaraju, S. Hyaluronidases, a neglected class of glycosidases from snake venom. Beyond a spreading factor. In *Handbook of Venoms and Toxins of Reptiles*; Mackessy, S.P., Ed.; CRC Press: Boca Raton, FL, USA, 2010; pp. 237–258.

36. Ownby, C.L.; Bjarnason, J.; Tu, A.T. Hemorrhagic toxins from rattlesnake (*Crotalus atrox*) venom. Pathogenesis of hemorrhage induced by three purified toxins. *Am. J. Pathol.* **1978**, *93*, 201–218. [PubMed]

37. Moreira, L.; Borkow, G.; Ovadia, M.; Gutiérrez, J.M. Pathological changes induced by BaH1, a hemorrhagic proteinase isolated from *Bothrops asper* (Terciopelo) snake venom, on mouse capillary blood vessels. *Toxicon* **1994**, *32*, 976–987. [CrossRef]

38. Halfter, W.; Oertle, P.; Monnier, C.A.; Camenzind, L.; Reyes-Lua, M.; Hu, H.; Candiello, J.; Labilloy, A.; Balasubramani, M.; Henrich, P.B.; et al. New concepts in basement membrane biology. *FEBS J.* **2015**, *282*, 4466–4479. [CrossRef] [PubMed]

39. Girish, K.S.; Jagadeesha, D.K.; Rajeev, K.B.; Kemparaju, K. Snake venom hyaluronidase: An evidence for isoforms and extracellular matrix degradation. *Mol. Cell. Biochem.* **2002**, *240*, 105–110. [CrossRef] [PubMed]

40. Escalante, T.; Rucavado, A.; Pinto, A.F.M.; Terra, R.M.S.; Gutiérrez, J.M.; Fox, J.W. Wound exudate as a proteomic window to reveal different mechanisms of tissue damage by snake venom toxins. *J. Proteome Res.* **2009**, *8*, 5120–5131. [CrossRef] [PubMed]

41. Herrera, C.; Macêdo, J.K.A.; Feoli, A.; Escalante, T.; Rucavado, A.; Gutiérrez, J.M.; Fox, J.W. Muscle tissue damage induced by the venom of *Bothrops asper*: Identification of early and late pathological events through proteomic analysis. *PLoS Negl. Trop. Dis.* **2016**, *10*. [CrossRef] [PubMed]

42. Fernandez, M.L.; Broadbent, J.A.; Shooter, G.K.; Malda, J.; Upton, Z. Development of an enhanced proteomic method to detect prognostic and diagnostic markers of healing in chronic wound fluid. *Br. J. Dermatol.* **2008**, *158*, 281–290. [CrossRef] [PubMed]

43. Broadbent, J.; Walsh, T.; Upton, Z. Proteomics in chronic wound research: Potentials in healing and health. *Proteomics Clin. Appl.* **2010**, *4*, 204–214. [CrossRef] [PubMed]

44. Eming, S.A.; Koch, M.; Krieger, A.; Brachvogel, B.; Kreft, S.; Bruckner-Tuderman, L.; Krieg, T.; Shannon, J.D.; Fox, J.W. Differential proteomic analysis distinguishes tissue repair biomarker signatures in wound exudates obtained from normal healing and chronic wounds. *J. Proteome Res.* **2010**, *9*, 4758–4766. [CrossRef] [PubMed]

45. Crutchfield, C.A.; Thomas, S.N.; Sokoll, L.J.; Chan, D.W. Advances in mass spectrometry-based clinical biomarker discovery. *Clin. Proteomics* **2016**, *13*. [CrossRef] [PubMed]

46. L'Imperio, V.; Smith, A.; Chinello, C.; Pagni, F.; Magni, F. Proteomics and glomerulonephritis: A complementary approach in renal pathology for the identification of chronic kidney disease related markers. *Proteomics Clin. Appl.* **2016**, *10*, 371–383. [CrossRef] [PubMed]

47. Randles, M.J.; Humphries, M.J.; Lennon, R. Proteomic definitions of basement membrane composition in health and disease. *Matrix Biol.* **2016**. [CrossRef] [PubMed]

48. Rucavado, A.; Escalante, T.; Shannon, J.; Gutiérrez, J.M.; Fox, J.W. Proteomics of wound exudate in snake venom-induced pathology: Search for biomarkers to assess tissue damage and therapeutic success. *J. Proteome Res.* **2011**, *10*, 1987–2005. [CrossRef] [PubMed]

49. Rucavado, A.; Escalante, T.; Shannon, J.D.; Ayala-Castro, C.N.; Villalta, M.; Gutiérrez, J.M.; Fox, J.W. Efficacy of IgG and F(ab')$_2$ antivenoms to neutralize snake venom-induced local tissue damage as assessed by the proteomic analysis of wound exudate. *J. Proteome Res.* **2012**, *11*, 292–305. [CrossRef] [PubMed]

50. Gutiérrez, J.M.; Núñez, J.; Escalante, T.; Rucavado, A. Blood flow is required for rapid endothelial cell damage induced by a snake venom hemorrhagic metalloproteinase. *Microvasc. Res.* **2006**, *71*, 55–63. [CrossRef] [PubMed]

51. Paes Leme, A.F.; Sherman, N.E.; Smalley, D.M.; Sizukusa, L.O.; Oliveira, A.K.; Menezes, M.C.; Fox, J.W.; Serrano, S.M.T. Hemorrhagic activity of HF3, a snake venom metalloproteinase: Insights from the proteomic analysis of mouse skin and blood plasma. *J. Proteome Res.* **2012**, *11*, 279–291. [CrossRef] [PubMed]

52. Gutiérrez, J.M.; Rucavado, A.; Escalante, T.; Díaz, C. Hemorrhage induced by snake venom metalloproteinases: Biochemical and biophysical mechanisms involved in microvessel damage. *Toxicon* **2005**, *45*, 997–1011. [CrossRef] [PubMed]

53. Escalante, T.; Rucavado, A.; Fox, J.W.; Gutiérrez, J.M. Key events in microvascular damage induced by snake venom hemorrhagic metalloproteinases. *J. Proteomics* **2011**, *74*, 1781–1794. [CrossRef] [PubMed]

54. McKay, D.G.; Moroz, C.; De Vries, A.; Csavossy, I.; Cruse, V. The action of hemorrhagin and phospholipase derived from *Vipera palestinae* venom on the microcirculation. *Lab. Investig.* **1970**, *22*, 387–399. [PubMed]

55. Ownby, C.L.; Geren, C.R. Pathogenesis of hemorrhage induced by hemorrhagic proteinase IV from timber rattlesnake (*Crotalus horridus horridus*) venom. *Toxicon* **1987**, *25*, 517–526. [CrossRef]

56. Osaka, A.; Just, M.; Habermann, E. Action of snake venom hemorrhagic principles on isolated glomerular basement membrane. *Biochim. Biophys. Acta* **1973**, *323*, 415–428. [CrossRef]

57. Timpl, R. Macromolecular organization of basement membranes. *Curr. Opin. Cell Biol.* **1996**, *8*, 618–624. [CrossRef]

58. Yurchenco, P.D.; Amenta, P.S.; Patton, B.L. Basement membrane assembly, stability and activities observed through a developmental lens. *Matrix Biol.* **2004**, *22*, 521–538. [CrossRef] [PubMed]

59. Yurchenco, P.D.; Patton, B.L. Developmental and pathogenic mechanisms of basement membrane assembly. *Curr. Pharm. Des.* **2009**, *15*, 1277–1294. [CrossRef] [PubMed]

60. Yurchenco, P.D. Basement membranes: Cell scaffoldings and signaling platforms. *Cold Spring Harb. Perspect. Biol.* **2011**, *3*. [CrossRef] [PubMed]

61. Iozzo, R. Basement membrane proteoglycans: From cellar to ceiling. *Nat. Rev. Mol. Cell Biol.* **2005**, *6*, 646–656. [CrossRef] [PubMed]

62. Rodrigues, V.M.; Soares, A.M.; Guerra-Sá, R.; Rodrigues, V.; Fontes, M.R.; Giglio, J.R. Structural and functional characterization of neuwiedase, a nonhemorrhagic fibrin(ogen)olytic metalloprotease from *Bothrops neuwiedi* snake venom. *Arch. Biochem. Biophys.* **2000**, *381*, 213–224. [CrossRef] [PubMed]

63. Rucavado, A.; Flores-Sánchez, E.; Franceschi, A.; Magalhaes, A.; Gutiérrez, J.M. Characterization of the local tissue damage induced by LHF-II, a metalloproteinase with weak hemorrhagic activity isolated from *Lachesis muta muta* snake venom. *Toxicon* **1999**, *37*, 1297–1312. [CrossRef]

64. Kühn, K. Basement membrane (type IV) collagen. *Matrix Biol.* **1995**, *14*, 439–445. [CrossRef]

65. Khoshnoodi, J.; Pedchenko, V.; Hudson, B.G. Mammalian collagen IV. *Microsc. Res. Tech.* **2008**, *71*, 357–370. [CrossRef] [PubMed]

66. Moura-da-Silva, A.M.; Ramos, O.H.P.; Baldo, C.; Niland, S.; Hansen, U.; Ventura, J.S.; Furlan, S.; Butera, D.; Della-Casa, M.S.; Tanjoni, I.; et al. Collagen binding is a key factor for the hemorrhagic activity of snake venom metalloproteinases. *Biochimie* **2008**, *90*, 484–492. [CrossRef] [PubMed]

67. Tanjoni, I.; Evangelista, K.; Della-Casa, M.S.; Butera, D.; Magalhães, G.S.; Baldo, C.; Clissa, P.B.; Fernandes, I.; Eble, J.; Moura-da-Silva, A.M. Different regions of the class P-III snake venom metalloproteinase jararhagin are involved in binding to alpha2beta1 integrin and collagen. *Toxicon* **2010**, *55*, 1093–1099. [CrossRef] [PubMed]

68. Miyoshi, S.; Nakazawa, H.; Kawata, K.; Tomochika, K.; Tobe, K.; Shinoda, S. Characterization of the hemorrhagic reaction caused by *Vibrio vulnificus* metalloprotease, a member of the thermolysin family. *Infect. Immun.* **1998**, *66*, 4851–4855. [PubMed]

69. Gould, D.B.; Phalan, F.C.; Breedveld, G.J.; van Mil, S.E.; Smith, R.S.; Schimenti, J.C.; Aguglia, U.; van der Knaap, M.S.; Heutink, P.; John, S.W.M. Mutations in COL4A1 cause perinatal cerebral hemorrhage and porencephaly. *Science* **2005**, *308*, 1167–1171. [CrossRef] [PubMed]

70. Gould, D.B.; Phalan, F.C.; van Mil, S.E.; Sundberg, J.P.; Vahedi, K.; Massin, P.; Bousser, M.G.; Heutink, P.; Miner, J.H.; Tournier-Lasserve, E.; et al. Role of COL4A1 in small-vessel disease and hemorrhagic stroke. *N. Engl. J. Med.* **2006**, *354*, 1489–1496. [CrossRef] [PubMed]

71. Federico, A.; Di Donato, I.; Bianchi, S.; Di Palma, C.; Taglia, I.; Dotti, M.T. Hereditary cerebral small vessel diseases: A review. *J. Neurol. Sci.* **2012**, *322*, 25–30. [CrossRef] [PubMed]

72. Costell, M.; Gustafsson, E.; Aszódi, A.; Mörgelin, M.; Bloch, W.; Hunziker, E.; Addicks, K.; Timpl, R.; Fässler, R. Perlecan maintains the integrity of cartilage and some basement membranes. *J. Cell Biol.* **1999**, *147*, 1109–1122. [CrossRef] [PubMed]

73. Stratman, A.N.; Davis, G.E. Endothelial cell-pericyte interactions stimulate basement membrane matrix assembly: Influence on vascular tube remodeling, maturation, and stabilization. *Microsc. Microanal.* **2012**, *18*, 68–80. [CrossRef] [PubMed]

74. Gustafsson, E.; Almonte-Becerril, M.; Bloch, W.; Costell, M. Perlecan maintains microvessel integrity in vivo and modulates their formation in vitro. *PLoS ONE* **2013**, *8*. [CrossRef] [PubMed]

75. Kiritsi, D.; Has, C.; Bruckner-Tuderman, L. Laminin 332 in junctional epidermolysis bullosa. *Cell Adhes. Migr.* **2013**, *7*, 135–141. [CrossRef] [PubMed]

76. Sugawara, K.; Tsuruta, D.; Ishii, M.; Jones, J.C.R.; Kobayashi, H. Laminin-332 and -511 in skin. *Exp. Dermatol.* **2008**, *17*, 473–480. [CrossRef] [PubMed]

77. Morrissey, M.A.; Sherwood, D.R. An active role for basement membrane assembly and modification in tissue sculpting. *J. Cell Sci.* **2015**, *128*, 1661–1668. [CrossRef] [PubMed]

78. Cescon, M.; Gattazzo, F.; Chen, P.; Bonaldo, P. Collagen VI at a glance. *J. Cell Sci.* **2015**, *128*, 3525–3531. [CrossRef] [PubMed]

79. Kuo, H.J.; Maslen, C.L.; Keene, D.R.; Glanville, R.W. Type VI collagen anchors endothelial basement membranes by interacting with type IV collagen. *J. Biol. Chem.* **1997**, *272*, 26522–26529. [CrossRef] [PubMed]

80. Tillet, E.; Wiedemann, H.; Golbik, R.; Pan, T.C.; Zhang, R.Z.; Mann, K.; Chu, M.L.; Timpl, R. Recombinant expression and structural and binding properties of alpha 1(VI) and alpha 2(VI) chains of human collagen type VI. *Eur. J. Biochem.* **1994**, *221*, 177–185. [CrossRef] [PubMed]

81. Camacho Vanegas, O.; Bertini, E.; Zhang, R.Z.; Petrini, S.; Minosse, C.; Sabatelli, P.; Giusti, B.; Chu, M.L.; Pepe, G. Ullrich scleroatonic muscular dystrophy is caused by recessive mutations in collagen type VI. *Proc. Natl. Acad. Sci. USA* **2001**, *98*, 7516–7521. [CrossRef] [PubMed]

82. Niiyama, T.; Higuchi, I.; Hashiguchi, T.; Suehara, M.; Uchida, Y.; Horikiri, T.; Shiraishi, T.; Saitou, A.; Hu, J.; Nakagawa, M.; et al. Capillary changes in skeletal muscle of patients with Ullrich's disease with collagen VI deficiency. *Acta Neuropathol.* **2003**, *106*, 137–142. [CrossRef] [PubMed]

83. Jöbsis, G.J.; Keizers, H.; Vreijling, J.P.; de Visser, M.; Speer, M.C.; Wolterman, R.A.; Baas, F.; Bolhuis, P.A. Type VI collagen mutations in Bethlem myopathy, an autosomal dominant myopathy with contractures. *Nat. Genet.* **1996**, *14*, 113–115. [CrossRef] [PubMed]

84. Merlini, L.; Martoni, E.; Grumati, P.; Sabatelli, P.; Squarzoni, S.; Urciuolo, A.; Ferlini, A.; Gualandi, F.; Bonaldo, P. Autosomal recessive myosclerosis myopathy is a collagen VI disorder. *Neurology* **2008**, *71*, 1245–1253. [CrossRef] [PubMed]

85. Paco, S.; Ferrer, I.; Jou, C.; Cusí, V.; Corbera, J.; Torner, F.; Gualandi, F.; Sabatelli, P.; Orozco, A.; Gómez-Foix, A.M.; et al. Muscle fiber atrophy and regeneration coexist in collagen VI-deficient human muscle: Role of calpain-3 and nuclear factor-κB signaling. *J. Neuropathol. Exp. Neurol.* **2012**, *71*, 894–906. [CrossRef] [PubMed]

86. Urciuolo, A.; Quarta, M.; Morbidoni, V.; Gattazzo, F.; Molon, S.; Grumati, P.; Montemurro, F.; Tedesco, F.S.; Blaauw, B.; Cossu, G.; et al. Collagen VI regulates satellite cell self-renewal and muscle regeneration. *Nat. Commun.* **2013**, *4*. [CrossRef] [PubMed]

87. Keene, D.R.; Engvall, E.; Glanville, R.W. Ultrastructure of type VI collagen in human skin and cartilage suggests an anchoring function for this filamentous network. *J. Cell Biol.* **1988**, *107*, 1995–2006. [CrossRef] [PubMed]

88. Gara, S.K.; Grumati, P.; Squarzoni, S.; Sabatelli, P.; Urciuolo, A.; Bonaldo, P.; Paulsson, M.; Wagener, R. Differential and restricted expression of novel collagen VI chains in mouse. *Matrix Biol.* **2011**, *30*, 248–257. [CrossRef] [PubMed]

89. Serrano, S.M.T.; Kim, J.; Wang, D.; Dragulev, B.; Shannon, J.D.; Mann, H.H.; Veit, G.; Wagener, R.; Koch, M.; Fox, J.W. The cysteine-rich domain of snake venom metalloproteinases is a ligand for von Willebrand factor A domains: Role in substrate targeting. *J. Biol. Chem.* **2006**, *281*, 39746–39756. [CrossRef] [PubMed]

90. Kadler, K.E.; Baldock, C.; Bella, J.; Boot-Handford, R.P. Collagens at a glance. *J. Cell Sci.* **2007**, *120*, 1955–1958. [CrossRef] [PubMed]

91. Mouw, J.K.; Ou, G.; Weaver, V.M. Extracellular matrix assembly: A multiscale deconstruction. *Nat. Rev. Mol. Cell Biol.* **2014**, *15*, 771–785. [CrossRef] [PubMed]

92. Thierry, L.; Geiser, A.S.; Hansen, A.; Tesche, F.; Herken, R.; Miosge, N. Collagen types XII and XIV are present in basement membrane zones during human embryonic development. *J. Mol. Histol.* **2004**, *35*, 803–810. [CrossRef] [PubMed]

93. Chiquet, M.; Birk, D.E.; Bönnemann, C.G.; Koch, M. Collagen XII: Protecting bone and muscle integrity by organizing collagen fibrils. *Int. J. Biochem. Cell Biol.* **2014**, *53*, 51–54. [CrossRef] [PubMed]

94. Scarr, G. Fascial hierarchies and the relevance of crossed-helical arrangements of collagen to changes in the shape of muscles. *J. Bodyw. Mov. Ther.* **2016**, *20*, 377–387. [CrossRef] [PubMed]

95. Listrat, A.; Lethias, C.; Hocquette, J.F.; Renand, G.; Ménissier, F.; Geay, Y.; Picard, B. Age-related changes and location of types I, III, XII and XIV collagen during development of skeletal muscles from genetically different animals. *Histochem. J.* **2000**, *32*, 349–356. [CrossRef] [PubMed]

96. Myers, J.C.; Dion, A.S.; Abraham, V.; Amenta, P.S. Type XV collagen exhibits a widespread distribution in human tissues but a distinct localization in basement membrane zones. *Cell Tissue Res.* **1996**, *286*, 493–505. [CrossRef] [PubMed]

97. Amenta, P.S.; Scivoletti, N.A.; Newman, M.D.; Sciancalepore, J.P.; Li, D.; Myers, J.C. Proteoglycan-collagen XV in human tissues is seen linking banded collagen fibers subjacent to the basement membrane. *J. Histochem. Cytochem.* **2005**, *53*, 165–176. [CrossRef] [PubMed]

98. Clementz, A.G.; Harris, A. Collagen XV: Exploring its structure and role within the tumor microenvironment. *Mol. Cancer Res.* **2013**, *11*, 1481–1486. [CrossRef] [PubMed]

99. Hägg, P.M.; Hägg, P.O.; Peltonen, S.; Autio-Harmainen, H.; Pihlajaniemi, T. Location of type XV collagen in human tissues and its accumulation in the interstitial matrix of the fibrotic kidney. *Am. J. Pathol.* **1997**, *150*, 2075–2086. [PubMed]

100. Eklund, L.; Piuhola, J.; Komulainen, J.; Sormunen, R.; Ongvarrasopone, C.; Fässler, R.; Muona, A.; Ilves, M.; Ruskoaho, H.; Takala, T.E.; et al. Lack of type XV collagen causes a skeletal myopathy and cardiovascular defects in mice. *Proc. Natl. Acad. Sci. USA* **2001**, *98*, 1194–1199. [CrossRef] [PubMed]

101. Rasi, K.; Piuhola, J.; Czabanka, M.; Sormunen, R.; Ilves, M.; Leskinen, H.; Rysä, J.; Kerkelä, R.; Janmey, P.; Heljasvaara, R.; et al. Collagen XV is necessary for modeling of the extracellular matrix and its deficiency predisposes to cardiomyopathy. *Circ. Res.* **2010**, *107*, 1241–1252. [CrossRef] [PubMed]

102. Anai, K.; Sugiki, M.; Yoshida, E.; Maruyama, M. Neutralization of a snake venom hemorrhagic metalloproteinase prevents coagulopathy after subcutaneous injection of *Bothrops jararaca* venom in rats. *Toxicon* **2002**, *40*, 63–68. [CrossRef]

103. Rucavado, A.; Escalante, T.; Teixeira, C.F.P.; Fernándes, C.M.; Diaz, C.; Gutiérrez, J.M. Increments in cytokines and matrix metalloproteinases in skeletal muscle after injection of tissue-damaging toxins from the venom of the snake *Bothrops asper*. *Mediat. Inflamm.* **2002**, *11*, 121–128. [CrossRef] [PubMed]

104. Saravia-Otten, P.; Robledo, B.; Escalante, T.; Bonilla, L.; Rucavado, A.; Lomonte, B.; Hernández, R.; Flock, J.I.; Gutiérrez, J.M.; Gastaldello, S. Homogenates of skeletal muscle injected with snake venom inhibit myogenic differentiation in cell culture. *Muscle Nerve* **2013**, *47*, 202–212. [CrossRef] [PubMed]

105. Shannon, J.D.; Baramova, E.N.; Bjarnason, J.B.; Fox, J.W. Amino acid sequence of a *Crotalus atrox* venom metalloproteinase which cleaves type IV collagen and gelatin. *J. Biol. Chem.* **1989**, *264*, 11575–11583. [PubMed]

106. Gutiérrez, J.M.; Ownby, C.L.; Odell, G.V. Pathogenesis of myonecrosis induced by crude venom and a myotoxin of *Bothrops asper*. *Exp. Mol. Pathol.* **1984**, *40*, 367–379. [CrossRef]

107. Gutiérrez, J.M.; Arce, V.; Brenes, F.; Chaves, F. Changes in myofibrillar components after skeletal muscle necrosis induced by a myotoxin isolated from the venom of the snake *Bothrops asper*. *Exp. Mol. Pathol.* **1990**, *52*, 25–36. [CrossRef]

108. Zuliani, J.P.; Fernandes, C.M.; Zamuner, S.R.; Gutiérrez, J.M.; Teixeira, C.F.P. Inflammatory events induced by Lys-49 and Asp-49 phospholipases A_2 isolated from *Bothrops asper* snake venom: Role of catalytic activity. *Toxicon* **2005**, *45*, 335–346. [CrossRef] [PubMed]

109. Vega, M.E.; Schwarzbauer, J.E. Collaboration of fibronectin matrix with other extracellular signals in morphogenesis and differentiation. *Curr. Opin. Cell Biol.* **2016**, *42*, 1–6. [CrossRef] [PubMed]

110. Sasaki, T.; Larsson, H.; Tisi, D.; Claesson-Welsh, L.; Hohenester, E.; Timpl, R. Endostatins derived from collagens XV and XVIII differ in structural and binding properties, tissue distribution and anti-angiogenic activity. *J. Mol. Biol.* **2000**, *301*, 1179–1190. [CrossRef] [PubMed]

111. Bix, G.; Iozzo, R.V. Matrix revolutions: "Tails" of basement-membrane components with angiostatic functions. *Trends Cell Biol.* **2005**, *15*, 52–60. [CrossRef] [PubMed]

112. Ramchandran, R.; Dhanabal, M.; Volk, R.; Waterman, M.J.; Segal, M.; Lu, H.; Knebelmann, B.; Sukhatme, V.P. Antiangiogenic activity of restin, NC10 domain of human collagen XV: Comparison to endostatin. *Biochem. Biophys. Res. Commun.* **1999**, *255*, 735–739. [CrossRef] [PubMed]

113. Maeshima, Y.; Manfredi, M.; Reimer, C.; Holthaus, K.A.; Hopfer, H.; Chandamuri, B.R.; Kharbanda, S.; Kalluri, R. Identification of the anti-angiogenic site within vascular basement membrane-derived tumstatin. *J. Biol. Chem.* **2001**, *276*, 15240–15248. [CrossRef] [PubMed]

114. Hamano, Y.; Zeisberg, M.; Sugimoto, H.; Lively, J.C.; Maeshima, Y.; Yang, C.; Hynes, R.O.; Werb, Z.; Sudhakar, A.; Kalluri, R. Physiological levels of tumstatin, a fragment of collagen IV alpha3 chain, are generated by MMP-9 proteolysis and suppress angiogenesis via alphaV beta3 integrin. *Cancer Cell* **2003**, *3*, 589–601. [CrossRef]

115. Poluzzi, C.; Iozzo, R.V.; Schaefer, L. Endostatin and endorepellin: A common route of action for similar angiostatic cancer avengers. *Adv. Drug Deliv. Rev.* **2016**, *97*, 156–173. [CrossRef] [PubMed]

116. Maquart, F.X.; Pasco, S.; Ramont, L.; Hornebeck, W.; Monboisse, J.C. An introduction to matrikines: Extracellular matrix-derived peptides which regulate cell activity. Implication in tumor invasion. *Crit. Rev. Oncol. Hematol.* **2004**, *49*, 199–202. [CrossRef] [PubMed]

117. Liu, A.; Mosher, D.F.; Murphy-Ullrich, J.E.; Goldblum, S.E. The counteradhesive proteins, thrombospondin 1 and SPARC/osteonectin, open the tyrosine phosphorylation-responsive paracellular pathway in pulmonary vascular endothelia. *Microvasc. Res.* **2009**, *77*, 13–20. [CrossRef] [PubMed]

118. Kazerounian, S.; Yee, K.O.; Lawler, J. Thrombospondins: From structure to therapeutics. *Cell. Mol. Life Sci.* **2008**, *65*, 700–712. [CrossRef] [PubMed]

119. Lopez-Dee, Z.; Pidcock, K.; Gutiérrez, L.S. Thrombospondin-1: Multiple paths to inflammation. *Mediat. Inflamm.* **2011**. [CrossRef] [PubMed]

120. Bonnefoy, A.; Moura, R.; Hoylaerts, M.F. The evolving role of thrombospondin-1 in hemostasis and vascular biology. *Cell. Mol. Life Sci.* **2008**, *65*, 713–727. [CrossRef] [PubMed]

121. Sweetwyne, M.T.; Murphy-Ullrich, J.E. Thrombospondin1 in tissue repair and fibrosis: TGF-β-dependent and independent mechanisms. *Matrix Biol.* **2012**, *31*, 178–186. [CrossRef] [PubMed]

122. Hynes, R.O. The extracellular matrix: Not just pretty fibrils. *Science* **2009**, *326*, 1216–1219. [CrossRef] [PubMed]

123. Arroyo, A.G.; Iruela-Arispe, M.L. Extracellular matrix, inflammation, and the angiogenic response. *Cardiovasc. Res.* **2010**, *86*, 226–235. [CrossRef] [PubMed]

124. Reilly, G.C.; Engler, A.J. Intrinsic extracellular matrix properties regulate stem cell differentiation. *J. Biomech.* **2010**, *43*, 55–62. [CrossRef] [PubMed]

125. Wells, R.G.; Discher, D.E. Matrix elasticity, cytoskeletal tension, and TGF-beta: The insoluble and soluble meet. *Sci. Signal.* **2008**, *1*. [CrossRef] [PubMed]
126. Raffetto, J.D.; Qiao, X.; Koledova, V.V.; Khalil, R.A. Prolonged increases in vein wall tension increase matrix metalloproteinases and decrease constriction in rat vena cava: Potential implications in varicose veins. *J. Vasc. Surg.* **2008**, *48*, 447–456. [CrossRef] [PubMed]
127. Kucukguven, A.; Khalil, R.A. Matrix metalloproteinases as potential targets in the venous dilation associated with varicose veins. *Curr. Drug Targets* **2013**, *14*, 287–324. [CrossRef] [PubMed]

Review

Hemorrhage Caused by Snake Venom Metalloproteinases: A Journey of Discovery and Understanding [†]

José María Gutiérrez [1],*, Teresa Escalante [1], Alexandra Rucavado [1] and Cristina Herrera [1,2]

[1]   Instituto Clodomiro Picado, Facultad de Microbiología, Universidad de Costa Rica, San José 11501-2060, Costa Rica; teresa.escalante@ucr.ac.cr (T.E.); alexandra.rucavado@ucr.ac.cr (A.R.); cristina.herreraarias@gmail.com (C.H.)

[2]   Facultad de Farmacia, Universidad de Costa Rica, San José 11501-2060, Costa Rica

*   Correspondence: jose.gutierrez@ucr.ac.cr; Tel.: +506-2511-7865

[†]   Dedicated to the memory of Akira Ohsaka for his pioneer contributions to our understanding of the mechanisms by which snake venoms induce hemorrhage.

Academic Editor: Bryan Grieg Fry
Received: 7 March 2016; Accepted: 18 March 2016; Published: 26 March 2016

Abstract: The historical development of discoveries and conceptual frames for understanding the hemorrhagic activity induced by viperid snake venoms and by hemorrhagic metalloproteinases (SVMPs) present in these venoms is reviewed. Histological and ultrastructural tools allowed the identification of the capillary network as the main site of action of SVMPs. After years of debate, biochemical developments demonstrated that all hemorrhagic toxins in viperid venoms are zinc-dependent metalloproteinases. Hemorrhagic SVMPs act by initially hydrolyzing key substrates at the basement membrane (BM) of capillaries. This degradation results in the weakening of the mechanical stability of the capillary wall, which becomes distended owing of the action of the hemodynamic biophysical forces operating in the circulation. As a consequence, the capillary wall is disrupted and extravasation occurs. SVMPs do not induce rapid toxicity to endothelial cells, and the pathological effects described in these cells *in vivo* result from the mechanical action of these hemodynamic forces. Experimental evidence suggests that degradation of type IV collagen, and perhaps also perlecan, is the key event in the onset of microvessel damage. It is necessary to study this phenomenon from a holistic, systemic perspective in which the action of other venom components is also taken into consideration.

Keywords: snake venom; viperids; metalloproteinases; hemorrhage; capillary vessels; basement membrane; type IV collagen

1. Introduction

Snakebite envenoming constitutes a highly relevant, albeit largely neglected, public health problem on a global basis, affecting primarily impoverished populations in the rural settings of Africa, Asia, Latin America and parts of Oceania [1–3]. Although accurate statistics on incidence and mortality due to snakebite envenoming are scarce, it has been estimated that between 1.2 and 5.5 million people suffer snakebite envenomings every year, with 25,000 to 125,000 deaths, and an estimated number of 400,000 victims left with permanent sequelae [4,5].

The clinical manifestations of envenomings vary depending on the species of snake causing the bite, and there is a large spectrum of pathophysiological effects induced by snake venoms, owing to the diverse arsenal described in their composition [6]. In addition, other factors play a key role in the severity of envenomings, such as the volume of venom injected, the anatomical site of the bite, and the physiological characteristics of the affected person [2]. Envenomings by snakes of the family Viperidae, and by some species of the family "Colubridae" (*sensu lato*), are characterized by prominent local and systemic hemorrhage. Blood vessel damage leading to extravasation, in turn, contributes to local tissue damage and poor muscle regeneration, and to massive systemic blood loss leading to hemodynamic disturbances and cardiovascular shock [2,7].

The study of the mechanism by which snake venoms induce hemorrhage and the characterization of hemorrhagic toxins and their mechanism of action has been a fascinating area of research within the field of Toxinology. The present contribution summarizes the journey of discovery and understanding that has led to our current view of snake venom-induced hemorrhage, and highlights some of the seminal discoveries and hypotheses generated during more than a century of scientific efforts.

2. Describing the Occurrence of Hemorrhage in Clinical and Experimental Viperid Snakebite Envenomings

Hemorrhage was recognized as a frequent and relevant manifestation early on in the description of the main clinical features of viperid snakebite envenomings, and bleeding in various organs was identified as one of the most serious consequences of these envenomings (see for example [8–10]). At the experimental level, J.B. de Lacerda, in 1884 [11], working in Brazil, described the effects induced by *Bothrops* sp. venoms in experimental animals. In his recount on local pathological effects, he reported that "*Le tissu cellulaire sous-cutané, est tout infiltré et présente de place en place des taches noirâtres, des points violacés et épars, des nuances livides diffuses et de nombreuses extravasions sanguines ... Le tissu cellulaire, qui cuvre les muscles, est également infiltré, il offre un aspect gélatineux et est plus ou moins imbibé du sang noir, en partie coagulé*". Two years later, Mitchel and Reichert, in the USA, described hemorrhagic events in animals treated with viperid snake venoms [12]. When reporting an experiment performed in a rabbit, the authors stated that "*On the peritoneum were placed a few small particles of the dried venom of* Crotalus adamanteus. *In two or three minutes some extravasations appeared immediately about the point of the application of the venom; a few moments later these extravasations were diffused over a considerable area and had run into each other to such an extent as to form a patch of bleeding surface*".

Other scientists investigated the hemorrhagic activity of viperid snake venoms in the first half of the 20th century by describing macroscopic observations in affected tissues, and by introducing the histological analysis of hemorrhagic lesions in organs. Likewise, researchers demonstrated the action of viperid venoms on the coagulation system and on platelets, and speculated on the contribution of these effects in the pathogenesis of venom-induced hemorrhage (reviewed in [13]). Thus hemorrhagic activity of snake venoms was a topic of interest for early clinical and experimental toxinologists owing to the relevance of hemorrhagic manifestations in the overall pathophysiology of viperid snakebite envenoming.

3. Devising Methods to Quantify Venom-Induced Hemorrhage

The experimental study of venom-induced hemorrhage and the identification of hemorrhagic components in snake venoms demanded the development of simple methods to quantify the extent of hemorrhagic lesions. One of the most significant advances towards this goal was achieved by the group of Akira Ohsaka in Japan [14]. A method was developed, consisting in the intracutaneous injection of various amounts of venom of *Trimeresurus* (now *Protobothrops*) *flavoviridis* into the depilated back skin of rabbits, followed by the measurement, 24 h later, of the size of the hemorrhagic spots in the inner side of the skin. The log dose–response curves showed that linear responses were obtained within a range of diameters from 10 mm to 18 mm. This skin method was then adapted for use in rats [15] and mice [16–18], with variations in the time lapse between venom injection and observation of hemorrhagic lesions. The hemorrhagic activity of a venom or toxin is expressed as the Minimum Hemorrhagic Dose (MHD), usually defined as the dose of venom or toxin that induces a hemorrhagic spot of 10 mm diameter [15,18]. This method has the limitation that it does not take into consideration the intensity of the hemorrhagic area, only its size. This has been circumvented by the quantification of the hemoglobin present in the hemorrhagic area of the skin [19].

Other methods for quantification of venom-induced hemorrhagic activity have been described [20], but modifications of the skin-based method of Kondo *et al.* [14] have been predominantly used in toxinological research owing to its simplicity and quantitative nature. It has been particularly useful in the assessment of hemorrhagic activity during fractionation of snake venoms for the purification of hemorrhagic components. On the other hand, methods have been established for the quantification of hemorrhagic activity by snake venoms in organs. The simplest one is based on the intramuscular or intravenous injection of venoms in mice followed by the euthanasia of animals and the sampling of affected organs. Tissues are then homogenized and the amount of hemoglobin in the tissue is quantified spectrophotometrically, by recording the absorbance at 540 nm [18,21]. For the quantification of hemorrhagic activity of venoms in the lungs, in addition to the method of Bonta *et al.* [20], the estimation of the Minimum Pulmonary Hemorrhagic Dose (MPHD) has been used. For this, groups of mice are injected intravenously with various doses of venom or toxin. After one hour, animals are euthanized and the thoracic cavity opened for observation of hemorrhagic spots in the surface of the lungs. The MPHD corresponds to the lowest venom dose which induces hemorrhagic lesions in the lungs in all mice injected [22].

4. Characterizing the Biochemical Properties of Hemorrhagic Toxins Present in Snake Venoms

Flexner and Noguchi [23] used the term "haemorrhagin" to denote the principle in snake venom responsible for hemorrhagic activity, and described it as *"the chief toxic constituent in Crotalus venom"*. By the middle of the 20th century, the hemorrhagic activity of viperid venoms was attributed by several researchers to the proteolytic activity of venoms [24], although the evidence for this hypothesis was mainly the fact that venoms with hemorrhagic activity also had high proteolytic action (see [13]). In addition, the inhibition of hemorrhagic activity by incubation with the chelating agent EDTA underscored the role of proteolytic activity in this effect [25]. However, it was not until modern chromatographic procedures were introduced in the study of snake venoms that this issue was properly addressed.

The early attempts to purify hemorrhagic toxins from snake venoms, by using electrophoretic and chromatographic methods, resulted in the isolation of proteinases devoid of hemorrhagic activity and of hemorrhagic toxins either having or lacking proteinase activity [26–28]. Thus, it was not completely clear at that time whether all viperid hemorrhagic toxins were proteinases or not. A source of confusion in these studies was the selection of the substrates for testing proteinase activity, as many of them followed the method of Kunitz [29], which uses casein as substrate, and assesses proteolytic activity by detecting acid-soluble peptides after precipitation of the enzyme-substrate mixture with trichloroacetic acid. Owing to the substrate specificity of snake venom hemorrhagic principles, this method is unable to detect proteolytic activity in some cases [30].

The advent of modern chromatographic procedures, especially after the decade of 1960, paved the way for the isolation of hemorrhagic components of good purity from a variety of viperid snake venoms. Among others, significant contributions in the purification of these components were performed by A. Ohsaka and colleagues in Japan [13], Grotto *et al.* in Israel [27], and F.R. Mandelbaum and coworkers in Brazil [31]. In 1978, Jon B. Bjarnason and Anthony T. Tu published a landmark study on the characterization of five hemorrhagic toxins from the venom of the rattlesnake *Crotalus atrox* [17]. Instead of employing the Kunitz's casein method for assaying proteolytic activity, they used dimethylcasein as substrate, which has a higher sensitivity. They demonstrated that these hemorrhagic toxins were metalloproteinases containing one mol of zinc per mol of toxin. Chelating agents eliminated both proteinase and hemorrhagic activities and, when zinc was removed, the toxins lost both actions to a similar degree. As there was not a correlation between the extent of hemorrhagic and proteinase activities, the authors suggested that these toxins have a highly selective substrate specificity for inducing hemorrhage. This study clarified a number of apparent contradictions in the characterization of snake venom hemorrhagic components, and paved the way for a new age in the biochemical characterization of hemorrhagic toxins.

Thereafter, many hemorrhagic proteinases, currently known as "snake venom metalloproteinases" (SVMPs), have been purified and characterized from snake venoms (see reviews [32,33]). All hemorrhagic SVMPs are zinc-dependent proteolytic enzymes that, together with the ADAMS ("*a* disintegrin *a*nd *m*etalloproteinase") belong to the M12B family of metalloproteinases which, in turn, are grouped within the Metzincin clan, characterized by a canonical sequence in the zinc-binding region at the catalytic site and by a Met-turn [34]. SVMPs have been classified in three classes (PI, PII, and PIII) on the basis of their domain composition. In the mature protein, PI SVMPs comprise the metalloproteinase domain only, whereas PII SVMPs present a disintegrin (Dis) domain in addition to the catalytic domain. On the other hand, PIII SVMPs present metalloproteinase, disintegrin-like (Dis-like), and cysteine-rich (Cys-rich) domains. In turn, several subclasses have been described within each class, depending on whether they are monomers or homo- or heterodimers, and also on the variable patterns of post-translational cleavage of several domains [35]. The variations in domain composition between different classes of SVMPs have implications in the mechanism of action of hemorrhagic toxins (see Section 9 of this review). The detailed analysis of the biochemical characterization of SVMPs, as well their molecular evolution after the recruitment of an ADAM-like gene early on in the course of advanced snakes diversification are beyond the scope of this review; readers are referred to excellent publications on these topics [33,35,36].

5. Exploring the Pathological Effects Induced by Venoms and SVMPs on the Microvasculature

One of the first histological analyses of snake venom-induced hemorrhage was published by Taube and Essex [37]. It was observed that rattlesnake venom affected the integrity of endothelial cells in vessels, inducing them to swell, and then burst and dissolve, leaving gaps in the vessel walls. Moreover, the authors described loss in blood coagulability. Further histological studies described hemorrhage, *i.e.*, presence of erythrocytes in the interstitial space of various tissues, as a consequence of injection of viperid venoms [38]. These authors also described pathological changes in the arterial walls. However, the characterization of the precise morphological alterations induced by venoms in the structure of microvessels had to wait for ultrastructural studies using the transmission electron microscope (TEM). Early TEM observations in tissues injected with a purified hemorrhagic toxin and the venom of *T. flavoviridis* revealed erythrocytes escaping through the intercellular junctions at the endothelial cell lining [39]. It was proposed that the basement membrane (BM) adjacent to the intercellular junctions has to be disrupted to allow the extravasation [13]. Moreover, by observing the mesentery of rats using cinematographic techniques at the microscope, Ohsaka *et al.* [40] described an initial vasoconstriction of arterioles, followed by vasodilation and then by the extravasation of erythrocytes one by one through holes formed in the capillaries, but without an overt rupture of the endothelium. Taken together, these findings suggested that hemorrhage occurs by extravasation of erythrocytes through openings in the intercellular endothelial cell junctions. As to the mechanisms of this opening, it was suggested that inflammatory mediators released by hemorrhagic toxins, such as histamine, serotonin and others, are responsible for this phenomenon [13].

A different picture emerged with a detailed characterization of ultrastructural alterations in capillary vessels after a subcutaneous injection of a hemorrhagic toxin from the venom of *Vipera palestinae* (currently *Daboia palestinae*) [41]. Ultrastructural lesions were described in capillary endothelial cells, and erythrocytes escaped through damaged endothelial cells and not through widened intercellular junctions. Similar and more detailed ultrastructural observations were carried out in muscle tissue injected with hemorrhagic SVMPs and crude venom of *Crotalus atrox* [42,43]. When examining the morphology of affected capillaries, endothelial cells showed dilatation of endoplasmic reticulum, formation of blebs, intracellular swelling, drop in the number of pinocytotic vesicles and, most importantly, disruption and formation of gaps within the cells through which erythrocytes escaped. The basal lamina was damaged in some portions and, interestingly, intercellular junctions between endothelial cells were not affected. Ownby *et al.* [42,43] named this mechanism *"hemorrhage per rhexis"*, in contrast with the mechanism proposed by Ohsaka and colleagues, which was named *"hemorrhage per diapedesis"*.

Afterwards, a number of groups investigated the morphological alterations in the microvasculature associated with the action of hemorrhagic SVMPs [22,44–47]. All of them described the extravasation of erythrocytes through lesions in capillary endothelial cells, and not through widened intercellular junctions, thus supporting the mechanism of hemorrhage *per rhexis*. (Figure 1).

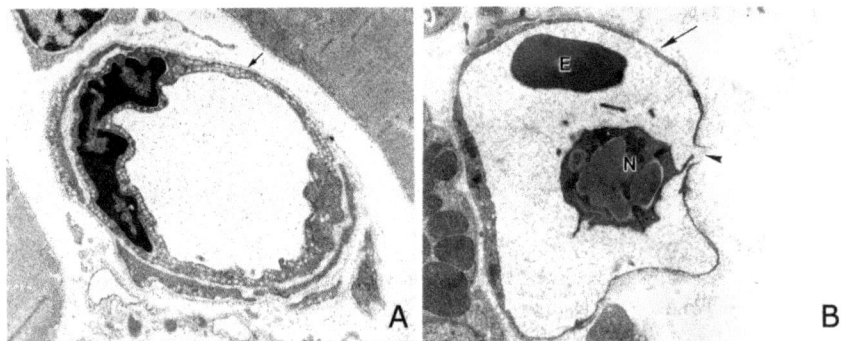

Figure 1. Pathological effects induced by a hemorrhagic snake venom metalloproteinase (SVMP) in capillary vessels. (**A**) Electron micrograph of a capillary vessel from muscle tissue injected with saline solution. Normal ultrastructure is observed in endothelial cell, including the presence of pynocytotic vesicles, and basement membrane (arrow). (**B**) Micrograph of a section of tissue injected with the hemorrhagic SVMP BaP1, from the venom of *Bothrops asper*. Notice prominent damage of endothelial cell, with loss of pynocytotic vesicles, distention and thinning of the cell, and rupture of cell integrity at one point (arrowhead). The basement membrane is absent along most of the periphery of the capillary (arrow). An erythrocyte (E) and a neutrophil (N) are observed inside the vessel. Magnification: $17,000\times$ (**A**); and $10,000\times$ (**B**). Reproduced by [47], Copyright 2006, Elsevier.

In contrast, Gonçalves and Mariano [48] described hemorrhage *per diapedesis* in the rat subcutaneous tissue injected with the venom of *Bothrops jararaca*. An explanation for this apparent discrepancy was proposed by Gutiérrez *et al.* [49]. It was suggested that the predominant mechanism of extravasation depends on the type of microvessel being observed. In electron micrographs of capillary vessels, endothelial cell disruption occurs and *per rhexis* mechanism predominates. In contrast, in the studies of Ohsaka [13] and Gonçalves and Mariano [48], the affected microvessels shown in the micrographs in which erythrocytes escape through widened intercellular junctions are venules. Venules react to several inflammatory mediators by contraction and widening of intercellular junctions, leading to an increment in vascular permeability [50]. It has been described that erythrocyte extravasation may take place by the intercellular route in cases of intense inflammation [51], as occurs in snakebite envenoming. Thus, both mechanisms of extravasation, *i.e.*, *per rhexis* and *per diapedesis*, are likely to be involved in tissues as a consequence of the action of hemorrhagic SVMPs. In the case of skeletal muscle, ultrastructural observations strongly indicate that *per rhexis* mechanism predominates, and that the main locus of action of SVMPs is the capillary network, and not the venular part of the microvasculature.

Intravital microscopy has been a useful tool to investigate the actions of venom and toxins in tissues from a dynamic perspective. When the venom of *Bothrops asper* or isolated SVMPs from this venom have been applied to the mouse cremaster muscle, microhemorrhagic lesions were observed in capillary vessels few minutes after application [52,53]. These microbleedings occurred in an explosive fashion and resulted in burst-shaped microhematomas. As time passed, more microhematomas appeared in the capillary network. These observations tend to support the *per rhexis* mechanism of hemorrhage, since such microvascular hemorrhagic bursts are more compatible with a rupture in the integrity of the capillary wall than with a discrete extravasation of erythrocytes through widened intercellular junctions.

6. Are Endothelial Cells Directly Damaged by Hemorrhagic SVMPs? A Two-Step Hypothesis and the Role of Biophysical Factors Operating *in Vivo*

The described ultrastructural observations of microvessels affected by SVMPs, occurring within few minutes of injection, highlight a prominent early damage to endothelial cells *in vivo*. An obvious corollary of these observations is that hemorrhagic SVMPs are directly cytotoxic to endothelial cells. However, when SVMPs are incubated with endothelial cells in culture, no cytotoxicity is observed during the first hours of incubation, in contrast to the very rapid damage observed in these cells *in vivo*, which occurs within minutes [53–56]. The most evident consequence of the action of SVMPs on endothelial cells in culture is a detachment of these cells, as a consequence of cleavage of proteins in the substrate of culture wells [53–55]. Interestingly, detached cells remain viable several hours until, eventually, they undergo cell death by apoptosis. A number of SVMPs have been described to induce apoptosis in endothelial cells in culture [57–60]. It was proposed that apoptosis occurs by anoikis, *i.e.*, as a consequence of the detachment of cells from their substrate [59,60], although the observation of SVMP-induced cytotoxicity in non-adherent cells in suspension argues for other mechanisms of cell death in addition to anoikis [61]. The cellular mechanisms involved in SVMP-induced endothelial cell apoptosis have been explored in detail by several groups (see for example [58–62]).

The use of endothelial cell monolayers in culture as a model to investigate the action of SVMPs has the drawback that, phenotypically, these cells bear differences with microvascular endothelial cells *in vivo*. Baldo *et al.* [63] approached this limitation by using two-dimensional and three-dimensional cultures of endothelial cells in an extracellular matrix scaffold, this model being closer to the *in vivo* conditions of vascular endothelial cells. It was observed that collagen and Matrigel substrates enhanced endothelial cell damage induced by a hemorrhagic SVMP, allowing the binding of this enzyme to focal adhesions, disruption of stress fibers, detachment and apoptosis. This study stressed the relevance of developing *in vitro* models of endothelial cells which more closely resemble their phenotype *in vivo*. Nevertheless, even in these conditions, cytotoxic effects on endothelial cells appear after several hours and, therefore, do not reproduce the very early necrotic damage occurring in the microvasculature *in vivo* after injection of hemorrhagic SVMPs.

A hypothesis was proposed by Gutiérrez *et al.* [49] to explain this apparent discrepancy, by taking into consideration the possible role of the hemodynamic biophysical forces operating in the circulation *in vivo*, *i.e.*, wall tension and shear stress. According to Laplace's law, wall tension is directly proportional to transmural pressure, *i.e.*, the difference between intravascular and extravascular hydrostatic pressures, and to the radius of the vessel [64]. In turn, shear stress is directly proportional to blood flow and viscosity, and inversely proportional to the vessel radius [65]. The distensibility of the vascular wall plays a key role in wall tension. In the case of capillaries, such distensibility is predominantly determined by the mechanical properties of their BM [66,67]. Thus, any effect in the mechanical stability of the BM would have a direct impact in the distensibility of the capillary wall. It was proposed by Gutiérrez *et al.* [49] that hemorrhagic SVMPs cause damage to the integrity of capillary vessels by a two-step mechanism: Initially, these enzymes hydrolyze key substrates at the BM surrounding endothelial cells in capillaries, causing a weakening in the mechanical stability of BM and increasing the distensibility of the microvessel wall. As a second step, the hemodynamic forces operating in the microcirculation, especially the hydrostatic force, induce a distention in the wall, which eventually culminates in the disruption of its integrity and in extravasation (Figure 2).

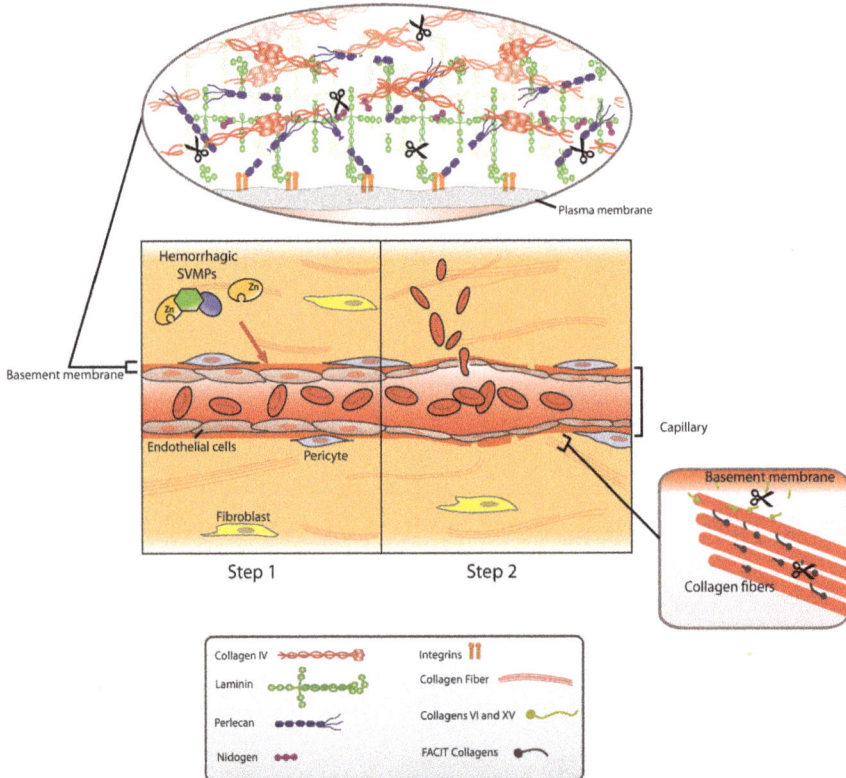

Figure 2. Two-step model to explain the mechanism of action of hemorrhagic SVMPs. The experimental evidence collected suggests that capillary vessel damage induced by hemorrhagic SVMPs occurs by a two-step mechanism. In the first step, SVMPs bind to and hydrolyze critical structural components of the basement membrane of capillary vessels, particularly type IV collagen and perlecan, and possibly other molecules that link the basement membrane to the fibrillar extracellular matrix. The cleavage of key peptide bonds of basement membrane components results in the mechanical weakening of this scaffold structure. As a consequence, in the second step, the biophysical hemodynamic forces normally operating in the microcirculation, *i.e.*, hydrostatic pressure, which largely determines wall tension, and shear stress, induce a distention of the vessel wall, until the capillary is eventually disrupted, with the consequent extravasation of blood. Reproduced by [68], Copyright 2011, Elsevier.

This hypothesis is compatible with ultrastructural alterations described above, *i.e.*, an increase in the capillary lumen associated with a thinning of endothelial cells and a drop in the number of pinocytotic vesicles, together with overt disruptions in the integrity of endothelial cells through which blood escapes to the interstitial space. One prediction arising from this hypothesis is that capillary damage should not occur in conditions of absence of blood flow. This was corroborated at the experimental level in muscle tissue of mice injected with a hemorrhagic SVMP in which blood flow was totally interrupted; no endothelial cell damage was observed in these conditions [47]. Thus, the rapid endothelial cell damage caused by hemorrhagic SVMPs *in vivo* is not due to a direct action of these enzymes in the cells, but instead to an indirect effect as a consequence of the weakening of the mechanical stability of the capillary wall owing to the effect of SVMP on the BM components. The obvious questions are, then: How do hemorrhagic SVMPs affect the BM? Why are some SVMP able to induce hemorrhage while others are not?

7. Exploring the Hydrolysis of Basement Membrane Components by SVMPs

The biochemical characterization of hemorrhagic components in snake venoms demonstrated that they are metalloproteinases. However, most of the early studies on purified SVMPs demonstrated proteolytic activity on a variety of substrates that do not correspond to the physiological targets of these enzymes [30,32]. Since microvessels are the target of hemorrhagic SVMPs, the study of degradation of BM components became a relevant task in understanding SVMPs' mechanism of action. Ohsaka *et al.* [69] first demonstrated that hemorrhagic SVMPs hydrolyzed proteins of isolated glomerular BM. In 1988, Bjarnason *et al.* [70] described the ability of SVMPs to hydrolyze proteins of Matrigel, a BM preparation from Engelbreth–Holm–Swarm (EHS) mouse sarcoma, and on isolated BM components. Since then, several studies have described the hydrolysis of BM components by a number of SVMPs [53,71–77]. Notwithstanding, several puzzling findings arose from these observations: (a) The time-course of SVMP-induced hydrolysis of BM components *in vitro*, as judged by SDS-PAGE, usually take several hours, whereas hemorrhage *in vivo* appears within minutes; this could be explained on the grounds that SDS-PAGE analysis may not reveal subtle, but significant, early cleavage of BM components, which would be relevant for the weakening of BM stability. (b) Several SVMPs devoid or having very low hemorrhagic activity have been shown to hydrolyze BM components *in vitro* [78–80]. This raises the possibility that a critical attack to the BM mechanical stability might be related to the selective hydrolysis of some BM components.

The BM is a complex and highly specialized extracellular matrix structure which plays a key scaffold role in capillary endothelial cells and in other cell types [81]. BM contains major components such as laminin, type IV collagen, nidogen/entactin, and heparin sulphate proteoglycan (perlecan), in addition to a number of minor components such as agrin, APARC/BM-40/osteopontin, fibulins and types XV and XVIII collagens [81–83]. In the context of this complexity, which are the structurally relevant components whose cleavage by hemorrhagic SVMPs weakens the mechanical stability of microvessels? When comparing the hydrolysis of Matrigel *in vitro* by PI hemorrhagic and non-hemorrhagic SVMPs, a significant difference was found on the hydrolysis of type IV collagen and perlecan, since the former cleaved these substrates, particularly type IV collagen, to a greater extent [80]. On these grounds, deciphering the sites of cleavage of these molecules might shed light on the molecular basis of BM destabilization. Jay W. Fox, Jon Bjarnason and colleagues have studied the cleavage patterns of *Crotalus atrox* hemorrhagic SVMP on type IV collagen [32,72]. Cleavage occurs at the α1 (IV) and α2 (IV) chains. The α1 (IV) chains are cleaved at a triplet interruption region of the triple helix, whereas the α2 (IV) chain is cleaved in the triple helical region near the NC2 domain. Owing to the role of these sites in the association of type IV collagen monomers into tetramers, these cleavages may have implications on the mechanical stability of the BM [32].

8. Exploring the Hydrolysis of BM Components by SVMPs *in Vivo*

The described *in vitro* observations of BM hydrolysis by SVMPs raised the need to explore the degradation of BM components *in vivo* as a consequence of injection of SVMPs. This has been approached by a combination of three methodologies, *i.e.*, immunohistochemistry in tissue sections, immunodetection by Western blot in tissue homogenates, and proteomics analysis of exudates collected in the vicinity of affected tissue. These complementary approaches have provided novel clues for understanding of the pathogenesis of SVMP-induced hemorrhage. Immunohistochemical staining of BM components as early as 15 min after injection of hemorrhagic SVMPs revealed loss of staining in microvessels of type IV collagen, laminin and nidogen [75,84], hence underscoring the rapid cleavage of these proteins. Such degradation was corroborated when performing immunoblots for detection of BM components on tissue homogenates and exudates [80,85]. Interestingly, when hemorrhagic and non-hemorrhagic SVMPs were compared, a similar pattern of hydrolysis was observed for laminin and nidogen, whereas the hemorrhagic enzyme induced a more extensive cleavage of perlecan and, especially, type IV collagen [80] (Figure 3).

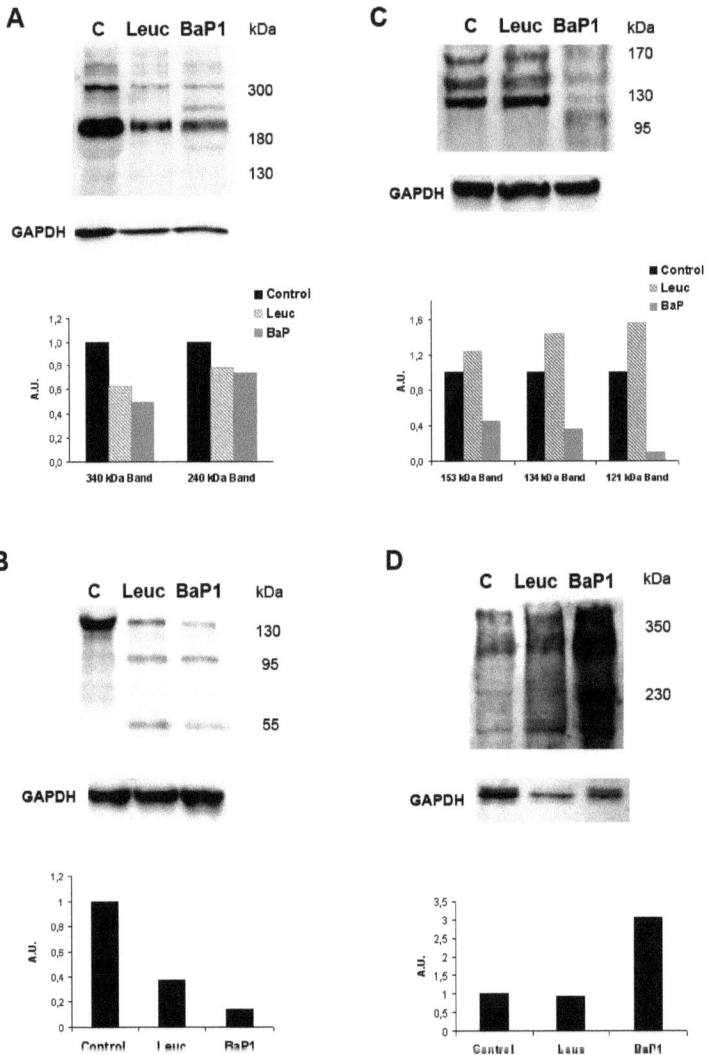

Figure 3. Differential hydrolysis of basement membrane components *in vivo* by hemorrhagic and non-hemorrhagic snake venom metalloproteinases (SVMPs). Non-hemorrhagic SVMP (leucurolysin a-leuc) and hemorrhagic SVMP (BaP1) were injected in the muscle of mice. After 15 min, animals were euthanized, and tissue was collected and homogenized. Supernatants of homogenates were separated by SDS-PAGE, and transferred to nitrocellulose membranes for immunodetection with either anti-laminin (**A**); anti-nidogen (**B**); anti-type IV collagen (**C**); and anti-endorepelin (perlecan) (**D**) antibodies, and with anti-GAPDH as loading control. (**C**) Control muscle injected with saline solution. A chemiluminiscent substrate was used to detect the reactions. Densitometric analysis was then performed. The molecular mass of various markers (in kDa) is shown at the right of the gels. A clear difference is observed between these SVMPs in the patterns of degradation of type IV collagen and endorepelin (perlecan). Reproduced by [80], Copyright 2011, PLOS.

Taken together, these findings strongly support the view that cleavage of type IV collagen, and probably perlecan as well, play a key role in the weakening of the mechanical stability of the BM of capillary vessels, and constitutes the first step in the pathogenesis of SVMP-induced microvessel damage [68,80]. When immunoblotting experiments were performed with muscle homogenates prepared from mice injected with a hemorrhagic SVMP in conditions where blood flow was interrupted, similar degradation patterns of BM components were observed (our unpublished results), but the integrity of capillary vessels was intact [47]. This supports the "two-step" hypothesis described above, since in conditions of no blood flow, the first step occurs, *i.e.*, enzymatic cleavage of BM components, but the second one, *i.e.*, distention and rupture of vessel integrity, does not, since the biophysical hemodynamic forces dependent on blood flow are not operating in these circumstances.

A significant step forward in the study of venom-induced tissue damage was the introduction of the proteomic analysis of exudate samples collected in the affected tissues [86], as well as of tissue homogenates [87]. This experimental approach allowed the identification of cleavage products of several extracellular matrix proteins, including BM components, as a consequence of the action of SVMPs [80,85,86]. In addition, fragments of type VI and type XV collagens were detected [80,85,86]. Interestingly, the amount of BM-specific heparan sulfate proteoglycan core protein, type VI collagen and type XV collagen were higher in exudates collected from tissue injected with a hemorrhagic SVMP than in samples injected with a non-hemorrhagic enzyme [80], thus suggesting that cleavage of these proteins may be related with the pathogenesis of hemorrhage. Type VI and XV collagens contribute to the stabilization of capillary vessels by connecting the BM with the surrounding extracellular matrix [88,89].

9. What Is the Basis for the Large Variation in the Hemorrhagic Potential of SVMPs?

When the hemorrhagic activity of isolated SVMPs is quantified, large differences are observed between enzymes. Moreover, some SVMPs are able to induce systemic hemorrhage, whereas others only cause local bleeding. The structural basis for these variations in hemorrhagic potential in enzymes that, otherwise, have similar cleavage patterns on protein substrates, has been deciphered as the structural features of SVMPs have become understood. In general, PIII SVMPs, which comprise Dis-like and Cys-rich domains in addition to the metalloproteinase domain, are more potent hemorrhagic toxins than PI SVMPs having only the catalytic domain [33,49]. Moreover, some of the few P-II SVMPs characterized to date are also potent hemorrhagic enzymes [90,91]. Thus, it is evident that the additional domains potentiate hemorrhagic activity, probably by the presence of relevant exosites in their sequences.

These exosites may contribute to hemorrhagic activity in various ways. They may promote the binding of SVMPs to specific substrates in the tissues at relevant targets, such as the capillary microvascular network. Indeed, immunohistochemical evidence has demonstrated that PII and PIII SVMPs preferentially bind to the microvasculature when applied in mouse tissues, and present a pattern of co-localization with type IV collagen in the vessel wall [84,85] (Figure 4). The same pattern of co-localization was observed for a fragment constituted by the Dis-like and Cys-rich domains (DC fragment) of a PIII SVMP, confirming that sequences in these domains are responsible for the targeting to microvessels [84]. In contrast, PI SVMPs, devoid of these additional domains, show a widespread localization in the extracellular matrix and are not concentrated in the vessels [84,85].

A   BaP1   BlatH1   CsH1

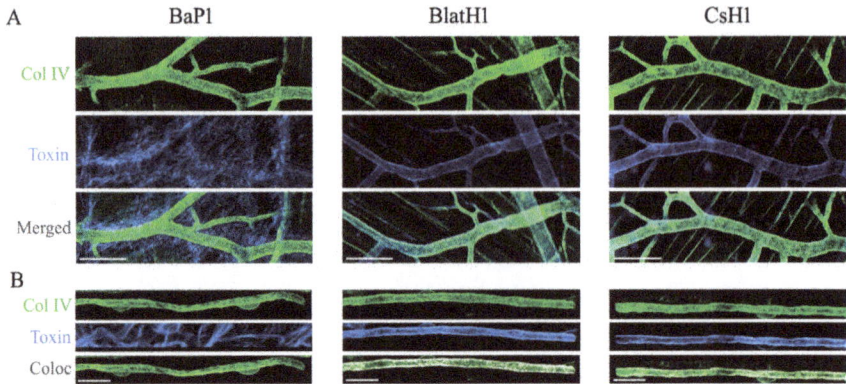

B

Figure 4. Immunolocalization of snake venom metalloproteinases (SVMPs) with vascular basement membrane on mouse cremaster muscle. Groups of mice were euthanized, and the cremaster muscle was dissected out. The isolated muscles were incubated for 15 min with equi-hemorrhagic amounts of three different SVMPs: BaP1 (PI SVMP, 30 µg), BlatH1 (PII SVMP, 3.5 µg) or CsH1 (PIII SVMP, 15 µg) labeled with Alexa Fluor® 647 (blue). Control tissues were incubated with the SVMPs without labeling and no fluorescence was detected. Whole tissues were fixed with 4% paraformaldehyde and immunostained with anti-collagen IV following the secondary antibody labeled with Alexa Fluor 488 (green). Tissues were visualized in a Zeiss LSM 5 Pascal laser-scanning confocal microscope. Three-dimensional reconstitution of the images was carried out using IMARIS ×64 7.4.2 software. (**A**) Distribution of the SVMPs in the cremaster muscle tissue. Scale bar represents 150 µm. (**B**) Localization of SVMPs in capillary vessels in the cremaster. Scale bar represents 20 µm. White areas represent co-localization of the SVMPs (blue) with collagen IV (green) of vascular basement membrane in capillaries. Notice the predominant localization of PII and PIII SVMPs in the vasculature, whereas PI SVMP localizes in a more widespread fashion in the tissue. Reproduced by [85], Copyright 2015, PLOS.

Biochemical studies with jararhagin, a hemorrhagic PIII SVMPs from the venom of *Bothrops jararaca*, showed that the Dis-like and Cys-rich domains mediate the binding to collagens and $\alpha_2\beta_1$ integrin, respectively [92,93]. Moreover, the Cys-rich domain of PIII SVMPs is responsible for the binding to type I collagen and several proteins presenting von Willebrand factor (vWF) A domains, such as vWF, various fibrillary-associated collagens with interrupted triple helices (FACITs) and matrylins [94–96]. In light of the pathological and proteomic observations described above, the binding to various fibrillar and non-fibrillar collagens are relevant, since hydrolysis of these substrates is likely to be causally related to the onset of microvessel damage [68]. The fact that PII SVMPs are also potent hemorrhagic toxins, and at least one of them co-localizes with type IV collagen in the vessels [85], demonstrates that exosites located in the Dis domain allow the targeting of these SVMPs to the microvasculature. Thus, by directing PII and PIII SVMPs to the microvessels, these additional domains are likely to position the catalytic sites of these enzymes nearby relevant substrates in the BM. In contrast, PI SVMPs are not targeted to these loci in the tissue and therefore are less effective in causing capillary damage.

It has been also described that the generalist plasma proteinase inhibitor α_2-macroglobulin (α_2M) effectively abrogates proteolytic and hemorrhagic activities of PI SVMPs, whereas it is ineffective to inhibit these activities in PII and PIII SVMPs [91,97–100]. The structural basis for this difference remains unknown, but it is likely that structural constraints in the Dis, Dis-like and Cys-rich domains impair the inhibitory action of α_2M. This finding is relevant in the light of the ability of PII and PIII SVMPs to induce systemic hemorrhage, since they are not inhibited upon their entrance into the circulation and their distribution to various organs. In contrast, PI SVMPs are readily inhibited by α_2M when they reach the bloodstream, thereby precluding their action at distant organs [49]. Moreover,

PII and PIII SVMPs are able to inhibit platelet aggregation, owing to the action of sequences of their Dis, Dis-like and Cys-rich domains [101], and also have the ability to degrade vWF [96,102]. However, these effects on hemostasis have been shown *in vitro*, and it remains to be investigated whether they also have an impact *in vivo*.

10. SVMP-Induced Hemorrhage Viewed from a Holistic Perspective: Studying the Action of SVMP in the Context of the Whole Venom

The field of Toxinology, as many other areas in biomedical research, is moving from a predominantly "reductionist" approach to a more "holistic" and integrated perspective [103]. In this context, the characterization of the structural and functional properties of hemorrhagic SVMPs has provided a rich body of knowledge that has expanded our understanding on the mechanism of action of these toxins in the tissues and on the structural determinants of their activity. Nevertheless, in actual snakebites, it is whole venom that is injected in the victim's tissues and, consequently, the action of toxins has to be analyzed in the context of venoms. Probably the best example of this need, in the case of hemorrhagic activity of SVMPs, is the potentiation of this activity by venom components that affect hemostasis. Viperid venoms contain an array of proteins acting on plasma coagulation factors and platelets [2,101,104,105]. Serine proteinases and SVMPs exert procoagulant effects by displaying thrombin-like activity, or by activating factors II, X or V of the coagulation cascade [106]. *In vivo*, these procoagulant effects result in defibrinogenation and incoagulability [2,105]. Although incoagulability *per se* does not necessarily result in significant bleeding, coagulopathy potentiates bleeding in the context of SVMP-induced vascular damage. In some cases, the action of procoagulant components in snake venoms or the damage to the endothelial lining cause regional thrombosis [107,108], with the consequent local or remote ischemic events, which contribute to the complexity of the pathophysiology of cardiovascular alterations in snakebite envenomings.

Likewise, various venom components, such as proteins of the C-type lectin-like family, and also serine proteinases, SVMPs, and disintegrins, exert effects that impair platelet function, either by causing thrombocytopenia or inhibition of platelet aggregation [101,109–112]. The role of venom-induced platelet disturbances in potentiating hemorrhagic activity has been demonstrated clinically [113] and at the experimental level, whereby mice rendered thrombocytopenic by a C-type lectin-like venom component developed stronger hemorrhage in the lungs after injection with a hemorrhagic SVMP [114].

Less evident instances of synergisms between SVMPs and other venom components have been also described. The co-administration of hyaluronidase and a hemorrhagic SVMP increased the hemorrhagic area as compared to the action of the SVMP alone, thus suggesting that hyaluronidase plays the role of a spreading factor that extends the action of SVMPs [115,116]. In this context, it is relevant to consider the possible involvement of non-hemorrhagic SVMPs since these enzymes, despite lacking hemorrhagic activity, degrade various extracellular matrix components [79]; this effect might contribute to the spreading and action of hemorrhagic SVMPs, a hypothesis that requires experimental analysis. More recently, a puzzling phenomenon was described whereby a non-toxic venom phospholipase A_2, which does not affect endothelial cells, potentiates the cell-detaching effect induced by a hemorrhagic SVMP on an endothelial cell line in culture [117]. Moreover, the inflammatory landscape generated by various components upon venom injection in the tissues is likely to promote vascular alterations that might potentiate the disruptive effect of hemorrhagic SVMPs in the microvasculature. The role of inflammation in the pathogenesis of systemic hemorrhage in organs such as the lungs or the brain remains to be explored.

11. Concluding Remarks

This summarized account of the journey followed by toxinologists for understanding the mechanisms by which SVMPs induce hemorrhage has provided an overview of some of the landmarks in the characterization of SVMPs and their action in the microvasculature. It is now clear that hemorrhage occurs mostly through the SVMP-induced degradation of key structural proteins at the BM and its surroundings. The mechanical weakening generated by this enzymatic degradation results in microvessel wall distention owing to the action of the hemodynamic forces operating in the circulation, with the eventual disruption of the capillary wall and extravasation. The wide variation in the hemorrhagic potential of SVMPs of different domain composition has been explained by the role of Dis, Dis-like and Cys-rich domains in positioning these enzymes at physiologically-relevant sites for exerting their hemorrhagic activity, *i.e.*, at the microvasculature. There is a need to explore, at deeper levels, which are the BM components critical for the disruption of microvessel integrity; in this regard, the use of immunoelectron microscopic techniques should provide novel information on the precise localization of SVMPs in the BM. Likewise, the proteomic analysis of samples of exudate and tissue collected after injection of SVMPs is likely to offer further details of the fine pathological alterations induced by these enzymes. The use of diverse, and complementary, methodological platforms has been helpful in generating valuable information on the mechanism of action of SVMPs, and should be strengthened and expanded in the future.

On the other hand, the action of hemorrhagic SVMPs in the organism has to be studied from a more holistic perspective, by understanding it in the overall context of envenoming. In this regard, the study of pathological alterations caused by SVMPs has to be complemented with the study of these effects induced by crude venoms and by combinations of various venom components, as well as by understanding the role of tissue responses to envenoming in the pathogenesis of hemorrhage and other pathological effects. The complexity of these tasks demands the integration of multidisciplinary research tools, including informatic resources. Toxinology is in need of renewed conceptual and experimental platforms aimed at reaching a more profound understanding of the highly complex pathophysiology of snakebite envenoming, including the pathogenesis of hemorrhage.

Acknowledgments: The authors thank our colleagues from Instituto Clodomiro Picado and other groups in various countries for many collaborative efforts and highly fruitful discussions in the study of SVMPs. Particular thanks are due to Jay W. Fox and his group at the University of Virginia (USA), Ana Moura-da-Silva and colleagues at Instituto Butantan (Brazil), R. David G. Theakston and Aura Kamiguti at the Liverpool School of Tropical Medicine (UK), and Michael Ovadia and Gadi Borkow at the University of Tel Aviv (Israel). Some of the studies reviewed in this work have been supported by Vicerrectoría de Investigación (Universidad de Costa Rica), the International Foundation for Science, US AID, NeTropica, and the Wellcome Trust.

Author Contributions: All the authors contributed to the writing and revision of this review article.

Conflicts of Interest: The authors declare no conflict of interest. The founding sponsors had no role in the design of the study, in the writing of the manuscript, and in the decision to publish the review.

References

1. Gutiérrez, J.M.; Theakston, R.D.G.; Warrell, D.A. Confronting the neglected problem of snake bite envenoming: The need for a global partnership. *PLoS Med.* **2006**, *3*. [CrossRef] [PubMed]
2. Warrell, D.A. Snake bite. *Lancet* **2010**, *375*, 77–88. [CrossRef]
3. Harrison, R.A.; Hargreaves, A.; Wagstaff, S.C.; Faragher, B.; Lalloo, D.G. Snake envenoming: A disease of poverty. *PLoS Negl. Trop. Dis.* **2009**, *3*. [CrossRef] [PubMed]
4. Kasturiratne, A.; Wickremasinghe, A.R.; de Silva, N.; Gunawardena, N.K.; Pathmeswaran, A.; Premaratna, R.; Savioli, L.; Lalloo, D.G.; de Silva, H.J. The global burden of snakebite: A literature analysis and modelling based on regional estimates of envenoming and deaths. *PLoS Med.* **2008**, *5*. [CrossRef] [PubMed]
5. Williams, D.; Gutiérrez, J.M.; Harrison, R.; Warrell, D.A.; White, J.; Winkel, K.D.; Gopalakrishnakone, P. The Global Snake Bite Initiative: An antidote for snake bite. *Lancet* **2010**, *375*, 89–91. [CrossRef]
6. Calvete, J.J. Proteomic tools against the neglected pathology of snake bite envenoming. *Expert Rev. Proteom.* **2011**, *8*, 739–758. [CrossRef] [PubMed]
7. Cardoso, J.L.C.; França, F.O.S.; Wen, F.H.; Málaque, C.M.S.; Haddad, V., Jr. *Animais Peçonhentos no Brasil. Biologia, Clínica e Terapêutica dos Acidentes*, 2nd ed.; Sarvier: São Paulo, Brazil, 2009. (In Portuguese)
8. Mitchel, S.W. *Researchers upon the Venom of the Rattlesnake: With an Investigation of the Anatomy and Physiology of the Organs Concerned*; Smithsonian Institution: Washington, DC, USA, 1860.
9. Brazil, V. *A Defesa Contra o Ophidismo*; Pocai & Weiss: São Paulo, Brasil, 1911. (In Portuguese)

10. Picado, C. Serpientes Venenosas de Costa Rica. Sus Venenos. In *Seroterapia Antiofídica*; Imprenta Alsina: San José, Costa Rica, 1931. (In Spanish)

11. De Lacerda, J.B. *Leçons sur le Venin des Serpents du Brésil*; Lombaerts: Rio de Janeiro, Brasil, 1884. (In French)

12. Mitchel, S.W.; Reichert, E.T. *Researches upon the Venoms of Poisonous Serpents*; Smithsonian Institution: Washington, DC, USA, 1886.

13. Ohsaka, A. Hemorrhagic, necrotizing and edema-forming effects of snake venoms. In *Handbook of Experimental Pharmacology*; Springer-Verlag: Berlin, Germany, 1979; Volume 52, pp. 480–546.

14. Kondo, H.; Kondo, S.; Ikezawa, H.; Murata, R.; Ohsaka, A. Studies on the quantitative method for the determination of hemorrhagic activity of Habu snake venom. *Jpn. J. Med. Sci. Biol.* **1960**, *13*, 43–51. [CrossRef] [PubMed]

15. Theakston, R.D.G.; Reid, H.A. Development of simple standard assay procedures for the characterization of snake venom. *Bull. World Health Organ.* **1983**, *61*, 949–956. [PubMed]

16. Tu, A.T.; Homma, M.; Hong, B.S. Hemorrhagic, myonecrotic, thrombotic and proteolytic activities of viper venoms. *Toxicon* **1969**, *6*, 175–178. [CrossRef]

17. Bjarnason, J.B.; Tu, A.T. Hemorrhagic toxins from Western diamondback rattlesnake (*Crotalus atrox*) venom: Isolation and characterization of five toxins and the role of zinc in hemorrhagic toxin e. *Biochemistry* **1978**, *17*, 3395–3404. [CrossRef] [PubMed]

18. Gutiérrez, J.M.; Gené, J.A.; Rojas, G.; Cerdas, L. Neutralization of proteolytic and hemorrhagic activities of Costa Rican snake venoms by a polyvalent antivenom. *Toxicon* **1985**, *23*, 887–893. [CrossRef]

19. Rucavado, A.; Escalante, T.; Teixeira, C.F.P.; Fernándes, C.M.; Díaz, C.; Gutiérrez, J.M. Increments in cytokines and matrix metalloproteinases in skeletal muscle after injection of tissue-damaging toxins from the venom of the snake *Bothrops asper*. *Mediat. Inflamm.* **2002**, *11*, 121–128. [CrossRef] [PubMed]

20. Bonta, I.L.; Vargaftig, V.V.; Bhargava, N.; de Vos, C.J. Method for study of snake venom induced hemorrhage. *Toxicon* **1970**, *8*, 3–10. [CrossRef]

21. Ownby, C.L.; Colberg, T.R.; Odell, G.V. A new method for quantitating hemorrhage induced by rattlesnake venoms: Ability of polyvalent antivenom to neutralize hemorrhagic activity. *Toxicon* **1984**, *22*, 227–233. [CrossRef]

22. Escalante, T.; Núñez, J.; Moura-da-Silva, A.M.; Theakston, R.D.G.; Gutiérrez, J.M. Pulmonary hemorrhage induced by jararhagin, a metalloproteinase from *Bothrops jararaca* snake venom. *Toxicol. Appl. Pharmacol.* **2003**, *193*, 17–28. [CrossRef]

23. Flexner, S.; Noguchi, H. The constitution of snake venom and snake sera. *J. Path. Bact.* **1903**, *8*, 379–410. [CrossRef]

24. Houssay, B.A. Classification des actions des venins de serpents sur l'organisme animal. *Comp. Rend. Soc. Biol. Paris* **1930**, *105*, 308–310. (In French).

25. Flowers, H.N.; Goucher, C.R. The effect of EDTA on the extent of tissue damage caused by the venoms of *Bothrops atrox* and *Agkistrodon piscivorus*. *Toxicon* **1965**, *2*, 221–224. [CrossRef]

26. Ohsaka, A.; Ikezawa, H.; Kondo, H.; Kondo, S.; Uchida, N. Haemorrhagic activities of Habu snake venom, and their relations to lethal toxicity, proteolytic activities and other pathological activities. *Br. J. Exp. Pathol.* **1960**, *41*, 478–486. [PubMed]

27. Grotto, L.; Moroz, C.; de Vries, A.; Goldblum, N. Isolation of *Vipera palestinae* hemorrhagin and distinction between its hemorrhagic and proteolytic activities. *Biochim. Biophys. Acta* **1967**, *133*, 356–362. [CrossRef]

28. Takahashi, T.; Ohsaka, A. Purification and some properties of two hemorrhagic principles (HR2a and HR2b) in the venom of *Trimeresurus flavoviridis*; complete separation of the principles from proteolytic activities. *Biochim. Biophys. Acta* **1970**, *207*, 65–75. [CrossRef]

29. Kunitz, M. Crystalline soybean trypsin inhibitor. II. General properties. *J. Gen. Physiol.* **1947**, *30*, 291–310. [CrossRef] [PubMed]

30. Bjarnason, J.B.; Fox, J.W. Hemorrhagic toxins from snake venoms. *J. Toxicol. Toxin Rev.* **1988–1989**, *7*, 121–209. [CrossRef]

31. Mandelbaum, F.R.; Reichl, A.P.; Assakura, M. Some physical and biochemical characteristics of HF2, one of the hemorrhagic factors in the venom of *Bothrops jararaca*. In *Animal, Plant, and Microbial Toxins*; Plenum Press: New York, NY, USA, 1976; Volume 1, pp. 111–121.

32. Bjarnason, J.B.; Fox, J.W. Hemorrhagic metalloproteinases from snake venoms. *Pharmacol. Ther.* **1994**, *62*, 325–372. [CrossRef]

33. Fox, J.W.; Serrano, S.M.T. Structural considerations of the snake venom metalloproteinases, key members of the M12 reprolysin family of metalloproteinases. *Toxicon* **2005**, *45*, 969–985. [CrossRef] [PubMed]

34. Bode, W.; Gomis-Rüth, F.X.; Stöckler, W. Astacins, serralysins, snake venom and matrix metalloproteinases exhibit identical zinc-binding environments (HEXXHXXGXXH and Met turn) and topologies and should be grouped into a common family, the "metzincins". *FEBS Lett.* **1993**, *331*, 134–140. [CrossRef]

35. Fox, J.W.; Serrano, S.M.T. Snake venom metalloproteinases. In *Handbook of Venoms and Toxins of Reptiles*; CRC Press: Boca Raton, FL, USA, 2010; pp. 95–113.

36. Casewell, N.R.; Sunagar, K.; Takacs, Z.; Calvete, J.J.; Jackson, T.N.W.; Fry, B.G. Snake venom metalloprotese enzymes. In *Venomous Reptiles and Their Toxins. Evolution, Pathophysiology and Discovery*; Oxford University Press: Oxford, UK, 2015; pp. 347–363.

37. Taube, H.N.; Essex, H.E. Pathologic changes in the tissue of the dog following injections of rattlesnake venom. *Arch. Pathol.* **1937**, *24*, 43–51.

38. Homma, M.; Tu, A.T. Morphology of local tissue damage in experimental snake envenomation. *Br. J. Exp. Pathol.* **1971**, *52*, 538–542. [PubMed]

39. Ohsaka, A.; Suzuki, K.; Ohashi, M. The spurting of erythrocytes through junctions of the vascular endothelium treated with snake venom. *Microvasc. Res.* **1975**, *10*, 208–213. [CrossRef]

40. Ohsaka, A.; Ohashi, M.; Tsuchiya, M.; Kamisaka, Y.; Fujishiro, Y. Action of *Trimeresurus flavoviridis* venom on the microcirculatory system of rat; dynamic aspects as revealed by cinematographic recording. *Jpn. J. Med. Sci. Biol.* **1971**, *24*, 34–39.

41. McKay, D.G.; Moroz, C.; de Vries, A.; Csavossy, I.; Cruse, V. The action of hemorrhagin and phospholipase derived from *Vipera palestinae* venom on the microcirculation. *Lab. Investig.* **1970**, *22*, 387–399. [PubMed]

42. Ownby, C.L.; Kainer, R.A.; Tu, A.T. Pathogenesis of hemorrhage induced by rattlesnake venom. An electron microscopic study. *Am. J. Pathol.* **1974**, *76*, 401–414. [PubMed]

43. Ownby, C.L.; Bjarnason, J.B.; Tu, A.T. Hemorrhagic toxins from rattlesnake (*Crotalus atrox*) venom. Pathogenesis of hemorrhage induced by three purified toxins. *Am. J. Pathol.* **1978**, *93*, 201–218. [PubMed]

44. Ownby, C.L.; Geren, C.R. Pathogenesis of hemorrhage induced by hemorrhagic proteinase IV from timber rattlesnake (*Crotalus horridus horridus*) venom. *Toxicon* **1987**, *25*, 517–526. [PubMed]

45. Moreira, L.; Borkow, G.; Ovadia, M.; Gutiérrez, J.M. Pathological changes induced by BaH1, a hemorrhagic proteinase isolated from *Bothrops asper* (Terciopelo) snake venom, on mouse capillary blood vessels. *Toxicon* **1994**, *32*, 976–987. [CrossRef]

46. Anderson, S.G.; Ownby, C.L. Pathogenesis of hemorrhage induced by proteinase H from eastern diamondback rattlesnake (*Crotalus adamanteus*) venom. *Toxicon* **1997**, *35*, 1291–1300. [CrossRef]

47. Gutiérrez, J.M.; Núñez, J.; Escalante, T.; Rucavado, A. Blood flow is required for rapid endothelial cell damage induced by a snake venom hemorrhagic metalloproteinase. *Microvasc. Res.* **2006**, *71*, 55–63. [PubMed]

48. Gonçalves, L.R.C.; Mariano, M. Local haemorrhage induced by *Bothrops jararaca* venom: Relationship to neurogenic inflammation. *Mediat. Inflamm.* **2000**, *9*, 101–107.

49. Gutiérrez, J.M.; Rucavado, A.; Escalante, T.; Díaz, C. Hemorrhage induced by snake venom metalloproteinases: Biochemical and biophysical mechanisms involved in microvessel damage. *Toxicon* **2005**, *45*, 997–1011. [CrossRef] [PubMed]

50. Gallin, J.I.; Snyderman, R. *Inflammation: Basic Principles and Clinical Correlates*; Lippincot: Philadelphia, PA, USA, 1999.

51. Malucelli, B.E.; Mariano, M. The haemorrhagic exudate and its possible relationship to neurogenic inflammation. *J. Pathol.* **1980**, *130*, 193–200. [PubMed]

52. Lomonte, B.; Lundgren, J.; Johansson, B.; Bagge, U. The dynamics of local tissue damage induced by *Bothrops asper* snake venom and myotoxin II on the mouse cremaster muscle. *Toxicon* **1994**, *32*, 41–55. [PubMed]

53. Rucavado, A.; Lomonte, B.; Ovadia, M.; Gutiérrez, J.M. Local tissue damage induced by BaP1, a metalloproteinase isolated from *Bothrops asper* (Terciopelo) snake venom. *Exp. Mol. Pathol.* **1995**, *63*, 186–199. [CrossRef] [PubMed]

54. Lomonte, B.; Gutiérrez, J.M.; Borkow, G.; Tarkowski, A.; Hanson, L.Å. Activity of hemorrhagic metalloproteinase BaH-1 and myotoxin II from *Bothrops asper* snake venom on capillary endothelial cells *in vitro*. *Toxicon* **1994**, *32*, 505–510. [CrossRef]

55. Borkow, G.; Gutiérrez, J.M.; Ovadia, M. *In vitro* activity of BaH1, the main hemorrhagic toxin of *Bothrops asper* snake venom on bovine endothelial cells. *Toxicon* **1995**, *33*, 1387–1391. [CrossRef]

56. Wu, W.B.; Chang, S.C.; Liau, M.Y.; Huang, T.F. Purification, molecular cloning and mechanism of action of graminelysin I, a snake-venom-derived metalloproteinase that induces apoptosis of human endothelial cells. *Biochem. J.* **2001**, *357*, 719–728. [CrossRef] [PubMed]

57. Masuda, S.; Ohta, T.; Kaji, K.; Fox, J.W.; Hayashi, H.; Araki, S. cDNA cloning and characterization of vascular apoptosis-inducing protein 1. *Biochem. Biophys. Res. Commun.* **2000**, *278*, 197–204. [CrossRef] [PubMed]

58. You, W.K.; Seo, H.J.; Chung, K.H.; Kim, D.S. A novel metalloprotease from *Gloydius halys* venom induces endothelial cell apoptosis through its protease and disintegrin-like domains. *J. Biochem.* **2003**, *134*, 739–749. [CrossRef] [PubMed]

59. Díaz, C.; Valverde, L.; Brenes, O.; Rucavado, A.; Gutiérrez, J.M. Characterization of events associated with apoptosis/anoikis induced by snake venom metalloproteinase BaP1 on human endothelial cells. *J. Cell. Biochem.* **2005**, *94*, 520–528. [CrossRef] [PubMed]

60. Tanjoni, I.; Weinlich, R.; Della-Casa, M.S.; Clissa, P.B.; Saldanha-Gama, R.F.; de Freitas, M.S.; Barja-Fidalgo, C.; Amarante-Mendes, G.P.; Moura-da-Silva, A.M. Jararhagin, a snake venom metalloproreinase, induces a specialized form of apoptosis (anoikis) selective to endothelial cells. *Apoptosis* **2005**, *10*, 851–861. [CrossRef] [PubMed]

61. Brenes, O.; Muñoz, E.; Roldán-Rodríguez, R.; Díaz, C. Cell death induced by *Bothrops asper* snake venom metalloproteinase on endothelial and other cell lines. *Exp. Mol. Pathol.* **2010**, *88*, 424–432. [CrossRef] [PubMed]

62. Wu, W.B.; Wang, T.F. Activation of MMP-2, cleavage of matrix proteins, and adherens junctions during a snake venom metalloproteinase-induced endothelial cell apoptosis. *Exp. Cell Res.* **2003**, *288*, 143–157. [CrossRef]

63. Baldo, C.; Lopes, D.S.; Faquim-Mauro, E.L.; Jacysyn, J.F.; Niland, S.; Eble, J.A.; Clissa, P.B.; Moura-da-Silva, A.M. Jararhagin disruption of endothelial cell anchorage is enhanced in collagen enriched matrices. *Toxicon* **2015**, *108*, 240–248. [CrossRef] [PubMed]

64. Milnor, W.R. Principles of hemodynamics. In *Medical Physiology*; Mosby: St. Louis, MO, USA, 1980; pp. 1017–1032.

65. Ballerman, B.J.; Dardik, A.; Eng, E.; Liu, A. Shear stress and the endothelium. *Kidney Int.* **1998**, *67*, S100–S108. [CrossRef]

66. Murphy, M.E.; Johnson, P.C. Possible contribution of basement membrane to the structural rigidity of blood capillaries. *Microvasc. Res.* **1975**, *9*, 242–245. [CrossRef]

67. Lee, J.; Schmid-Schönbein, G.W. Biomechanics of skeletal muscle capillaries: Hemodynamic resistance, endothelial distensibility, and pseudopod formation. *Ann. Biomed. Eng.* **1995**, *23*, 226–246. [CrossRef] [PubMed]

68. Escalante, T.; Rucavado, A.; Fox, J.W.; Gutiérrez, J.M. Key events in microvascular damage induced by snake venom hemorrhagic metalloproteinases. *J. Proteom.* **2011**, *74*, 1781–1794. [CrossRef] [PubMed]

69. Ohsaka, A.; Just, M.; Habermann, E. Action of snake venom hemorrhagic principles on isolated glomerular basement membrane. *Biochim. Biophys. Acta* **1973**, *323*, 415–438. [CrossRef]

70. Bjarnason, J.B.; Hamilton, D.; Fox, J.W. Studies on the mechanism of hemorrhage production by five proteolytic hemorrhagic toxins from *Crotalus atrox* venom. *Biol. Chem. Hoppe Seyler* **1988**, *369*, 121–129. [PubMed]

71. Baramova, E.N.; Shannon, J.D.; Bjarnason, J.B.; Fox, J.W. Degradation of extracellular matrix proteins by hemorrhagic metalloproteinases. *Arch. Biochem. Biophys.* **1989**, *275*, 63–71. [CrossRef]

72. Baramova, E.N.; Shannon, J.D.; Bjarnason, J.B.; Fox, J.W. Identification of the cleavage sites by a hemorrhagic metalloproteinase in type IV collagen. *Matrix* **1990**, *10*, 91–97. [CrossRef]

73. Baramova, E.N.; Shannon, J.D.; Fox, J.W.; Bjarnason, J.B. Proteolytic digestion of non-collagenous basement membrane proteins by the hemorrhagic metalloproteinase Ht-e from *Crotalus atrox* venom. *Biomed. Biochim. Acta* **1991**, *50*, 763–768. [PubMed]

74. Maruyama, M.; Sugiki, M.; Yoshida, E.; Shimaya, K.; Mihara, H. Broad substrate specificity of snake venom fibrinolytic enzymes: Possible role in haemorrhage. *Toxicon* **1992**, *30*, 1387–1397. [CrossRef]

75. Escalante, T.; Shannon, J.; Moura-da-Silva, A.M.; Gutiérrez, J.M.; Fox, J.W. Novel insights into capillary vessel basement membrane damage by snake venom hemorrhagic metalloproteinases: A biochemical and immunohistochemical study. *Arch. Biochem. Biophys.* **2006**, *455*, 144–153. [CrossRef] [PubMed]

76. Oliveira, A.K.; Paes-Leme, A.F.; Asega, A.F.; Camargo, A.C.; Fox, J.W.; Serrano, S.M.T. New insights into the structural elements involved in the skin haemorrhage induced by snake venom metalloproteinases. *Thromb. Haemost.* **2010**, *104*, 485–497. [CrossRef] [PubMed]

77. Bernardes, C.P.; Menaldo, D.L.; Camacho, E.; Rosa, J.C.; Escalante, T.; Rucavado, A.; Lomonte, B.; Gutiérrez, J.M.; Sampaio, S.V. Proteomic analysis of *Bothrops pirajai* snake venom and characterization of BpirMP, a new P-I metalloproteinase. *J. Proteom.* **2013**, *27*, 250–267. [CrossRef] [PubMed]

78. Rucavado, A.; Flores-Sanchez, E.; Franceschi, A.; Magalhaes, A.; Gutiérrez, J.M. Characterization of the local tissue damage induced by LHF-II, a metalloproteinase with weak hemorrhagic activity isolated from *Lachesis muta muta* snake venom. *Toxicon* **1999**, *37*, 1297–1312. [CrossRef]

79. Rodrigues, V.M.; Soares, A.M.; Guerra-Sá, R.; Rodrigues, V.; Fontes, M.R.; Giglio, J.R. Structural and functional characterization of neuwiedase, a nonhemorrhagic fibrin(ogen)olytic metalloprotease from *Bothrops neuwiedi* snake venom. *Arch. Biochem. Biophys.* **2000**, *381*, 213–224. [CrossRef] [PubMed]

80. Escalante, T.; Ortiz, N.; Rucavado, A.; Sanchez, E.F.; Richardson, M.; Fox, J.W.; Gutiérrez, J.M. Role of collagens and perlecan in microvascular stability: Exploring the mechanism of capillary vessel damage by snake venom metalloproteinases. *PLoS ONE* **2011**, *6*. [CrossRef] [PubMed]

81. Kalluri, R. Basement membranes: Structure, assembly and role in tumor angiogenesis. *Nat. Rev. Cancer* **2003**, *3*, 422–433. [CrossRef] [PubMed]

82. Timpl, R.; Brown, J.C. The laminins. *Matrix Biol.* **1994**, *14*, 275–281. [CrossRef]

83. Yurchenko, P.D.; Amenta, P.S.; Patton, B.L. Basement membrane assembly, stability and activities observed through a developmental lens. *Matrix Biol.* **2004**, *22*, 521–538. [CrossRef] [PubMed]

84. Baldo, C.; Jamora, C.; Yamanouye, N.; Zorn, T.M.; Moura-da-Silva, A.M. Mechanisms of vascular damage by hemorrhagic snake venom metalloproteinases: Tissue distribution and *in situ* hydrolysis. *PLoS Negl. Trop. Dis.* **2010**, *4*. [CrossRef] [PubMed]

85. Herrera, C.; Escalante, T.; Voisin, M.B.; Rucavado, A.; Morazán, D.; Macêdo, J.K.; Calvete, J.M.; Sanz, L.; Nourshargh, S.; Gutiérrez, J.M.; *et al.* Tissue localization and extracellular matrix degradation by PI, PII and PIII snake venom metalloproteinases: Clues on the mechanisms of venom-induced hemorrhage. *PLoS Negl. Trop. Dis.* **2015**, *9*. [CrossRef] [PubMed]

86. Escalante, T.; Rucavado, A.; Pinto, A.F.; Terra, R.M.; Gutiérrez, J.M.; Fox, J.W. Wound exudate as a proteomic window to reveal different mechanisms of tissue damage by snake venom toxins. *J. Proteome Res.* **2009**, *8*, 5120–5131. [CrossRef] [PubMed]

87. Paes-Leme, A.F.; Sherman, N.E.; Smalley, D.M.; Sizukusa, M.O.; Oliveira, A.K.; Menezes, M.C.; Fox, J.W.; Serrano, S.M.T. Hemorrhagic activity of HF3, a snake venom metalloproteinase: Insights from the proteomic analysis of mouse skin and blood plasma. *J. Proteome Res.* **2012**, *11*, 279–291. [CrossRef] [PubMed]

88. Kuo, H.J.; Maslen, C.L.; Keene, D.R.; Glanville, R.W. Type VI collagen anchors endothelial basement membranes by interacting with type IV collagen. *J. Biol. Chem.* **1997**, *272*, 26522–26529. [CrossRef] [PubMed]

89. Iozzo, R.V. Basement membrane proteoglycans: From cellar to ceiling. *Nat. Rev. Mol. Cell. Biol.* **2005**, *6*, 646–656. [CrossRef] [PubMed]

90. Nikai, T.; Taniguchi, K.; Komori, Y.; Masuda, K.; Fox, J.W.; Sugihara, H. Primary structure and functional characterization of bilitoxin-1, a novel dimeric P-II snake venom metalloproteinase from *Agkistrodon bilineatus* venom. *Arch. Biochem. Biophys.* **2000**, *378*, 6–15. [CrossRef] [PubMed]

91. Camacho, E.; Villalobos, E.; Sanz, L.; Pérez, A.; Escalante, T.; Lomonte, B.; Calvete, J.J.; Gutiérrez, J.M.; Rucavado, A. Understanding structural and functional aspects of PII snake venom metalloproteinases: Characterization of BlatH1, a hemorrhagic dimeric enzyme from the venom of *Bothriechis lateralis*. *Biochimie* **2014**, *101*, 145–155. [CrossRef] [PubMed]

92. Moura-da-Silva, A.M.; Ramos, O.H.; Baldo, C.; Niland, S.; Hansen, U.; Ventura, J.S.; Furlan, S.; Butera, D.; Della-Casa, M.S.; Tanjoni, I.; *et al.* Collagen binding is a key factor for the hemorrhagic activity of snake venom metalloproteinases. *Biochimie* **2008**, *90*, 484–492. [CrossRef] [PubMed]

93. Tanjoni, I.; Evangelista, K.; Della-Casa, M.S.; Butera, D.; Magalhães, G.S.; Baldo, C.; Clissa, P.B.; Fernandes, I.; Eble, J.; Moura-da-Silva, A.M. Different regions of the class P-III snake venom metalloproteinase jararhagin are involved in binding to α2β1 integrin and collagen. *Toxicon* **2010**, *55*, 1093–1099. [CrossRef] [PubMed]

94. Serrano, S.M.T.; Jia, L.G.; Wang, D.; Shannon, J.D.; Fox, J.W. Function of the cysteine-rich domain of the hemorrhagic metalloproteinase atrolysin A: Targeting adhesion proteins collagen I and von Willebrand factor. *Biochem. J.* **2005**, *391*, 69–76. [CrossRef] [PubMed]

95. Serrano, S.M.T.; Kim, J.; Wang, D.; Dragulev, B.; Shannon, J.D.; Mann, H.H.; Veit, G.; Wagener, R.; Koch, M.; Fox, J.W. The cysteine-rich domain of snake venom metalloproteinases is a ligand for von Willebrand factor A domains: Role in substrate targeting. *J. Biol. Chem.* **2006**, *281*, 39746–39756. [CrossRef] [PubMed]

96. Serrano, S.M.T.; Wang, D.; Shannon, J.D.; Pinto, A.F.; Polanowska-Grabowska, R.K.; Fox, J.W. Interaction of the cysteine-rich domain of snake venom metalloproteinases with the A1 domain of von Willebrand factor promotes site-specific proteolysis of von Willebrand factor and inhibition of von Willebrand factor-mediated platelet aggregation. *FEBS J.* **2007**, *274*, 3611–3621. [CrossRef] [PubMed]

97. Baramova, E.; Shannon, J.D.; Bjarnason, J.B.; Gonias, S.L.; Fox, J.W. Interaction of hemorrhagic metalloproteinases with human α2-macroglobulin. *Biochemistry* **1990**, *29*, 1069–1074. [CrossRef] [PubMed]

98. Kamiguti, A.S.; Desmond, H.P.; Theakston, R.D.G.; Hay, C.R.; Zuzel, M. Ineffectiveness of the inhibition of the main haemorrhagic metalloproteinase from *Bothrops jararaca* venom by its only plasma inhibitor, α2-macrogloblin. *Biochim. Biophys. Acta* **1994**, *1200*, 307–314. [CrossRef]

99. Estêvão-Costa, M.I.; Diniz, C.R.; Magalhães, A.; Markland, F.S.; Sanchez, E.F. Action of metalloproteinases mutalysin I and II on several components of the hemostatic and fibrinolytic systems. *Thromb. Res.* **2000**, *99*, 363–376. [CrossRef]

100. Escalante, T.; Rucavado, A.; Kamiguti, A.S.; Theakston, R.D.G.; Gutiérrez, J.M. *Bothrops asper* metalloproteinase BaP1 is inhibited by α_2-macroglobulin and mouse serum and does not induce systemic hemorrhage or coagulopathy. *Toxicon* **2004**, *43*, 213–217. [CrossRef] [PubMed]

101. Kamiguti, A.S. Platelets as targets of snake venom metalloproteinases. *Toxicon* **2005**, *45*, 1041–1049. [CrossRef] [PubMed]

102. Kamiguti, A.S.; Hay, C.R.; Theakston, R.D.G.; Zuzel, M. Insights into the mechanism of haemorrhage caused by snake venom metalloproteinases. *Toxicon* **1996**, *34*, 627–642. [CrossRef]

103. Gutiérrez, J.M.; Rucavado, A.; Escalante, T.; Lomonte, B.; Angulo, Y.; Fox, J.W. Tissue pathology induced by snake venoms: How to understand a complex pattern of alterations from a systems biology perspective? *Toxicon* **2010**, *55*, 166–170. [CrossRef] [PubMed]

104. Markland, F.S. Snake venoms and the hemostatic system. *Toxicon* **1998**, *36*, 1749–1800. [CrossRef]

105. White, J. Snake venoms and coagulopathy. *Toxicon* **2005**, *45*, 951–967. [CrossRef] [PubMed]

106. Mackessy, S.P. *Handbook of Venoms and Toxins of Reptiles*; CRC Press: Boca Raton, FL, USA, 2010.

107. Loría, G.D.; Rucavado, A.; Kamiguti, A.S.; Theakston, R.D.G.; Fox, J.W.; Alape, A.; Gutiérrez, J.M. Characterization of "basparin A", a prothrombin-activating metalloproteinase, from the venom of the snake *Bothrops asper* that inhibits platelet aggregation and induces defibrination and thrombosis. *Arch. Biochem. Biophys.* **2003**, *418*, 13–24. [CrossRef]

108. Malbranque, S.; Piercecchi-Marti, M.D.; Thomas, L.; Barbey, C.; Courcier, D.; Bucher, B.; Ridrch, A.; Smadja, D.; Warrell, D.A. Fatal diffuse thrombotic microangiopahy after a bite by the "Fer-de-Lance" pit viper (*Bothrops lanceolatus*) of Martinique. *Am. J. Trop. Med. Hyg.* **2008**, *78*, 856–861. [PubMed]

109. Santoro, M.L.; Sano-Martins, I.S. Platelet dysfunction during *Bothrops jararaca* snake envenomation in rabbits. *Thromb. Haemost.* **2004**, *92*, 369–383. [CrossRef] [PubMed]

110. Rucavado, A.; Soto, M.; Kamiguti, A.S.; Theakston, R.D.G.; Fox, J.W.; Escalante, T.; Gutiérrez, J.M. Characterization of aspercetin, a platelet aggregating component from the venom of the snake *Bothrops asper* which induces thrombocytopenia and potentiates metalloproteinase-induced hemorrhage. *Thromb. Haemost.* **2001**, *85*, 710–715. [PubMed]

111. Calvete, J.J.; Marcinkiewicz, C.; Monleón, D.; Esteve, V.; Celda, B.; Juárez, P.; Sanz, L. Snake venom disintegrins: Evolution of structure and function. *Toxicon* **2005**, *45*, 1063–1074. [CrossRef] [PubMed]

112. Lu, Q.; Navdaev, A.; Clemetson, J.M.; Clemetson, K.J. Snake venom C-type lectins interacting with platelet receptors. Structure-function relationships and effects on hemostasis. *Toxicon* **2005**, *45*, 1089–1098. [CrossRef] [PubMed]

113. Sano-Martins, I.S.; Santoro, M.L.; Castro, S.C.; Fan, H.W.; Cardoso, J.L.C.; Theakston, R.D.G. Platelet aggregation in patients bitten by the Brazilian snake *Bothrops jararaca*. *Thromb. Res.* **1997**, *87*, 183–195. [CrossRef]

114. Rucavado, A.; Soto, M.; Escalante, T.; Loría, G.D.; Arni, R.; Gutiérrez, J.M. Thrombocytopenia and platelet hypoaggregation induced by *Bothrops asper* snake venom. Toxins involved and their contribution to metalloproteinase-induced pulmonary hemorrhage. *Thromb. Haemost.* **2005**, *94*, 123–131. [CrossRef] [PubMed]

115. Tu, A.T.; Hendon, R.R. Characterization of lizard venom hyaluronidase and evidence for its action as a spreading factor. *Comp. Biochem. Physiol. B* **1983**, *76*, 377–383. [CrossRef]

116. Kemparaju, K.; Girish, K.S.; Nagaraju, S. Hyaluronidases, a neglected class of glycosidases from snake venom. Beyond a spreading factor. In *Handbook of Venoms and Toxins of Reptiles*; CRC Press: Boca Raton, FL, USA, 2010; pp. 237–258.

117. Bustillo, S.; García-Denegri, M.E.; Gay, C.; van de Velde, A.C.; Acosta, O.; Angulo, Y.; Lomonte, B.; Gutiérrez, J.M.; Leiva, L. Phospholipase A$_2$ enhances the endothelial cell detachment effect of a snake venom metalloproteinase in the absence of catalysis. *Chem. Biol. Interact.* **2015**, *240*, 30–36. [CrossRef] [PubMed]

toxins

MDPI

Review

Metalloproteases Affecting Blood Coagulation, Fibrinolysis and Platelet Aggregation from Snake Venoms: Definition and Nomenclature of Interaction Sites

R. Manjunatha Kini * and Cho Yeow Koh

Protein Science Laboratory, Department of Biological Sciences, Faculty of Science, 14 Science Drive 4, National University of Singapore, Singapore 117543, Singapore; choyeow@nus.edu.sg
* Correspondence: dbskinim@nus.edu.sg; Tel.: +65-6874-5235; Fax: +65-6779-2486

Academic Editors: Jay Fox and José María Gutiérrez
Received: 8 September 2016; Accepted: 22 September 2016; Published: 29 September 2016

Abstract: Snake venom metalloproteases, in addition to their contribution to the digestion of the prey, affect various physiological functions by cleaving specific proteins. They exhibit their activities through activation of zymogens of coagulation factors, and precursors of integrins or receptors. Based on their structure–function relationships and mechanism of action, we have defined classification and nomenclature of functional sites of proteases. These metalloproteases are useful as research tools and in diagnosis and treatment of various thrombotic and hemostatic conditions. They also contribute to our understanding of molecular details in the activation of specific factors involved in coagulation, platelet aggregation and matrix biology. This review provides a ready reference for metalloproteases that interfere in blood coagulation, fibrinolysis and platelet aggregation.

Keywords: procoagulant; anticoagulant; factor X activator; prothrombin activator; platelet aggregation; fibrinolytic; exosites in enzymes; allosteric sites

1. Introduction

Snake venoms are cocktails of pharmacologically active proteins and peptides. They are used as offensive weapons in immobilizing, killing and digesting the preys [1,2]. Some of these toxins exhibit various enzymatic activities, whereas others are nonenzymatic proteins. Most enzymes found in snake venoms are hydrolases that breakdown biological molecules including proteins, nucleic acids and phospholipids. In addition to their contribution to the digestion of the prey, a number of these hydrolases exhibit specific pharmacological effects. Snake venoms, particularly crotalid and viperid venoms, are rich sources of metalloproteases and serine proteases.

Snake venom metalloproteases (SVMPs) are Zn^{2+}-dependent, endoproteolytic enzymes that are classified into three different classes: P-I, P-II and P-III [3,4]. They are closely related to ADAM (a disintegrin and metalloprotease) family proteins and are included in the M12B clan [5]. SVMPs selectively cleave a small number of key proteins in the blood coagulation cascade and in platelet aggregation. Such limited proteolysis leads to either activation or inactivation of the protein involved in these processes, thus resulting in interference in blood coagulation and platelet aggregation (Figures 1 and 2). This review provides an overview on a number of metalloproteases that interfere in blood coagulation, fibrinolysis and platelet aggregation.

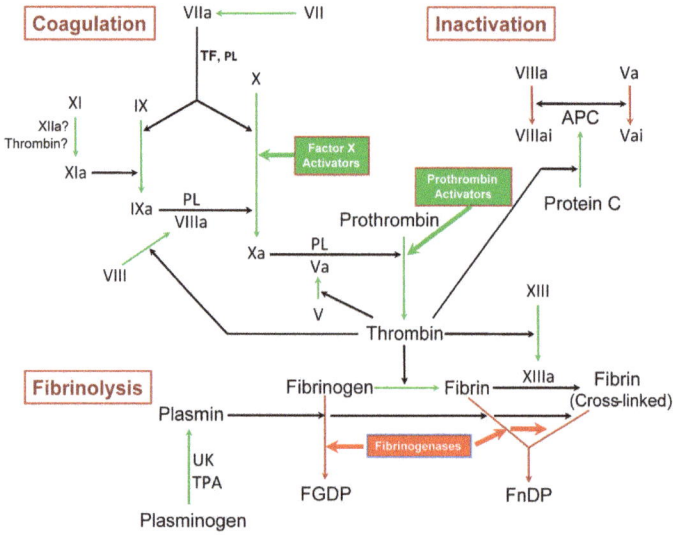

Figure 1. Snake venom metalloproteases affecting blood coagulation. Proteinases interfere by proteolysis of specific factors (thick arrow heads). Green boxes, procoagulant SVMPs; red box, fibrinogenases that cleave fibrinogen and fibrin; APC, activated protein C; FGDP, fibrinogen degradation products; FnDP, fibrin degradation products; PL, phospholipids; TF, tissue factor; TPA, tissue plasminogen activator; UK, urokinase.

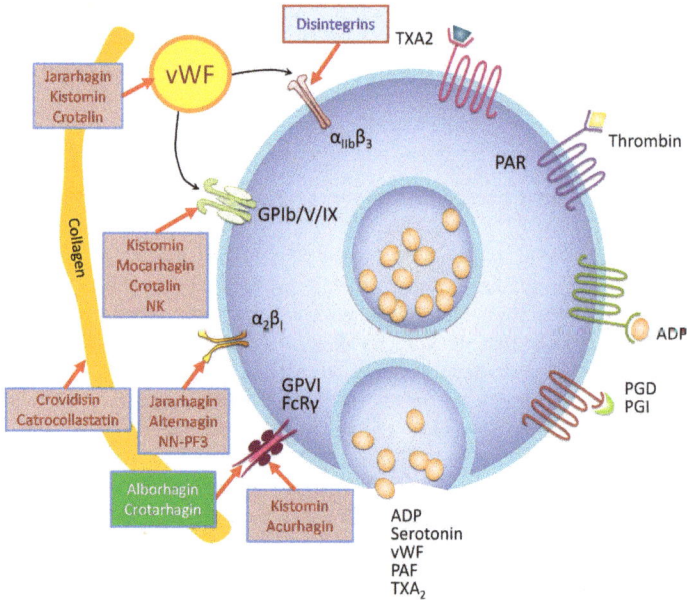

Figure 2. Snake venom metalloproteases affecting platelet aggregation. Proteinases that induce or inhibit platelet aggregation are shown in green or red boxes, respectively; Disintegrins that inhibit platelet aggregation are shown in blue box; PAF, platelet activating factor; PAR, protease activated receptor; PGD, prostaglandin D; PGI, prostaglandin I; TXA2, thromboxane A_2.

2. Procoagulant Proteases

Blood coagulation factors circulate as zymogens and they get activated through limited proteolytic cleavage during the breach of the blood vessel in a sequential manner leading to formation of fibrin clot that stops the blood leakage. All procoagulants from snake venoms characterized to date are proteases; they activate a zymogen of specific coagulation factors in the coagulation cascade and hasten clot formation. Unlike snake venom serine proteases, which activate various zymogens in the coagulation cascade (for reviews, see [6,7]), SVMPs activate only two key coagulation factors, factor X (FX) and prothrombin to exhibit their procoagulant effects.

2.1. Factor X Activators

Venoms from Viperidae, Crotalidae and Elapidae contain a variety of proteases capable of activating factor X (for reviews, see [8,9]). They are either metalloproteases or serine proteases. In general, metalloprotease FX activators are found in Viperidae and Crotalidae venoms [10–12], while serine protease FX activators are found in Elapidae venoms [13–15]. All the metalloprotease FX activators have two subunits held together by inter-subunit disulfide linkage; larger subunit is a P-III metalloprotease whereas the smaller subunit is a snaclec (snake C-type lectin-related proteins) with two chains covalently linked by an inter-chain disulfide bond. FX activator from Russell's viper (*Daboia russelli*) venom (RVV-X) is the well-characterized protein (for details, see [8]). As with other P-III enzymes, RVV-X possesses metalloprotease (M), disintegrin-like (D) and cysteine-rich (C) domains. The smaller subunit is a typical C-type lectin-related dimer and contributes to FX selectivity by binding to the γ-carboxy glutamate residues containing Gla domain of FX. Similar to physiological activators, intrinsic tenase (FIXa-FVIIIa) and extrinsic tenase (FVIIa-tissue factor) complexes, RVV-X activates FX by a proteolytic cleavage of Arg152-Ile153 bond resulting in the release of a 52-residue activation peptide and the activated FXaα [16,17]. *Bothrops atrox* activators, however, produce two other cleavages: one near the *N*-terminal end of the heavy chain of FX, generating FXμ, and a second one located at one extremity of the heavy chain of FXaα, generating FXaν [12].

Structural studies of RVV-X and other related P-III enzymes [18–23] help elucidate their structure–function relationship. The three domains of P-III SVMPs are arranged into a C-shaped configuration, with the *N*-terminal M domain interacting with C-terminal C domain (Figure 3A). One of the exceptions is kaouthiagin-like protease from *Naja atra*, which adopts a more elongated conformation due to the absence of a 17-residue segment and to a different disulfide bond pattern in the D domain [22] (Figure 3B). Other than variations in the peripheral loops, the structures of M domain among P-III [18–23], P-I [24–34] and P-II [35] enzymes are similar. M domains are folded as a five-stranded β-sheet interspersed with five α-helices into two subdomains flanking the catalytic cleft in which a zinc ion is localized. The conserved Zn^{2+}-binding HEXXHXXGXXHD motif is located at the bottom of the catalytic cleft. The catalytic Zn^{2+} ion is coordinated by the Nε atoms of three His side chains within the consensus motif (underlined) in addition to a solvent water molecule, which in turn is bound to the conserved Glu (italic). The identity of fourth ligand as water is ascertained by quantum mechanical and molecular mechanical simulations [36]. The D domain has two sub-domains named the "shoulder" (D_s) and the "arm" (D_a) (Figure 3). The bound Ca^{2+} ions and disulfide bonds in this domain are essential for the rigidity of the C-shaped since it lacks other secondary structural elements [37]. The D_a subdomain folds similar to disintegrin [38] with some variations in the RGD-containing disintegrin (D)-loop and the C-terminal region. Although the D-loop of disintegrin is thought to be involved in integrin-binding, it is not accessible for interaction in P-III enzymes as it packs against the C domain. The C domain of P-III SVMPs can be divided into two subdomains, the "wrist" (C_W) at the *N*-terminal, and the "hand" (C_h) towards the C-terminal. The C_w subdomain extends from D_s and D_a subdomains to form the C-shaped arm structure while the C_h subdomain forms a separated core of made of a unique α/β -fold structure (Figure 3). Within the C_h subdomain, a hyper-variable region (HVR) can be identified and may function in specific protein–protein interactions [18].

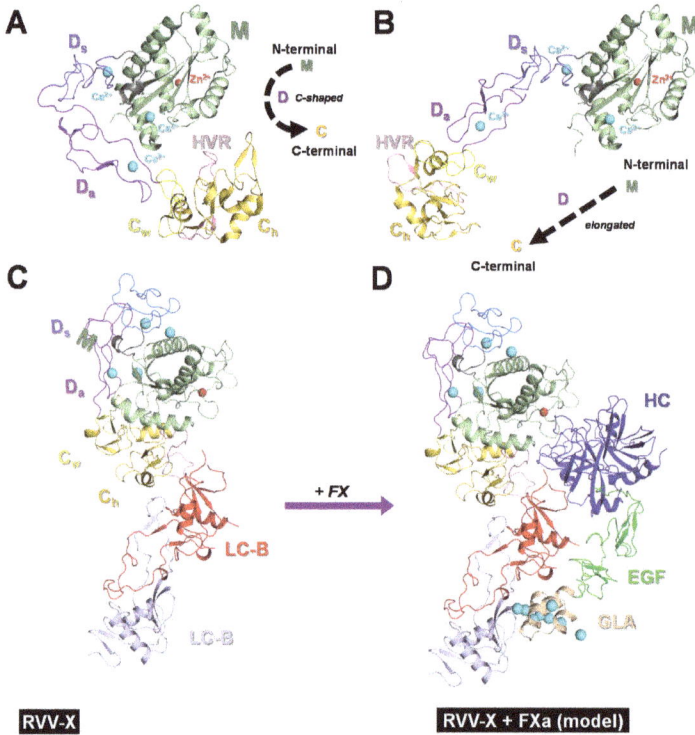

Figure 3. Structure of P-III snake venom metalloproteases. All proteins are shown as ribbon structures. Zn^{2+} and Ca^{2+} ions are shown as red and light blue spheres, respectively. Subdomains and segments are colored and named. (**A**) Catrocollastatin, an inhibitor of collagen-induced platelet aggregation prothrombin activator and a P-III SVMP, showing M, D and C domains, which form a C-shaped configuration (*inset*). (**B**) Kaouthiagin-like protease, in contrast exhibits straight configuration. The presence of "hinges" between the domains help P-III SVMPs to "open" and exhibit straighter configuration. (**C**) Ribbon structure of RVV-X, a P-IIId SVMP. Carinactivase and mutactivase, prothrombin activators, also belong to this class. (**D**) Docking model (prepared by Soichi Takeda) depicting the structural basis of FX activation.

P-IIId SVMPs is a subgroup that has additional subunits forming larger complexes. For example, RVV-X is a P-IIId complex [38] consisting of an MDC-containing heavy chain and two light chains of snaclec (Figure 3C). It has a hook-spanner-wrench-like architecture, in which the MD domains of the heavy chain resemble a hook, and the remainder of the molecule constitutes a handle [19]. A disulfide bridge between the Cys389 of heavy chain and Cys133 of light chain A links the two chains. Multiple hydrophobic interactions and hydrogen bonds further stabilize the interface. Light chains A and B are linked via a disulfide bond between Cys79 and Cys77 of the respective chains. The dimeric interface formed by the two snaclecs light chains is a concave structure similar to the ligand-binding site of factor IX/X binding protein [39]. This concave surface is likely to function as an exosite that binds to the gamma-carboxyglutamic acid-rich (Gla) domain of FX in the presence of Ca^{2+} [19]. A docking model indicates that the C_h/light chain portion may act as a scaffold to accommodate the elongated FX molecule. Ca^{2+} is likely to induce conformational changes in the Gla domain of FX, which might be necessary for the RVV-X recognition [17], consistent with the original proposal [8]. RVV-X is an example of venom complex that has evolved to target specific proteins in the blood coagulation cascade and to cause immediate toxicity to the vertebrate prey by coagulating its blood.

2.2. Prothrombin Activators

A large number of snake species contain prothrombin activators in their venoms (for an inventory, see [40], and for reviews, see [41–46]). Based on their structural properties, functional characteristics and cofactor requirements, they have been categorized into four groups [40,47,48]. Groups A and B prothrombin activators are metalloproteases and they convert prothrombin to meizothrombin. In contrast, groups C and D prothrombin activators are serine proteases and they convert prothrombin to mature thrombin. Here I will discuss some of the salient features of only groups A and B prothrombin activators. For more details, readers are advised to read recent reviews on prothrombin activators [44–46].

2.2.1. Group A Prothrombin Activators

These metalloproteases efficiently activate prothrombin without the requirement of any cofactors, such as Ca^{2+} ions, phospholipids or FVa [40,41]. They are found in several viper venoms and resistant to the natural endogenous coagulation inhibitors, such as serpins and antithrombin III [47]. They probably play the role of toxins in the venom. The best characterized Group A activator is ecarin, isolated from the venom of the saw-scale viper *Echis carinatus* [49]. The mature protein is a metalloprotease with 426 amino acids and shares 64% identity with the heavy chain of RVV-X [50]. Ecarin is also a P-III enzyme with MDC domains. In the disintegrin-like domain, the RGD tripeptide sequence is replaced by RDD sequence. Consequently, ecarin has no inhibitory effect on platelet aggregation. Ecarin is a highly efficient enzyme with a low Km for prothrombin and a high kcat. It cleaves the Arg_{320}–Ile_{321} bond in prothrombin and produces meizothrombin. Meizothrombin is ultimately converted to α-thrombin by autolysis. Ecarin can also activate descarboxyprothrombin that accumulates in plasma during warfarin therapy. Other prothrombin activators in this class [40,41], for example, those isolated from the *Bothrops* species [51], also have similar properties. In contrast, serine proteases that activate prothrombin (groups C and D) cleave at both Arg_{271}–Thr_{272} and Arg_{320}–Ile_{321} bonds of prothrombin [52–55], converting it to mature thrombin. Structural details of other Group A prothrombin activators are not available.

2.2.2. Group B Prothrombin Activators

In 1996, Yamada et al. [47] isolated and characterized carinactivase-1, another prothrombin activator from *E. carinatus* venom. In contrast to ecarin and other Group A prothrombin activators, this proteinase activity was Ca^{2+}-dependent. Similar to RVV-X, carinactivase-1 consists of two subunits held covalently through a disulfide bond: a 62 kDa P-III metalloprotease and a 25 kDa snaclec dimer linked by disulfide bridge. The snaclec subunit is homologous to the factor IX/X-binding protein from *Trimeresurus flavoviridis* venom [8,56]. Carinactivase-1 required millimolar concentrations of Ca^{2+} for its activity and had virtually no activity in the absence of Ca^{2+} ions. The light chains contribute to the specificity as well as Ca^{2+} dependency of Carinactivase-1. Therefore, unlike ecarin, Carinactivase-1 does not activate prothrombin derivatives, prethrombin-1 and descarboxyprothrombin, in which Ca^{2+}-binding has been perturbed. Based on this property, Yamada and Morita [57] developed a chromogenic assay for normal prothrombin in the plasma of warfarin-treated individuals. Functionally, the metalloprotease subunit by itself is similar to ecarin: it no longer requires Ca^{2+} for activity. Reconstitution of the snaclec subunit restores Ca^{2+} dependence. Prothrombin activation by carinactivase-1 is inhibited by prothrombin fragment 1, and the isolated snaclec subunit is capable of binding to fragment 1 in the presence of Ca^{2+} ions. Hence this protein recognizes the Ca^{2+}-bound conformation of the Gla domain in prothrombin via the 25 kDa regulatory subunit, and the subsequent conversion of prothrombin is catalyzed by the 62-kDa catalytic subunit. Subsequently, another prothrombin activator multactivase in *Echis multisquamatus* venom, which had very similar properties to carinactivase-1 was characterized [58]. Similar to Group A prothrombin activators, these enzymes also produce meizothrombin.

3. Fibrinolytic Enzymes

Fibrinogen is cleaved by both venom serine proteases and metalloproteases. Interestingly, serine proteases cleave the N-terminal end of the Aα or Bβ chains of fibrinogen releasing fibrinopeptide A or B, respectively, unlike thrombin, which releases both peptides [59,60]. These thrombin-like enzymes (TLEs) were isolated and characterized from venoms of pit vipers (*Agkistrodon*, *Bothrops*, *Lachesis* and *Trimeresurus*), true vipers (*Bitis* and *Cerastes*) and colubrids, *Dispholidus typus* (for an inventory and reviews, see [60–62]). Although classical serine protease inhibitors inhibit TLEs, most are not inhibited by thrombin inhibitors like antithrombin III and hirudin [59,60,63]. TLEs usually form friable and translucent clots presumably due to lack of crosslinking of fibrin by FXIIIa. In contrast, SVMPs selectively cleave the Aα chain of fibrinogen but not cleave Bβ and γ chains and thus classified as α-fibrinogenases [64–70]. They cleave at the C-terminal end of the Aα chain produce truncated fibrinogen, which is unable to form a stable fibrin clot, and thus inhibit blood coagulation. These SVMPs belong to all three classes, P-I, P-II and P-III. Unlike TLEs, these SVMPs also exhibit fibrinolytic activity. Thus, they may have clinical applications in the treatment of occlusive thrombi [71,72].

4. Platelet Aggregation Antagonists

Some α-fibrinogenases, described above, inhibit platelet aggregation [73,74]. Because of their ability to degrade fibrinogen, the antiplatelet effects of fibrinolytic enzymes were suggested to be caused by the formation of inhibitory fibrinogen degradation products [73,75,76]. Subsequent studies, however, showed that the degradation products of fibrinogen produced by either the α-fibrinogenase from *A. rhodostoma* venom or by plasmin do not show antiplatelet effects comparable to the protease [74,77]. Thus, the α-fibrinogenase was proposed to inhibit aggregation by elimination of the intact form of the adhesive molecule fibrinogen [74]. Interestingly, only a small number of but not all fibrinogenases inhibit platelet aggregation. Thus, the role of fibrinogen degradation in the inhibition of platelet aggregation by α-fibrinogenases was questionable. Our studies using F1-proteinase, an α-fibrinogenase from *Naja nigricollis* venom, showed that the degradation products of fibrinogen formed by this protease failed to inhibit platelet aggregation [78]. This SVMP inhibits platelet aggregation in washed platelets and in platelets that were reconstituted with defibrinogenated plasma. Thus, the inhibition of platelet aggregation by proteinase F1 is independent of its action on fibrinogen [78]. We speculated that the inhibition could be due to either binding to or hydrolysis of a plasma factor, or to accumulation of inhibitory peptides formed during the hydrolysis of a plasma factor other than fibrinogen.

In 1992, Huang et al. purified a P-I SVMP from *Agkistrodon rhodostoma* (=*Calloselsma rhodostoma*) venom that inhibited platelet aggregation [79]. It inhibited aggregation induced by low concentrations of thrombin (≤0.2 U/mL) with only slight effect on aggregation induced by high concentrations of thrombin (≥0.5 U/mL) [80]. This enzyme, named Kistomin, significantly inhibited cytosolic Ca^{2+} rise, completely blocked formation of thromboxane B2 and inositol phosphates in platelets stimulated by 0.1 U/mL of thrombin. In contrast, it inhibited significantly thromboxane but not inositol phosphates formation of platelets stimulated by a high concentration of thrombin (1 U/mL). They showed that incubation of platelets with kistomin resulted in a selective cleavage of platelet membrane glycoprotein Ib (GPIb) [80]. These results suggested that (a) kistomin is a highly selective SVMP that cleaves GPIb; and (b) thrombin activates platelets at least through two receptors; GPIb and a second receptor. Intact GPIb plays critical role in the extent and rate of platelet aggregation stimulated by low concentrations of thrombin [80]. Kistomin cleaves platelet GPIbα at two distinct sites releasing 45- and 130-kDa soluble fragments and specifically inhibits von Willebrand factor- (vWF-) induced platelet aggregation [81]. Kistomin also cleaves vWF resulting in the formation of low-molecular-mass multimers. It inhibits GPIbα agonist-induced platelet aggregation, and prolongs the occlusion time in mesenteric microvessels and tail-bleeding time in mice [81]. Kistomin also inhibits platelet aggregation induced by collagen and convulxin (Glycoprotein VI (GPVI) [82]. It cleaves GPVI but not integrins $\alpha_2\beta_1$ and $\alpha_{IIb}\beta_3$. The release of 25- and 35-kDa fragments from GPVI suggests that kistomin cleaved GPVI near the mucin-like region. Hsu et al. identified that kistomin cleaves Glu_{205}-Ala_{206} and Val_{218}-Phe_{219} peptide bonds using synthetic peptides [82]. Thus, P-I SVMP kistomin specifically targets receptors GPIbα and GPVI on platelets and vWF in the plasma to exhibit its effects on platelet aggregation. Kistomin may be useful for studying metalloprotease-substrate interactions and has a potential being developed as an antithrombotic agent. Huang and colleagues also characterized crotalin, a P-I SVMP from venom of *Crotalus atrox* that also cleaves vWF and GPIbα [83].

Mocarhagin, a 55-kDa SVMP from *Naja mocambique mocambique* (=*Naja mossabica*) cleaves GPIbα [84]. The GPIbα fragment cleaved by this SVMP, His_1-Glu_{282} was useful in identifying the thrombin-binding site; the sulfated tyrosine/anionic segment $Y_{276}DYYPEE_{282}$ are important for the binding of thrombin and the botrocetin-dependent binding of vWF [84]. Interestingly, mocarhagin cleaves a 10-amino acid residue peptide from the N-terminus of P-selectin glycoprotein ligand receptor (PSGL-1) expressed on neutrophils to abolish P-selectin binding on endothelial cells and prevents rolling of neutrophils [85]. In both cases, mocarhagin targets mucin-like substrates (GPIbα and PSGL-1) within anionic amino acid sequences containing sulfated tyrosines. Brendt and colleagues showed the presence of SVMPs that are immunologically and functionally similar to mocarhagin in *N. kaouthia* (*N. siamensis*), *N. nivea* (*N.* flava), *N. nigricollis crawshawii*, *N. nigricollis pallida*, *N. nigricollis nigricollis*, *N. atra*, *N. haje*, *N. naja*, *N. melanoleuca* and *N. oxiana*, but not in *N. sputatrix* venoms [86]. They also developed a simple method for purification of SVMPs using Ni^{2+}-agarose column and purified Nk from *Naja kaouthia* venom that cleaves GPIbα [87]. During the subsequent studies, same group found out that nerve growth factor (NGF) binds to Ni^{2+}-agarose column and NGF is co-purified with SVMPs [88]. They showed venom NGF and human NGF inhibits both SVMPs and human MPs.

Interestingly, another distinct P-III SVMP, NN-PF3, that inhibits platelet aggregation was purified and characterized from *Naja naja* venom [89]. NN-PF3, unlike the above *Naja* SVMPs, fails to inhibit ristocetin-induced platelet aggregation. Instead, it inhibits collagen-induced aggregation of washed platelets [89]. Western blot using anti-integrin $\alpha_2\beta_1$ mAb 6F1 suggested that NN-PF3 binds to $\alpha_2\beta_1$ integrin in a sequence-dependent manner only but does not cleave $\alpha_2\beta_1$ integrin. However, there is a drastic reduction in several intracellular signaling [89]. Further mechanistic details and structure–function relationships of NN-PF3 may help delineate the differences in the targeting of *Naja* SVMPs.

Jararhagin from *Bothrops jararaca* (Brazilian pit-viper) venom is a P-III SVMP with MDC domains [90]. The RGD tripeptide sequence in the D domain is replaced by ECD sequence. Jararhagin cleaves the C-terminal part of fibrinogen Aα chains, resulting in the removal of a 23 kDa fragment while leaving the β and γ chains unaffected [91]. The cleaved fibrinogen molecule is still fully functional in both platelet aggregation responses to ADP and adrenalin and in its ability to clot plasma by thrombin. However, the fibrin polymerization is abnormal [91]. Jararhagin inhibits both ristocetin- and collagen-induced platelet aggregations. The inhibition of ristocetin-induced platelet aggregation is attributed to a direct cleavage of vWF rather than its receptor GPIb-IX-V [92]. The cleavage vWF occurs in the N-terminal half, which contains the binding site for the GPIb receptor, the AI domain. Hydrolysis of vWF leads to the disappearance of the high molecular size multimeric structure of vWF and loss of platelet responses [92]. Ivaska et al. designed a series of eight short cyclic peptides corresponding to hydrophilic and charged regions along the protein sequence to identify the α2I binding site [93]. The peptide spanning $C^*_{241}TRKKHDNAQ_{249}C^*$ (*Cys residues form the disulfide bond) binds to α2I domain and interferes with the interaction between α2I domain and collagen. Using Ala scanning method, they identified the importance of RKK tripeptide sequence for this interaction [93]. Finally they developed a shorter, more potent version of this peptide C*TRKKHDC* which inhibits α2I domain and collagen interaction with an IC_{50} of 1.3 mM. These peptides bind near the metal ion-dependent adhesion site of the human integrin α_2I-domain [94]. The peptide C*TRKKHDC* competes for the collagen-binding site of α_2I but does not induce a large scale conformational rearrangement of the I domain [95].

In contrast, the inhibition of collagen-induced aggregation is driven by interference with the $\alpha_2\beta_1$ integrin, but not GPVI receptor [96]. However, treatment of platelets with jararhagin drastically reduces $\alpha_2\beta_1$ integrin on the platelet surface [92,97]. The effect was attributed both to binding to the α_2I domain [97] and to cleavage of the $\alpha_2\beta_1$ integrin [92,98]. The degradation of the β_1 subunit of $\alpha_2\beta_1$ by jararhagin results in the loss of $pp72^{syk}$ phosphorylation and thus β_1 subunit appears to be critically involved in collagen-induced platelet signaling [99]. Using recombinant fragments and monoclonal antibodies, Tanjoni et al. showed that jararhagin binding to collagen and $\alpha_2\beta_1$ integrin occurs by two independent motifs, which are located on D and C domains, respectively [99]. The roles of non-enzymatic domains in platelet aggregation are discussed below.

In addition to jararhagin (described above), several other P-III SVMPS, such as atrolysin A from *Crotalus atrox* venom [100], catrocollastatin from *Crotalus atrox* venom [101], crovidisin from *Crotalus viridis* venom [102], alternagin from *Bothrops alternatus* venom [103], acurhagin from *Agkistrodon acutus* venom [104], halydin (D domain from a P-III) from *Gloydius halys* venom [105] and kaouthiagin from *Naja kaouthia* venom [106] inhibit collagen-induced platelet aggregation. Mechanistically, these SVMPs bind and/or proteolytically cleave vWF, collagen, GPVI or $\alpha_2\beta_1$. Interestingly, acurhagin (87% identity with jararhagin) selectively inhibits platelet aggregation induced by collagen and suppresses tyrosine phosphorylation of several signaling proteins in convulxin-stimulated platelets [104]. Thus, acurhagin exhibits its function mainly through its binding to GPVI and collagen, instead of binding to $\alpha_2\beta_1$, or cleaving platelet membrane glycoproteins [104]. Recently, a P-I SVMP from *Bothrops barnetti* venom that inhibits platelet aggregation induced by vWF *plus* ristocetin and collagen was characterized [107]. It presumably cleaves both vWF and GPIb and thus, inhibits vWF-induced platelet aggregation. It also cleaves the collagen-binding α_2A domain of $\alpha_2\beta_1$ integrin and thus, inhibits collagen-induced platelet aggregation [107]. Despite the missing D and C domains, this P-I SVMP has similar properties compared jararhagin, a P-III SVMP. Such examples will help us understand subtleties in structure–function relationships.

5. Platelet Aggregation Agonists

A small number of SVMPs have been shown to induce platelet aggregation. Alborhagin, a P-III SVMP isolated from *Trimeresurus albolabris* venom activates platelets through a mechanism involving GPVI [108]. It induces similar tyrosine phosphorylation pattern [108] to convulxin, a GPVI agonist [109–111]. Interestingly, alborhagin has minimal effect on convulxin binding to GPVI-expressing cells, suggesting that these proteins may recognize distinct binding sites on GPVI. Both alborhagin and crotarhagin from *Crotalus horridus horridus* venom induce platelet aggregation [112]. They induce ectodomain shedding of GPVI by a mechanism that involves activation of endogenous platelet metalloproteases. This shedding of 55-kDa soluble GPVI fragment required GPVI-dependent platelet activation [112].

6. Role of Non-Enzymatic Domains and Subunits

In snake venoms, three distinct classes of SVMPs, P-I, P-II and P-III, are produced [3,4]. These enzymes exhibit various pharmacological effects by binding to specific target proteins. In most cases, the cleavage of the target proteins through their Zn^{2+}-dependent proteolytic activity leads to either destruction of the receptor or release of new ligands. Thus, M domain plays critical role in most of the pharmacological activities exhibited by SVMPs. However, in a significant number of instances, SVMPs exhibit their functions by non-enzymatic mechanisms through selective binding to key proteins. In such cases, non-enzymatic domains, such as D and C domains, as well as non-enzymatic subunits, such as snaclecs, play important roles. At times, these domains are proteolytically "processed" and exhibit independent pharmacological effects [3,4,113]. It is important to note that in some cases proteolytic activity is essential for the biological effects, while in others just physical binding and steric interference is sufficient for the function (although cleavage may still occur in any case). In this section, we will highlight the roles of these non-enzymatic domains and subunits in specific binding to the target proteins and inducing pharmacological effects.

As mentioned above, precursor of SVMPs are "processed" into various proteolytic products [113,114]. Accordingly, "processing" of P-II SVMPs lead to separation of M and D domains

(P-I-like SVMPs and disntegrins, respectively), while "processing" of P-III SVMPs lead to separation of M and DC domains. In 1994, Usami et al. isolated jararhagin-C, a 28 kDa protein containing the DC domain of jararhagin [115]. Jararhagin-C inhibits collagen- and ADP-induced platelet aggregation in high nanomolar concentrations [115]. Interestingly, phenanthroline-inactivated jararhagin inhibits collagen-induced platelet-aggregation with similar potency [96]. Native jararhagin is only 3- to 4-times more active than inactive jararhagin. These results suggest that there is a significant contribution of non-enzymatic mechanism to the inhibition of platelet aggregation and the small difference is due to proteolytic activity (enzymatic component) of jararhagin. Similarly, native and recombinant DC domains of alternagin, catrocollastatin and atrolysin A inhibit collagen-induced platelet aggregation [100,103,116]. In contrast, leberagin-C, DC domain containing protein from *Macrovipera labetina transmediterranea* venom inhibits platelet aggregation induced by thrombin and arachidonic acid with IC_{50} of 40 and 50 nM, respectively [117]. It inhibits the adhesion of melanoma tumor cells on fibrinogen and fibronectin, by interfering with the function of $\alpha_v\beta_3$ and, to a lesser extent, with $\alpha_v\beta_6$ and $\alpha_v\beta_1$ integrins. It does not bind to $\alpha_2\beta_1$ integrin. These studies support the importance of DC domains in the inhibition of platelet aggregation through non-enzymatic mechanisms. Structure–function relationships of these DC domains will help in determining the integrin selectivity and binding.

As with DC domains, "processed" D domains were also isolated as disintegrins from crotalid and viperid venoms. Disintegrins are among the potent inhibitors of platelet aggregation peptides [118–124]. These polypeptides, ranging from 49 to 84 amino acid residues, are isolated from crotalid and viperid snake venoms. They have a RGD/KGD tripeptide sequence in a 13-residue β-loop structure (dubbed as RGD loop), which is responsible for their biological activity. The active tripeptide RGD is located at the apex of a mobile loop protruding 14–17 Å from the protein core [125,126] and plays key role in the interaction of the disintegrins with the platelet integrin $\alpha_{IIb}\beta_3$ [127,128]. These disintegrins are derived by the processing of the D domains from P-II SVMP precursors [113]. Disintegrins with RGD sequence show different levels of binding affinity and selectivity towards $\alpha_{IIb}\beta_3$, $\alpha_v\beta_3$ and $\alpha_5\beta_1$ integrins [129], while KGD-containing barbourin inhibits the $\alpha_{IIb}\beta_3$ integrin with a high degree of selectivity [130]. This RGD tripeptide is replaced by various sequences including VGD, MLD, MVD and KTS, resulting in distinct integrin selectivity [131] and references therein]. Despite the role of disintegrins in inhibiting platelet aggregation, we will not focus on this group of non-enzymatic polypeptides. Readers can obtain details on this group of fascinating molecules elsewhere [132–137].

In significant number of P-III SVMPs, RGD sequence is replaced by various tripeptide sequences (for example, see [131]) and at the apex of the loop is a Cys residue involved in forming a disulfide bond. Thus, this D domain is appropriately named as "disintegrin-like" domain [3,4]. As with disintegrins, "disintegrin-like" D domains play important role in the recognition of various target receptors or integrins and inhibit platelet aggregation [100,106,138–140]. Recombinantly expressed D domain of jararhagin inhibits platelet–collagen interaction [140]. Linear peptides based on this domain were shown to inhibit the release of 5-hydroxytryptamine (5-HT) from collagen-stimulated platelets [140]. This selective inhibition of the secretion-dependent phase by jararhagin and its peptides is due to the defective phosphorylation of pleckstrin, which is involved in dense granule secretion [140]. Cyclic peptides that cover the loop inhibit platelet aggregation [139] as well as bind to collagen [93,95]. These studies indicate that the non-enzymatic D domains indeed plays critical role in recognition and binding of target receptor or integrin.

Thus far, polypeptides containing only C domain have not been isolated from snake venoms. Therefore, C domains are recombinantly expressed to evaluate their role in platelet functions. C domain of atrolysin A potently inhibits collagen- but not ADP-stimulated platelet aggregation [141]. These studies suggested that the C domain interacts with the collagen receptor $\alpha_2\beta_1$ integrin on the platelet surface. Using overlapping peptides from C domains of atrolysin A and jararhagin, Kamiguti et al. identified two peptides each corresponding to identical segments [142]. These peptides inhibit collagen-induced aggregation, but not convulxin-induced. Thus, they interact with $\alpha_2\beta_1$ integrin and not through GPVI. VKC-jararaca, but not VKC-atrox, induced a rapidly reversible weak aggregation [142]. Pinto et al. identified two regions, $_{365}$PCAPEDVKCG$_{374}$ and $_{372}$KCGRLYCK$_{379}$ in C domain of jararhagin which could bind to vWF [143]. They ruled out the latter region using molecular modeling and docking experiments. The C domain of atrolysin A not only bound directly to vWF and collagen I, but also blocked the collagen–vWF interaction [144]. The interaction of the C domain with the A1 domain of vWF promotes vWF proteolysis and inhibition of vWF-mediated platelet aggregation [145,146]. Similarly, C domain plays crucial role in ADAMTS-13, a vWF-cleaving protease; removal of this domain leads to a remarkable reduction of its ability to cleave vWF [147]. These studies strongly support the importance of the C domain in the non-enzymatic mechanism of inhibition of platelet aggregation.

Thus far, only snaclecs are found to be associated with P-III SVMPs [10–12,47,58]. As discussed above, these subunits are covalently linked through P-III SVMPs by interchain disulfide bond. As with other snaclecs, these subunits are heterodimeric proteins with two chains linked by an interchain disulfide bond. The concave dimeric interface forms the ligand-binding site of FX and prothrombin [17,47,56,58]. Respective Gla domains bind to these subunits in a Ca^{2+}-dependent manner and provide excellent selectivity. Thus, these non-enzymatic subunits impart to distinct properties. Correctly modified and folded Gla domain is important for optimal activity. It defines the Ca^{2+}-dependence, as Ca^{2+} ions are required for proper folding of Gla domain. Carinactivase-1 and multactivase fail to activate prethrombin-1 and descarboxyprothrombin in which Ca^{2+}-binding has been perturbed. On the other hand, Ecarin, which does not have this subunit, activates prothrombin, prethrombin-1 and descarboxyprothrombin with equal efficiency. This functional difference helps in measuring normal prothrombin versus descarboxyprothrombin in the plasma of warfarin-treated individuals [57]. Thus, these non-enzymatic regulatory subunits play critical role in substrate recognition and selectivity.

7. Definition and Nomenclature for Interaction Sites in Proteases

Proteases recognize and interact with specific substrates by binding them through various functional residues distributed among different sites. Each of these sites plays a specific role in the overall function of the enzyme. Our understanding of the chemical and biophysical interactions of various substrates with their respective enzymes has helped us to define these sites. Based on the interactions of SVMPs with various substrates, receptors and integrins (discussed above), we would like to propose new definitions of additional functional sites. We will also provide distinguishing features of these new sites in comparison with established functional sites.

7.1. Active Site

It is the region where substrate molecules bind (binding site) and undergo a chemical reaction (catalytic site). Binding site correctly orients the substrate for catalysis, while residues in the catalytic site play mechanistic role in lowering the activation energy to make the reaction proceed faster. Specific amino acid residues, cofactors and/or ions play critical roles in the catalytic mechanisms in protein enzymes. For example, each residue in the catalytic triad (Ser, His and Asp/Glu) plays a role in catalysis in serine proteases. The Acid–Base–Nucleophile triad generates a nucleophilic residue for covalent catalysis [148]. The residues form a charge-relay network to polarize and activate the nucleophile, which attacks the substrate and forms a covalent intermediate, which is then hydrolyzed to regenerate free enzyme. The nucleophile in serine proteases is a Ser; Cys, and occasionally Thr, also serve as nucleophile in other classes of proteases. Catalytic cleavage in SVMPs is through Zn^{2+} coordinated by three conserved His side chains and a water anchored to a conserved Glu [24,25]. This polarized water molecule acts as general base that catalyzes peptide bond cleavage.

Substrate binding site can be quite elaborate and complex; higher the complexity better is the substrate selectivity. The substrate binding site is divided into several subsites—the regions, which are on the enzyme surface that interact with individual amino acid residues on either side of the substrate cleavage site. The subsites on the amino side of the cleavage site are labeled as S1, S2, S3, etc.

(non-prime subsites), while those on the carboxyl side are labeled as S1′, S2′, S3′, etc. (prime subsites). Generally, these are discontinuous sites and thus, the residues forming these subsites are not contiguous in the protein sequence. P1 amino acid residue of the substrate associates with S1, P2 with S2, etc. Similarly, P1′ amino acid residue binds to S1′, P2′ with S2′, etc. P1-P1′ peptide bond of the substrate is proteolytically cleaved. Both non-prime and prime subsites could contribute to substrate selectivity and affinity. Paes Leme et al. [149] determined the amino acid preferences across the full P4 to P4′ range for the three P-I SVMPs, leucurolysin-a, atrolysin C, and BaP1, and one P-III SVMP, bothropasin, using high resolution mass spectrometric method and albumin-depleted plasma tryptic peptide library. All these SVMPs showed preferences (clear specificities) towards large, hydrophobic aliphatic residues at P1′, P2′ and P3′ sites [149].

7.2. Exosite

This is a secondary binding site, remote from the active site, on the enzyme. Exosites provide additional substrate (or inhibitor) selectivity. For example, thrombin (a serine protease) has two distinct electropositive surface regions, exosite I and exosite II, that contribute to the specificity of thrombin [150,151]. These exosites mediate the interactions of thrombin with its substrates, inhibitors and receptors. Exosite I is adjacent to the P′ side of the active site cleft and is the fibrinogen recognition exosite. Exosite II is more basic than exosite I and it binds to heparin. For details on the interaction of these exosites with substrates, receptors and inhibitors, see [150–152]. In SVMPs there is an exosite C^*_{241}TRKKHD$_{246}C^*$ (as numbered in jararhagin) that interacts with human integrin α_2I-domain [93–95]. Because of their importance in determining exquisite selectivity and specificity, the exosites are of immense interest in biomedical research as potential drug targets [153–158].

7.3. Allosteric Site

Small regulatory molecules interact with this site on the enzyme to activate or inhibit (positive or negative allosterism) the specific enzyme. In general, the non-covalent and reversible interaction of the allosteric effector often results in a conformational change. In homotropic allosterism, the modulator molecule is the substrate as well as the regulatory molecule for the target enzyme. It is typically an activator of the enzyme. In contrast, in heterotropic allosterism, modulator is not the enzyme's substrate. In this case, the modulator may be either an activator or an inhibitor. Although multimeric proteins (e.g., hemoglobin and ATPase) are considered to be prone to allosteric regulation, even monomeric proteins (e.g., myoglobin, human serum albumin, and human α-thrombin) exhibit heterotropic allosterism [159–161]. The rational design of specific antagonists targeting the active site to highly homologous enzymes is an extremely difficult task. As with exosites, allosteric sites are also used for designing drugs targeting specific enzymes [162,163]. For details on protein allosteric mechanisms, see [164].

7.4. Exosite versus Allosteric Site

Both exosite and allosteric site are on the surface of the enzyme or receptor. In the case of exosite, one part of the substrate or inhibitor interacts with the exosite while the other part interacts with the active site. Thus, exosite typically must be occupied first for optimal activity. In contrast, a substrate molecule (homotropic allosterism) or a ligand (heterotropic allosterism) interacts with the allosteric site and a second substrate molecule interacts with the active site. The binding at the allosteric site enhances or decreases the binding or catalysis at the active site. Thus far, no allosterism has been documented in SVMPs.

7.5. Classification of Exosites and Allosteric Sites (Figure 4)

Exosites and/or allosteric sites can be present in the same domain as the orthosteric site, such as active site (in enzymes) or agonist binding site (in receptors). These sites are thus closer to the orthosteric site and located on the enzymatic M domain and we name them as "p-exosite" (proximal-exosite) and "p-allosteric site" (proximal-allosteric site) (Figure 4A,B). The examples of p-exosites are exosite I and exosite II of thrombin [150,151] and C^*_{241}TRKKHD$_{246}$C* exosite of SVMPs [93–95]. In multi-domain enzymes and receptors, these regulatory sites may also be found in other domains. In such cases, we name them as "d-exosite" (distal-exosite) and "d-allosteric site" (distal-allosteric site) (Figure 4A,B). It is possible that these distal sites residing in different domains may be located physically closer to the orthosteric site in the tertiary structure of the proteins. The sites on D and C domains of SVMPs are excellent examples of d-exosites. A better understanding of the distance between orthrosteric site and the regulatory sites will be helpful in designing bifunctional ligands for the target enzyme or receptor.

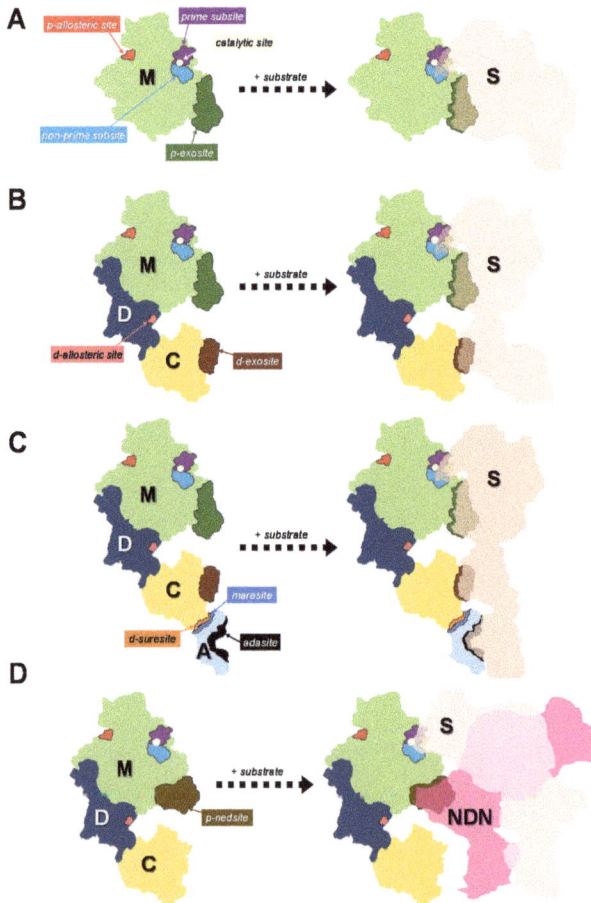

Figure 4. Nomenclature of interaction sites in snake venom metalloproteases. Left and right columns show free and respective substrate-bound protease. (**A**) **Left**: M domain showing catalytic site, prime and non-prime subsites, and proximal allosteric and exosites. **Right**: Substrate S interacts with the protease through active site and p-exosite. (**B**) **Left**: MDC domains showing distal allosteric and exosites. **Right**: Substrate S interacts with the protease through active site, p-exosite and d-exosite. (**C**) **Left**: MDC domains showing distal suresite and adaptor subunit, A showing interaction with MDC domain through distal maresite. A subunit also shows adasite. **Right**: Substrate S interacts with the protease through active site, p-exosite, d-exosite and adasite. (**D**) **Left**: MDC domains showing proximal nedsite. **Right**: Next-door neighbor (NDN) subunit interacts with p-nedsite, while the substrate S interacts with the protease through active site. See text for details.

Enzyme complexes, such as RVV-X, carinactivase-1 and multactivase [38,47,58], are heterodimers comprising a larger main subunit and smaller snaclec subunits. These enzymes use the concave dimeric interface of the snaclec subunits to bind to the substrate [8,47,58]. If these snaclec subunits of these SVMPs or the Gla domains of the substrates are removed, the substrate interaction is extremely poor. Thus, the concave dimeric interface of the adapted subunit acts as the exosite. Therefore, we named this site as "adasite" (adaptor exosite) (Figure 4C). In these cases, there are mutual recognition sites that form the interface between the SVMP and the snaclec subunit. These interaction sites are named as "maresite" (main subunit recognition site on the smaller subunit) and "suresite" (smaller subunit recognition site on the main subunit) (Figure 4C). As with exosite and allosteric sites, suresites can be either "p-suresite" (proximal-suresite, when located on the enzymatic M domain) or "d-suresite" (distal-suresite, when located in other domains). The finer definition and differentiation among various interaction sites will help improving the clarity in the field of SVMPs as well as other enzymes and receptors.

8. Unusual Behavior of Metalloproteases

During our analyses of the literature, we found two interesting, somewhat unusual behaviors of SVMPs. We have highlighted these observations as they will be useful in future research strategies in the field of SVMPs as well as other proteases.

8.1. Binding to Cell Surface Receptors

A key step in the identification of target receptor or acceptor on the cell surface is the characterization of specific binding and Scatchard plots [165,166]. Kamiguti et al. performed binding studies using ^{125}I-jararhagin to determine specific binding to platelets [96]. Their experiments showed no significant specific binding. Intelligently, they also studied the equilibrium binding of 1,10-phenanthroline-treated, catalytically inactive ^{125}I-jararhagin to platelets. The inactive jararhagin showed excellent specific binding to platelets (Figure 5). They had earlier determined that treatment of platelets with jararhagin drastically reduces $\alpha_2\beta_1$ integrin on the platelet surface [92,96]. These observations can be easily explained by the binding of active jararhagin to $\alpha_2\beta_1$ integrin and subsequent cleavage leading to the release of jararhagin from the platelet surface (Figure 5C,D). In contrast, inactive jararhagin continued to bind to $\alpha_2\beta_1$ integrin and stay bound to the platelet surface in the absence of proteolytic activity. Thus, the diligent strategy used by Kamiguti et al. makes an important contribution to specific binding studies of SVMPs. These strategies will also be extremely useful in studying specific binding of other proteases to cell surface receptors.

8.2. Unusual Cleavage of the $\alpha_2\beta_1$ Integrin

In general, proteases bind to a protein substrate and then cleave one or more peptide bonds of this substrate. Jararhagin and other SVMPs have an unusual behavior in cleavage of the $\alpha_2\beta_1$ integrin. They bind to α_2I domain of the α_2 integrin and cleave the β_1 subunit [97,98]. Thus, the binding and cleavage occur in two distinct protein subunits; these SVMPs bind to one protein subunit, but cleave the "next door neighbor" subunit. Such proteolytic cleavage away from the vicinity of the binding site may not be uncommon. The functional exosite that facilitates cleavage in the neighboring protein is named as "nedsite" (next door site) (Figure 4D). Nedsite can be further classified as either "p-nedsite" (proximal-nedsite, when located on the enzymatic M domain) or "d-nedsite" (distal-nedsite, when located in other domains). The p-exosite C^*_{241}TRKKHD$_{246}C^*$ of SVMPs [93–95] that binds to α_2I domain should be properly identified as a p-nedsite.

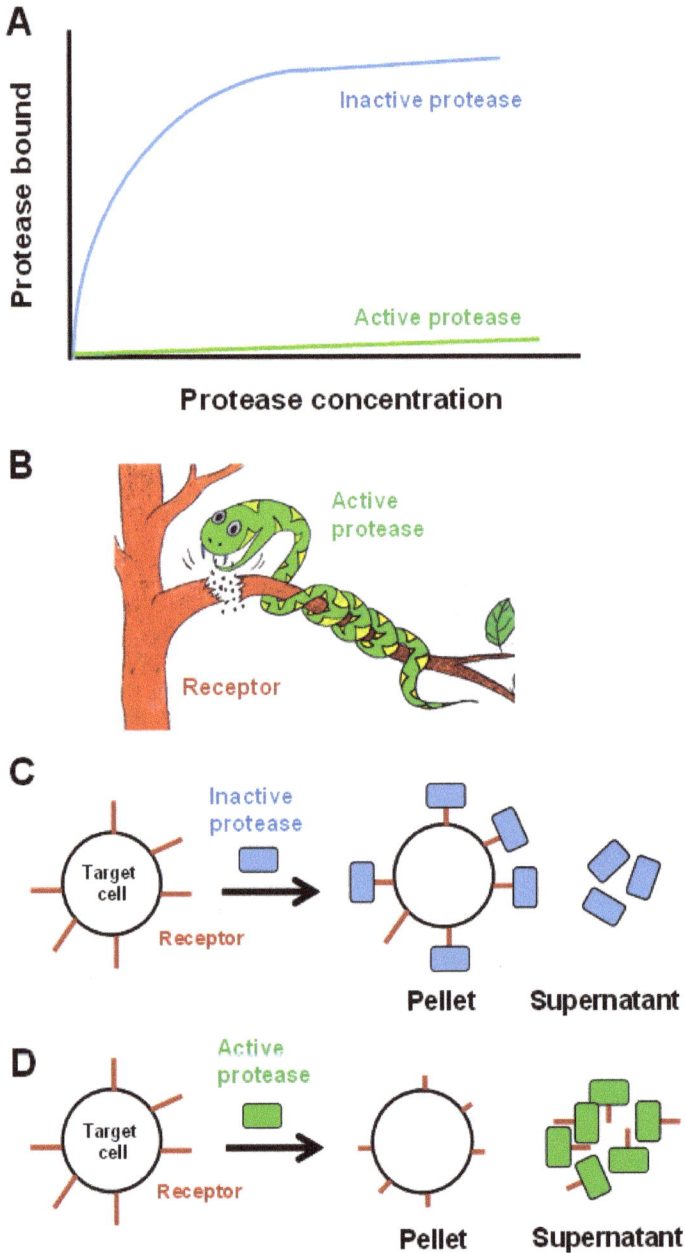

Figure 5. Unusual specific binding of snake venom metalloproteases with target receptor. (**A**) Schematic diagram showing specific binding of active and inactive SVMPs. Diagram is drawn based on the data published by Kamiguti et al. [96]. (**B**) Active protease cleaves the receptor and gets released into the solution. The picture was created by Cho Yeow Koh and Pol Zen Koh. (**C**) Inactive protease binds to receptors on the surface of the target cells and remains bound to the cells in the precipitate. (**D**) Active protease, on the other hand, cleaves the receptor and remains in the supernatant indicating low or no binding to cells in the precipitate.

9. Anticoagulant and Antiplatelet Activity in Hemorrhage

SVMPs frequently induce hemorrhage through the degradation of matrix proteins and basement membrane, resulting in the disruption of endothelial cell integrity in blood vessel walls [143,167–169]. This extra-vascular blood leakage is exacerbated by the disturbance of blood coagulation and platelet aggregation. A number of snake venom toxins have evolved to target various points along the blood coagulation cascade and platelet aggregation pathways. These toxins exhibit both pro- and anti-coagulation of blood or pro- and anti-platelet aggregation effects. Procoagulant toxins not only activate factor VII, factor X, factor V, and prothrombin but also act directly on fibrinogen [170,171]. In the whole animal, defibrogenating the blood and removing significant number of blood coagulation protein result in unclottable blood through consumptive coagulopathy [172]. In addition, a number of SVMPs interfere in blood coagulation and platelet aggregation (described above) and thus enhance hemorrhage. For example, Jararhagin affects hemostasis through fibrinogen degradation [91,173] and by the inhibition of platelet aggregation [92]. These effects significantly enhance its own as well as venom's hemorrhagic activity.

10. SVMPs as Research Tools, and Diagnostic and Therapeutic Agents

Due to high specificity and selectivity, SVMPs and their parts are used in various applications. Among them, their uses as diagnostic agents in hematology laboratories are well known. Stypven (Styptic venom) time is one the earliest one-step clotting time [174]. Russell's viper venom (capable of stopping the bleeding when applied to a wound and hence styptic venom) activates FX directly to initiate coagulation. The Stypven time is unaffected by deficiencies or abnormalities of factors VII, XII, XI, IX or VIII. However, it is abnormal in FV, prothrombin and in most cases of FX deficiency. Thus, it is used to detect hereditary deficiencies or abnormalities and disease- or drug-induced deficiencies. A modified version with limiting amounts of phospholipid and venom, dilute Russell viper venom time, is used for the detection of lupus anticoagulants [175,176]. The individuals with a lupus anticoagulant produce autoantibodies that bind to phospholipids. These antibodies prolong the clotting time by binding to phospholipids in dilute Russell viper venom time, a simple, reproducible, sensitive, and relatively specific method. The ecarin clotting time (ECT) allows us to carry out precise quantification of direct thrombin inhibitors [177]. Ecarin [49], a specific prothrombin activator, activates prothrombin to generate meizothrombin. The cleavage of a chromogenic substrate by meizothrombin is inhibited by direct thrombin inhibitors in a concentration-dependent fashion [178]. Various modifications of the ECT are important in both preclinical and clinical use, e.g., for biochemical investigations, as a point-of-care method and for cardiac surgery. For details of the advantages and disadvantages of these methods, see [177,178]. In CA-1 method, carinactivase-1 [47], a Ca^{2+}-dependent prothrombin activator, is used to activate prothrombin [57]. Since carinactivase-1 recognizes the carboxylated, fully folded Gla domain of prothrombin, CA-1 method measures only normal prothrombin and not descarboxyprothrombin (produced in warfarin-treated individuals). Thus, CA-1 method is a novel assay for monito ring coagulant activity in warfarin-treated individuals. For details on other snake venom proteins used as diagnostic agents, see [179,180].

SVMPs and their domains have also significantly contributed as research tools and also in the development of therapeutic leads. Although classical snake venom D and DC domains are proteolytically released from PII and PIII SVMPs [4,113,114], some heterodimeric disintegrins are encoded by separate genes [181–183]. Most common disintegrins with RGD motif bind selectively with high affinity to integrins including fibrinogen receptors ($\alpha_{IIb}\beta_3$), vitronectin receptors ($\alpha_v\beta_3$) and fibronectin receptor ($\alpha_5\beta_1$). Disintegrins with MLD motif are heterodimeric disintegrins and bind to $\alpha_4\beta_1$, $\alpha_4\beta_7$ and $\alpha_9\beta_1$ integrins. When their second subunit contains RGD, they bind to $\alpha_5\beta_1$ integrin [184]. Disintegrins with KTS/RTS motif bind to $\alpha_1\beta_1$ integrin [135,185–189]. The selectivity and potency strongly depends on the amino acid composition surrounding RGD/MLD/KTS/RTS motifs. For details on the selectivity of various disintegrins, see [134,136,137,184,186,190,191]. DC domains have a limited anti-integrin activity. Alternagin-C binds to collagen receptor, $\alpha_2\beta_1$ integrin through its RSECD sequence located in the D domain [102]. Leberagin-C binds to $\alpha_v \beta_3$ integrin [117]. However, specific integrin-binding motif was not evaluated. Because of their highly specific and selective interaction with various integrins, these disintegrins modulate cellular responses in platelets, neutrophils, T-lymphocytes, eosinophils and endothelial cells as well as various cancer cells (for details, see [134,137,190] and references therein). In addition, they also exhibit uniquely exclusive effects on smooth muscle cells [191,192], fibroblast-like cells [193,194], chondrocytes [195], osteoblasts [196], and neuronal progenitors [197]. Recent studies using obtustatin has shown that $\alpha_1 \beta_1$ integrin and integrin-linked kinase modulate angiotensin II effects in vascular smooth muscle cells and thus, are potential targets to the development of more effective therapeutic interventions in cardiovascular diseases [198,199]. Thus, D and C domains selectively target specific integrins and play critical role in our understanding of cell biology.

The high specificity, selectivity and affinity of D domains have helped the scientific community to design potent therapeutic agents for various human diseases. For example, RGD-disintegrins resulted in the successful design of two therapeutic drugs that inhibit $\alpha_{IIb} \beta_3$ integrin and are approved for the treatment of acute coronary ischemic disease and prevention of thrombotic complication in balloon angioplasty and stenting [200,201]. Integrilin (Eptifibatide, a synthetic cyclic heptapeptide) and tirofiban (Aggrastat, a non-peptide RGD mimic) were designed based on the structure of barbourin [121] and echistatin [119], respectively. Native or recombinant contortrostatin, a homodimeric RGD-disintegrin from *Agkistrodon contortrix contortrix* venom, exhibits potent antiangiogenic effects in in vitro and in vivo models [202–204]. Using liposomal delivery is effective as an anti-tumor agent in animal models of human breast, ovarian and prostate cancer [204,205]. A chimeric variant, Vicrostatin induces apoptosis and blocks tube formation in Matrigel [206]. Based on KTS-disintegrins, Vimocin and Vidapin (cyclic KTS peptides) that target $\alpha_1 \beta_1 / \alpha_2 \beta_1$ integrins are being developed as potent antagonists of angiogenesis for the treatment of angiogenesis disorders and cancer [207], whereas Vipegitide and Vipegitide-PEG2 (peptidomimetics) that target $\alpha_2 \beta_1$ integrin are being developed as another class of inhibitors of platelet aggregation for antithrombotic therapy [208]. Thus, research on the non-enzymatic D and C domains, which interact with integrins, have contributed significantly and appear to have tremendous future in basic cell biology as well as in biomedical applications [133,137,183,209,210].

SVMPs and their catalytically active M domains are also important in the development of therapeutic agents. A direct fibrinolytic enzyme from *Agkistrodon contortrix contortrix* venom, fibrolase and its recombinant analog, alfimeprase was developed as a clot-buster drug for myocardial infarction and stroke due to its thrombolytic properties [67,211–213]. Alfimeprase reached Phase 1 and Phase 2 clinical trials [214,215], but did not make it to the market. For details, see [216]. Despite the setback, there are several lessons learnt through their efforts. Dual antithrombotic therapy using hirudin (thrombin inhibitor) and S18886 (thromboxane A_2 receptor antagonist) were shown to improve reperfusion after thrombolysis with alfimeprase but not tissue plasminogen activator [217]. A careful strategy may help in developing this and related SVMPs as an alternative thrombolytic agent (clot buster) in clearing cardiovascular and cerebrovascular blockages in myocardial infarction and stroke.

11. Summary and Future Prospects

SVMPs, and their domains and complexes have evolved to bind to various integrins, receptors and extracellular matrix proteins. They activate or inactivate proteins through enzymatic or non-enzymatic mechanisms and interfere in blood coagulation and platelet aggregation, and contribute to venom toxicity, particularly to hemorrhagic activity and venom distribution in the prey or victim. The understanding of their structure–function relationships and mechanism of action has contributed significantly to basic sciences including protein chemistry, enzymology, hematology, angiogenesis and cancer biology, and also helped in the development of diagnostic and therapeutic agents. Further studies on this group of toxins will contribute to unlocking several complex physiological processes and pathological effects in blood coagulation, platelet aggregation, hemorrhage, matrix biology, angiogenesis and cancer biology. Their structure–function studies will enhance the potential in developing novel diagnostic and therapeutic agents.

Acknowledgments: We thank Soichi Takeda for providing the coordinates for the RVV-X and FX docking model and Pol Zen Koh for his help in preparing Figure 5B.

Author Contributions: R.M.K. and C.Y.K. wrote the review.

Conflicts of Interest: The authors declare no conflict of interest.

References

1. Harvey, A.L. *Snake Toxins*; Pergamon: New York, NY, USA, 1991.
2. Tu, A.T. *Reptile Venoms and Toxins*; Marcel Decker: New York, NY, USA, 1991.
3. Du, X.Y.; Clemetson, K.J. Snake venom L-amino acid oxidases. *Toxicon* **2002**, *40*, 659–665. [CrossRef]
4. Fox, J.W.; Serrano, S.M. Structural considerations of the snake venom metalloproteinases, key members of the M12 reprolysin family of metalloproteinases. *Toxicon* **2005**, *45*, 969–985. [CrossRef] [PubMed]
5. MEROPS. The Peptidase Database. Available online: http://merops.sanger.ac.uk/ (accessed on 26 September 2016).
6. Kini, R.M. Serine proteinases affecting blood coagulation and fibrinolysis from snake venom. *Pathophysiol. Haemost. Thromb.* **2005**, *34*, 200–204. [CrossRef] [PubMed]
7. Serrano, S.M. The long road of research on snake venom serine proteinases. *Toxicon* **2013**, *62*, 19–26. [CrossRef] [PubMed]
8. Morita, T. Proteases which activate factor X. In *Enzymes from Snake Venom*; Bailey, G.S., Ed.; Alaken Inc.: Fort Collins, CO, USA, 1998; pp. 179–208.
9. Tans, G.; Rosing, J. Snake venom activators of factor X: An overview. *Haemostasis* **2001**, *31*, 225–233. [CrossRef] [PubMed]
10. Kisiel, W.; Hermodson, M.A.; Davie, E.W. Factor X activating enzyme from Russell's viper venom: Isolation and characterization. *Biochemistry* **1976**, *15*, 4901–4906. [CrossRef] [PubMed]
11. Franssen, J.H.; Janssen-Claessen, T.; Van Dieijen, G. Purification and properties of an activating enzyme of blood clotting factor X from the venom of *Cerastes cerastes*. *Biochem. Biophys. Acta* **1983**, *747*, 186–190. [CrossRef]
12. Hofmann, H.; Bon, C. Blood coagulation induced by the venom of *Bothrops atrox*. 2. Identification, purification and properties of two factor X activators. *Biochemistry* **1987**, *26*, 780–787. [CrossRef] [PubMed]
13. Zhang, Y.; Lee, W.H.; Xiong, Y.L.; Wang, W.Y.; Zu, S.W. Characterization of OhS1, an arginine/lysine amidase from the venom of king cobra (*Ophiophagus hannah*). *Toxicon* **1994**, *32*, 615–623. [CrossRef]
14. Zhang, Y.; Xiong, Y.U.; Bon, C. An activator of blood coagulation factor X from the venom of *B. fasciatus*. *Toxicon* **1995**, *33*, 1277–1288. [CrossRef]
15. Khan, S.U.; Al-Saleh, S.S. Biochemical characterization of a factor X activator protein purified from *Walterinnesia aegyptia* venom. *Blood Coagul. Fibrinolysis* **2015**, *26*, 772–777. [CrossRef] [PubMed]
16. Fujikawa, K.; Coan, M.H.; Legaz, M.E.; Davie, E.W. The mechanism of activation of bovine factor X (Stuart factor) by intrinsic and extrinsic pathways. *Biochemistry* **1974**, *13*, 5290–5299. [CrossRef] [PubMed]
17. Takeya, H.; Nishida, S.; Miyata, T.; Kawada, S.; Saisaka, Y.; Morita, T.; Iwanaga, S. Coagulation factor X activating enzyme from Russell's viper venom (RVV-X). A novel metalloproteinase with disintegrin (platelet aggregation inhibitor)-like and C-type lectin-like domains. *J. Biol. Chem.* **1992**, *267*, 14109–14117. [PubMed]
18. Takeda, S.; Igarashi, T.; Mori, H.; Araki, S. Crystal structures of VAP1 reveal ADAMs' MDC domain architecture and its unique C-shaped scaffold. *EMBO J.* **2006**, *25*, 2388–2396. [CrossRef] [PubMed]
19. Takeda, S.; Igarashi, T.; Mori, H. Crystal structure of RVV-X: An example of evolutionary gain of specificity by ADAM proteinases. *FEBS Lett.* **2007**, *581*, 5859–5864. [CrossRef] [PubMed]
20. Igarashi, T.; Araki, S.; Mori, H.; Takeda, S. Crystal structures of catrocollastatin/VAP2B reveal a dynamic, modular architecture of ADAM/adamalysin/reprolysin family proteins. *FEBS Lett.* **2007**, *581*, 2416–2422. [CrossRef] [PubMed]

21. Muniz, J.R.; Ambrosio, A.L.; Selistre-de-Araujo, H.S.; Cominetti, M.R.; Moura-da-Silva, A.M.; Oliva, G.; Garratt, R.C.; Souza, D.H. The three-dimensional structure of bothropasin, the main hemorrhagic factor from *Bothrops jararaca* venom: Insights for a new classification of snake venom metalloprotease subgroups. *Toxicon* **2008**, *52*, 807–816. [CrossRef] [PubMed]

22. Guan, H.H.; Goh, K.S.; Davamani, F.; Wu, P.L.; Huang, Y.W.; Jeyakanthan, J.; Wu, W.G.; Chen, C.J. Structures of two elapid snake venom metalloproteases with distinct activities highlight the disulfide patterns in the D domain of ADAMalysin family proteins. *J. Struct. Biol.* **2009**, *169*, 294–303. [CrossRef] [PubMed]

23. Zhu, Z.; Gao, Y.; Yu, Y.; Zhang, X.; Zang, J.; Teng, M.; Niu, L. Structural basis of the autolysis of AaHIV suggests a novel target recognizing model for ADAM/reprolysin family proteins. *Biochem. Biophys. Res. Commun.* **2009**, *386*, 159–164. [CrossRef] [PubMed]

24. Gomis-Ruth, F.X.; Kress, L.F.; Bode, W. First structure of a snake venom metalloproteinase: A prototype for matrix metalloproteinases/collagenases. *EMBO J.* **1993**, *12*, 4151–4157. [PubMed]

25. Gomis-Ruth, F.X.; Kress, L.F.; Kellermann, J.; Mayr, I.; Lee, X.; Huber, R.; Bode, W. Refined 2.0 Å X-ray crystal structure of the snake venom zinc-endopeptidase adamalysin II. Primary and tertiary structure determination, refinement, molecular structure and comparison with astacin, collagenase and thermolysin. *J. Mol. Biol.* **1994**, *239*, 513–544. [CrossRef] [PubMed]

26. Kumasaka, T.; Yamamoto, M.; Moriyama, H.; Tanaka, N.; Sato, M.; Katsube, Y.; Yamakawa, Y.; Omori-Satoh, T.; Iwanaga, S.; Ueki, T.J. Crystal structure of H2-proteinase from the venom of *Trimeresurus flavoviridis*. *Biochemistry* **1996**, *119*, 49–57. [CrossRef]

27. Cirilli, M.; Gallina, C.; Gavuzzo, E.; Giordano, C.; Gomis-Rüth, F.X.; Gorini, B.; Kress, L.F.; Mazza, F.; Paradisi, M.P.; Pochetti, G.; et al. 2 Å X-ray structure of adamalysin II complexed with a peptide phosphonate inhibitor adopting a retro-binding mode. *FEBS Lett.* **1997**, *418*, 319–322. [CrossRef]

28. Gong, W.; Zhu, X.; Liu, S.; Teng, M.; Niu, L. Crystal structures of acutolysin A, a three-disulfide hemorrhagic zinc metalloproteinase from the snake venom of *Agkistrodon acutus*. *J. Mol. Biol.* **1998**, *283*, 657–668. [CrossRef] [PubMed]

29. Zhu, X.; Liu, S.; Teng, M.; Niu, L. Structure of acutolysin-C, a haemorrhagic toxin from the venom of *Agkistrodon acutus*, providing further evidence for the mechanism of the pH-dependent proteolytic reaction of zinc metalloproteinases. *Acta Crystallogr. D Biol. Crystallogr.* **1999**, *55*, 1834–1841. [CrossRef] [PubMed]

30. Watanabe, L.; Shannon, J.D.; Valente, R.H.; Rucavado, A.; Alape-Girón, A.; Kamiguti, A.S.; Theakston, R.D.; Fox, J.W.; Gutiérrez, J.M.; Arni, R.K. Amino acid sequence and crystal structure of BaP1, a metalloproteinase from *Bothrops asper* snake venom that exerts multiple tissue-damaging activities. *Protein Sci.* **2003**, *12*, 2273–2281. [CrossRef] [PubMed]

31. Lou, Z.; Hou, J.; Liang, X.; Chen, J.; Qiu, P.; Liu, Y.; Li, M.; Rao, Z.; Yan, G. Crystal structure of a non-hemorrhagic fibrin(ogen)olytic metalloproteinase complexed with a novel natural tri-peptide inhibitor from venom of *Agkistrodon acutus*. *J. Struct. Biol.* **2005**, *152*, 195–203. [CrossRef] [PubMed]

32. Lingott, T.; Schleberger, C.; Gutiérrez, J.M.; Merfort, I. High-resolution crystal structure of the snake venom metalloproteinase BaP1 complexed with a peptidomimetic: Insight into inhibitor binding. *Biochemistry* **2009**, *48*, 6166–6174. [CrossRef] [PubMed]

33. Akao, P.K.; Tonoli, C.C.; Navarro, M.S.; Cintra, A.C.; Neto, J.R.; Arni, R.K.; Murakami, M.T. Structural studies of BmooMPalpha-I, a non-hemorrhagic metalloproteinase from *Bothrops moojeni* venom. *Toxicon* **2010**, *55*, 361–368. [CrossRef] [PubMed]

34. Chou, T.L.; Wu, C.H.; Huang, K.F.; Wang, A.H. Crystal structure of a *Trimeresurus mucrosquamatus* venom metalloproteinase providing new insights into the inhibition by endogenous tripeptide inhibitors. *Toxicon* **2013**, *71*, 140–146. [CrossRef] [PubMed]

35. Souza, D.H.; Selistre-de-Araujo, H.S.; Moura-da-Silva, A.M.; Della-Casa, M.S.; Oliva, G.; Garratt, R.C. Crystallization and preliminary X-ray analysis of jararhagin, a metalloproteinase/disintegrin from *Bothrops jararaca* snake venom. *Acta Crystallogr. D Biol. Crystallogr.* **2001**, *57*, 1135–1137. [CrossRef] [PubMed]

36. Wu, E.L.; Wong, K.Y.; Zhang, X.; Han, K.; Gao, J. Determination of the structure form of the fourth ligand of zinc in Acutolysin A using combined quantum mechanical and molecular mechanical simulation. *J. Phys. Chem. B* **2009**, *113*, 2477–2485. [CrossRef] [PubMed]

37. Fujii, Y.; Okuda, D.; Fujimoto, Z.; Horii, K.; Morita, T.; Mizuno, H. Crystal structure of trimestatin, a disintegrin containing a cell adhesion recognition motif RGD. *J. Mol. Biol.* **2003**, *332*, 1115–1122. [CrossRef]
38. Fox, J.W.; Serrano, S.M. Insights into and speculations about snake venom metalloproteinase (SVMP) synthesis, folding and disulfide bond formation and their contribution to venom complexity. *FEBS J.* **2008**, *275*, 3016–3030. [CrossRef] [PubMed]
39. Mizuno, H.; Fujimoto, Z.; Atoda, H.; Morita, T. Crystal structure of an anticoagulant protein in complex with the Gla domain of factor X. *Proc. Natl. Acad. Sci. USA* **2001**, *98*, 7230–7234. [CrossRef] [PubMed]
40. Rosing, J.; Tans, G. Inventory of exogenous prothrombin activators. *Thromb. Haemost.* **1991**, *65*, 627–630. [PubMed]
41. Rosing, J.; Tans, G. Structural and functional properties of snake venom prothrombin activators. *Toxicon* **1992**, *30*, 1515–1527. [CrossRef]
42. Tans, G.; Rosing, J. Prothrombin activation by snake venom proteases. *J. Toxicol.* **1993**, *12*, 155–173. [CrossRef]
43. Petrovan, R.; Tans, G.; Rosing, J. Proteases activating prothrombin. In *Enzymes from Snake Venom*; Bailey, G.S., Ed.; Alaken Inc.: Fort Collins, CO, USA, 1998; pp. 227–252.
44. Kini, R.M.; Joseph, J.S.; Rao, V.S. Prothrombin activators from snake venoms. In *Perspectives in Molecular Toxinology*; Mènez, A., Ed.; John Wiley: Chichester, UK, 2002; pp. 341–355.
45. Joseph, J.S.; Kini, R.M. Snake venom prothrombin activators similar to blood coagulation factor Xa. *Curr. Drug Targets Cardiovasc. Haematol. Dis.* **2004**, *4*, 397–416. [CrossRef]
46. Kini, R.M. The intriguing world of prothrombin activators from snake venoms. *Toxicon* **2005**, *45*, 1133–1145. [CrossRef] [PubMed]
47. Yamada, D.; Sekiya, F.; Morita, T. Isolation and characterization of carinactivase, a novel prothrombin activator in *Echis carinatus* venom with a unique catalytic mechanism. *J. Biol. Chem.* **1996**, *271*, 5200–5207. [PubMed]
48. Kini, R.M.; Morita, T.; Rosing, J. Classification and nomenclature of prothrombin activators isolated from snake venoms. *Thromb. Haemost.* **2001**, *85*, 710–711.
49. Kornalik, F.; Blomback, B. Prothrombin activation induced by Ecarin-a prothrombin converting enzyme from *Echis carinatus* venom. *Thromb. Res.* **1975**, *6*, 57–63. [CrossRef]
50. Nishida, S.; Fujita, T.; Kohno, N.; Atoda, H.; Morita, T.; Takeya, H.; Kido, I.; Paine, M.J.; Kawabata, S.; Iwanaga, S. cDNA cloning and deduced amino acid sequence of prothrombin activator (ecarin) from Kenyan *Echis carinatus* venom. *Biochemistry* **1995**, *34*, 1771–1778. [CrossRef] [PubMed]
51. Hofmann, H.; Bon, C. Blood coagulation induced by the venom of *Bothrops atrox*. 1. Identification, purification, and properties of a prothrombin activator. *Biochemistry* **1987**, *26*, 772–780. [CrossRef] [PubMed]
52. Speijer, H.; Govers-Riemslag, J.W.P.; Zwaal, R.F.A.; Rosing, J. Prothrombin activation by an activator from the venom of *Oxyuranus scutellatus* (Taipan snake). *J. Biol. Chem.* **1986**, *261*, 13258–13267. [PubMed]
53. Joseph, J.S.; Chung, M.C.M.; Jeyaseelan, K.; Kini, R.M. Amino acid sequence of trocarin, a prothrombin activator from *Tropidechis carinatus* venom: Its structural similarity to coagulation factor Xa. *Blood* **1999**, *94*, 621–631. [PubMed]
54. Rao, V.S.; Joseph, J.S.; Kini, R.M. Group D prothrombin activators from snake venom are structural homologues of mammalian blood coagulation factor Xa. *Biochem. J.* **2003**, *369*, 635–642. [CrossRef] [PubMed]
55. Rao, V.S.; Kini, R.M. Pseutarin C, a prothrombin activator from *Pseudonaja textilis* venom: Its structural and functional similarity to mammalian coagulation factor Xa-Va complex. *Thromb. Haemost.* **2002**, *88*, 611–619. [PubMed]
56. Atoda, H.; Hyuga, M.; Morita, T. The primary structure of coagulation factor IX/factor X-binding protein isolated from the venom of *Trimeresurus flavoviridis*. Homology with asialoglycoprotein receptors, proteoglycan core protein, tetranectin, and lymphocyte Fc epsilon receptor for immunoglobulin E. *J. Biol. Chem.* **1991**, *266*, 14903–14911. [PubMed]
57. Yamada, D.; Morita, T. CA-1 method, a novel assay for quantification of normal prothrombin using a Ca^{2+}-dependent prothrombin activator, carinactivase-1. *Thromb. Res.* **1999**, *94*, 221–226. [CrossRef]
58. Yamada, D.; Morita, T. Purification and characterization of a Ca^{2+}-dependent prothrombin activator, multactivase, from the venom of *Echis multisquamatus*. *J. Biochem.* **1997**, *122*, 991–997. [CrossRef] [PubMed]
59. Aronson, D.L. Comparison of the actions of thrombin and the thrombin-like venom enzymes ancrod and batroxobin. *Thromb. Haemost.* **1976**, *36*, 1–13.
60. Bell, W.R., Jr. Defibrinogenating enzymes. *Drugs* **1997**, *54*, 18–30. [CrossRef] [PubMed]

61. Pirkle, H.; Stocker, K. Thrombin-like enzymes from snake venoms; an inventory. *Thromb. Haemost.* **1991**, *65*, 444–450. [PubMed]
62. Pirkle, H.; Theodor, I. Thrombin-like enzymes. In *Enzymes from Snake Venom*; Bailey, G.S., Ed.; Alaken Inc.: Fort Collins, CO, USA, 1998; pp. 39–69.
63. Hutton, R.A.; Warrell, D.A. Action of snake venom components on the haemostatic system. *Blood Rev.* **1993**, *7*, 176–189. [CrossRef]
64. Ouyang, C.; Teng, C.M.; Chen, Y.C. Physicochemical properties of α- and β-fibrinogenases of *Trimeresurus mucrosquamatus* venom. *Biochim. Biophys. Acta* **1977**, *481*, 622–630. [CrossRef]
65. Ouyang, C.; Huang, T.F. α and β-fibrinogenases from *Trimeresurus gramineus* snake venom. *Biochim. Biophys. Acta* **1979**, *571*, 270–283. [CrossRef]
66. Ouyang, C.; Hwang, L.J.; Huang, T.F. α-Fibrinogenase from *Agkistrodon rhodostoma* (Malayan pit viper) snake venom. *Toxicon* **1983**, *21*, 25–33. [CrossRef]
67. Markland, F.S. Fibrolase, an active thrombolytic enzyme in arterial and venous thrombosis model systems. *Adv. Exp. Med. Biol.* **1996**, *391*, 427–438. [PubMed]
68. Tu, A.T.; Baker, B.; Wongvibulsin, S.; Willis, T. Biochemical characterization of atroxase and nucleotide sequence encoding the fibrinolytic enzyme. *Toxicon* **1996**, *34*, 1295–1300. [CrossRef]
69. Terada, S.; Hori, J.; Fujimura, S.; Kimoto, E. Purification and amino acid sequence of brevilysin L6, a non-hemorrhagic metalloprotease from *Agkistrodon halys brevicaudus* venom. *J. Biochem.* **1999**, *125*, 64–69. [CrossRef] [PubMed]
70. Siigur, J.; Tonismagi, K.; Trummal, K.; Aaspollu, A.; Samel, M.; Vija, H.; Subbi, J.; Kalkkinen, N.; Siigur, E. *Vipera lebetina* venom contains all types of snake venom metalloproteases. *Pathophysiol. Haemost. Thromb.* **2005**, *34*, 209–214. [CrossRef] [PubMed]
71. Swenson, S.; Toombs, C.F.; Pena, L.; Johansson, J.; Markland, F.S., Jr. α-Fibrinogenases. *Curr. Drug Targets Cardiovasc. Haematol. Disord.* **2004**, *4*, 417–435. [CrossRef] [PubMed]
72. Markland, F.S.; Swenson, S. Fibrolase and its evolution to clinical trials: A long and winding road. In *Toxins and Hemostasis: From Bench to Bedside*; Kini, R.M., Clemetson, K.J., Markland, F.S., McLane, M.A., Morita, T., Eds.; Springer: Dordrecht, The Netherlands, 2010; pp. 409–427.
73. Ouyang, C.; Teng, C.M.; Chen, Y.C. Properties of fibrinogen degradation products produced by α- and β- fibrinogenases of *Trimeresurus mucrosquamatus* snake venom. *Toxicon* **1979**, *17*, 121–126. [CrossRef]
74. Ouyang, C.; Hwang, L.J.; Huang, T.F. Inhibition of rabbit platelet aggregation by alpha-fibrinogenase purified from *Calloselasma rhodostoma* (Malayan pit viper) venom. *J. Formos. Med. Assoc. (Taiwan Yi Xue Hui Za Zhi)* **1985**, *84*, 1197–1206.
75. Roschlau, W.H.; Gage, R. The effects of brinolase (fibrinolytic enzyme from *Aspergillus Oryzae*) on platelet aggregation of dog and man. *Thromb. Diath. Haemorrh.* **1972**, *28*, 31–48. [PubMed]
76. Boffa, M.C.; Boffa, G.A. Correlations between the enzymatic activities and the factors active on blood coagulation and platelet aggregation from the venom of *Vipera aspis*. *Biochim. Biophys. Acta* **1974**, *354*, 275–290. [CrossRef]
77. Miller, J.L.; Katz, A.J.; Feinstein, M.B. Plasmin inhibition of thrombin-induced platelet aggregation. *Thromb. Diath. Haemorrh.* **1975**, *33*, 286–309. [PubMed]
78. Kini, R.M.; Evans, H.J. Inhibition of platelet aggregation by a fibrinogenase from *Naja nigricollis* venom is independent of fibrinogen degradation. *Biochim. Biophys. Acta* **1991**, *1095*, 117–121. [CrossRef]
79. Huang, T.F.; Chang, M.C.; Peng, H.C.; Teng, C.M. A novel α-type fibrinogenase from *Agkistrodon rhodostoma* snake venom. *Biochim. Biophys. Acta* **1992**, *1160*, 262–268. [CrossRef]
80. Huang, T.F.; Chang, M.C.; Peng, H.C.; Teng, C.M. Antiplatelet protease, kistomin, selectively cleaves human platelet glycoprotein Ib. *Biochim. Biophys. Acta* **1993**, *1158*, 293–299. [CrossRef]
81. Hsu, C.C.; Wu, W.B.; Chang, Y.H.; Kuo, H.L.; Huang, T.F. Antithrombotic effect of a protein-type I class snake venom metalloproteinase, kistomin, is mediated by affecting glycoprotein Ib-von Willebrand factor interaction. *Mol. Pharmacol.* **2007**, *72*, 984–992. [CrossRef] [PubMed]
82. Hsu, C.C.; Wu, W.B.; Huang, T.F. A snake venom metalloproteinase, kistomin, cleaves platelet glycoprotein VI and impairs platelet functions. *J. Thromb. Haemost.* **2008**, *6*, 1578–1585. [CrossRef] [PubMed]
83. Wu, W.B.; Peng, H.C.; Huang, T.F. Crotalin, a vWF and GP Ib cleaving metalloproteinase from venom of *Crotalus atrox*. *Thromb. Haemost.* **2001**, *86*, 1501–1511. [PubMed]

84. Ward, C.M.; Andrews, R.K.; Smith, A.I.; Berndt, M.C. Mocarhagin, a novel cobra venom metalloproteinase, cleaves the platelet von Willebrand factor receptor glycoprotein Ibα. Identification of the sulfated tyrosine/anionic sequence Tyr-276-Glu-282 of glycoprotein Ibα as a binding site for von Willebrand factor and α-thrombin. *Biochemistry* **1996**, *35*, 4929–4938. [PubMed]

85. De Luca, M.; Dunlop, L.C.; Andrews, R.K.; Flannery, J.V., Jr.; Ettling, R.; Cumming, D.A.; Veldman, G.M.; Berndt, M.C. A novel cobra venom metalloproteinase, mocarhagin, cleaves a 10-amino acid peptide from the mature N terminus of P-selectin glycoprotein ligand receptor, PSGL-1, and abolishes P-selectin binding. *J. Biol. Chem.* **1995**, *270*, 26734–26737. [CrossRef] [PubMed]

86. Ward, C.M.; Vinogradov, D.V.; Andrews, R.K.; Berndt, M.C. Characterization of mocarhagin, a cobra venom metalloproteinase from *Naja mocambique mocambique*, and related proteins from other Elapidae venoms. *Toxicon* **1996**, *34*, 1203–1206. [CrossRef]

87. Wijeyewickrema, L.C.; Gardiner, E.E.; Shen, Y.; Berndt, M.C.; Andrews, R.K. Fractionation of snake venom metalloproteinases by metal ion affinity: A purified cobra metalloproteinase, Nk, from Naja kaouthia binds Ni^{2+}-agarose. *Toxicon* **2007**, *50*, 1064–1072. [CrossRef] [PubMed]

88. Wijeyewickrema, L.C.; Gardiner, E.E.; Gladigau, E.L.; Berndt, M.C.; Andrews, R.K. Nerve growth factor inhibits metalloproteinase-disintegrins and blocks ectodomain shedding of platelet glycoprotein VI. *J. Biol. Chem.* **2010**, *285*, 11793–11799. [CrossRef] [PubMed]

89. Kumar, M.S.; Girish, K.S.; Vishwanath, B.S.; Kemparaju, K. The metalloprotease, NN-PF3 from *Naja naja* venom inhibits platelet aggregation primarily by affecting $\alpha_2\beta_1$ integrin. *Ann. Hematol.* **2011**, *90*, 569–577. [CrossRef] [PubMed]

90. Paine, M.J.I.; Desmond, H.P.; Theakston, R.D.G.; Crampton, J.M. Purification, cloning and molecular characterization of a high molecular weight hemorrhagic metalloprotease, jararhagin, from *Bothrops jararaca* venom. Insights into the disintegrin-like gene family. *J. Biol. Chem.* **1992**, *267*, 22869–22876. [PubMed]

91. Kamiguti, A.S.; Slupsky, J.R.; Zuzel, M.; Hay, C.R. Properties of fibrinogen cleaved by jararhagin, a metalloproteinase from the venom of *Bothrops jararaca*. *Thromb. Haemost.* **1994**, *72*, 244–249. [PubMed]

92. Kamiguti, A.S.; Hay, C.R.M.; Theakston, R.D.G.; Zuzel, M. Insights into the mechanism of hemorrhage caused by the snake venom metalloproteinases. *Toxicon* **1996**, *34*, 627–642. [CrossRef]

93. Ivaska, J.; Käpylä, J.; Pentikainen, O.; Hoffren, A.M.; Hermonen, J.; Huttunen, P.; Johnson, M.S.; Heino, J. A peptide inhibiting the collagen binding function of integrin α_2I domain. *J. Biol. Chem.* **1999**, *274*, 3513–3521. [CrossRef] [PubMed]

94. Pentikainen, O.; Hoffren, A.M.; Ivaska, J.; Käpylä, J.; Nyrönen, T.; Heino, J.; Johnson, M.S. "RKKH" peptides from the snake venom metalloproteinase of *Bothrops jararaca* bind near the metal ion-dependent adhesion site of the human integrin α_2I-domain. *J. Biol. Chem.* **1999**, *274*, 31493–31505. [CrossRef] [PubMed]

95. Lambert, L.J.; Bobkov, A.A.; Smith, J.W.; Marassi, F.M. Competitive interactions of collagen and a jararhagin-derived disintegrin peptide with the integrin α_2-I domain. *J. Biol. Chem.* **2008**, *283*, 16665–16672. [CrossRef] [PubMed]

96. Kamiguti, A.S.; Hay, C.R.M.; Zuzel, M. Inhibition of collagen-induced platelet aggregation as the result of cleavage of $\alpha_2\beta_1$-integrin by the snake venom metalloproteinase jarararhagin. *Biochem. J.* **1996**, *320*, 635–641. [CrossRef] [PubMed]

97. De Luca, M.; Ward, C.M.; Ohmori, K.; Andrews, R.K.; Berndt, M.C. Jararhagin and jaracetin: Novel snake venom inhibitors of the integrin collagen receptor, $\alpha_2\beta_1$. *Biochem. Biophys. Res. Commun.* **1995**, *206*, 570–576. [CrossRef] [PubMed]

98. Kamiguti, A.S.; Markland, F.S.; Zhou, Q.; Laing, G.D.; Theakston, R.D.; Zuzel, M. Proteolytic cleavage of the β_1 subunit of platelet $\alpha_2\beta_1$ integrin by the metalloproteinase jararhagin compromises collagen-stimulated phosphorylation of pp72^{syk}. *J. Biol. Chem.* **1997**, *272*, 32599–32605. [CrossRef] [PubMed]

99. Tanjoni, I.; Evangelista, K.; Della-Casa, M.S.; Butera, D.; Magalhaes, G.S.; Baldo, C.; Clissa, P.B.; Fernandes, I.; Eble, J.A.; Moura-da-Silva, A.M. Different regions of the class P-III snake venom metalloproteinase jararhagin are involved in binding to $\alpha_2\beta_1$ integrin and collagen. *Toxicon* **2010**, *55*, 1093–1099. [CrossRef] [PubMed]

100. Jia, L.G.; Wang, X.M.; Shannon, J.D.; Bjarnason, J.B.; Fox, J.W. Function of disintegrin-like/cysteine-rich domains of atrolysin A. Inhibition of platelet aggregation by recombinant protein and peptide antagonists. *J. Biol. Chem.* **1997**, *272*, 13094–13102. [CrossRef] [PubMed]

101. Zhou, Q.; Smith, J.B.; Grossman, M.H. Molecular cloning and expression of catrocollastatin, a snake-venom protein from *Crotalus atrox* (western diamondback rattlesnake) which inhibits platelet adhesion to collagen. *Biochem. J.* **1995**, *307*, 411–417. [CrossRef] [PubMed]

102. Liu, C.Z.; Huang, T.F. Crovidisin, a collagen-binding protein isolated from snake venom of *Crotalus viridis*, prevents platelet-collagen interaction. *Arch. Biochem. Biophys.* **1997**, *337*, 291–299. [CrossRef] [PubMed]

103. Souza, D.H.; Iemma, M.R.; Ferreira, L.L.; Faria, J.P.; Oliva, M.L.; Zingali, R.B.; Niewiarowski, S.; Selistre-de-Araujo, H.S. The disintegrin-like domain of the snake venom metalloprotease alternagin inhibits $\alpha_2\beta_1$ integrin-mediated cell adhesion. *Arch. Biochem. Biophys.* **2000**, *384*, 341–350. [CrossRef] [PubMed]

104. Wang, W.J.; Shih, C.H.; Huang, T.F. Primary structure and antiplatelet mechanism of a snake venom metalloproteinase, acurhagin, from *Agkistrodon acutus* venom. *Biochimie* **2005**, *87*, 1065–1077. [CrossRef] [PubMed]

105. You, W.K.; Jang, Y.J.; Chung, K.H.; Kim, D.S. A novel disintegrin-like domain of a high molecular weight metalloprotease inhibits platelet aggregation. *Biochem. Biophys. Res. Commun.* **2003**, *309*, 637–642. [CrossRef] [PubMed]

106. Hamako, J.; Matsui, T.; Nishida, S.; Nomura, S.; Fujimura, Y.; Ito, M.; Ozeki, Y.; Titani, K. Purification and characterization of kaouthiagin, a von Willebrand factor-binding and -cleaving metalloproteinase from *Naja kaouthia* cobra venom. *Thromb. Haemost.* **1998**, *80*, 499–505. [PubMed]

107. Sanchez, E.F.; Richardson, M.; Gremski, L.H.; Veiga, S.S.; Yarleque, A.; Niland, S.; Lima, A.M.; Estevao-Costa, M.I.; Eble, J.A. A novel fibrinolytic metalloproteinase, barnettlysin-I from *Bothrops barnetti* (Barnett's pitviper) snake venom with anti-platelet properties. *Biochim. Biophys. Acta* **2016**, *1860*, 542–556. [CrossRef] [PubMed]

108. Andrews, R.K.; Gardiner, E.E.; Asazuma, N.; Berlanga, O.; Tulasne, D.; Nieswandt, B.; Smith, A.I.; Berndt, M.C.; Watson, S.P. A novel viper venom metalloproteinase, alborhagin, is an agonist at the platelet collagen receptor GPVI. *J. Biol. Chem.* **2001**, *276*, 28092–28097. [CrossRef] [PubMed]

109. Jandrot-Perrus, M.; Lagrue, A.H.; Okuma, M.; Bon, C. Adhesion and activation of human platelets induced by convulxin involve glycoprotein VI and integrin $\alpha_2\beta_1$. *J. Biol. Chem.* **1997**, *272*, 27035–27041. [CrossRef] [PubMed]

110. Polgar, J.; Clemetson, J.M.; Kehrel, B.E.; Weidemann, M.; Magnenat, E.M.; Wells, T.N.C.; Clemetson, K.J. Platelet activation and signal transduction by convulxin, a C-type lectin from Crotalus durissus terrificus (tropical rattlesnake) venom via the p62/GPVI collagen receptor. *J. Biol. Chem.* **1997**, *272*, 13576–13583. [CrossRef] [PubMed]

111. Leduc, M.; Bon, C. Cloning of subunits of convulxin, a collagen-like platelet-aggregating protein from *Crotalus durissus terrificus* venom. *Biochem. J.* **1998**, *333*, 389–393. [CrossRef] [PubMed]

112. Wijeyewickrema, L.C.; Gardiner, E.E.; Moroi, M.; Berndt, M.C.; Andrews, R.K. Snake venom metalloproteinases, crotarhagin and alborhagin, induce ectodomain shedding of the platelet collagen receptor, glycoprotein VI. *Thromb. Haemost.* **2007**, *98*, 1285–1290. [CrossRef] [PubMed]

113. Kini, R.M.; Evans, H.J. Structural domains in venom proteins: Evidence that metalloproteinases and nonenzymatic aggregation inhibitors (disintegrins) from snake venoms are derived by proteolysis from a common precursor. *Toxicon* **1992**, *30*, 265–293. [CrossRef]

114. Au, L.C.; Chou, J.S.; Chang, K.J.; Teh, G.W.; Lin, S.B. Nucleotide sequence of a full-length cDNA encoding a common precursor of platelet aggregation inhibitor and hemorrhagic protein from *Calloselasma rhodostoma* venom. *Biochim. Biophys. Acta* **1993**, *1173*, 243–245. [CrossRef]

115. Usami, Y.; Fujimura, Y.; Miura, S.; Shima, H.; Yoshida, E.; Yoshioka, A.; Hirano, K.; Suzuki, M.; Titani, K. A 28 kDa-protein with disintegrin-like structure (jararhagin-C) purified from *Bothrops jararaca* venom inhibits collagen- and ADP-induced platelet aggregation. *Biochem. Biophys. Res. Commun.* **1994**, *201*, 331–339. [CrossRef] [PubMed]

116. Shimokawa, K.; Shannon, J.D.; Jia, L.G.; Fox, J.W. Sequence and biological activity of catrocollastatin-C: A disintegrin-like/cysteine-rich two-domain protein from *Crotalus atrox* venom. *Arch. Biochem. Biophys.* **1997**, *343*, 35–43. [CrossRef] [PubMed]

117. Limam, I.; Bazaa, A.; Srairi-Abid, N.; Taboubi, S.; Jebali, J.; Zouari-Kessentini, R.; Kallech-Ziri, O.; Mejdoub, H.; Hammami, A.; El Ayeb, M.; et al. Leberagin-C, A disintegrin-like/cysteine-rich protein from *Macrovipera lebetina transmediterranea* venom, inhibits $\alpha_v\beta_3$ integrin-mediated cell adhesion. *Matrix Biol.* **2010**, *29*, 117–126. [CrossRef] [PubMed]

118. Huang, T.F.; Holt, J.C.; Lukasiewicz, H.; Niewiarowski, S. Trigramin. A low molecular weight peptide inhibiting fibrinogen interaction with platelet receptors expressed on glycoprotein IIb-IIIa complex. *J. Biol. Chem.* **1987**, *262*, 16157–16163. [PubMed]

119. Gan, Z.R.; Gould, R.J.; Jacobs, J.W.; Friedman, P.A.; Polokoff, M.A. Echistatin. A potent platelet aggregation inhibitor from the venom of the viper, *Echis carinatus*. *J. Biol. Chem.* **1988**, *263*, 19827–19832. [PubMed]

120. Huang, T.F.; Holt, J.C.; Kirby, E.P.; Niewiarowski, S. Trigramin: Primary structure and its inhibition of von Willebrand factor binding to glycoprotein IIb/IIIa complex on human platelets. *Biochemistry* **1989**, *28*, 661–666. [CrossRef] [PubMed]

121. Scarborough, R.M.; Rose, J.W.; Naughton, M.A.; Phillips, D.R.; Nannizzi, L.; Arfsten, A.; Campbell, A.M.; Charo, I.F. Characterization of the integrin specificities of disintegrins isolated from American pit viper venoms. *J. Biol. Chem.* **1993**, *268*, 1058–1065. [PubMed]

122. Gould, R.J.; Polokoff, M.A.; Friedman, P.A.; Huang, T.F.; Holt, J.C.; Cook, J.J.; Niewiarowski, S. Disintegrins: A family of integrin inhibitory proteins from viper venoms. *Proc. Soc. Exp. Biol. Med.* **1990**, *195*, 168–171. [CrossRef] [PubMed]

123. Dennis, M.S.; Henzel, W.J.; Pitti, R.M.; Lipari, M.T.; Napier, M.A.; Deisher, T.A.; Bunting, S.; Lazarus, R.A. Platelet glycoprotein IIb-IIIa protein antagonists from snake venoms: Evidence for a family of platelet-aggregation inhibitors. *Proc. Natl. Acad. Sci. USA* **1989**, *87*, 2471–2475. [CrossRef]

124. Niewiarowski, S.; McLane, M.A.; Kloczewiak, M.; Stewart, G.J. Disintegrins and other naturally antagonists of platelet fibrinogen receptors. *Semin. Hematol.* **1994**, *31*, 289–300. [PubMed]

125. Adler, M.; Lazarus, R.A.; Dennis, M.S.; Wagner, G. Solution structure of kistrin, a potent platelet aggregation inhibitor and GPIIb–IIIa antagonist. *Science* **1991**, *253*, 445–448. [CrossRef] [PubMed]

126. Saudek, V.; Atkinson, R.A.; Pelton, J.T. Three-dimensional structure of echistatin, the smallest active RGD protein. *Biochemistry* **1991**, *30*, 7369–7372. [CrossRef] [PubMed]

127. Dennis, M.S.; Carter, P.; Lazarus, R.A. Binding interactions of kistrin with platelet glycoprotein IIb-IIIa: Analysis by site-directed mutagenesis. *Proteins* **1993**, *15*, 312–321. [CrossRef] [PubMed]

128. Lazarus, R.A.; McDowell, R.S. Structural and functional aspects of RGD-containing protein antagonists of glycoprotein IIb-IIIa. *Curr. Opin. Biotechnol.* **1993**, *4*, 438–443. [CrossRef]

129. Marcinkiewicz, C.; Vijay-Kumar, S.; McLane, M.A.; Niewiarowski, S. Significance of the RGD loop and C-terminal domain of echistatin for recognition of $\alpha_{IIb}\beta_3$ and $\alpha_v\beta_3$ integrins and expression of ligand-induced binding sites. *Blood* **1997**, *90*, 1565–1575. [PubMed]

130. Scarborough, R.M.; Rose, J.W.; Hsu, M.A.; Phillips, D.R.; Fried, V.A.; Campbell, A.M.; Nannizzi, L.; Charo, I.F. Barbourin. A GPIIb–IIIa-specific integrin antagonist from the venom of *Sistrurus m. barbouri*. *J. Biol. Chem.* **1991**, *266*, 9359–9362. [PubMed]

131. Calvete, J.J.; Moreno-Murciano, M.P.; Theakston, R.D.G.; Kisiel, D.G.; Marcinkiewicz, C. Snake venom disintegrins: Novel dimeric disintegrins and structural diversification by disulphide bond engineering. *Biochem. J.* **2003**, *372*, 725–734. [CrossRef] [PubMed]

132. Swenson, S.; Ramu, S.; Markland, F.S. Anti-angiogenesis and RGD-containing snake venom disintegrins. *Curr. Pharm. Des.* **2007**, *13*, 2860–2871. [CrossRef] [PubMed]

133. McLane, M.A.; Joerger, T.; Mahmoud, A. Disintegrins in health and disease. *Front. Biosci.* **2008**, *13*, 6617–6637. [CrossRef] [PubMed]

134. Selistre-de-Araujo, H.S.; Pontes, C.L.; Montenegro, C.F.; Martin, A.C. Snake venom disintegrins and cell migration. *Toxins* **2010**, *2*, 2606–2621. [CrossRef] [PubMed]

135. Marcinkiewicz, C. Applications of snake venom components to modulate integrin activities in cell-matrix interactions. *Int. J. Biochem. Cell Biol.* **2013**, *45*, 1974–1986. [CrossRef] [PubMed]

136. Calvete, J.J. The continuing saga of snake venom disintegrins. *Toxicon* **2013**, *62*, 40–49. [CrossRef] [PubMed]

137. Arruda Macêdo, J.K.; Fox, J.W.; de Souza Castro, M. Disintegrins from snake venoms and their applications in cancer research and therapy. *Curr. Protein Pept. Sci.* **2015**, *16*, 532–548. [CrossRef] [PubMed]

138. Zhou, Q.; Dangelmaier, C.; Smith, J.B. The hemorrhagin catrocollastatin inhibits collagen-induced platelet aggregation by binding to collagen via its disintegrin-like domain. *Biochem. Biophys. Res. Commun.* **1996**, *219*, 720–726. [CrossRef] [PubMed]

139. Moura-da-Silva, A.M.; Línica, A.; Della-Casa, M.S.; Kamiguti, A.S.; Ho, P.L.; Crampton, J.M.; Theakston, R.D.G. Jararhagin ECD-disintegrin-like domain: Expression in Escherichia coli and inhibition of the platelet-collagen interaction. *Arch. Biochem. Biophys.* **1999**, *369*, 295–301. [CrossRef] [PubMed]

140. Kamiguti, A.S.; Moura-da-Silva, A.M.; Laing, G.D.; Knapp, T.; Zuzel, M.; Crampton, J.M.; Theakston, R.D.G. Collagen-induced secretion-dependent phase of platelet aggregation is inhibited by the snake venom metalloproteinase jararhagin. *Biochim. Biophys. Acta* **1997**, *1335*, 209–217. [CrossRef]

141. Jia, L.G.; Wang, X.M.; Shannon, J.D.; Bjarnason, J.B.; Fox, J.W. Inhibition of platelet aggregation by the recombinant cysteine-rich domain of the hemorrhagic snake venom metalloproteinase, atrolysin A. *Arch. Biochem. Biophys.* **2000**, *373*, 281–286. [CrossRef] [PubMed]

142. Kamiguti, A.S.; Gallagher, P.; Marcinkiewicz, C.; Theakston, R.D.G.; Zuzel, M.; Fox, J.W. Identification of sites in the cysteine-rich domain of the class P-III snake venom metalloproteinases responsible for inhibition of platelet function. *FEBS Lett.* **2003**, *549*, 129–134. [CrossRef]

143. Pinto, A.F.; Terra, R.M.; Guimaraes, J.A.; Fox, J.W. Mapping von Willebrand factor A domain binding sites on a snake venom metalloproteinase cysteine-rich domain. *Arch. Biochem. Biophys.* **2007**, *457*, 41–46. [CrossRef] [PubMed]

144. Serrano, S.M.T.; Jia, L.G.; Wang, D.; Shannon, J.D.; Fox, J.W. Function of the cysteine-rich domain of the haemorrhagic metalloproteinase atrolysin A: Targeting adhesion proteins collagen I and von Willebrand factor. *Biochem. J.* **2005**, *391*, 69–76. [CrossRef] [PubMed]

145. Serrano, S.M.; Kim, J.; Wang, D.; Dragulev, B.; Shannon, J.D.; Mann, H.H.; Veit, G.; Wagener, R.; Koch, M.; Fox, J.W. The cysteine-rich domain of snake venom metalloproteinases is a ligand for von Willebrand factor A domains: Role in substrate targeting. *J. Biol. Chem.* **2006**, *281*, 39746–39756. [CrossRef] [PubMed]

146. Serrano, S.M.T.; Wang, D.; Shannon, J.D.; Pinto, A.F.; Polanowska-Grabowska, R.K.; Fox, J.W. Interaction of the cysteine-rich domain of snake venom metalloproteinases with the A1 domain of von Willebrand factor promotes site-specific proteolysis of von Willebrand factor and inhibition of von Willebrand factor-mediated platelet aggregation. *FEBS J.* **2007**, *274*, 3611–3621. [CrossRef] [PubMed]

147. Soejima, K.; Matsumoto, M.; Kokame, K.; Yagi, H.; Ishizashi, H.; Maeda, H.; Nozaki, C.; Miyata, T.; Fujimura, Y.; Nakagaki, T. ADAMTS-13 cysteine-rich/spacer domains are functionally essential for von Willebrand factor cleavage. *Blood* **2003**, *102*, 3232–3237. [CrossRef] [PubMed]

148. Dodson, G.; Wlodawer, A. Catalytic triads and their relatives. *Trends Biochem. Sci.* **1998**, *23*, 347–352. [CrossRef]

149. Paes Leme, A.F.; Escalante, T.; Pereira, J.G.; Oliveira, A.K.; Sanchez, E.F.; Gutiérrez, J.M.; Serrano, S.M.; Fox, J.W. High resolution analysis of snake venom metalloproteinase (SVMP) peptide bond cleavage specificity using proteome based peptide libraries and mass spectrometry. *J. Proteom.* **2011**, *74*, 401–410. [CrossRef] [PubMed]

150. Huntington, J.A. Molecular recognition mechanisms of thrombin. *J. Thromb. Haemost.* **2005**, *3*, 1861–1872. [CrossRef] [PubMed]

151. Bode, W. The structure of thrombin: A janus-headed proteinase. *Semin. Thromb. Hemost.* **2006**, *32* (Suppl. 1), 16–31. [CrossRef] [PubMed]

152. Koh, C.Y.; Kini, R.M. Thrombin inhibitors from haematophagous animals. In *Toxins and Hemostasis: From Bench to Bedside*; Kini, R.M., Clemetson, K.J., Markland, F.S., Jr., McLane, M.A., Morita, T., Eds.; Springer: Dordrecht, The Netherlands, 2010; pp. 239–254.

153. Maraganore, J.M.; Bourdon, P.; Jablonski, J.; Ramachandran, K.L.; Fenton, J.W. Design and characterization of hirulogs: A novel class of bivalent peptide inhibitors of thrombin. *Biochemistry* **1990**, *29*, 7095–7101. [CrossRef] [PubMed]

154. Skrzypczak-Jankun, E.; Carperos, V.E.; Ravichandran, K.G.; Tulinsky, A.; Westbrook, M.; Maraganore, J.M. Structure of the hirugen and hirulog 1 complexes of α-thrombin. *J. Mol. Biol.* **1991**, *221*, 1379–1393. [CrossRef]

155. Koh, C.Y.; Kazimirova, M.; Trimnell, A.; Takac, P.; Labuda, M.; Nuttall, P.A.; Kini, R.M. Variegin, a novel fast and tight binding thrombin inhibitor from the tropical bont tick. *J. Biol. Chem.* **2007**, *282*, 29101–29113. [CrossRef] [PubMed]

156. Fields, G.B. New strategies for targeting matrix metalloproteinases. *Matrix Biol.* **2015**, *44–46*, 239–246. [CrossRef] [PubMed]

157. Tykvart, J.; Schimer, J.; Jančařík, A.; Bařinková, J.; Navrátil, V.; Starková, J.; Šrámková, K.; Konvalinka, J.; Majer, P.; Šácha, P. Design of highly potent urea-based, exosite-binding inhibitors selective for glutamate carboxypeptidase II. *J. Med. Chem.* **2015**, *58*, 4357–4363. [CrossRef] [PubMed]

158. Xue, S.; Javor, S.; Hixon, M.S.; Janda, K.D. Probing BoNT/A protease exosites: Implications for inhibitor design and light chain longevity. *Biochemistry* **2014**, *53*, 6820–6824. [CrossRef] [PubMed]

159. Ascenzi, P.; Bocedi, A.; Bolli, A.; Fasano, M.; Notari, S.; Polticelli, F. Allosteric Modulation of Monomeric Proteins. *Biochem. Mol. Biol. Educ.* **2005**, *33*, 169–176. [CrossRef] [PubMed]

160. Ascenzi, P.; Fasano, M. Allostery in a monomeric protein: The case of human serum albumin. *Biophys. Chem.* **2010**, *148*, 16–22. [CrossRef] [PubMed]

161. Ascenzi, P.; Marino, M.; Polticelli, F.; Coletta, M.; Gioia, M.; Marini, S.; Pesce, A.; Nardini, M.; Bolognesi, M.; Reeder, B.J.; et al. Non-covalent and covalent modifications modulate the reactivity of monomeric mammalian globins. *Biochim. Biophys. Acta* **2013**, *1834*, 1750–1756. [CrossRef] [PubMed]

162. Udi, Y.; Fragai, M.; Grossman, M.; Mitternacht, S.; Arad-Yellin, R.; Calderone, V.; Melikian, M.; Toccafondi, M.; Berezovsky, I.N.; Luchinat, C.; et al. Unraveling hidden regulatory sites in structurally homologous metalloproteases. *J. Mol. Biol.* **2013**, *425*, 2330–2346. [CrossRef] [PubMed]

163. Christopoulos, A. Allosteric binding sites on cell-surface receptors: Novel targets for drug discovery. *Nat. Rev. Drug Discov.* **2002**, *1*, 198–210. [CrossRef] [PubMed]

164. Tsai, C.J.; Del Sol, A.; Nussinov, R. Protein allostery, signal transmission and dynamics: A classification scheme of allosteric mechanisms. *Mol. Biosyst.* **2009**, *5*, 207–216. [CrossRef] [PubMed]

165. De Jong, L.A.; Uges, D.R.; Franke, J.P.; Bischoff, R. Receptor-ligand binding assays: Technologies and applications. *J. Chromatogr. B* **2005**, *829*, 1–25. [CrossRef] [PubMed]

166. Hulme, E.C.; Trevethick, M.A. Ligand binding assays at equilibrium: Validation and interpretation. *Br. J. Pharmacol.* **2010**, *161*, 1219–1237. [CrossRef] [PubMed]

167. Ohsaka, A. Hemorrhagic, necrotizing and edema-forming effects of snake venoms. In *Handbook of Experimental Pharmacology*; Lee, C.Y., Ed.; Springer: Berlin, Germany, 1979; Volume 52, pp. 480–546.

168. Bjarnason, J.B.; Fox, J.W. Hemorrhagic metalloproteinases from snake venoms. *Pharmacol. Ther.* **1994**, *62*, 325–372. [CrossRef]

169. Kini, R.M. Toxins in thrombosis and haemostasis: Potential beyond imagination. *J. Thromb. Haemost.* **2011**, *9* (Suppl. 1), 195–208. [CrossRef] [PubMed]

170. McCleary, R.J.R.; Kini, R.M. Snake bites and hemostasis/thrombosis. *Thromb. Res.* **2013**, *132*, 642–646. [CrossRef] [PubMed]

171. Berling, I.; Isbister, G.K. Hematologic effects and complications of snake envenoming. *Transfus. Med. Rev.* **2015**, *29*, 82–89. [CrossRef] [PubMed]

172. Markland, F.S. Snake venoms and the hemostatic system. *Toxicon* **1998**, *36*, 1749–1800. [CrossRef]

173. Hougie, C. Effect of Russell's viper venom (Stypven) on Stuart clotting defect. *Proc. Soc. Exp. Biol. Med.* **1956**, *98*, 570–573. [CrossRef]

174. Thiagarajan, P.; Pengo, V.; Shapiro, S.S. The use of the dilute Russell viper venom time for the diagnosis of lupus anticoagulants. *Blood* **1986**, *68*, 869–874. [PubMed]

175. Court, E.L. Lupus anticoagulants: Pathogenesis and laboratory diagnosis. *Br. J. Biomed. Sci.* **1997**, *54*, 287–298. [PubMed]

176. Nowak, G. The ecarin clotting time, a universal method to quantify direct thrombin inhibitors. *Pathophysiol. Haemost. Thromb.* **2003**, *33*, 173–183. [CrossRef] [PubMed]

177. Lange, U.; Nowak, G.; Bucha, E. Ecarin chromogenic assay–a new method for quantitative determination of direct thrombin inhibitors like hirudin. *Pathophysiol. Haemost. Thromb.* **2003**, *33*, 184–191. [CrossRef] [PubMed]

178. Marsh, N.; Williams, V. Practical applications of snake venom toxins in haemostasis. *Toxicon* **2005**, *45*, 1171–1181. [CrossRef] [PubMed]

179. Schoni, R. The use of snake venom-derived compounds for new functional diagnostic test kits in the field of haemostasis. *Pathophysiol. Haemost. Thromb.* **2005**, *34*, 234–240. [CrossRef] [PubMed]

180. Vija, H.; Samel, M.; Siigur, E.; Aaspõllu, A.; Tonismagi, K.; Trummal, K.; Subbi, J.; Siigur, J. VGD and MLD motifs containing heterodimeric disintegrin viplebedin-2 from *Vipera lebetina* snake venom. Purification and cDNA cloning. *Comp. Biochem. Physiol. B Biochem. Mol. Biol.* **2009**, *153*, 572–580. [CrossRef] [PubMed]

181. Okuda, D.; Koike, H.; Morita, T. A new gene structure of the disintegrin family: A subunit of dimeric disintegrin has a short coding region. *Biochemistry* **2002**, *41*, 14248–14254. [CrossRef] [PubMed]

182. Bazaa, A.; Marrakchi, N.; El Ayeb, M.; Sanz, L.; Calvete, J.J. Snake venomics: Comparative analysis of the venom proteomes of the Tunisian snakes *Cerastes cerastes*, *Cerastes vipera* and *Macrovipera lebetina*. *Proteomics* **2005**, *5*, 4223–4235. [CrossRef] [PubMed]

183. Walsh, E.M.; Marcinkiewicz, C. Non-RGD-containing snake venom disintegrins, functional and structural relations. *Toxicon* **2011**, *58*, 355–362. [CrossRef] [PubMed]

184. Kisiel, D.G.; Calvete, J.J.; Katzhendler, J.; Fertala, A.; Lazarovici, P.; Marcinkiewicz, C. Structural determinants of the selectivity of KTS-disintegrins for the $\alpha_1\beta_1$ integrin. *FEBS Lett.* **2004**, *577*, 478–482. [CrossRef] [PubMed]

185. Marcinkiewicz, C. Functional characteristic of snake venom disintegrins: Potential therapeutic implication. *Curr. Pharm. Des.* **2005**, *11*, 815–827. [CrossRef] [PubMed]

186. Olfa, K.Z.; Luis, J.; Daoud, S.; Bazaa, A.; Srairi-Abid, N.; Andreotti, N.; Lehmann, M.; Zouari, R.; Mabrouk, K.; Marvaldi, J.; et al. Lebestatin, a disintegrin from *Macrovipera* venom, inhibits integrin-mediated cell adhesion, migration and angiogenesis. *Lab. Investig.* **2005**, *85*, 1507–1516. [CrossRef] [PubMed]

187. Sanz, L.; Chen, R.Q.; Perez, A.; Hilario, R.; Juarez, P.; Marcinkiewicz, C.; Monleón, D.; Celda, B.; Xiong, Y.L.; Pérez-Payá, E.; et al. cDNA cloning and functional expression of jerdostatin, a novel RTS-disintegrin from *Trimeresurus jerdonii* and a specific antagonist of the $\alpha_1\beta_1$ integrin. *J. Biol. Chem.* **2005**, *208*, 40714–40722. [CrossRef] [PubMed]

188. Brown, M.C.; Eble, J.A.; Calvete, J.J.; Marcinkiewicz, C. Structural requirements of KTS-disintegrins for inhibition of $\alpha_1\beta_1$ integrin. *Biochem. J.* **2009**, *417*, 95–101. [CrossRef] [PubMed]

189. Calvete, J.J.; Marcinkiewicz, C.; Sanz, L. KTS and RTS-disintegrins: Anti-angiogenic viper venom peptides specifically targeting the alpha 1 beta 1 integrin. *Curr. Pharm. Des.* **2007**, *13*, 2853–2859. [CrossRef] [PubMed]

190. Calvete, J.J.; Marcinkiewicz, C.; Monleón, D.; Esteve, V.; Celda, B.; Juárez, P.; Sanz, L. Snake venom disintegrins: Evolution of structure and function. *Toxicon* **2005**, *45*, 1063–1074. [CrossRef] [PubMed]

191. Davenpeck, K.L.; Marcinkiewicz, C.; Wang, D.; Niculescu, R.; Shi, Y.; Martin, J.L.; Zalewski, A. Regional differences in integrin expression: Role of $\alpha_5\beta_1$ in regulating smooth muscle cell functions. *Circ. Res.* **2001**, *88*, 352–358. [CrossRef] [PubMed]

192. Sekimoto, H.; Eipper-Mains, J.; Pond-Tor, S.; Boney, C.M. $\alpha_v\beta_3$ integrins and Pyk2 mediate insulin-like growth factor I activation of Src and mitogen-activated protein kinase in 3T3-L1 cells. *Mol. Endocrinol.* **2005**, *19*, 1859–1867. [CrossRef] [PubMed]

193. Surazynski, A.; Sienkiewicz, P.; Wolczynski, S.; Palka, J. Differential effects of echistatin and thrombin on collagen production and prolidase activity in human dermal fibroblasts and their possible implication in β_1-integrin-mediated signaling. *Pharmacol. Res.* **2005**, *51*, 217–221. [CrossRef] [PubMed]

194. Pechkovsky, D.V.; Scaffidi, A.K.; Hackett, T.L.; Ballard, J.; Shaheen, F.; Thompson, P.J.; Thannickal, V.J.; Knight, D.A. Transforming growth factor β_1 induces $\alpha_v\beta_3$ integrin expression in human lung fibroblasts via a β_3 integrin-, c-Src-, and p38 MAPK-dependent pathway. *J. Biol. Chem.* **2008**, *283*, 12898–12908. [CrossRef] [PubMed]

195. Belisario, M.A.; Tafuri, S.; Pontarelli, G.; Staiano, N.; Gionti, E. Modulation of chondrocyte adhesion to collagen by echistatin. *Eur. J. Cell Biol.* **2005**, *8410*, 833–842. [CrossRef] [PubMed]

196. Long, R.K.; Nishida, S.; Kubota, T.; Wang, Y.; Sakata, T.; Elalieh, H.Z.; Halloran, B.P.; Bikle, D.D. Skeletal unloading induced insulin-like growth factor 1 (IGF-1) nonresponsiveness is not shared by platelet-derived growth factor: The selective role of integrins in IGF-1 signaling. *J. Bone Miner. Res.* **2011**, *26*, 2948–2958. [CrossRef] [PubMed]

197. Harper, M.M.; Ye, E.A.; Blong, C.C.; Jacobson, M.L.; Sakaguchi, D.S. Integrins contribute to initial morphological development and process outgrowth in rat adult hippocampal progenitor cells. *J. Mol. Neurosci.* **2010**, *40*, 269–283. [CrossRef] [PubMed]

198. Moraes, J.A.; Frony, A.C.; Dias, A.M.; Renovato-Martins, M.; Rodrigues, G.; Marcinkiewicz, C.; Assreuy, J.; Barja-Fidalgo, C. $\alpha_1\beta_1$ and integrin-linked kinase interact and modulate angiotensin II effects in vascular smooth muscle cells. *Atherosclerosis* **2015**, *243*, 477–485. [CrossRef] [PubMed]

199. Moraes, J.A.; Frony, A.C.; Dias, A.M.; Renovato-Martins, M.; Rodrigues, G.; Marcinkiewicz, C.; Assreuy, J.; Barja-Fidalgo, C. Data in support of $\alpha_1\beta_1$ and integrin-linked kinase interact and modulate angiotensin II effects in vascular smooth muscle cells. *Data Brief* **2015**, *6*, 330–340. [CrossRef] [PubMed]

200. Zeymer, U.; Wienbergen, H. A review of clinical trials with eptifibatide in cardiology. *Cardiovasc. Drug Rev.* **2007**, *25*, 301–315. [CrossRef] [PubMed]

201. Krotz, F.; Sohn, H.Y.; Klauss, V. Antiplatelet drugs in cardiological practice: Established strategies and new developments. *Vasc. Health Risk Manag.* **2008**, *4*, 637–675. [PubMed]

202. Zhou, Q.; Nakada, M.T.; Arnold, C.; Shieh, K.Y.; Markland, F.S., Jr. Contortrostatin, a dimeric disintegrin from *Agkistrodon contortrix contortrix*, inhibits angiogenesis. *Angiogenesis* **1999**, *3*, 259–269. [CrossRef] [PubMed]
203. Markland, F.S.; Shieh, K.; Zhou, Q.; Golubkov, V.; Sherwin, R.P.; Richters, V.; Sposto, R.A. Novel snake venom disintegrin that inhibits human ovarian cancer dissemination and angiogenesis in an orthotopic nude mouse model. *Haemostasis* **2001**, *31*, 183–191. [CrossRef] [PubMed]
204. Swenson, S.; Costa, F.; Ernst, W.; Fujii, G.; Markland, F.S. Contortrostatin, a snake venom disintegrin with anti-angiogenic and anti-tumor activity. *Pathophysiol. Haemost. Thromb.* **2005**, *34*, 169–176. [CrossRef] [PubMed]
205. Swenson, S.; Costa, F.; Minea, R.; Sherwin, R.P.; Ernst, W.; Fujii, G.; Yang, D.; Markland, F.S. Intravenous liposomal delivery of the snake venom disintegrin contortrostatin limits breast cancer progression. *Mol. Cancer Ther.* **2004**, *3*, 499–511. [PubMed]
206. Minea, R.O.; Helchowski, C.M.; Zidovetzki, S.J.; Costa, F.K.; Swenson, S.D.; Markland, F.S., Jr. Vicrostatin—An anti-invasive multi-integrin targeting chimeric disintegrin with tumor anti-angiogenic and pro-apoptotic activities. *PLoS ONE* **2010**, *5*. [CrossRef] [PubMed]
207. Momic, T.; Katzehendler, J.; Benny, O.; Lahiani, A.; Cohen, G.; Noy, E.; Senderowitz, H.; Eble, J.A.; Marcinkiewicz, C.; Lazarovici, P. Vimocin and vidapin, cyclic KTS peptides, are dual antagonists of $\alpha_1\beta_1/\alpha_2\beta_1$ integrins with antiangiogenic activity. *J. Pharmacol. Exp. Ther.* **2014**, *350*, 506–519. [CrossRef] [PubMed]
208. Momic, T.; Katzehendler, J.; Shai, E.; Noy, E.; Senderowitz, H.; Eble, J.A.; Marcinkiewicz, C.; Varon, D.; Lazarovici, P. Vipegitide: A folded peptidomimetic partial antagonist of $\alpha_2\beta_1$ integrin with antiplatelet aggregation activity. *Drug Des. Dev. Ther.* **2015**, *9*, 291–304.
209. Wijeyewickrema, L.C.; Berndt, M.C.; Andrews, R.K. Snake venom probes of platelet adhesion receptors and their ligands. *Toxicon* **2005**, *45*, 1051–1061. [CrossRef] [PubMed]
210. Kamiguti, A.S. Platelets as targets of snake venom metalloproteinases. *Toxicon* **2005**, *45*, 1041–1049. [CrossRef] [PubMed]
211. Randolph, A.; Chamberlain, S.H.; Chu, H.L.; Retzios, A.D.; Markland, F.S., Jr.; Masiarz, F.R. Amino acid sequence of fibrolase, a direct-acting fibrinolytic enzyme from *Agkistrodon contortrix contortrix* venom. *Protein Sci.* **1992**, *1*, 590–600. [CrossRef] [PubMed]
212. Markland, F.S.; Friedrichs, G.S.; Pewitt, S.R.; Lucchesi, B.R. Thrombolytic effects of recombinant fibrolase or APSAC in a canine model of carotid artery thrombosis. *Circulation* **1994**, *90*, 2448–2456. [CrossRef] [PubMed]
213. Toombs, C.F. Alfimeprase: Pharmacology of a novel fibrinolytic metalloproteinase for thrombolysis. *Haemostasis* **2001**, *31*, 141–147. [CrossRef] [PubMed]
214. Ouriel, K.; Cynamon, J.; Weaver, F.A.; Dardik, H.; Akers, D.; Blebea, J.; Gruneiro, L.; Toombs, C.F.; Wang-Clow, F.; Mohler, M.; et al. A phase I trial of alfimeprase for peripheral arterial thrombolysis. *J. Vasc. Interv. Radiol.* **2005**, *16*, 1075–1083. [CrossRef] [PubMed]
215. Moll, S.; Kenyon, P.; Bertoli, L.; De Maio, J.; Homesley, H.; Deitcher, S.R. Phase II trial of alfimeprase, a novel-acting fibrin degradation agent, for occluded central venous access devices. *J. Clin. Oncol.* **2006**, *24*, 3056–3060. [CrossRef] [PubMed]
216. Markland, F.S.; Swenson, S. Fibrolase: Trials and tribulations. *Toxins* **2010**, *2*, 793–808. [CrossRef] [PubMed]
217. Lee, J.Y.; Markland, F.S.; Lucchesi, B.R. Hirudin and S18886 maintain luminal patency after thrombolysis with alfimeprase. *J. Cardiovasc. Pharmacol.* **2013**, *61*, 152–159. [CrossRef] [PubMed]

Review

Natural Inhibitors of Snake Venom Metalloendopeptidases: History and Current Challenges

Viviane A. Bastos [1,2], Francisco Gomes-Neto [1,2], Jonas Perales [1,2], Ana Gisele C. Neves-Ferreira [1,2] and Richard H. Valente [1,2,*]

[1] Laboratory of Toxinology, Oswaldo Cruz Foundation (FIOCRUZ), Rio de Janeiro 21040-900, Brazil; vivika.bastos@gmail.com (V.A.B.); gomes.netof@gmail.com (F.G.-N.); jperales@ioc.fiocruz.br or jonasperales@gmail.com (J.P.); anag@ioc.fiocruz.br or anagextra@gmail.com (A.G.C.N.-F.)

[2] National Institute of Science and Technology on Toxins (INCTTOX), CNPq, Brasilia 71605-001, Brazil

* Correspondance: richardhemmi@gmail.com; Tel.: +55-21-2562-1345

Academic Editors: José María Gutiérrez and Jay Fox

Received: 18 June 2016; Accepted: 15 August 2016; Published: 26 August 2016

Abstract: The research on natural snake venom metalloendopeptidase inhibitors (SVMPIs) began in the 18th century with the pioneering work of Fontana on the resistance that vipers exhibited to their own venom. During the past 40 years, SVMPIs have been isolated mainly from the sera of resistant animals, and characterized to different extents. They are acidic oligomeric glycoproteins that remain biologically active over a wide range of pH and temperature values. Based on primary structure determination, mammalian plasmatic SVMPIs are classified as members of the immunoglobulin (Ig) supergene protein family, while the one isolated from muscle belongs to the ficolin/opsonin P35 family. On the other hand, SVMPIs from snake plasma have been placed in the cystatin superfamily. These natural antitoxins constitute the first line of defense against snake venoms, inhibiting the catalytic activities of snake venom metalloendopeptidases through the establishment of high-affinity, non-covalent interactions. This review presents a historical account of the field of natural resistance, summarizing its main discoveries and current challenges, which are mostly related to the limitations that preclude three-dimensional structural determinations of these inhibitors using "gold-standard" methods; perspectives on how to circumvent such limitations are presented. Potential applications of these SVMPIs in medicine are also highlighted.

Keywords: cross-linking; hydrogen/deuterium exchange; mass spectrometry; metalloendopeptidase inhibitor; modeling; natural immunity; natural resistance; snake venom; structure; therapeutic application

1. Introduction

Snakes and their venoms have always driven the fascination and curiosity of mankind—including the desire to freely handle them without being harmed by the venomous effects of their bites. Members of some ancient tribes used to drink small amounts of venom seeking protection from future envenomation; curiously, individuals from other tribes were thought to be resistant as a consequence of having snake blood running through their veins [1].

The idea that snakes could be resistant to their own venom traces back to Greek philosophers and physicians. Galen of Pergamum (131–*ca.* 201 A.D.) described in his treatises *De antidotis* and *De Theriaca ad Pisonem* the recipe for a concoction named "Theriac of Andromachus", which consists of a variety of ingredients including viper's flesh. This theriac was believed to be, amongst other things, an antidote to snakebite [2,3]. Many centuries later, Felice Fontana (1730–1805), an abbot from Trentino (Italy) [4,5], inoculated the venom of the common European viper (*Vipera berus*) into the viper itself. He observed that the animal did not display any symptoms of envenomation, even after thirty six hours, leading to his celebrated aphorism "*Il veleno della Vipera non è veleno per la propria specie*" [6], later translated by Joseph Skinner to "The venom of the viper is not a poison to the viper itself" [7].

Based on the viper's resistance to its own venom, Fontana concluded that this phenomenon was limited to the same species. Later, Guyon (1861) inoculated the venom of the viper into different snake species and found this resistance to be inter-specific [8]. By the end of the nineteenth century, Calmette, Phisalix and Bertrand first described the natural resistance displayed by some mammals, such as the mongoose and the hedgehog, towards snake bite envenomation [9,10]. Reports of natural protection against snake venom pathophysiological effects were also published in relation to several snakes, such as rattlesnakes (*Crotalus* sp.) [11], *Thamnophis s. sirtalis*, *Pituophis s. sayi*, *Natrix taxipilota*, *T. sirtalis infernalis*, *Heterodon contortrix* [12], *Sistrurus c. catenatus* [13], *Lampropeltis getulus floridana* [14], *Pseudoboa cloelia* [15], *Crotalus atrox* [16], *C. adamanteus* [17], and several mammals from the Didelphidae [18,19]. For a comprehensive review on the early days of the natural resistance field, the reader is referred to the work by Domont et al. [20].

After discovering the phenomenon of natural resistance, researchers in the field began to investigate its underlying mechanism of action. It is now currently accepted that this resistance can be conferred through two non-mutually exclusive mechanisms. In the first type, the resistant animal displays mutation(s) in the receptor(s) targeted by the snake's toxin(s), which prevent(s) the deleterious effect(s). The second mechanism, on which this review will focus, involves the occurrence of serum proteins that neutralize the toxins by forming noncovalent complexes, rendering them unable to exert their pathophysiological effects [21]. These natural inhibitors are distributed in two major classes—the phospholipases A_2 inhibitors (PLIs), which effectively inhibit the neuro- and myotoxic effects of snake venoms (for comprehensive reviews see [22–24]), and the SVMPIs, which can suppress the hemorrhagic symptoms commonly associated with Viperidae envenomation. In 2002, it was proposed that such inhibitors may be an important feature of the innate immune system of those venom-resistant animals due to their structural similarity to other proteins that exert relevant functions in immunity, and for acting as ready-made soluble acceptors in the serum, thus constituting the first line of defense against snake venom toxins [25].

During the second half of the 20th century, a large portion of the research in this field has been devoted to the isolation of SVMPIs for further physicochemical and chemical characterizations, including primary structure determination. However, over the last 15 years, the main goal of natural resistance research shifted from protein purification to mechanistic studies in an attempt to understand the interaction between inhibitors and target toxins at the molecular level. This review does not intend to present all known SVMPIs and their determined characteristics; this information can be found by the reader in a historical series of reviews [20,21,24,26–29]. In fact, with this contribution, we aimed to summarize the available knowledge in the field of SVMPIs (Figure 1) and to discuss novel perspectives in this research area, especially on how to address the actual bottleneck due to the lack of information on the three-dimensional structures of SVMPIs (Figure 2).

Early days

1787
Felice Fontana
First experimental evidence on the natural resistance to snake venoms [6].

1861
J.Guyon
Identification of the inter-specific nature of the resistance to snakebite envenomation [8] .

1895
Albert Calmette, Césaire Phisalix, Gabriel Bertrand
First report of mammals presenting natural immunity to snake envenomation [9,10].

1945
J.Vellard
Description of the natural resistance from mammals of the *Didelphidae* family [18,19].

Last 40 years

1972
Tamotsu Omori-Satoh
First purification of an antihemorrhagic factor (HSF) from the serum of a snake, *Protobothrops flavoviridis* [44].

1981
Juan M. Menchaca and John C.Pérez
First purification of an antihemorrhagic factor from the serum of a mammal, *Didelphis virginiana* [71].

1992
Yoshio Yamakawa and Tamotsu Omori-Satoh
Elucidation of the primary sequence of HSF; the first metalloendopeptidase inhibitor identified as a member of the cystatin superfamily of proteins [46].
Joseph J. Catanese and Lawrence F. Kress
Purification of oprin, the first metalloendopeptidase inhibitor identified as a member of the immunoglobulin supergene family of proteins [72].

1996
Dietrich Mebs and colleagues
Purification of a high-molecular-weight antihemorrhagic factor (erinacin) from the muscle extract of the european hedgehog *Erinaceus europaeus* [95].

2000
Tamotsu Omori-Satoh and colleagues
Molecular characterization and aminoacid sequence analysis of erinacin and classification into the ficolin/opsonin P35 superfamily [98].

2002
Jonas Perales and Gilberto B. Domont
Classification proposal of the natural inhibitors of metalloendopeptidases,phospholipases A₂ and myotoxins into the innate immune system of these animals [25].

2012
Ana Gisele C. Neves-Ferreira and colleagues
First quantitative studies on the interaction between a metalloendopeptidase inhibitor (DM43) and a SVMP (jararhagin) [91].

Future directions

Structural elucidation of inhibitors and inhibitor:toxin complexes (XL-MS, HDX-MS, SAXS, molecular modeling and dynamics).

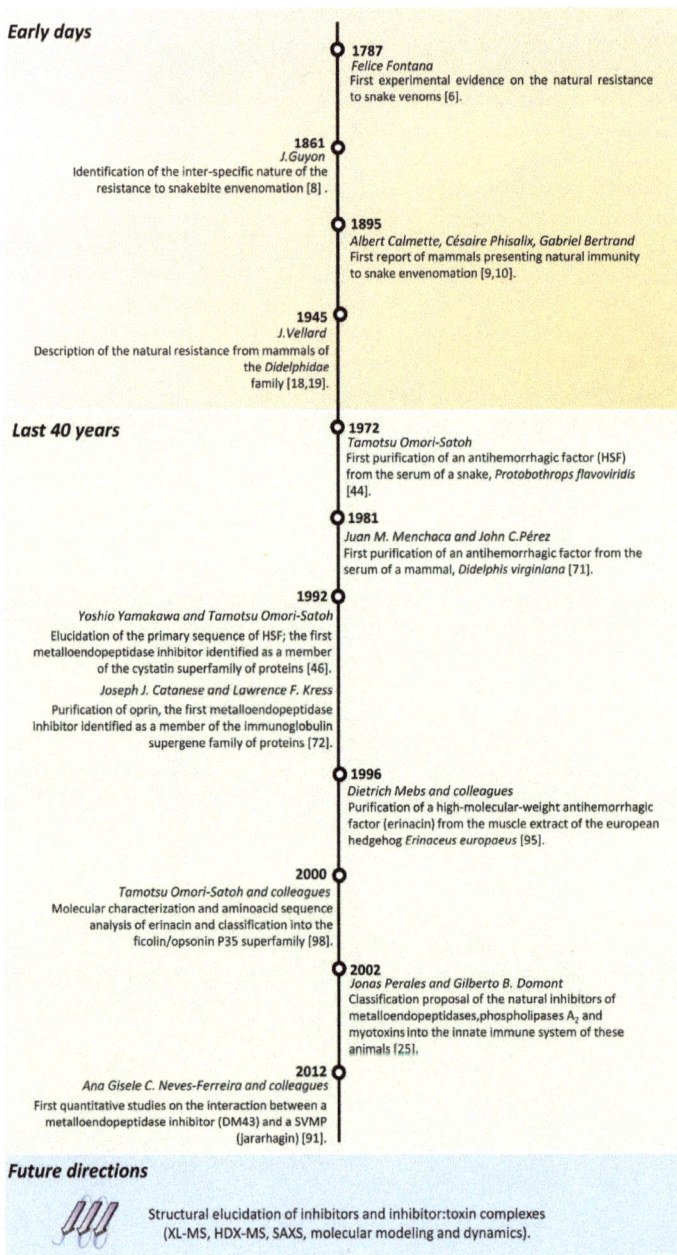

Figure 1. Research milestones on natural inhibitors of metalloendopeptidases. The investigation on the natural resistance that some animals presented to snake venoms began in the eighteenth century. Since Fontana's pioneering work, the field has grown considerably. Researchers have managed to purify several inhibitors from the sera of snakes and mammals and determined their relevant physicochemical properties. The challenges that lie ahead are the three-dimensional structure elucidation of these snake venom metalloendopeptidase inhibitors (SVMPIs) in their free and toxin-complexed forms in order to better understand the molecular dynamics of this interaction.

2. Biochemical Background

2.1. Snake Venom Metalloendopeptidases (Metalloproteinases)

In the early days of experimental research on the effects of viperid envenomation, hemorrhage was recognized as one of its main clinical features [30]. At that time, the mechanism of hemorrhage was largely unknown, and some authors referred to the principle in snake venom that caused hemorrhage as "hemorrhagin" [31].

In 1960, Japanese investigators were able to purify peptidases from *Trimeresurus (Protobothrops) flavoviridis* venom that displayed hemorrhagic activity. Their functional assays showed that both the proteolytic and hemorrhagic activities of these proteins were eliminated following EDTA addition, indicating that these molecules were most likely metallopeptidases [32–35]. The atomic absorption spectroscopy experiments conducted by Bjarnason and Tu in 1978 confirmed this hypothesis, demonstrating that hemorrhagins are zinc-dependent metallopeptidases, containing 1 mol of zinc ion per mol of enzyme [36].

According to their structural features, snake venom metalloendopeptidases (SVMPs) are currently grouped into three main classes: PI, PII, and PIII. SVMPs belonging to the PI class present only the metalloendopeptidase domain in their structure, whereas the enzymes belonging to the PII class present an additional disintegrin domain. Members of the PIII class present metalloendopeptidase, disintegrin-like, and cysteine-rich domains; eventually, a lectin-like domain may be present (PIIId) [37].

Upon venom injection, SVMPs primarily target the capillary vessels, hydrolyzing components of the basement membrane and promoting apoptosis of endothelial cells, leading to the extravasation of blood components [38,39]. Together with other snake toxins, SVMPs can also promote dermonecrosis and inflammatory reactions [40]. The local effects prompted by SVMPs occur shortly after the bite and contribute prominently to the high morbidity rates observed in snakebite envenoming [41,42]. Apart from their contribution to the important tissue damage frequently observed at the site of venom injection, SVMPs may also trigger systemic effects, being key toxins to the pathophysiology of snake envenomation [43].

2.2. SVMPIs Isolated from Snakes

2.2.1. Cystatin Superfamily (Fetuin-Like Proteins)

The first natural SVMPI purified from the sera of snakes was HSF (habu serum factor), from the serpent *Protobothrops (Trimeresurus) flavoviridis*. The purified protein has a molecular mass of 70 kDa, determined by ultracentrifugal sedimentation equilibrium, and an isoelectric point of 4.0. It inhibits the proteolytic activities of HR1 and HR2, two P-III class SVMPs isolated from the venom of this same snake, both in vivo and in vitro. Interestingly, no precipitin line was detected in immunodiffusion assays with the crude venom, HR1 or HR2, indicating that the neutralizing factor was not an immunoglobulin [44,45]. Indeed, HSF is a 323-amino acid-long glycoprotein that possesses two cystatin-like domains located in the *N*-terminal portion, followed by a C-terminal His-rich domain. MALDI-TOF MS (matrix-assisted laser desorption/ionization mass spectrometry) analysis of HSF showed that it has a molecular mass of 47,810 Da; compared to the native molecular mass (70 kDa), the results seem to indicate that HSF is homodimeric in solution [24,46,47]. Using molecular exclusion chromatography, Deshimaru and colleagues demonstrated that HSF binds the H6 protease from *Gloydius halys brevicaudus* venom at a 1:1 molar ratio; it also effectively inhibited several P-I, P-II and P-III class SVMPs from the venoms of *T. flavoviridis* (HR1A, HR1B, HR2a, HR2b, and H2) and *G. h. brevicaudus* (brevilysins H3, H4, H6, and L4), indicating that HSF has a broad inhibitory specificity, irrespective of the metalloendopeptidases' domain architecture [48]. Recently, HSF was shown to interact with small serum proteins (SSP), i.e., low-molecular mass proteins from *T. flavoviridis* serum with unknown functions [49]. The interaction of HSF with SSP-1 allowed the inhibition of HV1, a P-III class SVMP isolated from the venom of this same snake. Neither HSF nor SSP-1 alone could inhibit HV1; it was only through the ternary complex among HSF, SSP-1 and HV1 that the toxin's catalytic activity was abolished [50]. Currently, little is known about the regions of interaction between HSF and its target toxins. Aoki and co-workers have shown that the N-terminal half (residues 1–89) of the first cystatin-like domain of HSF is essential to its inhibitory activity. Additionally, molecular modeling analyses pinpointed a cluster of amino acid residues (Trp17, Trp48, Lys15, and Lys41) involved in the inhibition of SVMPs by HSF [51].

Another well-studied inhibitor from snake serum is BJ46a, isolated from *Bothrops jararaca*. BJ46a also presents two cystatin-like domains sharing 85% sequence identity with HSF. The glycoprotein's primary structure, confirmed by both Edman degradation and cDNA sequencing, consisted of 322 amino acids with 12 cysteine residues and four N-glycosylation sites (Asn76, Asn185, Asn263, and Asn274). By MALDI-TOF MS, BJ46a showed a molecular mass of 46,101 Da; by molecular exclusion chromatography and dynamic laser light scattering, it has a calculated mass of 79 kDa, suggesting a homodimeric structure. BJ46a effectively inhibited the proteolytic activity of the P-III class SVMP jararhagin and the P-I class atrolysin-C upon the fluorogenic peptide Abz-Ala-Gly-Leu-Ala-Nbz; titration experiments using molecular exclusion chromatography indicated that the inhibitor interacted with those SVMPs at a 1:2 (BJ46a monomer:toxin) molar ratio. SDS-PAGE (sodium dodecyl sulfate polyacrylamide gel electrophoresis) analyses, under reducing and nonreducing conditions, showed a noncovalent interaction between the inhibitor and each target toxin. Interestingly, BJ46a was not able to interact with jararhagin-C [52], a processed form of the SVMP jararhagin possessing only the disintegrin-like and cysteine-rich domains, even at a three-fold molar excess of the inhibitor, suggesting that the toxin's metalloendopeptidase domain is essential for BJ46a binding [53]. Upon complex formation, BJ46a dimer dissociated and each monomer noncovalently interacted with two molecules of metalloendopeptidase; thus, the inhibitor may have two toxin-binding sites for each monomer, a different stoichiometry than that reported for HSF [48]. A preliminary molecular modeling for BJ46a first cystatin domain was done, using HSF's model as template. The results indicated that, in addition to the cluster of residues Trp17, Trp48, Lys15, and Lys41 (also found in the three-dimensional model of HSF), BJ46a has a second cluster formed by the residues Trp52 and Lys58 in the first cystatin-like domain that could be involved in the binding of a second toxin molecule; another possibility is the involvement of the second cystatin-like domain in the interaction, although these assumptions remain to be experimentally verified [28]. Recently, Shi and colleagues expressed BJ46a in the methylotrophic yeast *Pichia pastoris* [54]. This recombinant BJ46a (rBJ46a) showed a molecular mass of 58 kDa, although after treatment with endoglycosidase H (for the removal of high mannose glycans) its mass was reduced to 38 kDa (corresponding to the protein moiety). rBJ46a was able to reduce the invasion of B16F10 melanoma cells and MHCC97H hepatocellular carcinoma cells in an in vitro trans-well migration assay. In subsequent in vivo assays, rBJ46a partially inhibited tissue colonization in a lung cancer model (C57BL/6 mice infected with B16F10 cells) and reduced the occurrence of metastasis in BALB/c nude mice infected with MHCC97H cells. The authors attributed these antitumoral activities to rBJ46a inhibitory capacity towards matrix metalloendopeptidases (MMPs) 2 and 9, even though no clear evidence was presented to support this claim [55].

From the sera of the Chinese (*Gloydius blomhoffi brevicaudus*) and the Japanese (*G. blomhoffi*) mamushis, two inhibitors were purified: cMSF and jMSF. Both proteins have a molecular mass of 40,500 Da by MALDI-TOF MS and were also classified as members of the cystatin superfamily, presenting sequence identities of 84% (cMSF) and 83% (jMSF) when compared to HSF. However, these new inhibitors presented a 17-residue deletion within their C-terminal His-rich domain. Despite this deletion, the inhibitor cMSF suppressed mamushi venom-induced hemorrhage in a dose-dependent manner and inhibited the proteolytic activities of the P-III class SVMPs HR1A and HR1B from *Protobothrops flavoviridis* venom. As for jMSF, it interacted with brevilysins H2, H3, H4, and H6 from *G. blomhoffi brevicaudus* venom but was unable to inhibit the SVMP HR2a from *P. flavoviridis* venom and brevilysin L6 from *Agkistrodon halys brevicaudus*. Even though the previously mentioned C-terminal deletion did not affect cMSF anti-hemorrhagic activity, the authors demonstrated that cMSF has a lower thermal stability limit (60 °C) when compared to HSF (100 °C) [56]. Shioi and colleagues also managed to purify SSPs from the serum of the Japanese mamushi that, as above discussed for HSF, interacted with jMSF in a yet to be described mechanism [57].

From the sera of *P. flavoviridis* and *G. blomhoffii brevicaudus*, Aoki and coworkers purified and characterized three proteins: habu HLP from *P. flavoviridis* and HLP-A and HLP-B from *G. b. brevicaudus*. All three proteins showed sequence homology to HSF but were devoid of antihemorrhagic activity; therefore, those proteins were named HLP, standing for habu-like proteins. One of these HLPs (HLP-B) was able to inhibit calcium phosphate precipitation, characterizing it as a *bona fide* (snake) fetuin, a protein class that is known to interact with calcium and prevent calcification [58]. To map the protein regions responsible for the maintenance of the antihemorrhagic activity, the sequences of all SVMPIs belonging to the cystatin superfamily of proteins (BJ46a, HSF, cMSF, and jMSF) were aligned with the deduced HLP sequences. The first cystatin-like domain showed approximately 60% identity between inhibitors and HLPs, whereas the second domain was conserved amongst all proteins (84% to 94% identity), indicating that the diversification process that originated SVMPIs and HLPs resulted from an alteration in amino acid sequences in the first cystatin-like domain [59]. Finally, the authors propose that these three snake blood proteins from the fetuin family (SVMPIs, HLPs, and true fetuin) evolved via gene duplication from a common ancestor to achieve different functions, including conferring resistance against the deleterious effects of envenomation [59].

2.2.2. Undetermined Protein Family

Additional inhibitors have been purified and partially characterized from the plasma/serum of venomous and non-venomous snakes, such as *Agkistrodon contortrix mokasen* [11,60], *Bothrops asper* [61], *Crotalus atrox* [62,63], *Dinodon semicarinatus* [64], *Natrix tesselata* [65], *Protobothrops mucrosquamatus* [66], and *Vipera palestinae* [67]. To date, none of them had their primary structure determined. However, the SVMPI isolated from *N. tesselata*, named NtAH, displayed unique structural characteristics. It is the only high-molecular-mass (880 kDa) metalloendopeptidase inhibitor isolated from snake blood displaying an oligomeric composition of three polypeptide chains of 150, 100, and 70 kDa in an unknown arrangement. NtAH inhibited BaH1, the main metalloendopeptidase from *Bothrops asper* venom [65].

2.3. SVMPIs Isolated from Mammals

The earliest reports of mammals with natural resistances to snake envenomation date back to the nineteenth century. In their experiments, Felix de Azara, Albert Calmette, Césaire Phisalix & Gabriel Bertrand described the immunities of the lutrine opossum (*Lutreolina*) [68], the mongoose (*Herpestes ichneumon*) [9], and the hedgehog (*Erinaceus europaeus*) [10], respectively.

Vellard [19], when studying the natural resistance that mammals of the family *Didelphidae* presented to snake venoms, proposed that such phenomenon should be an adaptation to prey on venomous snakes [69]. Based on his observations on the resistance of *Didelphis virginiana*, including the injection of a high dosage (15 mg/kg) of *Agkistrodon piscivorus* venom, Kilmon hypothesized that the only reason that this opossum could fight snakes and survive the venomous bites was the existence of a "unique and extremely efficient immune-response system" [70]. However, because there was no evidence of antibody involvement, the association with the immune system remained elusive.

In 1981, Menchaca and Pérez isolated an antihemorrhagic factor from *D. virginiana* serum, named AHF; this was the first antihemorrhagic factor to be purified from the serum of a mammal. AHF presented a molecular mass of 68 kDa, an isoelectric point of 4.1, thermal (0–37 °C) and pH (3–10) stabilities; no precipitin line formation was evident when AHF was incubated with rattlesnake venom, indicating that AHF did not interact with snake venoms in a classic antigen–antibody reaction [71].

2.3.1. Immunoglobulin Supergene Family

In 1992, Catanese and Kress purified another inhibitor from *D. virginiana* serum, which was named oprin. It showed sequence homology (36% identity) with α_1B-glycoprotein and was classified as a member of the immunoglobulin supergene family. Oprin was able to inhibit several snake venom metalloendopeptidases but failed to inhibit serine endopeptidases, MMPs or bacterial metalloendopeptidases; oprin interacted with *Crotalus atrox* α-proteinase. The authors proposed that oprin partially accounted for the natural resistance of *D. virginiana*, and that its serum would contain at least two inhibitors of metalloendopeptidases [72]. Other studies identified and characterized to different extents inhibitors belonging to the immunoglobulin supergene family from the plasma/serum of *Herpestes edwardsii* (AHF-1 to AHF-3) [73–75], *Lutreolina crassicaudata* [76], *Philander opossum* (PO41) [76,77], and *Didelphis albiventris* (DA2-II) [78].

Following the first studies on the resistance that *D. marsupialis* showed to snake venoms [79–81], Perales and colleagues optimized the purification of an antibothropic fraction (ABF) that effectively blocked the hemorrhagic and lethal effects of *Bothrops jararaca* venom in mice [82,83]. ABF was further fractionated yielding ABC (antibothropic complex), which was composed of two proteins, with apparent molecular masses of 48 kDa and 43 kDa, as determined by SDS-PAGE under reducing conditions. ABC inhibited the hemorrhagic, hyperalgesic and edematogenic effects of *Bothrops jararaca* venom [76,84,85]. In vitro, ABC inhibited the proteolytic activity of the venom upon fibrinogen, fibrin, collagen IV, laminin, and fibronectin [86]. ABC was also found in the opossum's milk, reinforcing neonatal protection against snakebite envenomation [87].

Neves-Ferreira and colleagues fractionated ABC, leading to the isolation of two SVMPIs: DM40 and DM43. Both of these factors are acidic glycoproteins with molecular masses of 40,318 Da for DM40 and 42,373–43,010 Da for DM43 by MALDI-TOF MS; by SDS-PAGE under reducing conditions, DM40 and DM43 molecular masses are 43 kDa and 48 kDa, respectively [88]. DM43 remains the most extensively studied inhibitor to date; it is a homodimeric glycoprotein bearing three immunoglobulin-like domains per monomer and is homologous to α1B-glycoprotein, a human serum protein [89]. The structural resemblance and the presence of a degenerate WSXWS sequon on each domain of DM43 (typically found in proteins bearing an Ig-like fold) classified DM43 into the immunoglobulin supergene family of proteins [89]. The analysis of its glycan moiety revealed that all *N*-glycosylation consensus sites (Asn23, Asn156, Asn160, and Asn175) were occupied with complex-type *N*-glycans containing the monosaccharides *N*-acetylglucosamine, mannose, galactose, and *N*-acetylneuraminic acid at a 4:3:2:2 molar ratio, which is compatible with biantennary glycan chains.

MALDI-TOF MS analyses of deglycosylated and native DM43 revealed that the glycan moiety corresponded to 21% of the average molecular mass of the inhibitor [89,90]. Similar to many glycoproteins, DM43 presents at least four glycoforms, which may result from glycan composition heterogeneity [91]. In vitro, DM43 inhibited the proteolytic activity of the SVMP jararhagin upon the fluorogenic substrate Abz-Ala-Gly-Leu-Ala-Nba and upon casein, fibrinogen, and fibronectin; in vivo, DM43 showed the same properties as the ABC in mice [88]. Titration experiments using molecular exclusion chromatography and electrophoresis in denaturing conditions demonstrated that DM43 interacted with snake venom metalloendopeptidases at a 1:1 (monomer of DM43:toxin) molar ratio, and that this interaction was maintained noncovalently [88]. Surface plasmon resonance analysis using a sensor chip with immobilized jararhagin indicated a high-affinity interaction, with an equilibrium dissociation constant (K_D) of 0.33 ± 0.06 nM [91]. The strength of the DM43-jararhagin binding was comparable to therapeutic monoclonal antibodies, which typically have K_D values in the range of 1 pM to 1 nM [92].

The current knowledge about the regions of interaction between DM43 and its target toxin is still very limited; DM43 does not bind jararhagin-C, indicating that the interaction between the inhibitor and target toxin involves the toxins' metalloendopeptidase domain [89]. Additionally, after partial deglycosylation with PNGase F under nondenaturing conditions, the inhibitory activity of DM43 was reduced to 50% compared to native DM43 [90]. It still remains to be verified whether the *N*-glycans were directly involved with the interaction between the inhibitor and the metalloendopeptidase or if the partial removal of the *N*-glycosylation induced a conformational change that hindered the formation of the toxin–antitoxin complex.

The interaction between DM43 and SVMPs from different snake venoms was explored through affinity chromatography, with the covalent immobilization of DM43 on a HiTrap NHS-activated column. The venoms of *Bothrops atrox*, *B. jararaca*, *B. insularis*, and *Crotalus atrox* were injected into the DM43-column and the unbound and bound protein fractions were collected and analyzed through two-dimensional protein electrophoresis (2D-PAGE) [93]. DM43 was able to interact with several metalloendopeptidases from those venoms, but the presence of SVMP spots in the 2D-PAGE gels of the unbound fractions indicated that DM43 was not a universal SVMP inhibitor. Accordingly, DM43 did not interact with HF3, a highly glycosylated P-III class SVMP from *B. jararaca* venom, suggesting that some SVMPs may have structural features that pose difficulties to DM43 binding [94]. On the other hand, DM43 was able to interact with MMPs from osteoarthritis synovial liquid and supernatants of MCF-7 cell cultures; Western blot analyses have shown that DM43 interacted with MMP-2, MMP-3, and MMP-9, outlining a promising application for DM43 in biotechnology [95].

2.3.2. Ficolin/Opsonin P35 Family

The antihemorrhagic factor erinacin was isolated from *Erinaceus europaeus* muscle extract. It is a high-molecular mass protein of 1040 kDa composed of two main subunits—α and β—at a molar ratio of 1α:2β. The α-subunit is a homodecamer of 370 kDa maintained by noncovalent bonds, and the β subunit is composed of ten polypeptide chains of 35 kDa interacting via covalent bonds [96]. When analyzed by electron microscopy, erinacin showed a molecular structure that resembled a flower bouquet, an arrangement typical of proteins from the ficolin/opsonin P35 superfamily, such as plasma ficolin and the Hakata antigen [97,98]. Amino acid sequencing revealed that both subunits of erinacin were composed of *N*-terminus, collagen- and fibrinogen-like domains homologous to proteins from this family [99]. In vitro, erinacin inhibited a metalloendopeptidase from the venom of *B. jararaca*, through the establishment of an equimolar complex; it did not inhibit serine endopeptidases such as trypsin or chymotrypsin, and the dissociation of erinacin into its subunits caused complete loss of its antihemorrhagic activity. Regarding the mechanism of metalloendopeptidase inhibition by erinacin, the authors suggested two possibilities: (a) the *C*-terminal region of the fibrinogen-like domain of erinacin could contribute to the metalloendopeptidase inhibition by recognizing an *N*-acetylglucosamine molecule, as reported for P35 lectin and plasma ficolin [97,100]; and (b) the collagen-like domain of erinacin would act as a "decoy" substrate for the SVMPs [99].

3. Possible Therapeutic Applications

SVMPs are members of the metzincin clan of metalloendopeptidases, together with ADAMs (a disintegrin and metalloendopeptidase), ADAMTS (ADAM with thrombospondin motifs), astacins, serralysins, and MMPs [101–103]. SVMPs are abundant toxins in Viperidae (and some Colubridae) venoms, being responsible for the onset of local (blistering, edema, inflammatory reactions, and dermonecrosis) and systemic (hemorrhage, coagulopathy, and myonecrosis) pathophysiological effects [43].

The current antiophidic therapy is based on intravenous administration of antivenom, which in turn relies on antibody specificity, affinity, and ability to reach SVMPs (and other snake venom toxins) to be effective. The application of antivenom soon after *B. jararaca* venom injection in mice was not able to fully reverse the local effects of envenomation due to impaired and delayed venom/antivenom interaction at the site of injury [104,105]. Therefore, one of the current initiatives for the improvement of antiophidic therapy is the local administration of inhibitors soon after the envenomation event to restrain the extent of tissue degradation, and thus lower the high morbidity rates associated with snakebite envenoming [106].

SVMPIs could be used as a scaffold for the rational development of peptidic inhibitors of metalloendopeptidases because the interaction between the inhibitors and their target toxins is specific and leads to a tight-binding complexation.

The concept of rational drug design has already been applied to proline-rich oligopeptides from *Bothrops jararaca* venom, known as bradykinin-potentiating peptides (BPPs), initially described by Ferreira and co-workers [107]. From this same venom, other authors were able to isolate and fully determine the primary structures of six BPPs [108], including <Glu-Trp-Pro-Arg-Pro-Gln-Ile-Pro-Pro, later named BPP-9a; a synthetic version of this peptide was named SC 20,881. Gavras et al. demonstrated that the parenteral administration of SC 20,881 to hypertensive patients led to a significant drop in arterial blood pressure [109]. However, due to its lack of oral activity, this peptide had limited clinical applicability [110]. In following studies, Ondetti and colleagues showed that BPPs were substrate analogs that bound competitively to the active site of angiotensin-converting enzyme (ACE), and the optimal inhibitory region of the sequence was composed by the tripeptide Phe-Ala-Pro [110]. Based on the Phe-Ala-Pro sequence and structural studies of carboxypeptidase A as a model for ACE, Cushman and Ondetti synthesized D-2-methylsuccinyl-L-proline. This molecule proved to be a specific inhibitor of ACE with an IC_{50} of 22 µM. A further substitution of a carboxyl to a sulfhydryl group enhanced the molecule's inhibitory activity by three orders of magnitude, yielding the compound SC 14,225, later named Captopril [111,112]. Captopril is widely used in the treatment of hypertension and paved the way for the development of many antihypertensive compounds [113].

Peptide drugs are an ever-growing branch of the pharmaceutical industry, with a market value estimated at more than 40 billion dollars per year; these pharmaceuticals offer high potency, high selectivity, high chemical diversity, lower toxicity, and lower accumulation in tissues [114]. Peptide inhibitors of metalloendopeptidases are currently approved for the therapeutic intervention of hypertension, periodontal disease, and osteoarthritis [115]. Hence, peptide drugs derived from the natural inhibitors of metalloendopeptidases could not only be used in the improvement of the antiophidic therapy but also for the treatment of many pathological conditions related to the abnormal expression of closely related molecules, such as the ADAMs, ADAMTS and MMPs. These last are associated with the spread of malignant tumors and chronic diseases (e.g., multiple sclerosis, arthritis, fibrosis, and inflammatory conditions), whereas ADAMs/ADAMTS are involved in interstitial pulmonary fibrosis, bronchial asthma, and neurodegenerative diseases [115,116].

Most inhibitors of metalloendopeptidases undergoing clinical trials are small molecules that possess zinc-binding groups, such as hydroxamate, that interact with side pockets of the catalytic site; classical representatives of these low selectivity inhibitors are marimastat and batimastat. However, both peptidic inhibitors have been discontinued during phase III of clinical trials for the treatment of invasive cancers because they displayed an excessive number of off-target effects [115,117].

The research history on natural SVMPIs described in this review envisage the possibility that they possess a different mechanism of inhibition than the one described for the previously mentioned artificial inhibitors, targeting different regions of the molecule with higher specificity, as shown for some TIMPs (tissue inhibitors of metalloendopeptidases) [118]. This potentially opens a new path for the treatment of pathological conditions related to the unbalanced expression of metalloendopeptidases. However, the current lack of knowledge regarding the tertiary and quaternary structures of these natural inhibitors, as well as their regions of interaction with SVMPs, is the bottleneck in this research area, and precludes further understanding of their mechanism(s) of action.

4. Status Quo and Perspectives on Three-Dimensional Structure Determination for SVMPIs

In this section, we will discuss experimental and computational approaches that could be used to further the knowledge of the tridimensional structures of SVMPIs (Figure 2) and the mapping of the regions of interaction between these inhibitors and SVMPs.

To date, no three-dimensional structures have been experimentally determined for any members of the different protein families related to SVMPIs (cystatin, immunoglobulin supergene, and ficolin/opsonin P35). Furthermore, molecular modeling attempts have only been performed for one member of the immunoglobulin supergene family—DM43 [89]—and one from the cystatin superfamily—HSF [51]. Due to the lack of literature on the subject, the discussion that follows will rely on some of our group's unpublished data, related to two well-characterized SVMPIs: BJ46a (isolated from *Bothrops jararaca*—cystatin superfamily) [53] and DM43 (from *Didelphis aurita*—immunoglobulin supergene family) [89].

3D structure elucidation of natural inhibitors of SVMP

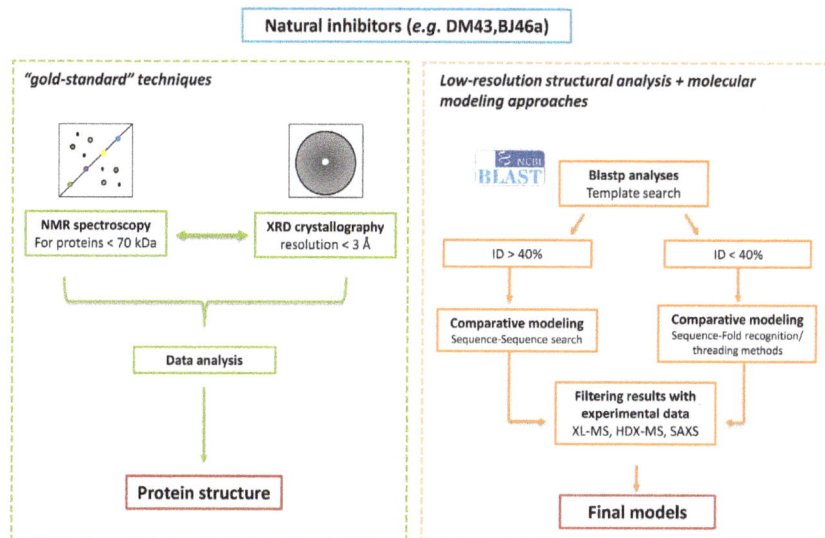

Figure 2. Strategies for a structural view of SVMPIs. (**Left**) The experimental methods for structure determination, NMR spectroscopy and XRD crystallography, are the "gold-standard" techniques in protein structure elucidation, providing atomic resolution of individual proteins and their complexes. The SVMPIs DM43 and BJ46a represent a challenge for these techniques. For NMR spectroscopy, due to the molecular size of both molecules, costly and time-consuming methods for sample labeling and analysis are required. For XRD crystallography, crystals of DM43 produced low-resolution diffraction pattern while BJ46a could not be crystallized, highlighting the limiting character of the crystallization step. Hence, modeling becomes an important tool for the structural studies of these molecules. (**Right**) In molecular modeling, the main step is the identification of a homologous protein, whose experimental structure has already been determined, to be used as a template structure. The identification in structure databases of sequences evolutionarily correlated with sequential identity greater than 40% is done by standard pairwise sequence search methods, allowing the generation of high accuracy models. However, below this sequence identity threshold the correlation between two structures is difficult to address. In this range, sequences are correlated directly with proteins of known structure (fold recognition). A drawback is that, due to the low evolutionary correlation and the low sensitivity in the sequence alignment building, the accuracy of the produced models is lower. On the other hand, the ensemble of models produced can be filtered according to their agreement with experimental data. In our proposed strategy, these data would come from XL-MS, HDX-MS and SAXS assays, leading to the selection of accurate models, and shedding some light on the three-dimensional structural characteristics of these SVMPIs. Consequently, the molecular basis of the interaction between the inhibitors and their target toxins could be established.

During the first 10 years after the primary structures of BJ46a and DM43 were published, our efforts were focused on applying standard X-ray diffraction (XRD) protocols to study the crystallized forms of these proteins. Multiple attempts at crystallizing the inhibitors BJ46a and DM43 have failed; after the eventual successful crystallization (DM43 only), the crystals showed a low-resolution diffraction pattern. A possible explanation is that these SVMPIs are glycoproteins whose glycan antennae show high conformational heterogeneity. These different states can be co-crystallized and interfere destructively in the diffraction pattern, decreasing its final resolution. Additionally, the absence of homologous proteins with an already determined crystallographic structure makes it impossible to solve the structure by molecular replacement, requiring more time and investment in producing heterologous proteins labeled with heavy atoms to solve the phase problem [119].

We have also evaluated if these proteins were good candidates for analysis by nuclear magnetic resonance (NMR) spectroscopy. However, DM43 (43 kDa monomer) and BJ46a (46 kDa monomer) are homodimeric in solution, exhibiting molecular masses outside the limit of standard protocols for NMR, requiring triple isotopic labeling (^{15}N, ^{13}C, and ^{2}H) in parallel with the selective/segmental labeling of specific amino acids, thus resulting in high production costs for NMR samples, and long data analysis times [120,121].

To overcome the absence of structural information, molecular modeling techniques can be used to produce models that could help us explain the mechanism of inhibition of SVMPIs. Molecular modeling is based on the assumption that proteins with similar primary sequences (defined by an identity threshold) should display matching three-dimensional structures and biological functions [122,123]. The limiting step in protein modeling is the identification of template sequences (homologous sequences) whose experimentally (XRD or NMR) determined structures are available. After the identification, the two sequences (target and template) are aligned. The coordinates and geometrical parameters of the template structure (in the aligned regions) are applied to the target sequence to generate the new model. Thus, the quality of the template/target alignment is essential to produce a biologically relevant model. Ideally, these two sequences must display a minimum of 40% sequence identity, with long aligned regions, and a low number of sequence alignment gaps [124].

The first application of modeling for SVMPI structure determination was done for DM43 [89]. This member of the Ig supergene family is composed of three Ig-like domains (D0, D1 and D2) for a total of 291 amino acid residues. At that time, the best template available was the inhibitory receptor (p58-cl42) for human natural killer cells, a two-domain protein whose Protein Data Bank identifier (PDB ID) is 1NKR. The overall sequence identity (taking only domains D1 and D2 into account) is 25.9% (Table 1). The low sequence identity level led to a difficult modeling process that required manual interference at all steps. The model allowed the prediction of the third domain (domain D2) as the one interacting with the metalloendopeptidases. However, the detailed SVMP interacting regions proposed by the model are not supported by low-resolution structural data (crosslinking resolved by mass spectrometry (XL-MS), hydrogen/deuterium exchange monitored by mass spectrometry (HDX-MS), and small angle X-ray scattering (SAXS)) recently acquired by our group.

Recent searches in the PDB database for structures analogous to DM43 now revealed PDB ID 5EIQ (Table 1), human OSCAR ligand-binding domain, as the best match. Released 17 years later than PDB ID 1NKR, the structure 5EIQ shows an increased identity level, and similar number of positive matches for the same DM43 region. Even though the expected value level (*E*-value) of alignment for DM43/5EIQ points to a good match (7×10^{-18}), it is still limited to domains D1 and D2, in the same fashion as for the alignment DM43/1NKR (original model). Nevertheless, the alignment is still below the 40% sequential identity threshold, suggesting that no new structural information is present in the protein structure database that could suggest a new direction for DM43 molecular modeling.

We used the same analysis for the SVMPIs HSF and BJ46a, belonging to the cystatin superfamily and displaying 85% sequence identity between themselves. As can be seen from Table 1, the sequence pairwise search results yielded sequence identity levels below the homology-modeling threshold of 40%, a low number of aligned residues, and high e-values. Altogether, these results suggest that, for the time being, SVMPIs are still a challenge for the application of standard modeling techniques.

Table 1. Template search for the SVMPIs DM43, HSF, and BJ46a. All structures identified as possible templates are below the threshold of 40% of sequence identity.

Feature	Ig Supergene Family		Cystatin Superfamily	
	DM43 (291)		HSF (323)	BJ46a (322)
Template (PDB ID)	1NKR	5EIQ	2KZX	2WBK
Release date	11 November 1998	25 November 2015	15 February 2012	4 April 2014
Number of aligned residues	193	192	98	55
E-value	4×10^{-5}	7×10^{-18}	1×10^{-1}	5×10^{-1}
Identity	(50/193) 26%	(71/192) 37%	(26/98) 27%	(15/55) 33%
Positives	(78/193) 40%	(91/192) 47%	(42/98) 42%	(29/55) 52%
Gaps	(15/193) 7%	(14/192) 7%	(5/98) 5%	(1/55) 1%
Aligned region	9–196	1–188	14–110	207–261

The complete primary structures of DM43, HSF, and BJ46a are composed of 291, 323, and 322 amino acids, respectively (numbers in parentheses). PSI-Blast search with default parameters (Expect threshold 10, Word size 3, Matrix BLOSUM62, Gap Costs Existence 11, Extension 1, PSI-BLAST threshold 0.005) were done against the PDB database. Template is the best hit, identified by its PDB ID number. Release date is the structure's publication date in the database. E-value is the expected number of chances that the match is random. Three percentages are calculated relatively to the number of aligned residues: identity (exact match residues), positives (exact + homology match residues), and gaps (inserted spaces to allow the alignment). 1NKR: inhibitory receptor (p58-cl42) for human natural killer cells. 5EIQ: human OSCAR ligand-binding domain. 2KZX: A3DHT5 from *Clostridium thermocellum*, Northeast Structural Genomics Consortium Target CmR116. 2WBK: beta-mannosidase, Man2A.

Another methodology for modeling was independently proposed with the seminal papers of Bowie, Jones, and Zhang [125–127]. This method is able to correlate two sequences that are evolutionarily distant (low sequence identity), based on the concept that the folding, and consequently the function, is more conserved than the primary structure. These authors introduced the term "threading", a method where the target sequence is fitted onto the backbone coordinates of a known protein structure (the template). The fitting is scored by an energy potential, and the lowest energy fitting corresponds to the best model.

There is one report in the literature describing this structural modeling approach for HSF [51]. This paper used sequence-to-structure methods to thread the HSF sequence into template PDB ID 1G96. Applying the algorithm DELTA-BLAST [128], we were able to trace the new structures available since then (Table 2). As can be seen, since 2001 (release date of structure 1G96), three more structures that are structurally related to HSF were determined. All selected structures are members of the cystatin superfamily, in agreement with the prediction from HSF's primary structure. Moreover, entry 4LZI is a convergent choice between several search algorithms (data not shown).

Table 2. New HSF-correlated structures in the PDB database, using DELTA-Blast.

Feature	HSF (323)			
Template (PDB ID)	4LZI	3PS8	1R4C	1G96
Release date	26 February 2014	21 December 2011	21 September 2004	6 April 2001
Number of aligned residues	222	115	107	115
E-value	4×10^{-55}	6×10^{-35}	8×10^{-36}	2×10^{-34}
Identity	(31/222) 14%	(15/115) 13%	(17/107) 16%	(15/115) 13%
Positives	(67/222) 30%	(39/115) 33%	(32/107) 29%	(39/115) 33%
Gaps	(55/222) 24%	(8/115) 6%	(4/107) 3%	(8/115) 6%

DELTA-Blast search followed by PSI-Blast, with the default parameters described in Table 1. Template is the best hit, identified by its PDB ID number. Release date is the structure's publication date in the database. E-value is the expected number of chances that the match is random. Three percentages are calculated relatively to the number of aligned residues: identity (exact match residues), positives (exact + homology match residues), and gaps (inserted spaces to allow the alignment). Despite the intermediate sequential identity value, the structure 4LZI shows the best sequence coverage (number of aligned residues) and positive matches, being the best template since structure 1G96. 4LZI: *Solanum tuberosum* multicystatin. 3PS8: L68V mutant of human cystatin C. 1R4C: N-truncated human cystatin C, dimeric form. 1G96: human cystatin C, dimeric form.

Finally, the same analysis was done for BJ46a (85% sequential identity with HSF), and, as expected, the results were very similar (Table 3). Three out of four structures selected as template to model HSF were also found in this case. However, even though the results are equivalent, there was a significant difference in the alignment with the secondary structure elements calculated using the threading algorithm (data not shown).

Table 3. BJ46a-correlated structures in PDB database, using DELTA-Blast.

Feature	BJ46a (322)		
Template (PDB ID)	4LZI	3PS8	1G96
Release date	26 February 2014	21 December 2011	6 April 2001
Number of aligned residues	226	115	115
E-value	3×10^{-33}	3×10^{-27}	6×10^{-27}
Identity	(30/226) 13%	(14/115) 12%	(14/115) 12%
Positives	(65/226) 28%	(41/115) 35%	(41/115) 35%
Gaps	(55/226) 25%	(8/115) 6%	(8/115) 6%

DELTA-Blast search followed by PSI-Blast, with default parameters described in Table 1. Template Ⅰ is the best hit, identified by its PDB ID number. Release date is the structure's publication date Ⅰ in the database. E-value is the expected number of chances that the match is random. Three percentages are calculated relatively to the number of aligned residues: identity (exact match residues), positives (exact + homology match residues), and gaps (inserted spaces to allow the alignment). Despite the intermediate sequential identity value, the structure 4LZI shows the best sequence coverage (number of aligned residues) and positive matches, being the best template since structure 1G96. 4LZI: *Solanum tuberosum* multicystatin. 3PS8: L68V mutant of human cystatin C. 1G96: human cystatin C dimeric form.

In summary, sequence-based methods (i.e., comparative homology modeling) can only produce good quality alignments and high accuracy models for closely related sequences (>40% identity). Below this identity level, threshold sequence-to-structure methods (i.e., fold recognition modeling or 3D-threading) show better performance. However, the low quality of the alignments may compromise the accuracy of the generated models. These limitations led to the development of hybrid strategies, which combine search algorithms based on sequence-profiling methods, and the energy potentials derived from threading methods. The new generation of fully automated servers for protein structure prediction is based on this hybrid strategy (genThreader, PSIPred, and i-Tasser), allowing the structure prediction of proteins at a proteome scale [129–131].

Hence, in order obtain confident structural models for these SVMPIs, we advocate that the overall strategy should be to apply sequence-to-structure methods to produce large ensembles of models, followed by validation against experimental data generated by XL-MS [132], HDX-MS [133], and SAXS [134–136] (Figure 2, Right panel).

5. Conclusions

The field of natural inhibitors of snake venom toxins has advanced considerably since the amazing phenomenon of innate venom resistance was first described more than two centuries ago. Currently, the physicochemical characteristics of the antiophidic proteins are known, but the molecular bases underlying their neutralizing properties are not quite well understood. For instance, translating this scientific knowledge into novel effective therapies (e.g., preventing snake envenomation morbidity) necessarily requires a deep understanding of the structure–function relationship. To tackle this challenge, greater emphasis should be placed on the concerted use of emerging structural biology techniques that are complementary to traditional approaches.

Acknowledgments: Viviane A. Bastos is a Ph.D. fellow from *Coordenação de Aperfeiçoamento de Pessoal de Nível Superior* (CAPES)(grant AuxPE 1214/2011) enrolled in the Biochemistry Graduate Program at *Universidade Federal do Rio de Janeiro* (Rio de Janeiro, Brazil). Jonas Perales is a *Conselho Nacional de Desenvolvimento Científico e Tecnológico* (CNPq) fellow (grant 312311/2013-3) and a *Fundação de Amparo à Pesquisa do Estado do Rio de Janeiro* (FAPERJ) fellow (grant E26/202-960/2015). Ana Gisele C. Neves-Ferreira is a CNPq fellow (grant 311539/2015-7).

Richard H. Valente was supported by CAPES (grant BEX 17666/12-0), CNPq (grant Universal 471439/2011-8), FAPERJ (grant APQ1 E26/111.781/2012), and *Fundação Oswaldo Cruz* (grant PAPES VI 407611/2012-6).

Author Contributions: R.H.V. and V.A.B. wrote the main body of this review with general contributions from A.G.C.N.-F. and J.P., and specific contributions (modeling strategies) from F.G.N.

Conflicts of Interest: The authors declare no conflicts of interest.

Abbreviations

The following abbreviations are used in this manuscript:

ABC	AntiBothropic Complex
ABF	AntiBothropic Fraction
ADAM	A Disintegrin And Metalloendopeptidase
ADAMTS	ADAM with ThromboSpondin motifs
BPP	Bradikynin-Potentiating Peptide
HDX-MS	Hydrogen/Deuterium eXchange MS
HLP	Habu-Like Protein
HSF	Habu Serum Factor
Ig	Immunoglobulin
MALDI TOF	Matrix-Assisted Laser/Desorption Ionization
MS	Mass Spectrometry
MMP	Matrix MetalloendoPeptidase
PDB ID	Protein Data Bank IDentifier
PLI	PhosphoLipase A_2 Inhibitor
NMR	Nuclear Magnetic Resonance
SAXS	Small-Angle X-ray Scattering
SDS-PAGE	Sodium Dodecyl Sulfate PolyAcrylamide Gel Electrophoresis
SSP	Small Serum Protein
SVMP	Snake Venom MetalloendoPeptidase
SVMPI	SVMP Inhibitor
TIMP	Tissue Inhibitor of MetalloendoPeptidase
2D-PAGE	Two-Dimensional PolyAcrilamide Gel Electrophoresis
XL-MS	Cross-Linking MS
XRD	X-ray Diffraction

References

1. Fraser, T. *Serpent's Venom: Artificial and Natural Immunity; Antidotal Properties of the Blood-Serum of Immunized Animals and of Venomous Serpents;* Neill and Company: Edinburgh, UK, 1895; p. 42.
2. Karaberopoulos, D.; Karamanou, M.; Androutsos, G. The theriac in antiquity. *Lancet* **2012**, *379*, 1942–1943. [CrossRef]
3. Parojcic, D.; Stupar, D.; Mirica, M. Theriac: Medicine and antidote. *Vesalius* **2003**, *9*, 28–32. [PubMed]
4. Garrison, F.H. Felice Fontana: A forgotten physiologist of the Trentino. *Bull. N. Y. Acad. Med.* **1935**, *11*, 117–122. [PubMed]
5. Hawgood, B.J. Abbé Felice Fontana (1730–1805): Founder of Modern Toxinology. *Toxicon* **1995**, *33*, 591–601. [CrossRef]
6. Fontana, F. *Ricerche Fisiche Sopra il Veleno della Vipera con Alcune Offervazioni Sopra la Anguillette dele Grano Sperone;* Jacopo Giusti: Lucca, Italy, 1767; p. 170. (In Italian)
7. Fontana, F. *Treatise on the Venom of the Viper; on the American Poisons; and on the Cherry Laurel; and Some Other Vegetable Poisons;* Jamie Murray: London, UK, 1787; Volume 1, p. 447.
8. Guyon, J. Le Venin de Serpents Exerce-t-il sur Eux-mêmes l'Action Qu'il Exerce sur D'autres Animaux? In *Animaux Venimeux et Venins: La Fonction Venimeuse chez tous les Animaux; les Appareils Venimeux; les Venins et leurs Propriétés; les Fonctions et Usages des Venins; L'envenimation et son Traitement;* Phisalix, M., Ed.; Masson & Cie.: Paris, France, 1861; Volume 2, p. 744. (In French)

9. Calmette, A. *Le Venin des Serpents: Physiologie de l'Envenimation, Traitement des Morsures Venimeuses*; Société d'Éditions Scientifiques: Paris, France, 1896; Volume 1, p. 90. (In French)

10. Phisalix, C.; Bertrand, G. Recherches sur l'immunité du hérisson contre le venin de vipère. *C. R. Soc. Biol.* **1895**, *47*, 639–641. (In French)

11. Gloyd, H.K. On the effects of the mocassin venom upon a rattlesnake. *Science* **1933**, *78*, 13–14. [CrossRef] [PubMed]

12. Keegan, H.L.; Andrews, T.F. Effects of crotalid venom on North American snakes. *Copeia* **1942**, *4*, 251–254. [CrossRef]

13. Swanson, P.L. Effects of snake venoms on snakes. *Copeia* **1946**, *4*, 242–249. [CrossRef]

14. Philpot, V.B.; Smith, R.G. Neutralization of pit viper venom by king snake serum. *Proc. Soc. Exp. Biol. Med.* **1950**, *75*, 521–523. [CrossRef]

15. Abalos, J.W. The Ophiofagus habits of *Pseudoboa cloelia*. *Toxicon* **1963**, *1*, 90–91. [CrossRef]

16. Straight, R.; Glenn, J.L.; Snyder, C.C. Antivenom activity of rattlesnake blood plasma. *Nature* **1976**, *261*, 259–260. [CrossRef] [PubMed]

17. Clark, W.C.; Voris, H.K. Venom neutralization by rattlesnake serum albumin. *Science* **1969**, *164*, 1402–1404. [CrossRef] [PubMed]

18. Vellard, J. Resistencia de los "Didelphis" (Zarigueya) a los venenos ofidicos (Nota Previa). *Rev. Bras. Biol.* **1945**, *5*, 463–467. (In Spanish) [PubMed]

19. Vellard, J. Investigaciones sobre imunidad natural contra los venenos de serpientes. *I. Publ. Mus. Hist. Nat. Lima Ser. A Zool.* **1949**, *1*, 1–61. (In Spanish)

20. Domont, G.B.; Perales, J.; Moussatché, H. Natural anti-snake venom proteins. *Toxicon* **1991**, *29*, 1183–1194. [CrossRef]

21. Perales, J.; Neves-Ferreira, A.G.C.; Valente, R.H.; Domont, G.B. Natural inhibitors of snake venom hemorrhagic metalloproteinases. *Toxicon* **2005**, *45*, 1013–1020. [CrossRef] [PubMed]

22. Fortes-Dias, C.L. Endogenous inhibitors of snake venom phospholipases A(2) in the blood plasma of snakes. *Toxicon* **2002**, *40*, 481–484. [CrossRef]

23. Lizano, S.; Domont, G.; Perales, J. Natural phospholipase A_2 myotoxin inhibitor proteins from snakes mammals and plants. *Toxicon* **2003**, *42*, 963–977. [CrossRef] [PubMed]

24. Neves-Ferreira, A.G.C.; Valente, R.H.; Perales, J.; Domont, G.B. Natural Inhibitors—Innate Immunity to Snake Venoms. In *Handbook of Venoms and Toxins of Reptiles*; Mackessy, S.P., Ed.; CRC Press: Boca Raton, FL, USA, 2010; Volume 1, pp. 259–284.

25. Perales, J.; Domont, G.B. Are Inhibitors of Metalloproteinases, Phospholipases A_2 and Myotoxins Members of the Innate Immune System? In *Perspectives in Molecular Toxinology*; Ménez, A., Ed.; John Wiley & Sons: Chichester, UK, 2002; pp. 435–456.

26. Thwin, M.M.; Gopalakrishnakone, P. Snake envenomation and protective natural endogenous proteins: A mini review of the recent developments (1991–1997). *Toxicon* **1998**, *36*, 1471–1482. [CrossRef]

27. Perez, J.C.; Sanchez, E.E. Natural protease inhibitors to hemorrhagins in snake venoms and their potential use in medicine. *Toxicon* **1999**, *37*, 703–728. [CrossRef]

28. Valente, R.H.; Neves-Ferreira, A.G.C.; Caffarena, E.R.; Domont, G.B.; Perales, J. Snake Venom Metalloproteinase Inhibitors—An Overview And Future Perspectives. In *Animal Toxins—State of the Art. Perspectives in Health and Biotechnology*; Lima, M.E., Ed.; Editora UFMG: Belo Horizonte, Brazil, 2009; pp. 547–558.

29. Neves-Ferreira, A.G.C.; Valente, R.H.; Domont, G.B.; Perales, J. Natural Inhibitors of Snake Venom Metallopeptidases. In *Toxins and Drug Discovery*; Gopalakrishnakone, P., Ed.; Springer: Dordrecht, The Netherlands, 2016; Volume 1, pp. 1–23.

30. De Lacerda, J.B. *Leçons sur le Venin des Serpents du Brèsil et sur la Méthode de Traitement des Morsures Venimeuses par le Permanganate de Potasse*; Livraria Lombaerts & C.: Rio de Janeiro, Brazil, 1884; p. 194. (In French)

31. Flexner, S.; Noguchi, H. The constitution of snake venom and snake sera. *J. Pathol. Bacteriol.* **1903**, *8*, 379–410. [CrossRef]

32. Maeno, H.; Mitsuhashi, S.; Sato, R. Studies on habu snake venom. 2c. Studies on H beta-proteinase of habu venom. *Jpn. J. Microbiol.* **1960**, *4*, 173–180. [CrossRef] [PubMed]

33. Ohsaka, A. Fractionation of habu snake venom by chromatography on cm-cellulose with special reference to biological activities. *Jpn. J. Med. Sci. Biol.* **1960**, *13*, 199–205. [CrossRef] [PubMed]

34. Ohsaka, A.; Ikezawa, H.; Kondo, H.; Kondo, S.; Uchida, N. Haemorrhagic activities of habu snake venom, and their relations to lethal toxicity, proteolytic activities and other pathological activities. *Br. J. Exp. Pathol.* **1960**, *41*, 478–486. [PubMed]

35. Okonogi, T.; Hoshi, S.; Honma, M.; Mitsuhashi, S.; Maeno, H.; Sawai, Y. Studies on the habu snake venom. 3–2. A comparative study of histopathological changes caused by crude venom, purified Habu-proteinase and other proteinases. *Jpn. J. Microbiol.* **1960**, *4*, 189–192. [CrossRef] [PubMed]

36. Bjarnason, J.B.; Tu, A.T. Hemorrhagic toxins from Western diamondback rattlesnake (*Crotalus atrox*) venom: Isolation and characterization of five toxins and the role of zinc in hemorrhagic toxin e. *Biochemistry* **1978**, *17*, 3395–3404. [CrossRef] [PubMed]

37. Fox, J.W.; Serrano, S.M. Insights into and speculations about snake venom metalloproteinase (SVMP) synthesis, folding and disulfide bond formation and their contribution to venom complexity. *FEBS J.* **2008**, *275*, 3016–3030. [CrossRef] [PubMed]

38. Gutierrez, J.M.; Rucavado, A. Snake venom metalloproteinases: Their role in the pathogenesis of local tissue damage. *Biochimie* **2000**, *82*, 841–850. [CrossRef]

39. Tanjoni, I.; Weinlich, R.; Della-Casa, M.S.; Clissa, P.B.; Saldanha-Gama, R.F.; de Freitas, M.S.; Barja-Fidalgo, C.; Amarante-Mendes, G.P.; Moura-da-Silva, A.M. Jararhagin, a snake venom metalloproteinase, induces a specialized form of apoptosis (anoikis) selective to endothelial cells. *Apoptosis* **2005**, *10*, 851–861. [CrossRef] [PubMed]

40. Gutierrez, J.M.; Escalante, T.; Rucavado, A.; Herrera, C. Hemorrhage caused by snake venom metalloproteinases: A journey of discovery and understanding. *Toxins* **2016**, *8*. [CrossRef] [PubMed]

41. Farsky, S.H.; Goncalves, L.R.; Cury, Y. Characterization of local tissue damage evoked by *Bothrops jararaca* venom in the rat connective tissue microcirculation: An intravital microscopic study. *Toxicon* **1999**, *37*, 1079–1083. [CrossRef]

42. Warrell, D.A. Snake bite. *Lancet* **2010**, *375*, 77–88. [CrossRef]

43. Gutiérrez, J.M.; Rucavado, A.; Escalante, T. Snake Venom Metalloproteinases—Biological Roles and Participation in the Pathophysiology of Envenomation. In *Handbook of Venoms and Toxins of Reptiles*; Mackessy, S.P., Ed.; CRC Press: Boca Raton, FL, USA, 2010; pp. 115–138.

44. Omori-Satoh, T.; Sadahiro, S.; Ohsaka, A.; Murata, R. Purification and characterization of an antihemorrhagic factor in the serum of *Trimeresurus flavoviridis*, a crotalid. *Biochim. Biophys. Acta* **1972**, *285*, 414–426. [CrossRef]

45. Omori-Satoh, T. Antihemorrhagic factor as a proteinase inhibitor isolated from the serum of *Trimeresurus flavoviridis*. *Biochim. Biophys. Acta* **1977**, *495*, 93–98. [CrossRef]

46. Yamakawa, Y.; Omori-Satoh, T. Primary structure of the antihemorrhagic factor in serum of the Japanese habu: A snake venom metalloproteinase inhibitor with a double-headed cystatin domain. *J. Biochem.* **1992**, *112*, 583–589. [PubMed]

47. Deshimaru, M.; Tanaka, C.; Tokunaga, A.; Goto, M.; Terada, S. Efficient Purification of an Antihemorrhagic factor in the serum of Japanese Habu (*Trimeresurus flavoviridis*). *Fukuoka Univ. Sci. Rep.* **2003**, *33*, 45–53.

48. Deshimaru, M.; Tanaka, C.; Fujino, K.; Aoki, N.; Terada, S.; Hattori, S.; Ohno, M. Properties and cDNA cloning of an antihemorrhagic factor (HSF) purified from the serum of *Trimeresurus flavoviridis*. *Toxicon* **2005**, *46*, 937–945. [CrossRef] [PubMed]

49. Shioi, N.; Narazaki, M.; Terada, S. Novel function of antihemorrhagic factor HSF as an SSP-binding protein in habu (*Trimeresurus flavoviridis*) serum. *Fukuoka Univ. Sci. Rep.* **2011**, *41*, 177–184.

50. Shioi, N.; Ogawa, E.; Mizukami, Y.; Abe, S.; Hayashi, R.; Terada, S. Small serum protein-1 changes the susceptibility of an apoptosis-inducing metalloproteinase HV1 to a metalloproteinase inhibitor in habu snake (*Trimeresurus flavoviridis*). *J. Biochem.* **2012**, *153*, 121–129. [CrossRef] [PubMed]

51. Aoki, N.; Deshimaru, M.; Terada, S. Active fragments of the antihemorrhagic protein HSF from serum of habu (*Trimeresurus flavoviridis*). *Toxicon* **2007**, *49*, 653–662. [CrossRef] [PubMed]

52. Usami, Y.; Fujimura, Y.; Miura, S.; Shima, H.; Yoshida, E.; Yoshioka, A.; Hirano, K.; Suzuki, M.; Titani, K. A 28 kDa-protein with disintegrin-like structure (jararhagin-C) purified from *Bothrops jararaca* venom inhibits collagen- and ADP-induced platelet aggregation. *Biochem. Biophys. Res. Commun.* **1994**, *201*, 331–339. [CrossRef] [PubMed]

53. Valente, R.H.; Dragulev, B.; Perales, J.; Fox, J.W.; Domont, G.B. BJ46a, a snake venom metalloproteinase inhibitor. Isolation, characterization, cloning and insights into its mechanism of action. *Eur. J. Biochem.* **2001**, *268*, 3042–3052. [CrossRef] [PubMed]

54. Shi, Y.; Ji, M.-K.; Xu, J.-W.; Lin, X.; Lin, J.-Y. High-level expression, purification, characterization and structural prediction of a snake venom metalloproteinase inhibitor in *Pichia pastoris*. *Protein J.* **2012**, *31*, 212–221. [CrossRef] [PubMed]

55. Ji, M.-K.; Shi, Y.; Xu, J.-W.; Lin, X.; Lin, J.-Y. Recombinant snake venom metalloproteinase inhibitor BJ46A inhibits invasion and metastasis of B16F10 and MHCC97H cells through reductions of matrix metalloproteinases 2 and 9 activities. *Anticancer Drugs* **2013**, *24*, 461–472. [CrossRef] [PubMed]

56. Aoki, N.; Tsutsumi, K.; Deshimaru, M.; Terada, S. Properties and cDNA cloning of antihemorrhagic factors in sera of Chinese and Japanese mamushi (*Gloydius blomhoffi*). *Toxicon* **2008**, *51*, 251–261. [CrossRef] [PubMed]

57. Shioi, N.; Deshimaru, M.; Terada, S. Structural analysis and characterization of new small serum proteins from the serum of a venomous snake (*Gloydius blomhoffii*). *Biosci. Biotechnol. Biochem.* **2014**, *78*, 410–419. [CrossRef] [PubMed]

58. Jahnen-Dechent, W.; Schinke, T.; Trindl, A.; Muller-Esterl, W.; Sablitzky, F.; Kaiser, S.; Blessing, M. Cloning and targeted deletion of the mouse fetuin gene. *J. Biol. Chem.* **1997**, *272*, 31496–31503. [CrossRef] [PubMed]

59. Aoki, N.; Deshimaru, M.; Kihara, H.; Terada, S. Snake fetuin: Isolation and structural analysis of new fetuin family proteins from the sera of venomous snakes. *Toxicon* **2009**, *54*, 481–490. [CrossRef] [PubMed]

60. Weinstein, S.A.; Lafaye, P.J.; Smith, L.A. Observations on a venom neutralizing fraction isolated from the serum of the Northern Copperhead. *Agkistrodon Contortrix Mokasen Copeia* **1991**, *5*, 777–786. [CrossRef]

61. Borkow, G.; Gutierrez, J.M.; Ovadia, M. Isolation, characterization and mode of neutralization of a potent antihemorrhagic factor from the serum of the snake *Bothrops asper*. *Biochim. Biophys. Acta* **1995**, *1245*, 232–238. [CrossRef]

62. Weissenberg, S.; Ovadia, M.; Fleminger, G.; Kochva, E. Antihemorrhagic factors in the blood serum of the western diamondback rattlesnake *Crotalus atrox*. *Toxicon* **1991**, *29*, 807–818. [CrossRef]

63. Weissenberg, S.; Ovadia, M.; Kochva, E. Inhibition of the proteolytic activity of hemorrhagin-e from *Crotalus atrox* serum by antihemorrhagins from homologous serum. *Toxicon* **1992**, *30*, 591–597. [CrossRef]

64. Tomihara, Y.; Kawamura, Y.; Yonaha, K.; Nozaki, M.; Yamakawa, M.; Yoshida, C. Neutralization of hemorrhagic snake venoms by sera of *Trimeresurus flavoviridis* (habu), *Herpestes edwardsii* (mongoose) and *Dinodon semicarinatus* (akamata). *Toxicon* **1990**, *28*, 989–991. [CrossRef]

65. Borkow, G.; Gutierrez, J.M.; Ovadia, M. A potent antihemorrhagin in the serum of the non-poisonous water snake *Natrix tessellata*: Isolation, characterization and mechanism of neutralization. *Biochim. Biophys. Acta* **1994**, *1201*, 482–490. [CrossRef]

66. Huang, K.-F.; Chow, L.-P.; Chiou, S.-H. Isolation and characterization of a novel proteinase inhibitor from the snake serum of Taiwan Habu (*Trimeresurus mucrosquamatus*). *Biochem. Biophys. Res. Commun.* **1999**, *263*, 610–616. [CrossRef] [PubMed]

67. Ovadia, M. Purification and characterization of an antihemorrhagic factor from the serum of the snake *Vipera palaestinae*. *Toxicon* **1978**, *16*, 661–672. [CrossRef]

68. Azara, F.D. *Apuntamientos para la Historia Natural de los Quadrúpedos del Paragüay y Río de la Plata*; Imprenta de la Viuda de Ibarra: Madrid, Spain, 1802; p. 340. (In Spanish)

69. Voss, R.S.; Jansa, S.A. Snake-venom resistance as a mammalian trophic adaptation: Lessons from didelphid marsupials. *Biol. Rev.* **2012**, *87*, 822–837. [CrossRef] [PubMed]

70. Kilmon, J.A. High tolerance to snake venom by the Virginia opossum, *Didelphis virginiana*. *Toxicon* **1976**, *14*, 337–340. [CrossRef]

71. Menchaca, J.M.; Perez, J.C. The purification and characterization of an antihemorrhagic factor in opossum (*Didelphis virginiana*) serum. *Toxicon* **1981**, *19*, 623–632. [CrossRef]

72. Catanese, J.J.; Kress, L.F. Isolation from opossum serum of a metalloproteinase inhibitor homologous to human alpha 1B-glycoprotein. *Biochemistry* **1992**, *31*, 410–418. [CrossRef] [PubMed]

73. Tomihara, Y.; Yonaha, K.; Nozaki, M.; Yamakawa, M.; Kamura, T.; Toyama, S. Purification of three antihemorrhagic factors from the serum of a mongoose (*Herpestes edwardsii*). *Toxicon* **1987**, *25*, 685–689. [CrossRef]

74. Qi, Z.-Q.; Yonaha, K.; Tomihara, Y.; Toyama, S. Characterization of the antihemorrhagic factors of mongoose (*Herpestes edwardsii*). *Toxicon* **1994**, *32*, 1459–1469. [CrossRef]

75. Qi, Z.Q.; Yonaha, K.; Tomihara, Y.; Toyama, S. Isolation of peptides homologous to domains of human alpha 1B-glycoprotein from a mongoose antihemorrhagic factor. *Toxicon* **1995**, *33*, 241–245. [CrossRef]

76. Perales, J.; Moussatché, H.; Marangoni, S.; Oliveira, B.; Domont, G.B. Isolation and partial characterization of an anti-bothropic complex from the serum of the south american Didelphidae. *Toxicon* **1994**, *32*, 1237–1249. [CrossRef]

77. Jurgilas, P.B.; Neves-Ferreira, A.G.; Domont, G.B.; Perales, J. PO41, a snake venom metalloproteinase inhibitor isolated from *Philander opossum* serum. *Toxicon* **2003**, *42*, 621–628. [CrossRef] [PubMed]

78. Farah, M.D.F.L.; One, M.; Novello, J.C.; Toyama, M.H.; Perales, J.; Moussatché, H.; Domont, G.B.; Oliveira, B.; Marangoni, S. Isolation of protein factors from opossum (*Didelphis albiventris*) serum which protect against *Bothrops jararaca* venom. *Toxicon* **1996**, *34*, 1067–1071. [CrossRef]

79. Moussatché, H.; Yates, A.; Leonardi, F.; Borche, L. Mechanisms of resistance of the opossum to some snake venoms. *Toxicon* **1979**, *17*, 130.

80. Moussatché, H.; Yates, A.; Leonardi, F.; Borche, L. Obtención de una fracción del suero de *Didelphis* activa contra la acción tóxica del veneno de *B. jararaca*. *Acta Cient. Venez.* **1980**, *31*, 104. (In Spanish)

81. Moussatché, H.; Leonardi, F.; Mandelbaum, F. Inhibición por una proteína aislada del suero de *D. marsupialis* a la acción hemorrágica por una fracción del veneno de *Bothrops jararaca*. *Acta Cient. Venez.* **1981**, *32*, 173. (In Spanish)

82. Perales, J.; Munoz, R.; Moussatché, H. Isolation and partial characterization of a protein fraction from the opossum (*Didelphis marsupialis*) serum, with protecting property against *Bothrops jararaca* venom. *An. Acad. Bras. Cienc.* **1986**, *58*, 155–162. [PubMed]

83. Moussatché, H.; Perales, J. Factors underlying the natural resistance of animals against snake venoms. *Mem. Inst. Oswaldo Cruz* **1989**, *84*, 391–394. [CrossRef]

84. Perales, J.; Amorim, C.Z.; Rocha, S.L.G.; Domont, G.B.; Moussatché, H. Neutralization of the oedematogenic activity of *Bothrops jararaca* venom on the mouse paw by a antibothropic fraction isolated from opossum (*Didelphis marsupialis*) serum. *Agents Actions* **1992**, *37*, 250–259. [CrossRef] [PubMed]

85. Rocha, S.L.; Frutuoso, V.S.; Domont, G.B.; Martins, M.A.; Moussatche, H.; Perales, J. Inhibition of the hyperalgesic activity of *Bothrops jararaca* venom by an antibothropic fraction isolated from opossum (*Didelphis marsupialis*) serum. *Toxicon* **2000**, *38*, 875–880. [CrossRef]

86. Neves-Ferreira, A.G.C.; Perales, J.; Ovadia, M.; Moussatché, H.; Domont, G.B. Inhibitory properties of the antibothropic complex from the south american opossum (*Didelphis marsupialis*) serum. *Toxicon* **1997**, *35*, 849–863. [CrossRef]

87. Jurgilas, P.B.; Neves-Ferreira, A.G.C.; Domont, G.B.; Moussatché, H.; Perales, J. Detection of an antibothropic fraction in opossum (*Didelphis marsupialis*) milk that neutralizes *Bothrops jararaca* venom. *Toxicon* **1999**, *37*, 167–172. [CrossRef]

88. Neves-Ferreira, A.G.C.; Cardinale, N.; Rocha, S.L.G.; Perales, J.; Domont, G.B. Isolation and characterization of DM40 and DM43, two snake venom metalloproteinase inhibitors from *Didelphis marsupialis* serum. *Biochim. Biophys. Acta* **2000**, *1474*, 309–320. [CrossRef]

89. Neves-Ferreira, A.G.C.; Perales, J.; Fox, J.W.; Shannon, J.D.; Makino, D.L.; Garratt, R.C.; Domont, G.B. Structural and functional analyses of DM43, a snake venom metalloproteinase inhibitor from *Didelphis marsupialis* serum. *J. Biol. Chem.* **2002**, *277*, 13129–13137. [CrossRef] [PubMed]

90. Léon, I.R.; Neves-Ferreira, A.G.C.; Rocha, S.L.G.; Trugilho, M.R.O.; Perales, J.; Valente, R.H. Using mass spectrometry to explore the neglected glycan moieties of the antiophidic proteins DM43 and DM64. *Proteomics* **2012**, *12*, 2753–2765. [CrossRef] [PubMed]

91. Brand, G.D.; Salbo, R.; Jorgensen, T.J.; Bloch, C., Jr.; Boeri Erba, E.; Robinson, C.V.; Tanjoni, I.; Moura-da-Silva, A.M.; Roepstorff, P.; Domont, G.B.; et al. The interaction of the antitoxin DM43 with a snake venom metalloproteinase analyzed by mass spectrometry and surface plasmon resonance. *J. Mass Spectrom.* **2012**, *47*, 567–573. [CrossRef] [PubMed]

92. Drake, A.W.; Myszka, D.G.; Klakamp, S.L. Characterizing high-affinity antigen/antibody complexes by kinetic- and equilibrium-based methods. *Anal. Biochem.* **2004**, *328*, 35–43. [CrossRef] [PubMed]

93. Rocha, S.L.; Neves-Ferreira, A.G.; Trugilho, M.R.; Chapeaurouge, A.; Leon, I.R.; Valente, R.H.; Domont, G.B.; Perales, J. Crotalid snake venom subproteomes unraveled by the antiophidic protein DM43. *J. Proteome Res.* **2009**, *8*, 2351–2360. [CrossRef] [PubMed]

94. Asega, A.F.; Oliveira, A.K.; Menezes, M.C.; Neves-Ferreira, A.G.; Serrano, S.M. Interaction of *Bothrops jararaca* venom metalloproteinases with protein inhibitors. *Toxicon* **2014**, *80*, 1–8. [CrossRef] [PubMed]

95. Jurgilas, P.B.; de Meis, J.; Valente, R.H.; Neves-Ferreira, A.G.C.; da Cruz, D.A.M.; de Oliveira, D.A.F.; Savino, W.; Domont, G.B.; Perales, J. Use of DM43 and Its Fragments as Matrix Metalloproteinases Inhibitor. U.S. Patent 20080249005 A1, 9 October 2008.

96. Mebs, D.; Omori-Satoh, T.; Yamakawa, M.; Nagaoka, Y. Erinacin, an antihemorrhagic factor from the european hedgehog, *Erinaceus europaeus*. *Toxicon* **1996**, *34*, 1313–1316. [CrossRef]

97. Ohashi, T.; Erickson, H.P. Two oligomeric forms of plasma ficolin have differential lectin activity. *J. Biol. Chem.* **1997**, *272*, 14220–14226. [CrossRef] [PubMed]

98. Sugimoto, R.; Yae, Y.; Akaiwa, M.; Kitajima, S.; Shibata, Y.; Sato, H.; Hirata, J.; Okochi, K.; Izuhara, K.; Hamasaki, N. Cloning and characterization of the Hakata antigen, a member of the ficolin/opsonin p35 lectin family. *J. Biol. Chem.* **1998**, *273*, 20721–20727. [CrossRef] [PubMed]

99. Omori-Satoh, T.; Yamakawa, Y.; Mebs, D. The antihemorrhagic factor, erinacin, from the European hedgehog (*Erinaceus europaeus*), a metalloprotease inhibitor of large molecular size possessing ficolin/opsonin P35 lectin domains. *Toxicon* **2000**, *38*, 1561–1580. [CrossRef]

100. Matsushita, M.; Endo, Y.; Taira, S.; Sato, Y.; Fujita, T.; Ichikawa, N.; Nakata, M.; Mizuochi, T. A novel human serum lectin with collagen- and fibrinogen-like domains that functions as an opsonin. *J. Biol. Chem.* **1996**, *271*, 2448–2454. [CrossRef] [PubMed]

101. Rawlings, N.D.; Barrett, A.J. Evolutionary families of peptidases. *Biochem. J.* **1993**, *290*, 205–218. [CrossRef] [PubMed]

102. Fox, J.W.; Serrano, S.M. Snake Venom Metalloproteinases. In *Handbook of Venoms and Toxins of Reptiles*; Mackessy, S.P., Ed.; CRC Press: Boca Raton, FL, USA, 2010; pp. 96–109.

103. Khokha, R.; Murthy, A.; Weiss, A. Metalloproteinases and their natural inhibitors in inflammation and immunity. *Nat. Rev. Immunol.* **2013**, *13*, 649–665. [CrossRef] [PubMed]

104. Battellino, C.; Piazza, R.; Silva, A.M.M.; Cury, Y.; Farsky, S.H.P. Assessment of efficacy of bothropic antivenom therapy on microcirculatory effects induced by *Bothrops jararaca* snake venom. *Toxicon* **2003**, *41*, 583–593. [CrossRef]

105. Gutierrez, J.M.; Leon, G.; Lomonte, B. Pharmacokinetic-pharmacodynamic relationships of immunoglobulin therapy for envenomation. *Clin. Pharmacokinet.* **2003**, *42*, 721–741. [CrossRef] [PubMed]

106. Gutierrez, J.M.; Lomonte, B.; Leon, G.; Rucavado, A.; Chaves, F.; Angulo, Y. Trends in snakebite envenomation therapy: Scientific, technological and public health considerations. *Curr. Pharm. Des.* **2007**, *13*, 2935–2950. [CrossRef] [PubMed]

107. Ferreira, S.H.; Bartelt, D.C.; Greene, L.J. Isolation of bradykinin-potentiating peptides from *Bothrops jararaca* venom. *Biochemistry* **1970**, *9*, 2583–2593. [CrossRef] [PubMed]

108. Ondetti, M.A.; Williams, N.J.; Sabo, E.F.; Pluscec, J.; Weaver, E.R.; Kocy, O. Angiotensin-converting enzyme inhibitors from the venom of *Bothrops jararaca*. Isolation, elucidation of structure, and synthesis. *Biochemistry* **1971**, *10*, 4033–4039. [CrossRef] [PubMed]

109. Gavras, H.; Brunner, H.R.; Laragh, J.H.; Sealey, J.E.; Gavras, I.; Vukovich, R.A. An angiotensin converting-enzyme inhibitor to identify and treat vasoconstrictor and volume factors in hypertensive patients. *N. Engl. J. Med.* **1974**, *291*, 817–821. [CrossRef] [PubMed]

110. Cushman, D.W.; Ondetti, M.A. Design of angiotensin converting enzyme inhibitors. *Nat. Med.* **1999**, *5*, 1110–1113. [CrossRef] [PubMed]

111. Cushman, D.W.; Cheung, H.S.; Sabo, E.F.; Ondetti, M.A. Design of potent competitive inhibitors of angiotensin-converting enzyme. Carboxyalkanoyl and mercaptoalkanoyl amino acids. *Biochemistry* **1977**, *16*, 5484–5491. [CrossRef] [PubMed]

112. Ondetti, M.A.; Rubin, B.; Cushman, D.W. Design of specific inhibitors of angiotensin-converting enzyme: New class of orally active antihypertensive agents. *Science* **1977**, *196*, 441–444. [CrossRef] [PubMed]

113. Acharya, K.R.; Sturrock, E.D.; Riordan, J.F.; Ehlers, M.R. ACE revisited: A new target for structure-based drug design. *Nat. Rev. Drug Discov.* **2003**, *2*, 891–902. [CrossRef] [PubMed]

114. Craik, D.J.; Fairlie, D.P.; Liras, S.; Price, D. The future of peptide-based drugs. *Chem. Biol. Drug Des.* **2013**, *81*, 136–147. [CrossRef] [PubMed]

115. Abbenante, G.; Fairlie, D.P. Protease inhibitors in the clinic. *Med. Chem.* **2005**, *1*, 71–104. [CrossRef] [PubMed]

116. Shiomi, T.; Lemaitre, V.; D'Armiento, J.; Okada, Y. Matrix metalloproteinases, a disintegrin and metalloproteinases, and a disintegrin and metalloproteinases with thrombospondin motifs in non-neoplastic diseases. *Pathol. Int.* **2010**, *60*, 477–496. [CrossRef] [PubMed]

117. Drag, M.; Salvesen, G.S. Emerging principles in protease-based drug discovery. *Nat. Rev. Drug Discov.* **2010**, *9*, 690–701. [CrossRef] [PubMed]

118. Batra, J.; Robinson, J.; Soares, A.S.; Fields, A.P.; Radisky, D.C.; Radisky, E.S. Matrix metalloproteinase-10 (MMP-10) interaction with tissue inhibitors of metalloproteinases TIMP-1 and TIMP-2: Binding studies and crystal structure. *J. Biol. Chem.* **2012**, *287*, 15935–15946. [CrossRef] [PubMed]

119. Wlodawer, A.; Minor, W.; Dauter, Z.; Jaskolski, M. Protein crystallography for non-crystallographers, or how to get the best (but not more) from published macromolecular structures. *FEBS J.* **2008**, *275*, 1–21. [CrossRef] [PubMed]

120. Skrisovska, L.; Schubert, M.; Allain, F.H. Recent advances in segmental isotope labeling of proteins: NMR applications to large proteins and glycoproteins. *J. Biomol. NMR* **2010**, *46*, 51–65. [CrossRef] [PubMed]

121. Pervushin, K.; Riek, R.; Wider, G.; Wuthrich, K. Attenuated T2 relaxation by mutual cancellation of dipole-dipole coupling and chemical shift anisotropy indicates an avenue to NMR structures of very large biological macromolecules in solution. *Proc. Natl. Acad. Sci. USA* **1997**, *94*, 12366–12371. [CrossRef] [PubMed]

122. Sander, C.; Schneider, R. Database of homology-derived protein structures and the structural meaning of sequence alignment. *Proteins* **1991**, *9*, 56–68. [CrossRef] [PubMed]

123. Brenner, S.E.; Chothia, C.; Hubbard, T.J. Assessing sequence comparison methods with reliable structurally identified distant evolutionary relationships. *Proc. Natl. Acad. Sci. USA* **1998**, *95*, 6073–6078. [CrossRef] [PubMed]

124. Rost, B. Twilight zone of protein sequence alignments. *Protein Eng.* **1999**, *12*, 85–94. [CrossRef] [PubMed]

125. Bowie, J.U.; Luthy, R.; Eisenberg, D. A method to identify protein sequences that fold into a known three-dimensional structure. *Science* **1991**, *253*, 164–170. [CrossRef] [PubMed]

126. Jones, D.T.; Taylor, W.R.; Thornton, J.M. A new approach to protein fold recognition. *Nature* **1992**, *358*, 86–89. [CrossRef] [PubMed]

127. Zhang, Y. Progress and challenges in protein structure prediction. *Curr. Opin. Struct. Biol.* **2008**, *18*, 342–348. [CrossRef] [PubMed]

128. Boratyn, G.M.; Schaffer, A.A.; Agarwala, R.; Altschul, S.F.; Lipman, D.J.; Madden, T.L. Domain enhanced lookup time accelerated BLAST. *Biol. Direct.* **2012**, *7*. [CrossRef] [PubMed]

129. Jones, D.T. Protein secondary structure prediction based on position-specific scoring matrices. *J. Mol. Biol.* **1999**, *292*, 195–202. [CrossRef] [PubMed]

130. Jones, D.T. GenTHREADER: An efficient and reliable protein fold recognition method for genomic sequences. *J. Mol. Biol.* **1999**, *287*, 797–815. [CrossRef] [PubMed]

131. Yang, J.; Yan, R.; Roy, A.; Xu, D.; Poisson, J.; Zhang, Y. The I-TASSER Suite: Protein structure and function prediction. *Nat. Methods* **2015**, *12*, 7–8. [CrossRef] [PubMed]

132. Kahraman, A.; Herzog, F.; Leitner, A.; Rosenberger, G.; Aebersold, R.; Malmstrom, L. Cross-link guided molecular modeling with ROSETTA. *PLoS ONE* **2013**, *8*, e73411. [CrossRef] [PubMed]

133. Zhang, Y.; Majumder, E.L.; Yue, H.; Blankenship, R.E.; Gross, M.L. Structural analysis of diheme cytochrome c by hydrogen-deuterium exchange mass spectrometry and homology modeling. *Biochemistry* **2014**, *53*, 5619–5630. [CrossRef] [PubMed]

134. Schneidman-Duhovny, D.; Kim, S.J.; Sali, A. Integrative structural modeling with small angle X-ray scattering profiles. *BMC Struct. Biol.* **2012**, *12*. [CrossRef] [PubMed]

135. Zheng, W.; Doniach, S. Protein structure prediction constrained by solution X-ray scattering data and structural homology identification. *J. Mol. Biol.* **2002**, *316*, 173–187. [CrossRef] [PubMed]

136. Schneidman-Duhovny, D.; Hammel, M.; Tainer, J.A.; Sali, A. Accurate SAXS profile computation and its assessment by contrast variation experiments. *Biophys. J.* **2013**, *105*, 962–974. [CrossRef] [PubMed]

Review

Snake Genome Sequencing: Results and Future Prospects

Harald M. I. Kerkkamp [1], R. Manjunatha Kini [2], Alexey S. Pospelov [3], Freek J. Vonk [4], Christiaan V. Henkel [1] and Michael K. Richardson [1,*]

[1] Institute of Biology, University of Leiden, Leiden 2300 RA, The Netherlands; h.m.i.kerkkamp@biology.leidenuniv.nl (H.M.I.K.); henkel.c@hsleiden.nl (C.V.H.)

[2] Department of Biological Science, National University of Singapore, Singapore 117543, Singapore; dbskinim@nus.edu.sg

[3] Department of Biosciences and Neuroscience Center, University of Helsinki, Helsinki 00014, Finland; apospelo@mappi.helsinki.fi

[4] Naturalis Biodiversity Center, Darwinweg 2, Leiden 2333 CR, The Netherlands; freek.vonk@naturalis.nl

* Correspondence: m.k.richardson@biology.leidenuniv.nl

Academic Editors: Jay Fox and José María Gutiérrez

Received: 2 November 2016; Accepted: 25 November 2016; Published: 1 December 2016

Abstract: Snake genome sequencing is in its infancy—very much behind the progress made in sequencing the genomes of humans, model organisms and pathogens relevant to biomedical research, and agricultural species. We provide here an overview of some of the snake genome projects in progress, and discuss the biological findings, with special emphasis on toxinology, from the small number of draft snake genomes already published. We discuss the future of snake genomics, pointing out that new sequencing technologies will help overcome the problem of repetitive sequences in assembling snake genomes. Genome sequences are also likely to be valuable in examining the clustering of toxin genes on the chromosomes, in designing recombinant antivenoms and in studying the epigenetic regulation of toxin gene expression.

Keywords: snake; genome; genomics; king cobra; reptile; Malayan pit viper

1. Introduction

The sequencing of animal genomes is uncovering a treasure trove of biological information. Genomes can be defined in various ways. Functional definitions based on concepts of information-encoding and transfer tend to ignore the role of extra-genomic (epigenetic) mechanisms in these processes [1]. Therefore, we shall simply assume the genome to comprise the nucleotide sequence of all nuclear and mitochondrial DNA of an organism. The genome may be sequenced in its entirety via whole genome sequencing [2,3]. It may be more practical for some research questions to sequence only the region of interest, using a 'targeted capture' approach [4]. Targeted approaches include the selective sequencing of bacterial artificial chromosome (BAC) libraries [5].

Genome sequencing has tended to focus on *Homo sapiens* and there are reportedly plans to sequence 2 million human genomes for biomedical research objectives including personalized medicine [6]. Further, the genomes of many animal species used as models in biomedical research, or reared in agriculture, have also been sequenced. The genomes of non-model species have received far less attention although there are plans to sequence many thousands of vertebrate genome in the near future [7].

1.1. Why Snakes Are Interesting

Snake genomics is a neglected topic, as can be seen by the relatively modest number of published genomes and projects in the pipeline (Table 1). Nonetheless, it is a topic that is attracting increasing interest from a biologists in several sub-disciplines [8]. This interest in snake genomes stems from the medical importance of snakebite in many developing countries [9], the potential for finding novel drugs and other bioactive compounds in venoms [10] and, from the perspective of fundamental research, the extraordinary array of evolutionary novelties found in snakes [11,12].

Table 1. Snake genome projects published or in progress.

Trivial Name	Scientific Name	Family	Notes
Prong-snouted blind snake	*Anilios bituberculatus*	Typhlopidae	F.J. Vonk et al., in progress
Texas blind snake	*Rena dulcis*	Leptotyphlopidae	T.A. Castoe et al., in progress
Boa constrictor	*Boa constrictor*	Boidae	Ref. [13]; GenB: PRJNA210004
Boa constrictor	*Boa constrictor*	Boidae	Ref. [14]
Burmese python	*Python bivittatus*	Pythonidae	Published [2]; GenB: AEQU00000000
Garter snake	*Thamnophis sirtalis*	Colubridae	GenB: LFLD00000000
	Thamnophis elegans	Colubridae	Ref. [13]; GenB: PRJNA210004
Corn snake	*Pantherophis guttatus*	Colubridae	Ref. [15]; GenB: JTLQ01000000
Corn snake	*Pantherophis guttatus*	Colubridae	Targeted sequencing: 5′ hox genes [16]
King cobra	*Ophiophagus hannah*	Elapidae	Published [3]; GenB: AZIM00000000
Malayan pit viper	*Calloselasma rhodostoma*	Viperidae	F.J. Vonk et al., in progress
Five-pacer viper	*Deinagkistrodon acutus*	Viperidae	Ref. [17]
European adder	*Vipera berus berus*	Viperidae	Baylor College of Medicine, Human Genome Sequencing Center; GenB: JTGP00000000
Habu	*Protobothrops flavoviridis*	Viperidae	H. Shibata et al., in progress
Brown spotted pit viper	*Protobothrops mucrosquamatus*	Viperidae	A.S. Mikheyev et al., in progress; GenB: PRJDB4386
Prairie rattlesnake	*Crotalus viridis viridis*	Viperidae	T.A. Castoe et al., in progress
Western diamond-backed rattlesnake	*Crotalus atrox*	Viperidae	Ref. [5]
Timber rattlesnake	*Crotalus horridus*	Viperidae	GenB: LVCR00000000.1
Speckled rattlesnake	*Crotalus mitchellii pyrrhus*	Viperidae	Ref. [18]; GenB: JPMF01000000
Western Diamondback rattlesnake, Mojave rattlesnake and Eastern Diamondback rattlesnake	*Crotalus atrox, C. scutulatus, and C. adamanteus*	Viperidae	Targeted sequencing of bacterial artificial chromosome (BAC) clones containing phospholipase A₂ genes.
Pygmy rattlesnake	*Sistrurus miliarius*	Viperidae	Ref. [13]; GenB: PRJNA210004
Temple pit viper	*Tropidolaemus wagleri*	Viperidae	R.M. Kini et al., in progress

This list is not necessarily exhaustive. Abbreviation: GenB, GenBank accession number. Taxonomy according to the Pubmed Taxonomy database [19].

Snakes (Serpentes) are represented by around 3000 extant species [20]. They show a suite of adaptations common to many lineages of vertebrates that have independently evolved long, thin bodies. This suite includes limb reduction or loss, axial elongation, increase in vertebral count and asymmetry of paired viscera. Extant snakes have completely lost all traces of the forelimb and pectoral girdle. In most species there is also loss of the hindlimb and pelvic girdle [11]. Exceptions include the femoral and pelvic girdle remnants found on each side in Leptotyphlopidae (reviewed in Ref. [21]); and the single pelvic element on each side in Typhlopidae [22]. Pelvic vestiges are also present in Aniliidae, Cylindrophiidae and Anomochilidae; and in boas and pythons. There are both pelvic and femoral vestiges, the latter often tipped with a horny spur [22,23]. Compared to ancestral squamates, snakes show elongation of the primary axis with a high vertebral number and poor demarcation of the vertebral regions [24]. The left lung is reduced in size or absent [25].

Other adaptations include jaw modifications and metabolic adaptations associated with swallowing prey whole [2]; the presence of a venom delivery system [26], consisting of the venom glands and fangs; and heat-sensing "pit organs". Venom delivery systems are found in approximately 600 species in the Elapidae, Viperidae, Colubridae and Atractaspididae [27]. Heat-sensitive pit organs are represented by the loreal pit of Crotalinae [28], and the labial pits of some pythons and boas [29].

In the context of this special journal issue, the most relevant of these adaptations is venom, and the peptide or protein toxins that it contains. Venom can be defined as any glandular secretion produced by a metazoan that can be introduced into the tissues of another animal through a puncture (inflicted by the venomous animal for that purpose) and which incapacitates prey or deters attackers by virtue of its potent bioactivity [30]. Venom, and an associated venom delivery system (a gland connected to a puncturing device), has evolved independently in many animal clades [31,32].

1.2. What Genomes Can Tell Us

As has been pointed out [33], toxin evolution has been studied for many years using traditional sequencing and proteomics approaches; but new tools for genomics, transcriptomics and proteomics are greatly advancing the field. For biomedical research in general, there are many advantages in having genomic sequence data and we now summarize just a few of these advantages. A whole genome sequence allows, in principle, the prediction of all translated genes (the exome) by means of ab initio gene prediction algorithms [34] and homology searches using reference sequences [35]. Because of the paucity of genome sequences and the apparent frequent duplication of toxin-encoding genes, the use of transcriptome data from the same species, or one closely related, makes this task easier. The genes predicted may include genes for translated proteins as well as microRNAs (mRNAs) and other non-coding genes.

Predicted gene sequences can then be used in a host of applications and analyses, ranging from the design of probes for in situ hybridization [26], to searches for genes under positive or negative selection, as inferred by the d_N/d_S ratio [36]. The latter analyses have shown that multiple genes are under selection in snakes, or in clades within the snakes, including some venom toxin genes [3] and developmental genes possibly connected to development of the serpentiform body plan [2].

With genome sequences, evolutionary gene loss can be more confidently asserted than by looking at the transcriptome alone. Hypotheses about gene loss, or the degeneration of functional genes into pseudogenes, can be more easily tested because non-coding pseudogenes can be identified in the genome sequence on the basis of sequence homology or synteny [37]. Synteny refers to the location of loci on the same chromosome, or the order and orientation of neighboring genes, especially when compared across species [37].

Analysis of genome sequences shows that several visual pigment genes have been lost in snakes compared with other squamates [2]. This may be related to the putative fossorial (burrowing) lifestyle of an ancestral snake which might have had reduced eyes [38]. Genomics also reveals that some neurotoxin genes have been lost in the lineage leading to the Western and Eastern diamond-backed rattlesnakes (*Crotalus atrox* and *C. adamanteus*, respectively) [5].

Using genome sequences, it is possible to look for candidate regulatory regions; this in turn may allow genomic regulatory blocks to be identified [39]. In the context of toxinology, it will be especially interesting to examine whether duplicated toxin genes of the same toxin family are clustered [5] and functionally part of a common regulatory landscape—in a way analogous, perhaps, to the well-studied *hox* developmental genes [16]. Genome sequences allow the identification of structural variations, including inversions, insertions, deletions and tandem duplications and other large rearrangements [40]. It is also possible to look for transposable elements and other repetitive sequences [41].

Genomic data can be used in phylogeny reconstruction (which is one aspect of the discipline phylogenomics) although this endeavor is not without difficulties [37,42]. One such difficulty is that the evolution of nucleotide sequences effectively overwrites the ancestral sequence making homologies (orthologues) more difficult to identify [37,42]. Horizontal gene transfer (between species), gene loss and genome duplications [43] can further obscure the phylogenetic relationships among species (reviewed in Ref. [37]).

1.3. Aims and Objectives of This Review

Our aim here is to review some of the biological results yielded, to date, by snake genomes; and to consider some of the research questions that may one day be solved by the analysis of snake genomes. We will focus mainly on the evolution of venom toxins, but discuss also some questions related to snake morphological and physiological adaptations that are being illuminated by genomics.

2. Status of Snake Genome Sequencing Projects

The sequencing of snake genomes is very much in its infancy. The first draft genomes of snakes to be published were those of the Boa constrictor (*Boa constrictor*) [13,14], Burmese python (*Python molurus bivittatus*) [2] and the king cobra (*Ophiophagus hannah*) [3], followed by a high coverage (238 ×) assembly of the first viper genome (*Crotalus mitchellii*) [18]. Some key data on the first two of these draft genomes are summarized in Table 2. The status of some other snake genome projects known to us, including studies based on targeted capture, is summarized in Table 1. As can be seen, the genome sizes of the Burmese python [2] and king cobra [3] are 1.44 and 1.36–1.59 Gbp, respectively. This is roughly half the size of the human genome and closer to the smaller genomes of some other sauropsids such as the chicken and the anole lizard (Table 2).

Table 2. Selected data from the Burmese python and king cobra draft genomes and comparison with genomes of other species.

Species	Coding Genes (k)	Genome Size (Gb)	Repeats (%)
Burmese python	25 [2]	1.44 [2]	31.8–59.4 [2]
King cobra	21.19 [3]	1.36–1.59 [3]	35.2–60.4 [2]
Chicken	20–23 * [44]	1.05 [44]	4.3–8.0 [45]; 9.4 [44]
Human	20.4 ¶; 19 [46]	3.54 ¶	>66–69 [47]
Anolis	18.5 †	1.70 †	30% ‡ [48]

* v. 85.4 in ensembl.org gives the number of coding genes in the chicken genome as 15,508; ¶ Human genome, build 38; ensembl.org; † GenBank Assembly ID GCA_000090745.1; ‡ Mobile elements.

3. Genome Data in the Reconstruction of Toxin Evolution

3.1. Overview of Possible Mechanisms of Toxin Evolution

Toxin evolution is reviewed in Ref. [32]. Waglerin toxins in Wagler's viper (*Tropidolaemus wagleri*) may well have arisen de novo since no orthologues have been found [49]. This is an exceptional case and in general, the evolution of genes de novo is thought to be comparatively rare. Thus, in the human genome, entirely new genes (i.e., those not found in other primates) are very few in number, and tend not to be expressed in the proteome, suggesting that they function as non-protein-coding genes [46]. In fact, the likelihood of a gene being expressed in the human proteome at all was found to be related to the age of evolutionary origin of that gene [46].

Cysteine-rich secretory protein (CRISP) and kallikrein toxins in Wagler's viper are suggested to have become toxic simply as a result of evolutionary changes in the coding sequence of existing salivary proteins [49]. Indeed, another study concluded that not just a few, but in fact most, snake venom toxins evolved from proteins expressed ancestrally in salivary glandular tissue [50]. In any case, it is clear most venom toxins share close sequence similarity, at least in their functional domains, with known, non-venom genes (physiological or body genes) [49].

Alternative splicing can result in both physiological and toxin isoforms being generated from the same gene in different tissues. This appears to be the case with acetylcholinesterase gene of *Bungarus fasciatus* [51].

3.2. Moonlighting: The Strange Case of Nerve Growth Factor

Nerve growth factor (NGF) is a component of venoms in many snakes. At first sight, it may seem to be nothing more than an innocuous neurotrophin apparently occurring in the venom for no good reason. However, NGF is an extremely potent inducer of mast cell degranulation; thus it is possible that it may produce increased local vascular permeability and toxin absorption; it may also produce or enhance anaphylaxis [52,53]. The possibility that venom nerve growth factor may contribute to the toxicity of venom is further suggested by the fact that, like other true venom toxins, it is under positive selection in at least some snakes [52,53]. Nerve growth factor may also play an ancillary (non-toxic) role while the venom is stored in the venom gland by inhibiting metalloprotease-mediated degradation [54]. Since a single isoform is present in *Bothrops jararaca* [55] it is possible that nerve growth factor may be 'moonlighting' as a venom component—that is, taking on functions in the venom additional to those of its function as a neurotropin (the concept of moonlighting is discussed in Ref. [56]). However, arguing against moonlighting is the fact that nerve growth factor is present in at least two copies in the king cobra genome [3] and in other cobras (reviewed in Ref. [50]; see also Table 3 in the current article).

Table 3. Number of copies (paralogues) of toxin genes in the king cobra (*Ophiophagus hannah*) genome; data from Ref. [3].

Venom Toxin or Toxin Family	Number of Paralogues
3FTx (three-finger toxin) *	21
PLA2 (phospholipase A2) *	12
Lectin *	11
Kunitz *	10
Waprins *	6
Cystatin	5
CRISP (cysteine-rich secretory protein)	3
CVF (cobra venom factor)	3
Kallikrein	3
SVMP (snake venom metalloproteinase)	3
LAAO (L-amino acid oxidase)	2
NGF (nerve growth factor)	2
NP (natriuretic peptide) *	2
Acetylcholinesterase	1
Hyaluronidase	1
PLB (phospholipase-B)	1
VEGF (vascular endothelial growth factors)	1
Vespryn	1

Key: (*) estimated number of paralogues; the current genome assembly is not sufficiently well-scaffolded to allow the number of paralogues to be determined with certainty.

3.3. Gene Duplication

Gene duplication may be important in the evolution of venom toxins at two levels: (i) in the origin of the toxin gene from its ancestral counterpart and (ii) in subsequent expansion of the established toxin gene into a multigene family.

Some toxin genes may have undergone an initial duplication event, after which one copy came to be relatively highly expressed in the venom gland by some change in tissue-specific regulation [32]. The nascent toxin gene could then, in principle, undergo sequence evolution independently of its non-venom paralogue to evolve a new function. This process is called neo-functionalization [57,58]. One problem with neo-functionalization as a mechanism is that mutations are more likely to be deleterious than beneficial [59]. An alternative model of gene evolution after duplication is sub-functionalization. This phenomenon can account for the survival of both paralogues because, weakened in function by deleterious mutations, the two copies will need to be retained in the genome in order to make up, together, the full ancestral function by virtue of their complementary effects [60]. This has been called the duplication-degeneration-complementation (DDC) hypothesis [60].

There is no predictable pattern of duplication events in toxin evolution, as can be readily seen, for example, in the highly variable number of different toxin paralogues in the king cobra genome (Table 3). The number varies from one (hyaluronidase) to 21 (three-finger toxins). It is possible that some genes have undergone what we have referred to as 'hijacking' [3]; that is, sequence modification without duplication (Figure 1 in the current article). Comparative analysis of synteny (Figure 1) suggests that the ancestral PLBD1 gene may have evolved into the king cobra venom phospholipase-B (PLB), and that the HYALP1 gene similarly gave rise to venom-expressed hyaluronidase (HYAL).

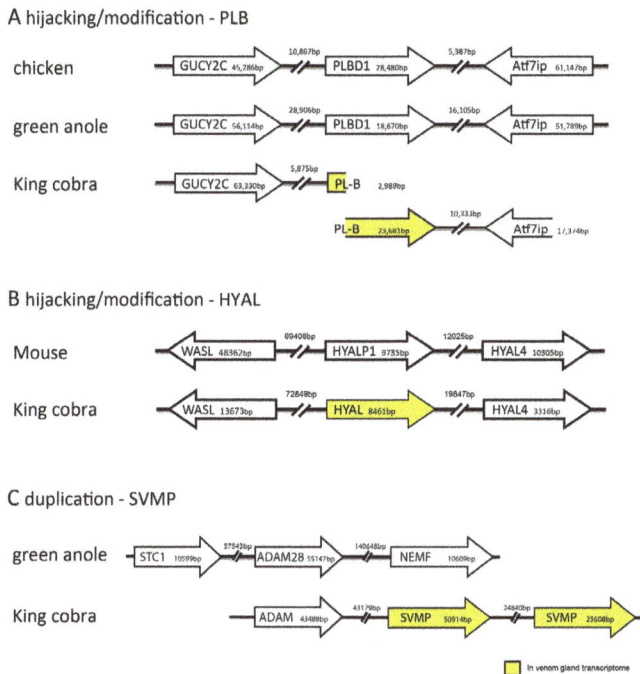

Figure 1. Syntenic comparisons of venom genes in the king cobra with other vertebrates revealing toxin recruitment by hijacking/modification and gene duplication. (**A**) Modification of PLBD1 gene found in the green anole lizard (*Anolis carolinensis*) and the chicken (*Gallus gallus*) results in the venom gland expressed phospholipase-B (PLB). Note that PLB is found split across two king cobra genome scaffolds; (**B**) Modification of HYALP1 gene found in the mouse (*Mus musculus*) results in the venom gland expressed hyaluronidase (HYAL); (**C**) Duplication of the non-venom gland expressed ADAM gene in the king cobra results in a venom gland expressed snake venom metalloproteinase (SVMP) gene. The ADAM gene in the green anole is flanked on both sides by non-SVMP genes, demonstrating the absence of gene duplication in this species. Note that subsequent downstream duplication of the SVMP gene in the king cobra results in multiple venom gland expressed SVMP isoforms. Based on Figure S5 from [3].

An example of a gene that has undergone duplication is the ADAM gene which underwent duplication and subsequently these duplicates evolved into a venom-expressed snake venom metalloproteinase (SVMP) gene (Figure 1). Other toxins that have undergone multiple rounds of duplication to produce multigene families include phospholipase A_2 in rattlesnakes [5]. In that gene family, some paralogues subsequently disappeared from the genome in different lineages, possibly because of a change in prey type [5]. The origin of genes by duplication, and the subsequent loss of some paralogues in this way, is consistent with the birth-and-death model of the evolution of multigene families [61,62].

3.4. Possible Selective Advantage of Possessing Multigene Toxin Families

A preliminary analysis of the king cobra genome (Figure 2) suggests one possible selective advantage of duplication in the evolution of multigene toxin families. There is a tendency for paralogues that have undergone recent expansion to be more highly expressed in the venom gland transcriptome. More work is required to confirm this hypothesis although it is consistent, for example, with the relationship between amylase abundance and mRNA abundance in mice [63].

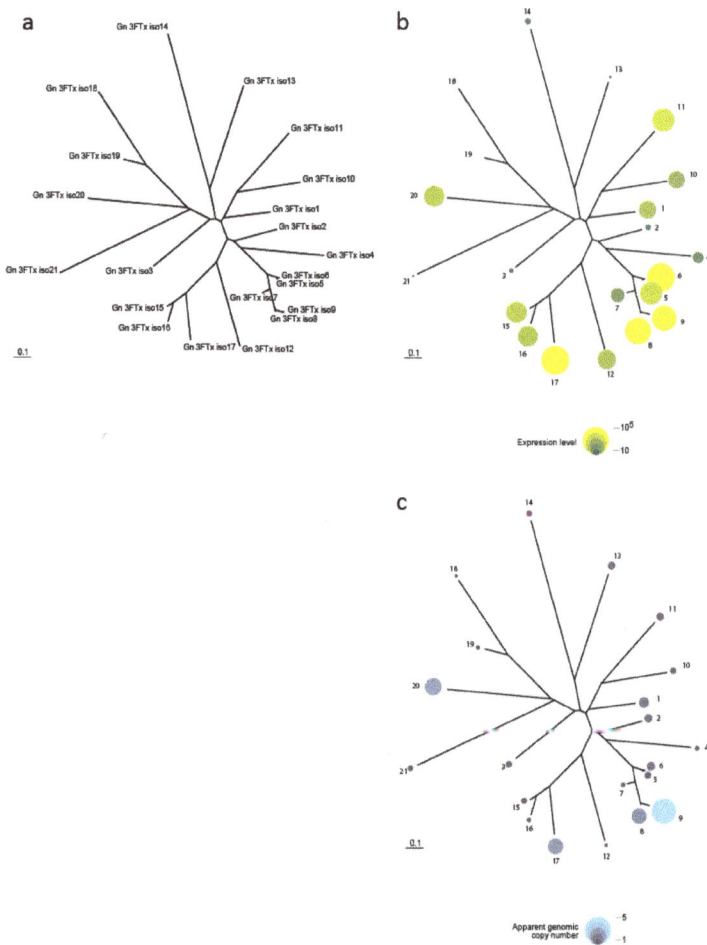

Figure 2. Preliminary analysis of three finger toxin isoforms in the king cobra genome. (**a**) Phylogeny showing isoform numbers; (**b**) Expression level (transcript abundance) in the venom gland; (**c**) Apparent copy number in genome. One hypothesis consistent with the figure is that the more recently expanded paralogues tend to be more highly expressed. The figure is an unpublished analysis by one of us (Christiaan Henkel) based on data in Ref. [3]. See Table 4 for corresponding genome sequencing and accession codes of the three finger toxin isoforms.

Other explanations for the evolution of multiple isoforms of the same toxin is that they might provide broad spectrum toxicity against a range of prey species. Presumably, this is more likely to be advantageous for generalists, than for specialists such as the king cobra. Multiple isoforms might also provide potentiation, so that the toxin complex is more potent than a single toxin. Potentiation is known, for example, in the cone snails [64]. The possession of multiple gene copies might make it more difficult for prey to evolve resistance.

Table 4. King cobra three finger toxin genome sequencing and accession codes. Isoforms correspond with the ones referred to in Figure 2.

3FTX Isoform	Nucleotide Sequence	Accession Code Genbank
Iso1	GATACACCTTGACATGTCTAACACATGAATCATTATTTTTTGAAACCACTGAGAC TTGTTCAGATGGGCAGAACCTATGCTATGCAAAATGGTTTGCAGTTTTTCCAGGTG	AZIM01011044.1
Iso2	GATACACCAGGATATGCCACAAATCTTCTTTTATCTCTGAGACTTGTCCAGATGG GCAGAACCTATGCTATTTAAAATCGTCGGTGTGTGACATTTTTT	AZIM01016929.1
Iso3	GATACACCTTGACATGCATCACATCTGCTCGTAACTTTGAGACTTGTCCACCTGG GCAGAACCTATGCTTTTTAAAATCATGGTATGAAGCTTCAT	AZIM01214498.1
Iso4	TACAAAACCGGTGAACGTATTATTTCTGAGACTTGTCCCCCTGGGCAGGACCTAT GCTATATGAAGACTTGGTGTGACGTTTTTT	AZIM01146344.1
Iso5	GATACACCATGACATGTTACACACAGTACTCATTGTCTCCTCCAACCACTAAGAC TTGTCCAGATGGGCAGAACCTATGCTATAAAAGGTGATTTGCGTTTATTCCACATG	AZIM01015434.1
Iso6	GATACACCACGAAATGCTACGTAACACCTGATGCTACCTCTCAGACTTGTCCAG ATGGGGAGAACATATGCTATACAAAGTCTTGGTGTGACGGTTTTT	AZIM01133918.1
Iso7	GATACACCACGAAATGCTATGTAACACCTGATGCTACCTCTCAGACTTGTCCAGA TGGGGAGAACATATGCTATACAAAGTCTTGGTGTGACGTTTTTT	AZIM01229389.1
Iso8	GATACACCACGAAATGCTACATAACACCTGATGTGAAGTCTCAGACTTGTCCAG ATGGGGAGAACATATGCTATACAAAGACTTGGTGTGATGTTTGGT	AZIM01229389.1
Iso9	GATACACCACGAAATGCTACGTAACACCTGATGTTAAGTCTGAGACTTGTCCAG ATGGGCAGGACATATGCTATACAGAGACTTGGTGTGACGTTTGGT	AZIM01028336.1
Iso10	GATACACCACGAAATGCTACGTAACACCTGATGTTAAGTCTGAGACTTGTCCAG CTGGGCAGGACATATGCTATACAGAGACTTGGTGTGATGCTTGGT	AZIM01097792.1
Iso11	GACACACCAGGATATGTCTCACAGACTACTCAAAAGTTAGTGAAACCATTGAGA TTTGTCCAGATGGGCAGAACTTCTGCTTTAAAAAGTTTCCTAAGGGTATTCCATTTT	AZIM01006046.1
Iso12	GATACACCATGAAATGTCTCACAAAGTACTCCCGGGTTAGTGAAACCTCTCAGA CTTGTCACGTTTGGCAGAACCTATGTTTTAAAAAGTGGCAGAAGG	AZIM01011575.1
Iso13	GACACACCTTGATATGTGTCAAACAGTACACAATTTTTGGTGTAACCCCTGAGAT TTGCGCAGATGGGCAGAACCTATGCTATAAAAACATGGCATATGGTGTATCCAGGTG	AZIM01011969.1
Iso14	GATACACCACGAAATGTTACAACCACCAGTCAACGACTCCTGAAACCACTGAAA TTTGTCCAGATTCAGGGTACTTTTGCTATAAAAAGCTCTTGGATTGATGGACGTG	AZIM01034614.1
Iso15	GATACACCCTGATATGTCACCGAGTGCATGGACTTCAGACTTGTGAACCAGATG AGAAGTTTTGCTTTAGAAAGACGACAATGTTTTTTCCAAATC	AZIM01009352.1
Iso16	GATACACCAGGAAATGTCTCAACACACCGCTTCCTTTGATCTATANTTAAAATGA CTATTAAGAAGTTGCCATCTA	AZIM01009586.1
Iso17	NATACACCAGGATATGTTTAAAGCAAGAGCCATTTCAACCTGAAACCAGTACAA CTTGTCCAGATGGGGAAGATGCTTGCTATAGTACATTTTGGAGTGATAACC	AZIM01019523.1
Iso18	NATACACCAGGATATGTTTAAAGCAAGAGCCGTTTCAACCTGAAACCACTACAA CTTGTCCAGAAGGGGAGGATGCTTGCTATAATTTGTTTTGGAGTGATCACA	AZIM01052732.1
Iso19	GATACAGCTTGATATGTTTAACCAAGAGACGTATCGACCTGAAACCACTACAA CTTGTCCAGATGGGGAGGACACTTGCTATAGTACATTTTGGAATGATCACCATG	AZIM01009977.1
Iso20	CACAAACCAAGACATGTTACTCATGCACTGGAGCATTTTGTTCTAATCGTCAAAA ATGTTCGGGTGGGCAGGTCATATGCTTTAAAAGTTGGAAAAATACTCTTCTGATAT	AZIM01013260.1
Iso21	CACACACCCTGACATGTTACTCATGCAATGGATTATTATGTTCTGACCGTGAACA ATGTCCAGATGGGTAGGACATATGCTTTAAGAGATGGAATGATACTGATTGGTCAG	AZIM01013561.1
Iso22	GATACAGCTTGACATGTCTCAATTGCCCAGAACAGTATTGTAAAAGAATTCACA CTTGTCCAGATGGGGAGAACGTATGCTTTAAAAGGTTTTACGAGGGTAAACTATTAT	AZIM01071124.1
Iso23	GATACACTCTGTTGTGTGTTGCAAATGCAATCAAACGGTTTGTGATCTCAATTCGTAT TGTTCAGCAGGCAAGAACCAATGCTATATATTGCAGAATAATA	AZIM01008565.1

3.5. The Selective Expression of Toxin Genes, or Their Ancestral Orthologues, in the Venom Gland

Given that many toxins have arisen by duplication from genes whose ancestral function was something other than that of a venom toxin [5,49,55,65], how did one or more of the duplicates (paralogues) come to be selectively expressed in the venom gland?

3.5.1. Recruitment and Neo-Functionalisation Hypothesis

One scenario for the selective expression of toxin genes in the venom gland is as follows [49]. One of the copies of the ancestral gene underwent a change in tissue-specific regulation so as to become expressed de novo in the venom gland [66]. This paralogue then underwent evolution of its coding sequence so as to become more effective as a venom toxin [3]. Such adaptive changes in the coding sequence of the newly-recruited gene represent "neo-functionalization"—the evolution of a function not related to the ancestral function (reviewed in Ref. [57,58]). Changes in the coding sequence may be accompanied by additional changes in the regulation of toxin gene transcription, as well as in translation and post-translational modification of the protein [67]. Finally, there is evidence that a toxin gene may ultimately undergo a further change in tissue-specific regulation and revert to being expressed in a tissue or organ other than the venom gland [55,68]. While this hypothesis has been disputed [69], recent work comparing toxin expression in multiple different tissues of *B. jararaca* provided additional evidence for reverse recruitment [55].

3.5.2. Restriction and Sub-Functionalisation Hypothesis

The hypothesis of duplication and neo-functionalisation outlined above has been questioned [50] on the grounds that gene duplication in vertebrate genomes is an extremely rare event, and that persuasive examples of neo-functionalisation have rarely been described in any context. Furthermore, since new transcriptional regulatory relations have to be established for a gene to become highly expressed in the venom gland, the whole scenario is argued to be improbable [50].

An alternative hypothesis is that the ancestral gene was expressed in a wide range of tissues, including the venom gland; it then underwent duplication, with one paralogue becoming restricted in expression to the venom gland and losing expression in the other tissues [50]. Thus, while the recruitment and neo-functionalisation hypothesis is critically dependent on the acquisition of new regulatory regions (for the novel expression of a paralogue in the venom gland), the recruitment and sub-functionalisation depends on the loss of regulatory regions (that ancestrally drove expression in tissues other than the venom gland).

3.5.3. Testing the Recruitment and Restriction Hypotheses

It may well prove to be difficult or impossible to test these hypotheses using comparative transcriptomic data only. An essential pre-requisite will be the availability of multiple snake genomes that provide appropriate taxon sampling, together with the identification of regulatory regions that control the tissue-specific expression of toxin genes and their ancestral paralogues. Putative regulatory sequences will also have to be tested functionally. Progress is being made in this area, as we shall now discuss.

3.6. Mechanisms of Transcriptional Regulation That Might Have Led to Selective Expression of Toxin Genes in the Venom-Gland

3.6.1. Non-Coding RNA Genes

RPTLN are long, non-coding RNA genes that may have been involved in the evolution of snake venom metalloproteinases (SVMPs) from a disintegrin and metalloproteinase (ADAM) gene. According to one hypothesis [66], RPTLN was under the control of a venom gland promotor and its signal sequence became fused with the extracellular domain of the one copy of the ancestrally physiological ADAM gene (the latter having previously undergone tandem duplication). After thus being activated in the venom gland, the ADAM gene evolved into an SVMP [66]. The authors note that their hypothesis can be tested as soon as genome builds for the relevant snake species are available (see also this issue: see Ref. [70] for more information).

3.6.2. Transposable Elements

Another intriguing possibility is that *CR1 LINE* transposable elements, which are much more abundant in advanced than in basal snakes, may have played a role in toxin gene recruitment [2]. *CR1 LINEs* are abundant in the genome of the copperhead (*Agkistrodon contortrix*)—much more abundant than they are in the Burmese python (*Python molurus bivittatus*) genome [41]. We discuss transposable elements and other repetitive sequences in more detail below.

3.6.3. VERSE

It has previously been shown that the gene sequences of TroD (venom prothrombin activator gene) and TrFX (blood coagulation factor X gene) are highly similar, except for promoter and intron 1 regions, indicating that TroD probably evolved by duplication of its plasma counterpart [22]. The insertion, in the promoter of TroD, of a VERSE sequence (VEnom Recruitment/Switch Element) accounts for elevated, but not tissue-specific, expression [23].

3.6.4. AG-Rich Motifs

More recently, it was found that AG-rich motifs, in the first intron, silence gene expression in non-venom gland tissues [71]. These AG-rich motifs are promotor-independent silencers, and such cis-elements are also found in some snake toxin genes, but not in housekeeping genes. Several polycomb group proteins (transcription factors) were identified to bind these motifs to regulate expression. Genome sequences will help in identifying regulatory elements that control tissue-specific expression of toxin genes in venom glands as well as expression of cognate genes in respective tissues.

3.7. Evolution of Toxin Resistance in Snakes as Studied with Genomic Data

Genome sequences have cast light on the resistance by snakes to the toxins of their prey. Thus, the Eastern hog-nosed snake (*Heterodon platirhinos*) is resistant to the tetrodotoxin of *Notophthalmus viridescens* a newt on which it preys [72], and the garter snake *Thamnophis sirtalis* likewise shows resistance to the neurotoxic tetrodotoxin of newts in the genus *Taricha* [73]. Tetrodotoxin resistance in *Thamnophis* is due to modification of the amino acid sequence of tetrodotoxin targets: sodium ion channels on skeletal muscles ($Na_V1.4$) and peripheral neurons ($Na_V1.6$ and $Na_V1.7$; reviewed in Refs. [73,74]). Analysis of snake genomic data and partial sequences, suggests that tetrodotoxin resistance in *Thamnophis* arose stepwise over a long period of evolutionary time, with the ancient modifications of the sodium channels in nerves providing the necessary conditions for evolution of resistance in skeletal muscle sodium-channels [73].

4. Transposable Elements and Other Repetitive Sequences in Snake Genomes

Repetitive elements (repeats) are DNA sequences present in many copies in a genome; they can be classified into tandem repeats and transposable elements [75]. They are relatively abundant in snake genomes, especially the genomes of advanced snakes.

Studies of snake genomes have shown how transposable elements have accumulated in the Hox complex of developmental genes. The Hox complex consists of transcription factor genes that have important roles in regulating embryonic pattern formation. Di-Poï and colleagues selectively sequenced the 5′ regions of the Hox clusters of different species [6]. In the squamates studied, including the corn snake (*Pantherophis guttatus*) the clusters had become expanded in size due the accumulation of numerous transposable elements. These transposons include retrotransposons and DNA transposons, and occur mainly in the introns and intergenic regions [6]. The availability of genomic sequences is also helping to uncover possible incidences of horizontal gene transfer. Thus, the long interspersed element (LINE) non-LTR retrotransposon BovB is suggested to have been transferred, by tics, from squamates to bovids [28].

5. Future Prospects in Snake Genomics

In the future, it may be possible to scan snake genomes for bioactive molecules; this could be done, for example, by comparison with a pharmacophore database. Drug discovery from venoms has already delivered drugs such as Prialt, Integrilin, Captopril and Byetta, and multiple candidates are now progressing in clinical trials [76,77]. Furthermore, peptides derived from venoms are valid pharmacological tools to study diseases [78]. For example, the study of the snake toxin α-bungarotoxin led to characterisation of the nicotinic acetylcholine receptor (nAChR) and a new understanding of the disease myasthenia gravis [79]. Genome sequences also provide the prospect of generating recombinant antivenoms [80]. Additionally, methylome sequencing will allow us to investigate the role of epigenetics in regulating toxin gene expression [81].

Another step forward in snake genome sequencing will be new sequencing techniques. Next- or second-generation sequencing, so-called because of the advance in Sanger sequencing, produces short reads with low error rates and high throughput [82]. Examples of next-generation platforms are Illumina and Roche 454. The newly-emerging third-generation sequencing platforms are able to provide reads many kB long [83,84]. These platforms include the so-called "PacBio" system—single molecule, real-time (SMRT) sequencing—from Pacific Biosciences; and MinION™ from Oxford Nanopore Technologies [82]. MinION™ has a higher error rate than PacBio [82] and both have higher error rates than second-generation sequencing. For a review of different sequencing platforms and their properties, see Ref. [84].

Given the relatively high percentage of repetitive sequences in advanced snake genomes, a hybrid approach is very promising, and has indeed proved useful for the tackling the same problem in the human genome [84]. This approach involves using the long reads of third-generation sequencing to bridge the gaps due to repeats; and combining them with reads from second-generation sequencing to ameliorate the problem of errors in the long reads [84]. We have used this approach to assemble a draft genome of the Malayan pit viper (unpublished data). However, as the error rate of long-read sequencing improves, the hybrid approach will likely lose favor.

Other interesting developments in sequencing include optical mapping [85], useful for examining the large-scale organization of genomic features, especially around large repetitive clusters; and single-cell RNA-seq [86]. The latter would be very useful in investigating the regulation of toxin production in the venom gland, as it provides a transcriptomic profile per individual cell (and thereby an overview of the different cell types in a gland).

Acknowledgments: We thank Nicholas Casewell for reading the manuscript. Any errors are ours and not his. We are also grateful to Todd Castoe, Naoko Oda-Ueda and Kim Worley for helpful discussions.

Author Contributions: All authors contributed ideas, comments and references to this article. H.M.I.K. and M.K.R. took the lead in writing and structuring the article. H.M.I.K. and C.V.H. made the figures.

Conflicts of Interest: The authors declare no conflicts of interest.

References

1. Goldman, A.D.; Landweber, L.F. What is a genome? *PLoS Genet.* **2016**, *12*. [CrossRef] [PubMed]
2. Castoe, T.A.; de Koning, A.P.; Hall, K.T.; Card, D.C.; Schield, D.R.; Fujita, M.K.; Ruggiero, R.P.; Degner, J.F.; Daza, J.M.; Gu, W.; et al. The burmese python genome reveals the molecular basis for extreme adaptation in snakes. *Proc. Natl. Acad. Sci. USA* **2013**, *110*, 20645–20650. [CrossRef] [PubMed]
3. Vonk, F.J.; Casewell, N.R.; Henkel, C.V.; Heimberg, A.M.; Jansen, H.J.; McCleary, R.J.; Kerkkamp, H.M.; Vos, R.A.; Guerreiro, I.; Calvete, J.J.; et al. The king cobra genome reveals dynamic gene evolution and adaptation in the snake venom system. *Proc. Natl. Acad. Sci. USA* **2013**, *110*, 20651–20656. [CrossRef] [PubMed]
4. Jones, M.R.; Good, J.M. Targeted capture in evolutionary and ecological genomics. *Mol. Ecol.* **2016**, *25*, 185–202. [CrossRef] [PubMed]
5. Dowell, N.L.; Giorgianni, M.W.; Kassner, V.A.; Selegue, J.E.; Sanchez, E.E.; Carroll, S.B. The deep origin and recent loss of venom toxin genes in rattlesnakes. *Curr Biol* **2016**, *26*, 2434–2445. [CrossRef] [PubMed]
6. Ledford, H. Astrazeneca launches project to sequence 2 million genomes. *Nature* **2016**, *532*. [CrossRef] [PubMed]
7. Koepfli, K.P.; Paten, B.; O'Brien, S.J. The genome 10K project: A way forward. *Annu. Rev. Anim. Biosci.* **2015**, *3*, 57–111. [CrossRef] [PubMed]

8. Schield, D.R.; Card, D.C.; Reyes-Velasco, J.; Andrew, A.L.; Modahl, C.A.; Mackessy, S.M.; Pollock, D.D.; Castoe, T.A. A role for genomics in rattlesnake research—Current knowledge and future potential. In *Rattlesnakes of Arizona*; Schuett, G.W., Porras, L.W., Reiserer, R.S., Eds.; Eco Books: Rodeo, NM, USA, in press.
9. World Health Organization. *Rabies and Envenomings: A Neglected Public Health Issue: Report of a Consultative Meeting, World Health Organization, Geneva, 10 January 2007*; World Health Organization: Geneva, Switzerland, 2007.
10. Vonk, F.J.; Jackson, K.; Doley, R.; Madaras, F.; Mirtschin, P.J.; Vidal, N. Snake venom: From fieldwork to the clinic: Recent insights into snake biology, together with new technology allowing high-throughput screening of venom, bring new hope for drug discovery. *Bioessays* **2011**, *33*, 269–279. [CrossRef] [PubMed]
11. Coates, M.; Ruta, M. Nice snake, shame about the legs. *Trends Ecol. Evol.* **2000**, *15*, 503–507. [CrossRef]
12. Greene, H.W. *Snakes: The Evolution of Mystery in Nature*; University of California Press: Berkeley, CA, USA, 1997; p. 351.
13. Vicoso, B.; Emerson, J.J.; Zektser, Y.; Mahajan, S.; Bachtrog, D. Comparative sex chromosome genomics in snakes: Differentiation, evolutionary strata, and lack of global dosage compensation. *PLoS Biol.* **2013**, *11*. [CrossRef] [PubMed]
14. Bradnam, K.R.; Fass, J.N.; Alexandrov, A.; Baranay, P.; Bechner, M.; Birol, I.; Boisvert, S.; Chapman, J.A.; Chapuis, G.; Chikhi, R.; et al. Assemblathon 2: Evaluating de novo methods of genome assembly in three vertebrate species. *Gigascience* **2013**, *2*. [CrossRef] [PubMed]
15. Ullate-Agote, A.; Milinkovitch, M.C.; Tzika, A.C. The genome sequence of the corn snake (*Pantherophis guttatus*), a valuable resource for evodevo studies in squamates. *Int. J. Dev. Biol.* **2014**, *58*, 881–888. [CrossRef] [PubMed]
16. Di-Poi, N.; Montoya-Burgos, J.I.; Miller, H.; Pourquie, O.; Milinkovitch, M.C.; Duboule, D. Changes in hox genes' structure and function during the evolution of the squamate body plan. *Nature* **2010**, *464*, 99–103. [CrossRef] [PubMed]
17. Yin, W.; Wang, Z.; Li, Q.; Lian, J.; Zhou, Y.; Lu, B.; Jin, L.; Qiu, P.; Zhang, P.; Zhu, W.; et al. Evolution trajectories of snake genes and genomes revealed by comparative analyses of five-pacer viper. *Nat. Commun.* **2016**, *7*. [CrossRef] [PubMed]
18. Gilbert, C.; Meik, J.M.; Dashevsky, D.; Card, D.C.; Castoe, T.A.; Schaack, S. Endogenous hepadnaviruses, bornaviruses and circoviruses in snakes. *Proc. R. Soc.* **2014**, *281*. [CrossRef] [PubMed]
19. Pubmed taxonomy database. Available online: https://www.Ncbi.Nlm.Nih.Gov/taxonomy (accessed on 30 November 2016).
20. Vidal, N.; Delmas, A.S.; David, P.; Cruaud, C.; Couloux, A.; Hedges, S.B. The phylogeny and classification of caenophidian snakes inferred from seven nuclear protein-coding genes. *C. R. Biol.* **2007**, *330*, 182–187. [CrossRef] [PubMed]
21. Pinto, R.R.; Martins, A.R.; Curcio, F.; Ramos, L.O. Osteology and cartilaginous elements of trilepida salgueiroi (amaral, 1954) (scolecophidia: Leptotyphlopidae). *Anat. Rec. (Hoboken)* **2015**, *298*, 1722–1747. [CrossRef] [PubMed]
22. Boulenger, G.A. *Catalogue of the Snakes in the British Museum (Natural History)*; British Museum (Natural History): London, UK, 1893; Volume 1, p. 448.
23. Cohn, M.J.; Tickle, C. Developmental basis of limblessness and axial patterning in snakes. *Nature* **1999**, *399*, 474–479. [CrossRef] [PubMed]
24. Head, J.J.; Polly, P.D. Evolution of the snake body form reveals homoplasy in amniote hox gene function. *Nature* **2015**, *520*, 86–89. [CrossRef] [PubMed]
25. Van Soldt, B.J.; Metscher, B.D.; Poelmann, R.E.; Vervust, B.; Vonk, F.J.; Muller, G.B.; Richardson, M.K. Heterochrony and early left-right asymmetry in the development of the cardiorespiratory system of snakes. *PLoS ONE* **2015**, *10*. [CrossRef] [PubMed]
26. Vonk, F.J.; Admiraal, J.F.; Jackson, K.; Reshef, R.; de Bakker, M.A.; Vanderschoot, K.; van den Berge, I.; van Atten, M.; Burgerhout, E.; Beck, A.; et al. Evolutionary origin and development of snake fangs. *Nature* **2008**, *454*, 630–633. [CrossRef] [PubMed]
27. Jackson, K. Evolution of the venom conducting fang in snakes. *Integr. Comp. Biol.* **2002**, *42*, 1249.

28. Hofstadler Deiques, C. The development of the pit organ of bothrops jararaca and crotalus durissus terrificus (serpentes, viperidae): Support for the monophyly of the subfamily crotalinae. *Acta Zool.* **2002**, *83*, 175–182. [CrossRef]

29. Gracheva, E.O.; Ingolia, N.T.; Kelly, Y.M.; Cordero-Morales, J.F.; Hollopeter, G.; Chesler, A.T.; Sanchez, E.E.; Perez, J.C.; Weissman, J.S.; Julius, D. Molecular basis of infrared detection by snakes. *Nature* **2010**, *464*, 1006–1011. [CrossRef] [PubMed]

30. Weinstein, S.A. Snake venoms: A brief treatise on etymology, origins of terminology, and definitions. *Toxicon* **2015**, *103*, 188–195. [CrossRef] [PubMed]

31. Fry, B.G.; Roelants, K.; Champagne, D.E.; Scheib, H.; Tyndall, J.D.; King, G.F.; Nevalainen, T.J.; Norman, J.A.; Lewis, R.J.; Norton, R.S.; et al. The toxicogenomic multiverse: Convergent recruitment of proteins into animal venoms. *Annu. Rev. Genom. Hum. Genet.* **2009**, *10*, 483–511. [CrossRef] [PubMed]

32. Casewell, N.R.; Wuster, W.; Vonk, F.J.; Harrison, R.A.; Fry, B.G. Complex cocktails: The evolutionary novelty of venoms. *Trends Ecol. Evol.* **2013**, *28*, 219–229. [CrossRef] [PubMed]

33. Reyes-Velasco, J.; Card, D.C.; Andrew, A.L.; Shaney, K.J.; Adams, R.H.; Schield, D.R.; Casewell, N.R.; Mackessy, S.P.; Castoe, T.A. Expression of venom gene homologs in diverse python tissues suggests a new model for the evolution of snake venom. *Mol. Biol. Evol.* **2015**, *32*, 173–183. [CrossRef] [PubMed]

34. Majoros, W.H.; Pertea, M.; Salzberg, S.L. Tigrscan and glimmerhmm: Two open source ab initio eukaryotic gene-finders. *Bioinformatics* **2004**, *20*, 2878–2879. [CrossRef] [PubMed]

35. Collins, J.E.; White, S.; Searle, S.M.; Stemple, D.L. Incorporating RNA-seq data into the zebrafish ensembl genebuild. *Genome Res.* **2012**, *22*, 2067–2078. [CrossRef] [PubMed]

36. Spielman, S.J.; Wan, S.; Wilke, C.O. A comparison of one-rate and two-rate inference frameworks for site-specific dn/ds estimation. *Genetics* **2016**, *24*, 2499–2511. [CrossRef] [PubMed]

37. Tekaia, F. Inferring orthologs: Open questions and perspectives. *Genom. Insights* **2016**, *9*, 17–28. [CrossRef] [PubMed]

38. Simoes, B.F.; Sampaio, F.L.; Jared, C.; Antoniazzi, M.M.; Loew, E.R.; Bowmaker, J.K.; Rodriguez, A.; Hart, N.S.; Hunt, D.M.; Partridge, J.C.; et al. Visual system evolution and the nature of the ancestral snake. *J. Evol. Biol.* **2015**, *28*, 1309–1320. [CrossRef] [PubMed]

39. Irimia, M.; Maeso, I.; Roy, S.W.; Fraser, H.B. Ancient cis-regulatory constraints and the evolution of genome architecture. *Trends Genet.* **2013**, *29*, 521–528. [CrossRef] [PubMed]

40. Tattini, L.; D'Aurizio, R.; Magi, A. Detection of genomic structural variants from next-generation sequencing data. *Front. Bioeng. Biotechnol.* **2015**, *3*. [CrossRef] [PubMed]

41. Castoe, T.A.; Hall, K.T.; Mboulas, M.L.G.; Gu, W.; de Koning, A.P.; Fox, S.E.; Poole, A.W.; Vemulapalli, V.; Daza, J.M.; Mockler, T.; et al. Discovery of highly divergent repeat landscapes in snake genomes using high-throughput sequencing. *Genome Biol. Evol.* **2011**, *3*, 641–653. [CrossRef] [PubMed]

42. Telford, M.J.; Copley, R.R. Improving animal phylogenies with genomic data. *Trends Genet.* **2011**, *27*, 186–195. [CrossRef] [PubMed]

43. Taylor, J.S.; Van de Peer, Y.; Braasch, I.; Meyer, A. Comparative genomics provides evidence for an ancient genome duplication event in fish. *Philos. Trans. R. Soc. Lond. Ser. B Biol. Sci* **2001**, *356*, 1661–1679. [CrossRef] [PubMed]

44. Hillier, L.W.; Miller, W.; Birney, E.; Warren, W.; Hardison, R.C.; Ponting, C.P.; Bork, P.; Burt, D.W.; Groenen, M.A.; Delany, M.E.; et al. Sequence and comparative analysis of the chicken genome provide unique perspectives on vertebrate evolution. *Nature* **2004**, *432*, 695–716. [CrossRef] [PubMed]

45. Wicker, T.; Robertson, J.S.; Schulze, S.R.; Feltus, F.A.; Magrini, V.; Morrison, J.A.; Mardis, E.R.; Wilson, R.K.; Peterson, D.G.; Paterson, A.H.; et al. The repetitive landscape of the chicken genome. *Genome Res.* **2005**, *15*, 126–136. [CrossRef] [PubMed]

46. Ezkurdia, I.; Juan, D.; Rodriguez, J.M.; Frankish, A.; Diekhans, M.; Harrow, J.; Vazquez, J.; Valencia, A.; Tress, M.L. Multiple evidence strands suggest that there may be as few as 19,000 human protein-coding genes. *Hum. Mol. Genet.* **2014**, *23*, 5866–5878. [CrossRef] [PubMed]

47. De Koning, A.P.; Gu, W.; Castoe, T.A.; Batzer, M.A.; Pollock, D.D. Repetitive elements may comprise over two-thirds of the human genome. *PLoS Genet.* **2011**, *7*. [CrossRef] [PubMed]

48. Alfoldi, J.; Di Palma, F.; Grabherr, M.; Williams, C.; Kong, L.; Mauceli, E.; Russell, P.; Lowe, C.B.; Glor, R.E.; Jaffe, J.D.; et al. The genome of the green anole lizard and a comparative analysis with birds and mammals. *Nature* **2011**, *477*, 587–591. [CrossRef] [PubMed]

49. Fry, B.G. From genome to "venome": Molecular origin and evolution of the snake venom proteome inferred from phylogenetic analysis of toxin sequences and related body proteins. *Genome Res.* **2005**, *15*, 403–420. [CrossRef] [PubMed]
50. Hargreaves, A.D.; Swain, M.T.; Hegarty, M.J.; Logan, D.W.; Mulley, J.F. Restriction and recruitment-gene duplication and the origin and evolution of snake venom toxins. *Genome Biol. Evol.* **2014**, *6*, 2088–2095. [CrossRef] [PubMed]
51. Cousin, X.; Bon, S.; Massoulie, J.; Bon, C. Identification of a novel type of alternatively spliced exon from the acetylcholinesterase gene of bungarus fasciatus. Molecular forms of acetylcholinesterase in the snake liver and muscle. *J. Biol. Chem.* **1998**, *273*, 9812–9820. [CrossRef] [PubMed]
52. Sunagar, K.; Fry, B.G.; Jackson, T.N.; Casewell, N.R.; Undheim, E.A.; Vidal, N.; Ali, S.A.; King, G.F.; Vasudevan, K.; Vasconcelos, V.; et al. Molecular evolution of vertebrate neurotrophins: Co-option of the highly conserved nerve growth factor gene into the advanced snake venom arsenalf. *PLoS ONE* **2013**, *8*. [CrossRef]
53. Kostiza, T.; Meier, J. Nerve growth factors from snake venoms: Chemical properties, mode of action and biological significance. *Toxicon* **1996**, *34*, 787–806. [CrossRef]
54. Wijeyewickrema, L.C.; Gardiner, E.E.; Gladigau, E.L.; Berndt, M.C.; Andrews, R.K. Nerve growth factor inhibits metalloproteinase-disintegrins and blocks ectodomain shedding of platelet glycoprotein vi. *J. Biol. Chem.* **2010**, *285*, 11793–11799. [CrossRef] [PubMed]
55. Junqueira-de-Azevedo, I.L.; Bastos, C.M.; Ho, P.L.; Luna, M.S.; Yamanouye, N.; Casewell, N.R. Venom-related transcripts from bothrops jararaca tissues provide novel molecular insights into the production and evolution of snake venom. *Mol. Biol. Evol.* **2015**, *32*, 754–766. [CrossRef] [PubMed]
56. Jeffery, C.J. Protein species and moonlighting proteins: Very small changes in a protein's covalent structure can change its biochemical function. *J. Proteom.* **2016**, *134*, 19–24. [CrossRef] [PubMed]
57. True, J.R.; Carroll, S.B. Gene co-option in physiological and morphological evolution. *Annu. Rev. Cell Dev. Biol.* **2002**, *18*, 53–80. [CrossRef] [PubMed]
58. Taylor, J.S.; Raes, J. Duplication and divergence: The evolution of new genes and old ideas. *Annu. Rev. Genet.* **2004**, *38*, 615–643. [CrossRef] [PubMed]
59. Loewe, L.; Hill, W.G. The population genetics of mutations: Good, bad and indifferent. *Philos. Trans. R. Soc. B Biol. Sci.* **2010**, *365*, 1153–1167. [CrossRef] [PubMed]
60. Force, A.; Lynch, M.; Pickett, F.B.; Amores, A.; Yan, Y.L.; Postlethwait, J. Preservation of duplicate genes by complementary, degenerative mutations. *Genetics* **1999**, *151*, 1531–1545. [PubMed]
61. Fry, B.G.; Wuster, W.; Kini, R.M.; Brusic, V.; Khan, A.; Venkataraman, D.; Rooney, A.P. Molecular evolution and phylogeny of elapid snake venom three-finger toxins. *J. Mol. Evol.* **2003**, *57*, 110–129. [CrossRef] [PubMed]
62. Nei, M.; Rooney, A.P. Concerted and birth-and-death evolution of multigene families. *Annu. Rev. Genet.* **2005**, *39*, 121–152. [CrossRef] [PubMed]
63. Meisler, M.H.; Antonucci, T.K.; Treisman, L.O.; Gumucio, D.L.; Samuelson, L.C. Interstrain variation in amylase gene copy number and mRNA abundance in three mouse tissues. *Genetics* **1986**, *113*, 713–722. [PubMed]
64. Olivera, B.M.; Seger, J.; Horvath, M.P.; Fedosov, A.E. Prey-capture strategies of fish-hunting cone snails: Behavior, neurobiology and evolution. *Brain Behav. Evol.* **2015**, *86*, 58–74. [CrossRef] [PubMed]
65. Fry, B.G.; Vidal, N.; Norman, J.A.; Vonk, F.J.; Scheib, H.; Ramjan, S.F.; Kuruppu, S.; Fung, K.; Hedges, S.B.; Richardson, M.K.; et al. Early evolution of the venom system in lizards and snakes. *Nature* **2006**, *439*, 584–588. [CrossRef] [PubMed]
66. Sanz-Soler, R.; Sanz, L.; Calvete, J.J. Distribution of rptln genes across reptilia: Hypothesized role for rptln in the evolution of svmps. *Integr. Comp. Biol.* **2016**. [CrossRef] [PubMed]
67. Casewell, N.R.; Wagstaff, S.C.; Wuster, W.; Cook, D.A.; Bolton, F.M.; King, S.I.; Pla, D.; Sanz, L.; Calvete, J.J.; Harrison, R.A. Medically important differences in snake venom composition are dictated by distinct postgenomic mechanisms. *Proc. Natl. Acad. Sci. USA* **2014**, *111*, 9205–9210. [CrossRef] [PubMed]
68. Casewell, N.R.; Huttley, G.A.; Wuster, W. Dynamic evolution of venom proteins in squamate reptiles. *Nat. Commun.* **2012**, *3*. [CrossRef] [PubMed]

69. Hargreaves, A.D.; Swain, M.T.; Logan, D.W.; Mulley, J.F. Testing the toxicofera: Comparative transcriptomics casts doubt on the single, early evolution of the reptile venom system. *Toxicon* **2014**, *92*, 140–156. [CrossRef] [PubMed]
70. Sanz, L.; Calvete, J.J. Insights into the evolution of a snake venom multi-gene family from the genomic organization of echis ocellatus svmp genes. *Toxins (Basel)* **2016**, *8*. [CrossRef] [PubMed]
71. Han, S.X.; Kwong, S.; Ge, R.; Kolatkar, P.R.; Woods, A.E.; Blanchet, G.; Kini, R.M. Regulation of expression of venom toxins: Silencing of prothrombin activator trocarin d by ag-rich motifs. *FASEB J.* **2016**, *30*, 2411–2425. [CrossRef] [PubMed]
72. Feldman, C.R.; Durso, A.M.; Hanifin, C.T.; Pfrender, M.E.; Ducey, P.K.; Stokes, A.N.; Barnett, K.E.; Brodie, E.D., 3rd; Brodie, E.D., Jr. Is there more than one way to skin a newt? Convergent toxin resistance in snakes is not due to a common genetic mechanism. *Heredity (Edinb)* **2016**, *116*, 84–91. [CrossRef] [PubMed]
73. McGlothlin, J.W.; Kobiela, M.E.; Feldman, C.R.; Castoe, T.A.; Geffeney, S.L.; Hanifin, C.T.; Toledo, G.; Vonk, F.J.; Richardson, M.K.; Brodie, E.D.; et al. Historical contingency in a multigene family facilitates adaptive evolution of toxin resistance. *Curr. Biol.* **2016**, *26*, 1616–1621. [CrossRef] [PubMed]
74. Soong, T.W.; Venkatesh, B. Adaptive evolution of tetrodotoxin resistance in animals. *Trends Genet.* **2006**, *22*, 621–626. [CrossRef] [PubMed]
75. Padeken, J.; Zeller, P.; Gasser, S.M. Repeat DNA in genome organization and stability. *Curr. Opin. Genet. Dev.* **2015**, *31*, 12–19. [CrossRef] [PubMed]
76. Marcinkiewicz, C. Functional characteristic of snake venom disintegrins: Potential therapeutic implication. *Curr. Pharm. Des.* **2005**, *11*, 815–827. [CrossRef] [PubMed]
77. Laing, G.D.; Moura-da-Silva, A.M. Jararhagin and its multiple effects on hemostasis. *Toxicon* **2005**, *45*, 987–996. [CrossRef] [PubMed]
78. McCleary, R.J.; Kini, R.M. Non-enzymatic proteins from snake venoms: A gold mine of pharmacological tools and drug leads. *Toxicon* **2013**, *62*, 56–74. [CrossRef] [PubMed]
79. Kini, R.M.; Doley, R. Structure, function and evolution of three-finger toxins: Mini proteins with multiple targets. *Toxicon* **2010**, *56*, 855–867. [CrossRef] [PubMed]
80. Wagstaff, S.C.; Laing, G.D.; Theakston, R.D.; Papaspyridis, C.; Harrison, R.A. Bioinformatics and multiepitope DNA immunization to design rational snake antivenom. *PLoS Med.* **2006**, *3*. [CrossRef] [PubMed]
81. Bird, A. Perceptions of epigenetics. *Nature* **2007**, *447*, 396–398. [CrossRef] [PubMed]
82. Karlsson, E.; Larkeryd, A.; Sjodin, A.; Forsman, M.; Stenberg, P. Scaffolding of a bacterial genome using minion nanopore sequencing. *Sci. Rep.* **2015**, *5*. [CrossRef] [PubMed]
83. Lu, H.; Giordano, F.; Ning, Z. Oxford nanopore minion sequencing and genome assembly. *Genom. Proteom. Bioinform.* **2016**, *31*, 265–279. [CrossRef] [PubMed]
84. Xiao, W.; Wu, L.; Yavas, G.; Simonyan, V.; Ning, B.; Hong, H. Challenges, solutions, and quality metrics of personal genome assembly in advancing precision medicine. *Pharmaceutics* **2016**, *8*. [CrossRef] [PubMed]
85. Levy-Sakin, M.; Ebenstein, Y. Beyond sequencing: Optical mapping of DNA in the age of nanotechnology and nanoscopy. *Curr. Opin. Biotechnol.* **2013**, *24*, 690–698. [CrossRef] [PubMed]
86. Saliba, A.E.; Westermann, A.J.; Gorski, S.A.; Vogel, J. Single-cell RNA-seq: Advances and future challenges. *Nucleic Acids Res.* **2014**, *42*, 8845–8860. [CrossRef] [PubMed]

Section 2:
Original Research

Article

Novel Catalytically-Inactive PII Metalloproteinases from a Viperid Snake Venom with Substitutions in the Canonical Zinc-Binding Motif

Erika Camacho [1], Libia Sanz [2], Teresa Escalante [1], Alicia Pérez [2], Fabián Villalta [1],
Bruno Lomonte [1], Ana Gisele C. Neves-Ferreira [3], Andrés Feoli [1], Juan J. Calvete [2,4],
José María Gutiérrez [1] and Alexandra Rucavado [1,*]

[1] Instituto Clodomiro Picado, Facultad de Microbiología, Universidad de Costa Rica, San José 11501,
Costa Rica; ecamachoum@gmail.com (E.C.); teresa.escalante@ucr.ac.cr (T.E.); fvillalta85@gmail.com (F.V.);
bruno.lomonte@ucr.ac.cr (B.L.); boch31@gmail.com (A.F.); jose.gutierrez@ucr.ac.cr (J.M.G.)
[2] Instituto de Biomedicina de Valencia, Consejo Superior de Investigaciones Científicas, Valencia 46010, Spain;
libia.sanz@ibv.csic.es (L.S.); aperez@ibv.csic.es (A.P.); jcalvete@ibv.csic.es (J.J.C.)
[3] Laboratório de Toxinologia, Instituto Oswaldo Cruz, Fiocruz, Rio de Janeiro 21040-900, Brazil;
anagextra@gmail.com
[4] Departamento de Biotecnología, Universidad Politécnica de Valencia, Valencia 46022, Spain
* Correspondence: alexandra.rucavado@ucr.ac.cr; Tel.: +506-25117876

Academic Editor: Nicholas R. Casewell
Received: 12 September 2016; Accepted: 30 September 2016; Published: 12 October 2016

Abstract: Snake venom metalloproteinases (SVMPs) play key biological roles in prey immobilization and digestion. The majority of these activities depend on the hydrolysis of relevant protein substrates in the tissues. Hereby, we describe several isoforms and a cDNA clone sequence, corresponding to PII SVMP homologues from the venom of the Central American pit viper *Bothriechis lateralis*, which have modifications in the residues of the canonical sequence of the zinc-binding motif HEXXHXXGXXH. As a consequence, the proteolytic activity of the isolated proteins was undetectable when tested on azocasein and gelatin. These PII isoforms comprise metalloproteinase and disintegrin domains in the mature protein, thus belonging to the subclass PIIb of SVMPs. PII SVMP homologues were devoid of hemorrhagic and in vitro coagulant activities, effects attributed to the enzymatic activity of SVMPs, but induced a mild edema. One of the isoforms presents the characteristic RGD sequence in the disintegrin domain and inhibits ADP- and collagen-induced platelet aggregation. Catalytically-inactive SVMP homologues may have been hitherto missed in the characterization of snake venoms. The presence of such enzymatically-inactive homologues in snake venoms and their possible toxic and adaptive roles deserve further investigation.

Keywords: snake venom metalloproteinases; PII SVMP homologues; disintegrin domain; zinc-binding motif; hemorrhagic activity; platelet aggregation; proteinase activity

1. Introduction

Snake venom metalloproteinases (SVMPs) are abundant components in the venoms of advanced snakes of the superfamily Colubroidea and play key roles in venom toxic and digestive actions [1–3]. Together with the ADAMs and ADAMTs, SVMPs are classified within the M12 family of metalloproteinases (subfamily M12B or reprolysin) [1], which, in turn, belong to the metzincin superfamily of these proteinases. This superfamily is characterized by a canonical zinc-binding sequence (HEXXHXXGXXH) followed by a Met-turn [4].

SVMPs have been classified into three classes on the basis of the domain composition (PI, PII and PIII). All contain a signal sequence, a prodomain and a metalloproteinase domain. In the case of the

PI class, the mature protein has only the metalloproteinase domain, while the PII class has, in addition to the catalytic domain, a disintegrin (Dis) domain, which may be proteolytically processed, giving rise to a free disintegrin, that in most cases hosts the canonical RGD motif [5]. In turn, PIII class SVMPs have disintegrin-like (Dis-like) and cysteine-rich (Cys-rich) domains following the metalloproteinase domain. In some PIII SVMPs, two additional C-type lectin-like subunits are covalently linked to the cysteine-rich domain. Within each one of these SVMP classes, there are several subclasses that vary in their post-translational processing and quaternary structure [1,6]. SVMPs play key biological roles for snakes, both as digestive enzymes and as toxicity factors, since they are responsible for local and systemic hemorrhage, blistering and dermonecrosis, myonecrosis and coagulopathies, in addition to exerting a pro-inflammatory role [2]. Most of the biological effects of SVMPs depend on their catalytic activity, particularly on their capacity to hydrolyze extracellular matrix components and coagulation factors. However, other functions are associated with the non-metalloproteinase domains, such as the action of disintegrins and disintegrin-like sequences on platelet aggregation [1,2,7,8].

SVMPs constitute a model of evolution and neofunctionalization of multigene families [9]. The evolutionary history of these toxins, which has been characterized by domain loss, gene duplication, positive selection and neofunctionalization, started with the recruitment of an ADAM (ADAM 7, ADAM 28 and/or ADAM decysin 1), followed by the loss of C-terminal domains characteristic of ADAMs, resulting in the formation of P-III SVMPs [10,11]. After that, a duplicated SVMP PIII gene evolved by positive selection into a PII SVMP via loss of the Cys-rich domain. Then, PI evolved by disintegrin domain loss, with further neofunctionalization of the metalloproteinase domain to play diverse biological roles [6,10,11]. PIII SVMPs are present in all advanced snake families, whereas PI and PII SVMPs occur only in the family Viperidae [2,6]. This accelerated and complex evolutionary process has generated a great functional diversity in this group of enzymes and the expression of multiple, structurally-distinct SVMP isoforms in snake venoms [6,12,13].

Many representatives of PII SVMPs undergo a post-translational modification by which the disintegrin domain is released from the precursor protein [1,6]. However, in two subclasses of PII SVMPs, i.e., PIIb and PIIc, this proteolytic cleavage does not occur, and the mature proteins are comprised by the metalloproteinase and the Dis domain held together [6]. In addition, PIIc SVMPs occur as dimers, whereas PIIb are monomers. Despite the great diversification of SVMPs, to the best of our knowledge, there are no reports on enzymes with mutations in the canonical zinc-binding motif of the active site, and all SVMPs described to date are catalytically active.

In the course of the characterization of SVMPs from the venom of the arboreal pit viper *Bothriechis lateralis*, a species distributed in Central America [14], a fraction containing internal sequences characteristic of SVMPs, but being devoid of proteinase and hemorrhagic activities, was isolated. The characterization of this fraction and the cDNA cloning of a related isoform showed the presence of modifications in the residues constituting the zinc-binding motif characteristic of the reprolysin group of metalloproteinases. This study describes the isolation and characterization of these SVMP variants and discusses some biological implications of their presence in snake venoms.

2. Results

2.1. Purification of SVMP BlatPII

A PII SVMP, hereby named BlatPII, was purified from the venom of *B. lateralis* by two chromatographic steps. First, ion-exchange chromatography on diethylaminoethyl (DEAE)-Sepharose yielded five protein fractions (Figure 1A). Then, Peak III, which showed hemorrhagic activity, was fractionated by phenyl sepharose chromatography (Figure 1B). Fraction II of the latter chromatography was devoid of hemorrhagic activity, but had sequences characteristic of SVMPs by mass spectrometric analysis. When run on SDS-PAGE, it showed bands of 36 kDa and 39 kDa, under non-reducing and reducing conditions, respectively (Figure 1C), indicating that it is a monomeric protein containing intramolecular disulfide bond(s). It stained positive in the carbohydrate detection test, i.e., it is a glycoprotein (not shown). Reversed-phase HPLC analysis demonstrated that the protein, apparently homogeneous on SDS-PAGE, can be separated into two peaks (Figure 1D) with molecular masses of 35,409 ± 248 Da and 36,715 ± 249 Da, estimated by MALDI-TOF mass spectrometry.

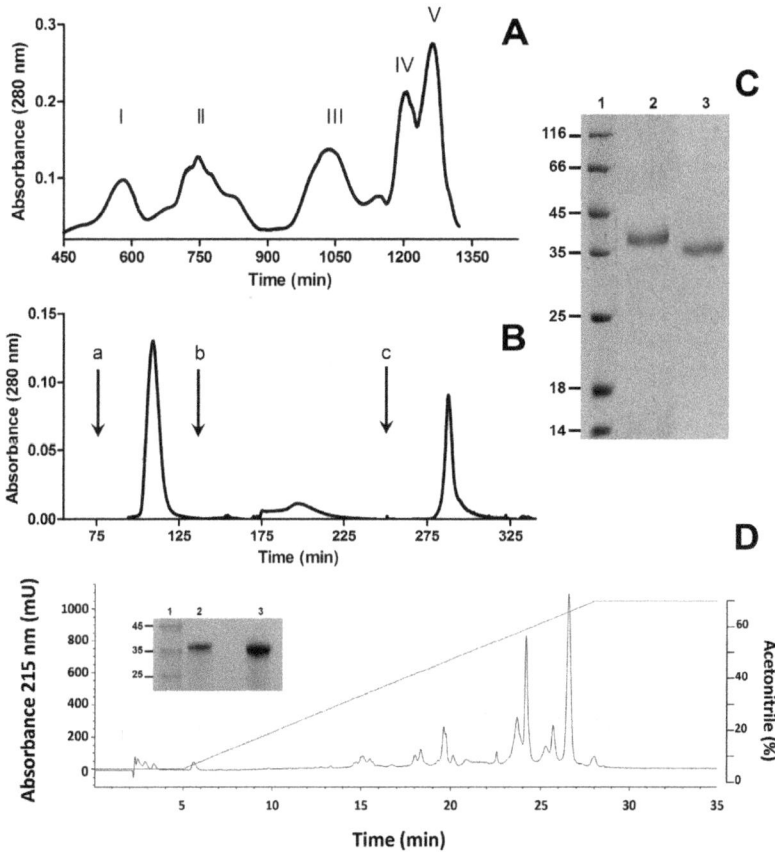

Figure 1. Isolation of BlatPII from *B. lateralis* venom. Venom was fractionated by ion-exchange chromatography on diethylaminoethyl (DEAE)-Sepharose (**A**); and hydrophobic interaction chromatography on phenyl sepharose (**B**); as described in the Materials and Methods. Arrows inserted in (**B**) correspond to the addition of 100 mL of 0.01 M phosphate buffer, pH 7.8, containing 1 M NaCl (a); 100 mL of 0.01 M phosphate buffer, pH 7.8 (b); and 100 mL of deionized water (c). Fraction II from the hydrophobic interaction chromatography is a SVMP devoid of hemorrhagic activity and was named BlatPII. (**C**) SDS-PAGE of the purified protein under reducing (Lane 2) and non-reducing (Lane 3) conditions. (**D**) BlatPII was separated into two main peaks by RP-HPLC. Red line corresponds to acetonitrile gradient (see the text for details). SDS-PAGE under reducing conditions of Peak 1 (Lane 2) and Peak 2 (Lane 3) is shown in the insert of (**D**). Lane 1 in gels of (**C**) and the insert of (**D**) corresponds to molecular mass standards (kDa).

2.2. Determination of Internal Peptide Sequences

Mass spectrometry analysis and Edman *N*-terminal degradation of peptides obtained by digestion of RP-HPLC Peaks 1 and 2 with different proteinases identified several sequences characteristic of the metalloproteinase and disintegrin domains of SVMPs (Table 1). Sequences having two different patterns of substitutions in the canonical sequence of the zinc-binding motif of SVMPs

(HEXXHXXGXXH) were found. One of them (HELGHNLGIHQ) has a Gln instead of a characteristic His, whereas the other (HDLGHNLCIDH) has two substitutions, i.e., an Asp instead of Glu and a Cys replacing a Gly. These sequences were found in peptides having variations in other internal sequences, hence evidencing the presence of several isoforms within these two basic patterns of substitutions in the zinc-binding motif (Table 1). Representative high-resolution MS/MS spectra of the above referred peptides are shown in the Supplementary Material (Figures S1–S5). All fragmentation patterns were automatically interpreted by PEAKS De Novo software (version 8, Bioinformatics Solutions Inc., Waterloo, Canada, 2016). Only sequence results with very high confidence (average local confidence (ALC) ≥99% and local confidence score for each amino acid ≥98%) were considered. The sequence RGD, characteristic of many disintegrins, was observed in a few peptides derived from proteins eluted in the HPLC Peak 2 (Table 1). The high-resolution MS/MS spectrum of one of these peptides is shown in the Supplementary Material (Figure S6). For this specific peptide, the de novo ALC was 91%, and the local confidence score for each amino acid ranged from 59% to 99%. All peptides derived from proteins from HPLC Peaks 1 and 2 identified by automatic de novo sequencing and showing ALC ≥ 99% were aligned against the BlatPII-c sequence using PepExporer, a similarity-driven tool (Figures S7 and S8). Considering a minimum identity of 75%, the sequence coverage obtained for BlatPII-c was 22.1% and 49.1%, respectively, for peptides from HPLC Peak 1 and Peak 2.

2.3. Cloning of a P-II SVMP

The PCR amplified a 1458-bp cDNA sequence (Figure 2), with a precursor protein sequence including a 17-residue *N*-terminal signal peptide, a pre-pro-domain (residues −18–−190, including the canonical cysteine switch motif (PKMCGVT)), followed by a mature protein comprising a 206-amino acid metalloproteinase domain, a short spacer sequence (residues 207–211) and a long PII disintegrin domain (212–297) containing the characteristic RGD motif. Noteworthy, the metalloproteinase domain contains the CIM sequence characteristic of the Met-turn of metzincins [4] and a mutated zinc-binding motif with the sequence HDLGHNLCIDH, instead of the canonical HEXXHXXGXXH. This sequence is identical to the mutated catalytic site of some of the internal sequences determined in proteins eluted in Peak 2 of the RP-HPLC separation (Table 1). When comparing the complete sequence of the clone with the peptide sequences identified in RP-HPLC peaks (Table 1), there were differences in some residues. Thus, this cDNA-deduced amino acid sequence is likely to correspond to another isoform having the same substitutions in the canonical catalytic site found in some of the isolated protein isoforms. The protein coded by the cDNA clone is hereby named BlatPII-c. The BlatPII-c clone presents two Cys residues, corresponding in the mature protein to residues at positions 217 and 236. The presence of these Cys residues identifies this domain as a member of the long-chain disintegrin subfamily [13,15] and has been suggested to underlay the lack of proteolytic processing of the disintegrin domain in some PII SVMPs. Therefore, the data suggest that BlatPII-c belongs to the subclass PIIb of SVMPs, i.e., monomeric PII containing metalloproteinase and disintegrin domains in the mature protein [16] (Figure 2). The sequence of this clone was deposited in GenBank under Accession Number KU885992.

2.4. Proteolytic Activity on Azocasein and Gelatin

Proteolytic activity on these two substrates was tested with the fraction BlatPII, which contains isoforms that can be separated only by RP-HPLC. Since the solvents used for this separation denature the enzymes, it was not possible to assess the proteolytic activity of the RP-HPLC peaks separately. Under the experimental conditions used, BlatPII did not show proteolytic activity on the two substrates tested. In contrast, both BaP1 and BlatH1 SVMPs, which are active proteinases, hydrolyzed azocasein and gelatin (Figure 3).

Table 1. Amino acid sequence of peptides obtained by digestion with trypsin, chymotrypsin or Glu-C endoproteinases from Peak 1 and Peak 2 of RP-HPLC, as determined by mass spectrometry (MALDI-TOF-TOF or nESI-MS/MS) or Edman N-terminal sequencing.

Peptide Sequence	HPLC Peak	m/z	z	Confidence [a]	Error [b] (ppm)	Method [c]	Domain
YIELVIVADHR	1	1327.8	1	MV	47.4	MALDI/DE NOVO	metalloproteinase
NLITPEEQRY	1	1149.6	1	MV	40.0	MALDI/DE NOVO	metalloproteinase
SRYHFVANR	1	1149.6	1	MV	7.8	MALDI/DE NOVO	metalloproteinase
YHFVANR	1	-	-	MV	-	EDMAN	metalloproteinase
MAHELGHNLGLHQDR	1	1727.8	1	MV	22.5	EDMAN	metalloproteinase
MAHELGHNLGLHQDR	1	576.6180	3	99	0.2	nESI/DE NOVO	metalloproteinase
DSCSCGSNSCIMSATVSNEPSSR	1	2493.0	1	MV	10.0	MALDI/DE NOVO	metalloproteinase
CIDNEPLR	1	1016.5	1	MV	16.7	MALDI/DE NOVO	metalloproteinase
NEPLRTDIVSPPFCGNYYPE *	1	2368.3	1	MV	88.2	MALDI/DE NOVO	metalloproteinase/disintegrin
LTTGSOCAEGLCCDQCR	1	2015.9	1	MV	47.6	MALDI/DE NOVO	disintegrin
FEGLCCDQCR	1	1344.4	1	MV	84.0	MALDI/DE NOVO	disintegrin
KTDIVSPPF **	1	1031.6	1	MV	46.5	MALDI/DE NOVO	disintegrin
YIELVIVADHR	2	1327.8	1	MV	47.4	MALDI/DE NOVO	metalloproteinase
VTLSADDTLDLFGTWR	2	-	-	MV	-	EDMAN	metalloproteinase
VALIGLEIWSSGELSK	2	1702.0	1	MV	34.0	MALDI/DE NOVO	metalloproteinase
YHFVANR	2	906.5	1	MV	46.3	MALDI/DE NOVO	metalloproteinase
MAHDLGHNLCIDHDR	2	1803.9	1	MV	54.8	MALDI/DE NOVO	metalloproteinase
MAHDLGHNLCLDHDR	2	902.4039	2	99	-0.4	nESI/DE NOVO	metalloproteinase
MAHDLGHNLCLDHDDR	2	959.9163	2	99	-1.5	nESI/DE NOVO	metalloproteinase
MAHELGHNLGLHQDDR	2	614.9611	3	99	1.2	nESI/DE NOVO	metalloproteinase
MAHELGHNLGLHQDR	2	864.4214	2	99	-2.1	nESI/DE NOVO	metalloproteinase
NKFPSETDIVSPPVCGNY **	2	2040.1	1	MV	79.4	MALDI/DE NOVO	metalloproteinase/disintegrin
CNGISAGCPRNPF **	2	1449.7	1	MV	44.1	MALDI/DE NOVO	disintegrin
RARGDDVNDYCNGISAGCPRNPFH *	2	2748.4	1	MV	68.7	MALDI/DE NOVO	disintegrin
ARGDDVNDYCNGISAGCPR	2	2097.1	1	MV	101.5	MALDI/DE NOVO	disintegrin
ARGDDVDYDCGLDAGCPR	2	709.6204	3	91	-1.7	nESI/DE NOVO	disintegrin
AATCKLTTGSOCADGLCDQCKFMRE *	2	3068.5	2	MV	71.0	MALDI/DE NOVO	disintegrin

[a] Confidence: MALDI/DE NOVO and EDMAN (MV, manual validation); nESI/DE NOVO (PEAKS 8 De Novo average local confidence); [b] Error: error of the precursor ion (in ppm) relative to the proposed sequence; [c] Method: MALDI/DE NOVO: MALDI-TOF-TOF and manual de novo sequencing; EDMAN: N-terminal sequencing; nESI/DE NOVO: nLC-nanoelectrospray MS/MS and automatic de novo sequencing using PEAKS 8 software; * peptides generated by Glu-C endoproteinase digestion; ** peptides generated by chymotrypsin digestion. Peptides without asterisks were obtained following trypsin digestion. All cysteines are carboxamidomethylated. Sequences colored in red correspond to the mutated catalytic site.

```
  1 atgatcccagttctcttggtaactatatgcttagcagcttttccttatcaagggagctct  60
    |------------- Fw primer ------------->
      M  I  P  V  L  L  V  T  I  C  L  A  A  F  P  Y  Q  G  S  S   -20
    <-------------------- Signal peptide sequence ---------->|---------
 61 ataatcctggaatctgggaacatgaatgattatgaagtagtgtatccacgaaaagtcact  120
      I  I  L  E  S  G  N  M  N  D  Y  E  V  V  Y  P  R  K  V  T   -40
    ------------------------------ Prodomain ------------->
121 gcattgccaaaaggagaagttcagccaaagtatgaagacaccatgcaatatgaatttaag  180
      A  L  P  K  G  E  V  Q  P  K  Y  E  D  T  M  Q  Y  E  F  K   -60
181 gtgaatggagagccagtggtccttcacctggaaaaaaataaaggacttttttcaaaagat  240
      V  N  G  E  P  V  V  L  H  L  E  K  N  K  G  L  F  S  K  D   -80
241 tacagcgagactcattattcccctgatggcagagaaattacaacataccctccggttgag  300
      Y  S  E  T  H  Y  S  P  D  G  R  E  I  T  T  Y  P  P  V  E   -100
301 gatcactgctattatcatggacgcatcgagaatgatgctgactcaactgaaagcatcagt  360
      D  H  C  Y  Y  H  G  R  I  E  N  D  A  D  S  T  E  S  I  S   -120
361 gcatgcaacggtttgaaaggacatttcaagcttcaaggggagatgtaccttattgaaccc  420
      A  C  N  G  L  K  G  H  F  K  L  Q  G  E  M  Y  L  I  E  P   -140
421 ttgaagctttccgacagtgaagcccatgcaatctacaaatatgaaaacgtagaaaaagag  480
      L  K  L  S  D  S  E  A  H  A  I  Y  K  Y  E  N  V  E  K  E   -160
481 gatgaggcccccaaaatgtgtggagtaaccgagactaattgggaatcatatgagcccatc  540
      D  E  A  P  K  M  C  G  V  T  E  T  N  W  E  S  Y  E  P  I   -180
541 aaaaaggcctctcagtcaaatcttactcctgaacaacaaagatttaacccttcaaatac  600
-181  K  K  A  S  Q  S  N  L  T  P  E  Q  Q  R  F  N  P  F  K  Y   10
                               |--------- Metalloproteinase ---
601 gttgagcttgttatagttgcggatcacagaatgtacacaaaatatgacggtgataaaact  660
 11   V  E  L  V  I  V  A  D  H  R  M  Y  T  K  Y  D  G  D  K  T   30
    ------------------------------------->
661 gagataagttcaataatatatgaaattgtcaacattctaactctgatttacagatctttg  720
 31   E  I  S  S  I  I  Y  E  I  V  N  I  L  T  L  I  Y  R  S  L   50
721 catattcgtgtagctctgattggcctagaaatttggtccagtggagaattgagtaaagtg  780
 51   H  I  R  V  A  L  I  G  L  E  I  W  S  S  G  E  L  S  K  V   70
781 acattatcagcagatgatactttggacttatttggaacctggagagagacagatttgctg  840
 71   T  L  S  A  D  D  T  L  D  L  F  G  T  W  R  E  T  D  L  L   90
841 aagcgcaaaaaacatgataatgctcagttactcacgggcatgatcttcaatgaaacaatt  900
 91   K  R  K  K  H  D  N  A  Q  L  L  T  G  M  I  F  N  E  T  I   110
901 gaaggaaggacttacaccagtggtatatgcgacccaaagcgttctgtaggaattgttcgg  960
111   E  G  R  T  Y  T  S  G  I  C  D  P  K  R  S  V  G  I  V  R   130
961 gatcatagaagtaaatatcatttgttgcaaatagaatggcccatgacctgggtcataat 1020
131   D  H  R  S  K  Y  H  F  V  A  N  R  M  A  H  D  L  G  H  N   150
1021 ctgtgcattgatcatgacagagattcctgtacttgcggtgctaacccatgcattatgtct 1080
151   L  C  I  D  H  D  R  D  S  C  T  C  G  A  N  P  C  I  M  S   170
1081 gcgacagtaagcgatcaaccttccagtcaattcagcgattgtagtgttgatcaatatttg 1140
171   A  T  V  S  D  Q  P  S  S  Q  F  S  D  C  S  V  D  Q  Y  L   190
1141 agaaatattattcattcttttacaacaaactgcctttacaataaaccctcggagacagat 1200
191   R  N  I  I  H  S  F  T  T  N  C  L  Y  N  K  P  S  E  T  D   210
                                        |-- Spacer --
1201 attgtttcacctccagtttgtggcaattactatcgggagatgggagaagagtgtgactgt 1260
211   I  V  S  P  P  V  C* G  N  Y  Y  R  E  M  G  E  E  C  D  C   230
    -->|----------------- Long Disintegrin ------------------->
1261 ggccctcctgcaaattgtcagaatccatgctgtgatgctgcaacgtgtaaactgacaaca 1320
231   G  P  P  A  N  C* Q  N  P  C  C  D  A  A  T  C  K  L  T  T   250
1321 gggtcacagtgtgcagacggactgtgttgtgaccagtgcagatttatgaaagaaggaaca 1380
251   G  S  Q  C  A  D  G  L  C  C  D  Q  C  R  F  M  K  E  G  T   270
1381 gtatgccggagagcaaggggtgatgacgtgaatgattactgcaatggcatatctgctgac 1440
                                             |----------
271   V  C  R  R  A  R  G  D  D  V  N  D  Y  C  N  G  I  S  A  D   290
             R  A  R  G  D  D  V  N  D  Y  C  N  G  I  S  A  G
             |----------------- MS/MS 2748.4 (1+) --------------------
1441 tgtcccagtaatggctaa                                          1458
    ---- Rev primer --->
291   C  P  S  N  G  *
      C  P  R  N  P  F  H                                         297
      -------------------------|
```

Figure 2. cDNA sequence and deduced amino acid sequence of BlatPII-c. The deduced amino acid sequence of the mature protein is depicted in bold letters and is preceded by the pro-domain, which contains the sequence PKMCGVT characteristic of the Cys switch (highlighted in green). Sequences corresponding to signal peptide, pro-domain, metalloproteinase domain, spacer region and the disintegrin domain are indicated. The sequences corresponding to the mutated zinc-binding motif and the sequence CIM (Cys-Ile-Met) of the Met-turn are highlighted in yellow, as well as the sequence RGD in the disintegrin domain. Cysteine residues proposed to be related to the lack of proteolytic processing of the disintegrin from the metalloproteinase domain are labeled with asterisks. Primers used to PCR-amplify the cDNA clone and the C-terminal amino acid sequence gathered by MS/MS analysis of a tryptic peptide are shown.

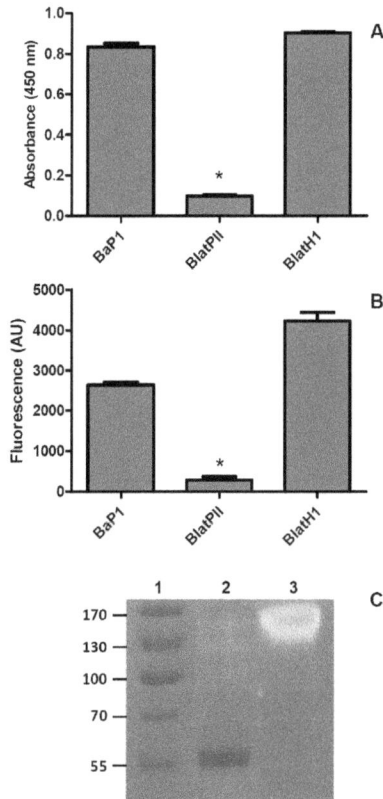

Figure 3. Evaluation of the proteolytic activity of BlatPII and other SVMPs on azocasein and gelatin. (**A**) BlatPII (5.5 µM) was incubated with azocasein (see the Materials and Methods for details). For comparison, an equimolar concentration of PI SVMP BaP1, from the venom of *Bothrops asper*, and of the hemorrhagic PII SVMP BlatH1, from the venom of *Bothriechis lateralis*, were also tested. PBS was used as a negative control; (**B**) Quantification of the gelatinolytic activity of the SVMP by a fluorescent commercial kit (EnzCheck® protocol Gelatinase/Collagenase Assay Kit, Molecular Probes, Life Technologies, Eugene, OR, USA). One-point-six micrograms of BlatPII were incubated with 20 µg of the fluorescent gelatin; equimolar quantities of BaP1 and BlatH1 were used as positive controls; samples were incubated at room temperature for 6 h. Fluorescence intensity was measured in the BioTek Synergy HT microplate reader using the absorption filter at 495 nm and the emission filter at 515 nm. * $p < 0.05$ when compared with BaP1 and BlatH1. In (**A**,**B**), the signal of BlatPII was not significantly different when compared to the controls without enzyme ($p > 0.05$); (**C**) Gelatin zymography: 2.5 µg of BlatPII (Lane 2) and 2.5 µg of BlatH1 (Lane 3) were separated on SDS-PAGE under non-reducing conditions. Lane 1 corresponds to molecular mass standards. The dark band observed in Lane 2 corresponds to BlatPII, whose migration in SDS-PAGE-gelatin gels is delayed as compared to SDS-PAGE gels without gelatin.

2.5. Hemorrhagic, Edema-Forming, Coagulant and Platelet Aggregation Inhibitory Activities

BlatPII did not exert local hemorrhagic activity when injected in the skin of mice at a dose of 100 µg, nor did it show coagulant activity on human plasma. Likewise, the protein did not induce pulmonary hemorrhage after i.v. injection of 100 µg in mice. On the other hand, BlatPII demonstrated a low edema-forming activity (Figure 4), and inhibited ADP- and collagen-mediated aggregation of human platelets in PRP, with estimated IC_{50}s of 0.25 and 0.33 μM, respectively (Figure 5A,B). In the case of the peaks obtained by RP-HPLC separation, Peak 1 did not induce the inhibition of platelet aggregation, while Peak 2 inhibited platelet aggregation using ADP and collagen as the agonists (Figure 5C).

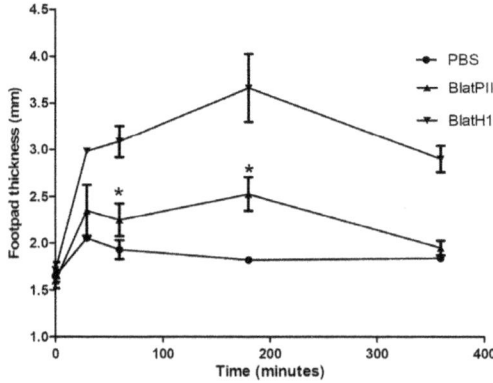

Figure 4. Edema-forming activity of BlatPII. Groups of four CD-1 mice were injected subcutaneously in the right footpad with 5 μg of BlatPII, in a volume of 50 μL PBS. Controls were injected with 50 μL PBS only. The paw thickness was measured using a low pressure spring caliper, as a quantitative index of edema, before injection and at 30, 60, 180 and 360 min after injection. Edema was calculated as the percentage of increase in the paw thickness of the right foot injected with SVMP as compared to the left foot; in parallel, 5 μg of BlatH1 were used as a positive control. BlatH1 induced significantly higher ($p < 0.05$) edema than BlatPII at all time intervals, whereas BlatPII induced a significant edema, as compared to the control, only at 60 and 180 min (* $p < 0.05$).

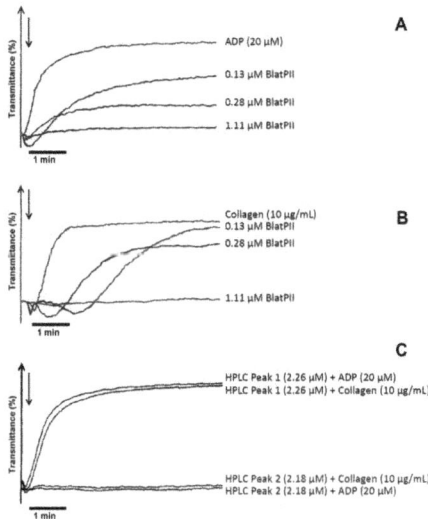

Figure 5. Inhibition of platelet aggregation by BlatPII and RP-HPLC Peaks 1 and 2. Various concentrations of BlatPII were tested using ADP (**A**) or collagen (**B**) as agonists. (**C**) Inhibition of platelet aggregation by HPLC Peak 1 and Peak 2. Results presented correspond to one experiment representative of three different independent experiments.

2.6. Three-Dimensional Model of the Active Site of BlatPII-c

As described above, the clone codes for the sequence HDLGHNLCIDH instead of the characteristic canonical site HEXXHXXGXXH at the enzyme active site. Three-dimensional structures, based on homology models, were generated using the Modeller method available in the Accelrys Discovery Studio 3.5 software [17]. The vascular apoptosis-inducing protein 1 (VAP-1) (PDB Code 2ERQ) was used as a template. After aligning the generated model of the BlatPII-c with the VAP-1 crystal structure, it was shown that the substitution of Glu for Asp causes an increase in the distance between residue Asp335 and the water molecule (4.203 Å), both of which play a key role in the hydrolysis of the peptide bond. In the case of VAP-1, a PIII SVMP from the venom of *Crotalus atrox*, which has the characteristic active site with Glu, the distance between the water molecule and Glu is 2.691 Å (Figure 6). A PIII SVMP (VAP-1) was used as a model, because no PII SVMP has been crystallized, and the sequence identity percentage between these toxins is acceptable (55%) for the modeling program.

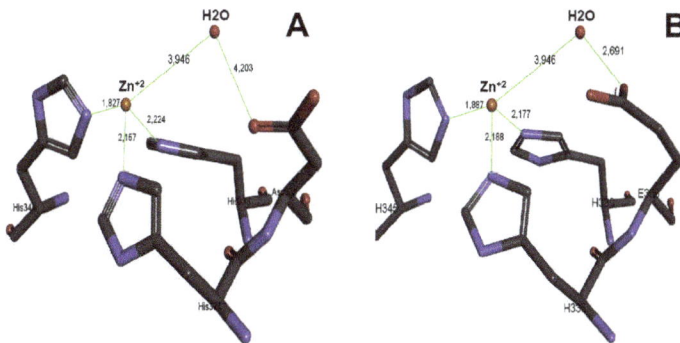

Figure 6. 3D model of the active site HDLGHNLCIDH. Using the Discovery Studio Program Version 3.5.0.12158 (Accelrys Software Inc, San Diego, CA, USA), a 3D model of the active site of BlatPII was generated (**A**). Determination of distances in the predicted model was made using the coordinates of the zinc atom and the water molecule from the crystallographic structure of VAP-1 from venom of *Crotalus atrox* (**B**).

3. Discussion

In contrast to the SVMPs described so far, the PII SVMP homologues described in the present study have the peculiarity of being devoid of proteinase activity. To the best of our knowledge, this constitutes the first case of a snake venom protein having a SVMP structural scaffold, but being devoid of proteolytic activity. Mass spectrometric and molecular analyses of the isoforms of BlatPII underscore the presence of mutations in the canonical sequence of the zinc-binding motif characteristic of the active site of SVMPs. In the canonical sequence (HEXXHXXGXXH), the catalytic zinc is coordinated by the three His residues, and a water molecule plays the role of a fourth ligand and is clamped between the catalytic Glu of the motif and the metal, forming a trigonal pyramidal coordination sphere around the zinc [4]. In the catalytic process, the water molecule is added to the peptidic bond by a nucleophilic attack, in which the carbonyl group of the scissile peptide bond is polarized by the active site zinc. In such a mechanism, the zinc-bound water molecule mediates between this carbonyl group and the catalytic Glu residue, giving rise to the acquisition of nucleophilicity by the water molecule, necessary for attacking the carbonyl carbon of the scissile peptide bond, resulting in a tetra-coordinate transition state. The catalytic Glu then presumably serves as a proton shuttle to the cleavage products [4,18,19]. Thus, it is likely that the substitutions of these highly-conserved catalytic residues occurring in the isoforms of BlatPII drastically affect one or more steps in this catalytic process. The presence of catalytically-inactive SVMP homologues in snake venoms needs to be further investigated since the isolation of SVMPs is usually followed by testing proteinase or toxic activities of the chromatographic fractions; this approach does not detect catalytically-inactive variants. It is therefore recommended that the search for this type of inactive SVMP homologues is based on either immunological detection of components cross-reacting with SVMPs or on the proteomic identification of these variants, as performed in our study.

In contrast to SVMPs, where no examples of enzymatically-inactive variants have been described before, mutations in the catalytic site have been reported in the case of the other branch of the reprolysin group of metalloproteinases, i.e., the ADAMs [1,20,21]. Some ADAMs have completely lost the sequence of the zinc-binding motif [1,21–24]. In other cases, such as ADAMDEC-1 [20] and ADAM-7 [21,25], substitutions in this canonical motif have been described, with variable impact on the proteolytic activity of these proteins. On the other hand, the detection of several mutated isoforms of BlatPII and the presence of variations in the internal sequences determined highlight the existence of several isoforms in this venom within the two basic mutated patterns of the zinc-binding motif.

Two basic patterns of substitutions in the canonical sequence of the zinc-binding motif were found in our study. In one case, an Asp substitutes the Glu and a Cys substitutes a Gly, whereas in the other, a Gln substitutes the third His in the canonical sequence. It is therefore proposed that these substitutions drastically affect the catalytic machinery in these PII SVMP homologues. A 3D modeling of the active site, using the sequence of BlatPII-c, showed that the substitution of Glu for Asp causes an increase of the distance between the Asp335 residue and the water molecule, as compared to the distance between Glu335 and water in a catalytically-active variant. The increment of this distance may affect the activation of the water molecule, thus precluding the nucleophilic attack of the peptide bond. On the other hand, the substitution of a Gly by Cys might also bear structural consequences that affect the catalytic machinery. The elucidation of the 3D structure of BlatPII-c will certainly provide an explanation for the described effect in proteolytic activity.

The lack of hemorrhagic and coagulant activities in BlatPII is probably related to the loss of enzymatic activity, since these effects have been associated with proteolysis in catalytically-active SVMPs. The observation that BlatPII induces mild edema is compatible with a study that described edema-forming activity of a fragment constituted by the Dis-like and Cys-rich domains [26]. HPLC peak 2 inhibits platelet aggregation induced by ADP and collagen, and as expected, sequences found in HPLC Peak 2 present the characteristic RGD motif in the disintegrin domain. Since this sequence is associated with the ability of many disintegrins to bind $\alpha_{IIB}\beta_3$ integrin in the platelet membrane, causing inhibition of aggregation [27–29], it is suggested that it is responsible for the effect observed in platelets.

The existence of catalytically-inactive PII SVMP homologues in the venom of *B. lateralis* is intriguing from the adaptive standpoint, since most of the described biological roles of SVMPs, associated with both prey immobilization and digestion, are dependent on the proteolytic degradation of diverse substrates, usually extracellular matrix components or plasma proteins [2]. BlatPII isoforms are devoid of hemorrhagic and coagulant effects, in agreement with their lack of proteinase activity. The only biological effects shown by BlatPII were a mild edema-forming effect and inhibition of platelet aggregation, which may be due to the presence of RGD in the disintegrin domain and may play a role in the overall toxicity of the venom by affecting this aspect of hemostasis, although this requires experimental demonstration. Alternatively, the presence of BlatPII in the venom may reveal a neutral evolutionary process by which these catalytically-inactive homologues are expressed in the venom gland without playing a significant trophic role. Many components have been described in snake venoms that have not been shown to play a biological role. Nevertheless, it would be relevant to explore the possible toxicity of these isoforms by assessing a wider set of effects and also by studying their toxicity in prey other than mice, since adult *B. lateralis* also feeds on birds and bats in addition to rodents [14].

In the wider context of hydrolytic enzymes, examples of catalytically-inactive homologues of phospholipases A$_2$ (PLA$_2$) and of serine proteinases have been described in snake venoms. In the case of PLA$_2$ homologues, substitutions of key residues in the catalytic machinery have resulted in the loss of enzymatic activity. However, many catalytically-inactive PLA$_2$ homologues exert toxic effects, such as myotoxicity and pro-inflammatory activities, owing to the appearance of 'toxic sites' in some regions of the surface of these molecules [30]. Likewise, serine proteinase homologues have been reported, which are characterized by substitutions in the catalytic triad His, Asp, Ser [31,32]. In a serine proteinase isolated from the venom of *Trimeresurus jerdonii*, the replacement of His-43 by Arg-43 at the catalytic triad has been associated with the loss of proteolytic activity [32].

In conclusion, cDNA encoding for a PII SVMP and several isoforms of PII SVMP homologues with a mutated catalytic site occur in the venom of the arboreal viperid snake *B. lateralis*, which contains a high percentage of SVMPs (55.1%) [33]. Such mutations in the canonical sequence of the zinc-binding motif of these enzymes are likely to be responsible for the lack of enzymatic activity observed. To the best of our knowledge, this is the first case of catalytically-inactive SVMP homologues described in snake venoms. The possible biological role, if any, of these SVMP variants remains unknown, although one of them is able to inhibit ADP- and collagen-induced platelet aggregation in vitro. These components expand the wide set of SVMP variants in viperid snake venoms. The search for other enzymatically-inactive SVMP homologues in venoms is a relevant task in order to ascertain how frequent they are and to discover new possible biological roles for this group of venom components.

4. Materials and Methods

4.1. Venom and Toxins

A pool of venom obtained from at least 20 adult specimens of *B. lateralis* collected in various locations of Costa Rica, and maintained at the Serpentarium of Instituto Clodomiro Picado, was used. Venom was lyophilized and stored at −20 °C. Venom solutions were prepared immediately before each experiment. In some experiments, BlatH1, a PII SVMP isolated from the venom of *B. lateralis* [34], and BaP1, a PI SVMP from the venom of *Bothrops asper* [35], were used for comparative purposes.

4.2. Purification of SVMP

One hundred milligrams of *B. lateralis* venom were dissolved in 5 mL of 0.01 M phosphate buffer, pH 7.8, and centrifuged at 500× *g* for 10 min. The supernatant was loaded onto a DEAE sepharose column (column 10 cm in length × 2 cm internal diameter) previously equilibrated with the same buffer. After eluting the unbound material, a linear NaCl gradient (0–0.4 M) was applied, and fractions were collected at a flow rate of 0.3 mL/min. The fraction containing SVMPs, identified on the basis of hemorrhagic activity in mice, was then adjusted to 1 M NaCl and applied to a phenyl sepharose hydrophobic interaction column (6 cm × 1.5 cm i.d.). Unbound proteins were eluted with 100 mL of 0.01 M phosphate buffer, pH 7.8, containing 1 M NaCl. Then, 100 mL of 0.01 M phosphate buffer, pH 7.8, were applied, and finally, the chromatographic separation was finished by the addition of 100 mL of deionized water. The eluted fractions were analyzed by SDS-PAGE, hemorrhagic activity and mass spectrometry analysis of tryptic peptides. A peak of 39 kDa devoid of hemorrhagic activity, but showing amino acid sequences corresponding to SVMPs, was isolated after the application of 100 mL of 0.01 M phosphate buffer, pH 7.8, and was further characterized in this study; it was named BlatPII. The peak obtained after the application of water has been previously characterized as a dimeric PII hemorrhagic SVMP, known as BlatH1 [34]. Homogeneity and molecular mass were determined by SDS-polyacrylamide gel electrophoresis (SDS-PAGE) under reducing and non-reducing conditions, using 12% polyacrylamide gels [36].

4.3. HPLC Separation of Isoforms

One hundred micrograms of BlatPII were loaded on a Thermo AQUASIL C4 column (150 mm × 4.6 mm i.d.), and the reversed-phase HPLC fractionation was made on an Agilent 1200 instrument (Agilent Technologies, Santa Clara, CA, USA) at a flow of 0.5 mL/min, with detection set at 215 nm. The solvent system was 0.1% trifluoroacetic acid in H_2O (Solvent A) and 0.1% trifluoroacetic acid in acetonitrile (Solvent B). The gradient program began with 0% Solvent B for 5 min and was then ramped to 70% Solvent B at 300 min, followed by 70% Solvent B for 5 min.

4.4. Mass Spectrometry Analyses

4.4.1. MALDI-TOF/TOF

Mass determination of the whole protein was performed by mixing 0.5 μL of saturated sinapinic acid (in 50% acetonitrile, 0.1% trifluoroacetic acid) and 0.5 μL of sample, which were spotted onto an OptiTOF-384 plate, dried and analyzed in positive linear mode on an Applied Biosystems 4800-Plus MALDI-TOF/TOF instrument (Applied Biosystems, Framingham, MA, USA). Mass spectra were acquired with delayed extraction in the m/z range of 20,000–130,000, at a laser intensity of 4200 and 500 shots per spectrum. Protein identification and amino acid sequencing of peptides obtained by digestion with trypsin, chymotrypsin or endoproteinase Glu-C were carried out by CID tandem mass spectrometry. Protein bands were excised from Coomassie Blue R-250-stained gels (12% polyacrylamide) following SDS-PAGE run under either reducing or non-reducing conditions and were reduced with dithiothreitol, alkylated with iodoacetamide and digested with sequencing-grade bovine trypsin (Sigma-Aldrich Co, St. Louis MO, USA), chymotrypsin (GBiosciences, St. Louis, MO, USA) or endoproteinase Glu-C (V8 protease) (GBiosciences, St. Louis, MO, USA), on an automated processor (ProGest, Digilab, Marlborough, MA, USA) overnight at 37 °C, according to the instructions of the manufacturer. The peptide mixtures were then analyzed by MALDI-TOF-TOF. Mixtures of 0.5 μL of sample and 0.5 μL of saturated α-cyano-4-hydroxycinnamic acid (in 50% acetonitrile, 0.1% trifluoroacetic acid) were spotted, dried and analyzed in positive reflector mode. Spectra were acquired using a laser intensity of 3000 and 1625 shots per spectrum, after external MS and MS/MS calibration with CalMix-5 standards (ABSciex, Framingham, MA, USA) spotted on the same plate. Up to 10 precursor peaks were selected from each MS spectrum for automated collision-induced dissociation MS/MS spectra acquisition at 2 kV (500 shots/spectrum, laser intensity of 3000). The spectra were analyzed using ProteinPilot v.4.0.8 and the Paragon® algorithm (ABSciex) against the UniProt/SwissProt database (Serpentes; downloaded 12 January 2016) for protein identification at a confidence level of 99%. Subsequently, sequences were manually inspected to confirm amino acid sequences de novo.

4.4.2. nLC-MS/MS

RP-HPLC Peaks 1 and 2 were dissolved in 0.4 M ammonium bicarbonate/8 M urea, reduced in 10 mM dithiothreitol at 37°C for 3 h, alkylated in 25 mM of iodoacetamide, followed by final quenching step in 4.5 mM DTT. The last two steps were performed at room temperature, in the dark, for 15 min each. After diluting the urea to 1 M, trypsin (Promega, Madison, WI, USA) was added (enzyme to substrate ratio 1:50 w/w) and the hydrolysis proceeded for 18 h at 37°C. Trifluoroacetic acid was then added to a final concentration of 1% (v/v), followed by sample desalting on Poros R2 microcolumns. Peptides were completely dried in a vacuum centrifuge, followed by re-suspension with 1% formic acid in water. Peptide quantitation (2 μL aliquots) was based on the UV absorbance at 280 nm on a NanoDrop 2000 spectrophotometer (Thermo Scientific, Waltham, MA, USA) (1 Absorbance unit = 1 mg/mL, considering 1 cm path length). Peptides (~1 μg) were submitted to a reversed phase nanochromatography (Ultimate 3000 system, Thermo Scientific Dionex, Sunnyvale, CA, USA) hyphenated to a quadrupole Orbitrap mass spectrometer (Q Exactive Plus, Thermo Scientific, Waltham, MA, USA). For desalting and concentration, samples were first loaded at 2 μL/min onto a home-made capillary guard column (2 cm × 100 μm i.d.) packed with 5 μm, 200 Å Magic C18 AQ matrix (Michrom Bioresources, Auburn, CA, USA). Peptide fractionation was performed at 200 nL/min on an analytical column (30 cm × 75 μm i.d.) with a laser pulled tip (~5 μm), packed with 1.9 μm ReproSil-Pur 120 C18-AQ (Dr. Maisch). The following mobile phases were used: (A) 0.1% formic acid in water; (B) 0.1%

formic acid in acetonitrile. Peptides were eluted with a gradient of 2%–40% B over 162 min, which was increased to 80% in 4 min, followed by a washing step at this concentration for 2 min before column re-equilibration.

Using data-dependent acquisition, up to 12 most intense precursor ions in each survey scan (excluding singly-charged ions and ions with unassigned charge states) were selected for higher energy collisional dissociation (HCD) fragmentation with 30% normalized collision energy. The following settings were used: (a) full MS (profile mode): 70,000 resolution (FWHM at m/z 200), Automatic Gain Control (AUC) AGC target 1×10^6, maximum injection time 100 ms, scan range 300–1500 m/z; (b) dd-MS/MS (centroid mode): 17,500 resolution, AGC target 5×10^4, maximum injection time 50 ms, isolation window 2.0 m/z (with 0.5 m/z offset), dynamic exclusion 60 s. The spray voltage was set to 1.9 kV with no sheath or auxiliary gas flow and with a capillary temperature of 250 °C. Data were acquired in technical triplicates using the Xcalibur software (Version 3.0.63, Thermo Fisher Scientific, Waltham, MA, USA). To avoid cross-contamination, two blank injections were run before the analysis of each biological replicate. The mass spectrometer was externally calibrated using a calibration mixture that was composed of caffeine, peptide MRFA and Ultramark 1621, as recommended by the instrument manufacturer.

PEAKS (version 8, Bioinformatics Solutions Inc., Waterloo, KW, Canada, 2016) was used for de novo sequencing analysis, using the following parameters: fixed cysteine modification (carbamidomethylation); 10 ppm peptide mass error tolerance; 0.02 Da fragment mass tolerance. Only de novo sequences showing ≥99% ALC (average of local confidence) were exported and submitted to sequence alignment against the mature BlatPII-c clone sequence using the algorithm PepExplorer [37]. The following parameters were used in the similarity-driven analysis: 75% minimum identity, minimum of 6 residues per peptide, substitution matrix PAM30MS.

4.5. Determination of Internal Peptide Sequences by Edman Degradation

The fractions obtained by reversed-phase HPLC were dried and reconstituted in 20 μL of 0.4 M NH$_4$HCO$_3$, 8 M urea. Then, 5 μL of 100 mM dithiothreitol were added, and the solution was incubated for 3 h at 37 °C. After that, 400 mM of iodoacetamide were added and incubated at room temperature for 15 min, protected from light. Then, 130 μL of deionized water were added, and digestion was performed with 2 μg trypsin (Sigma-Aldrich Co, St. Louis, MO, USA), overnight at 37 °C. The reaction was stopped with 20 μL of 1% (*v/v*) trifluoroacetic acid. Peptides were separated by reversed-phase HPLC on an Agilent 1200 instrument (Agilent Technologies, Santa Clara, CA, USA) in a Thermo AQUASIL C18 column (4.6 × 150 mm) at a flow of 0.5 mL/min, with detection at 215 nm. The solvent system was 0.1% trifluoroacetic acid in H$_2$O (Solvent A) and 0.1% trifluoroacetic acid in acetonitrile (Solvent B). The gradient program started with 5% Solvent B for 5 min and was then raised to 75% Solvent B at 50 min, and 75% Solvent B was maintained for 5 min. *N*-terminal sequences of isolated peptides were determined on a Shimadzu Biotech PPSQ 33A instrument according to the manufacturer's instruction. These tryptic peptides were also used for MALDI-TOF/TOF MS analysis.

4.6. Detection of Carbohydrates

Two-point-five micrograms of purified SVMP were separated by 12% SDS-PAGE under reducing and non-reducing conditions. The gel was stained using a Pro-Q Emerald 300 Glycoprotein Kit (Molecular Probes, Eugene, OR, USA), as described by the manufacturer.

4.7. PCR Amplification of SVMPs cDNA

An adult specimen of *B. lateralis*, kept at the Serpentarium of Instituto Clodomiro Picado, was sacrificed 3 days after venom extraction, and its venom glands were dissected out for total RNA isolation. After homogenization of the venom glands, the total RNA was extracted using the RNeasy Mini Kit (Qiagen, Hilden, Germany), following the manufacturer's instructions. The first strand of cDNA was reverse-transcribed using the RevertAid H Minus First Strand cDNA

Synthesis kit, according to the manufacturer's (ThermoScientific, Waltham, MA, USA) protocol, using the Qt-primer 5'-CCAGTGAGCAGAGTGACGAGGACTCGAGCTCAAGC(T)17-3'. cDNAs coding for PII SVMP sequences were amplified by PCR using the following primers: Forward (Fw) primer 5'-ATGATCCCAGTTCTCTTGGTAACTATATGCTTAGC-3' (corresponding to a conserved sequence in the pro-peptide of a metalloproteinase precursor; Figure 2), and reverse (Rev) primer, 5'-AGCCATTACTGGGACAGTCAGCAG-3', coding for the C-terminal amino acid sequence of BlatPII, obtained by mass spectrometry analysis (Figure 2). The PCR program was as follows: initial denaturation step (94 °C for 2 min), 35 cycles of denaturation (94 °C for 45 s), annealing (55 °C for 45 s), extension (72 °C for 120 s) and final extension (72 °C for 10 min). The PCR products were separated by 1% agarose electrophoresis, and the fragments with the expected molecular mass were purified using Illustra GFXTM PCR DNA and Gel Band Purification Kit (GE Healthcare, Life Sciences, Uppsala, Sweden). The purified sequences were cloned in a pGEM-T vector (Promega, Madison, WI, USA), which were used to transform *Escherichia coli* DH5α cells (Novagen, Darmstadt, Germany) by using an Eppendorf 2510 electroporator following the manufacturer's instructions. Positive clones, selected by growing the transformed cells in LB (Luria-Bertani medium) (Thermo Fisher Scientific, Waltham, MA, USA) broth containing 10 µg/mL of ampicillin, were confirmed by PCR-amplification using M13 primers of the pGEM-T vector. The inserts of positive clones were isolated using kit Wizard (Promega) and sequenced on an Applied Biosystems Model 377 DNA sequencer.

4.8. Proteolytic Activity on Azocasein

The proteinase activity of the SVMP on azocasein was assessed according to [38]. For comparative purposes, the PI SVMP BaP1 and the dimeric PII SVMP BlatH1 were used.

4.9. Proteolytic Activity on Gelatin

Proteolytic activity on gelatin was assessed by two methods: gelatin zymography and using a fluorescent commercial kit (EnzCheck® protocol Gelatinase/Collagenase Assay Kit, Molecular Probes, Life Technologies, Eugene, OR, USA). For the zymography, the method described by Herron et al. [39], as modified by Rucavado et al. [40], was used. Briefly, 2.5 µg of the SVMP were separated on a 10% SDS-PAGE containing 0.5 mg/mL of Type A gelatin (Sigma Chemical Co., St Louis, MO, USA). The same amount of BlatH1 was included as a positive control. After electrophoresis, gels were incubated with zymography substrate buffer overnight at 37 °C. Then, the gel was stained with 0.5% Coomassie Blue R-250 in acetic acid:isopropyl alcohol:water (1:3:6) and destained with distilled water to visualize gelatinolytic bands. In order to quantify the gelatinolytic activity of the SVMP, the commercial kit EnzCheck® was used following the manufacturer's instructions. Various concentrations of the SVMP, dissolved in 100 µL of reaction buffer, were incubated with 20 µg of the fluorescent gelatin substrate in a 96-well microplate. Samples were incubated at room temperature for 6 h and protected from light. Samples were analyzed in triplicate; a reagent blank was included. BlatH1 and BaP1 were included as positive controls. Fluorescence intensity was measured in the BioTek Synergy HT microplate reader using the absorption filter at 495 nm and the emission filter at 515 nm.

4.10. Local and Systemic Hemorrhagic Activities

Local hemorrhagic activity was determined by the mouse skin test [41]. Groups of four CD-1 mice (18–20 g) were intradermally (i.d.) injected in the ventral abdominal region with different doses of *B. lateralis* SVMP, dissolved in 100 µL of PBS. After 2 h, animals were sacrificed by CO_2 inhalation. The presence of hemorrhagic areas in the inner side of the skin was determined. The experimental protocols involving the use of animals in this study were approved by the Institutional Committee for the Care and Use of Laboratory Animals (CICUA) of the University of Costa Rica. Systemic hemorrhagic activity was tested as described by Escalante et al. [42]. Groups of four CD-1 mice were intravenously (i.v.) injected, in the tail vein, with various doses of the SVMP, dissolved in 100 µL PBS. After one hour, mice were sacrificed by an overdose of a ketamine/xylazine mixture.

The thoracic cavity was opened, and the presence of hemorrhagic spots was assessed by macroscopic observation of the lungs. Lungs of injected animals were dissected out and fixed in formalin solution for processing and embedding in paraffin. Then, sections were stained with hematoxylin-eosin for microscopic observation.

4.11. Edema-Forming Activity

The edema-forming activity was assessed according to the method of Lomonte et al. [43]. Groups of four CD-1 mice were injected subcutaneously (s.c.) in the right footpad with 5 µg of SVMP, dissolved in 50 µL PBS. In other mice, the SVMP BlatH1 was used as a positive control. Control mice were injected with PBS under otherwise identical conditions. The paw thickness was measured using a low pressure spring caliper before injection and at 0, 30, 60, 180 and 360 min after injection. Edema was calculated as the percentage of increase in the paw thickness of the right foot injected with SVMP as compared to the thickness before injection.

4.12. Coagulant Activity

Coagulant activity was determined as described by [44], with the following modifications: various doses of SVMP, dissolved in 25 µL PBS, were added to 0.25 mL of citrated human plasma, previously incubated for 5 min at 37 °C. Coagulant activity was monitored for 15 min at 37 °C. PBS was used as a negative control.

4.13. Platelet Aggregation Inhibitory Activity

Platelet-rich plasma (PRP) was obtained from adult healthy human volunteers, as described [45]. Two hundred twenty five microliters of PRP were pre-warmed at 37 °C for 5 min. The SVMP, dissolved in 25 µL of sterile saline solution, was added to PRP, and the mixture incubated for 5 min. After that, either ADP (20 µM final concentration) or collagen (10 µg/mL final concentration) (Helena Laboratories, Beaumont, TX, USA) was added to the PRP-SVMP mixture, and platelet aggregation was recorded for 5 min using an AggRAM analyzer (Helena Laboratories, Beaumont, Texas, TX, USA) at a constant spin rate of 600 rpm. Platelet aggregation was expressed as the percentage of transmittance, considering 100% response the aggregation induced by ADP or collagen alone.

4.14. Statistical Analysis

Significant differences between the mean values of experimental groups were assessed by one-way ANOVA, and a Tukey post-test was used in order to compare all pairs of means. Values of *p* lower than 0.05 were considered significant.

Supplementary Materials: The following are available online at www.mdpi.com/2072-6651/8/10/292/s1, Figure S1: High-resolution fragmentation spectrum of the m/z 576.6180 ion (z = 3; precursor error 0.2 ppm) from HPLC Peak 1 obtained on a QExactive Plus instrument (Thermo Scientific, Waltham, MA, USA) (**A**); the peptide sequence was interpreted by the PEAKS 8 de novo sequencing software. The average local confidence was set to 99%, and the local confidence score for each amino acid (i.e., "the likelihood of each amino acid assignment") was ≥98%. (**B**) Ion table showing the calculated mass of possible fragment ions. All fragment ions matching a peak within the mass error tolerance were colored blue (b-ion series) or red (y ion series). The lower graph shows the mass error distribution of all matched fragment ions (represented by blue and red dots); Figure S2: High-resolution fragmentation spectrum of the m/z 614.9611 ion (z = 3; precursor error 1.2 ppm) from HPLC Peak 2 obtained on a QExactive Plus instrument (**A**). The peptide sequence was interpreted by the PEAKS 8 de novo sequencing software. The average local confidence was set to 99%, and the local confidence score for each amino acid (i.e., "the likelihood of each amino acid assignment") was ≥98%. (**B**) Ion table showing the calculated mass of possible fragment ions. All fragment ions matching a peak within the mass error tolerance were colored blue (b-ion series) or red (y ion series). The lower graph shows mass error distribution of all matched fragment ions (represented by blue and red dots); Figure S3: High-resolution fragmentation spectrum of the m/z 864.4214 ion (z = 2; precursor error −2.1 ppm) from HPLC Peak 2 obtained on a QExactive Plus instrument (**A**). The peptide sequence was interpreted by the PEAKS 8 de novo sequencing software. The average local confidence was set to 99%, and the local confidence score for each amino acid (i.e., "the likelihood of each amino acid assignment") was ≥98%. (**B**) Ion table showing the calculated mass of possible fragment ions. All fragment ions matching a peak within the mass error tolerance were colored blue (b-ion series) or red (y ion series). The lower graph shows mass error distribution of all matched fragment ions (represented by blue and red dots); Figure S4: High-resolution fragmentation spectrum of the m/z 959.9163 ion ($z = 2$; precursor error -1.5 ppm) from HPLC Peak 2 obtained on a QExactive Plus instrument (**A**). The peptide sequence was interpreted by the PEAKS 8 de novo sequencing software. The average local confidence was set to 99%, and the local confidence score for each amino acid (i.e., "the likelihood of each amino acid assignment") was $\geq$ 98%. (**B**) Ion table showing the calculated mass of possible fragment ions. All fragment ions matching a peak within the mass error tolerance were colored blue (b-ion series) or red (y ion series). The lower graph shows mass error distribution of all matched fragment ions (represented by blue and red dots); Figure S5: High-resolution fragmentation spectrum of the m/z 902.4039 ion ($z = 2$; precursor error -0.4 ppm) from HPLC Peak 2 obtained on a QExactive Plus instrument (**A**). The peptide sequence was interpreted by the PEAKS 8 de novo sequencing software. The average local confidence was set to 99%, and the local confidence score for each amino acid (i.e., "the likelihood of each amino acid assignment") was $\geq$98%. (**B**) Ion table showing the calculated mass of possible fragment ions. All fragment ions matching a peak within the mass error tolerance were colored blue (b-ion series) or red (y ion series). The lower graph shows mass error distribution of all matched fragment ions (represented by blue and red dots); Figure S6: High-resolution fragmentation spectrum of the m/z 709.6204 ion ($z = 3$; precursor error -1.7 ppm) from HPLC Peak 2 obtained on a QExactive Plus instrument (**A**). The peptide sequence was interpreted by the PEAKS 8 de novo sequencing software. The average local confidence was 91%, and the local confidence score for each amino acid (i.e., "the likelihood of each amino acid assignment") is shown in the yellow inset. (**B**) Ion table showing the calculated mass of possible fragment ions. All fragment ions matching a peak within the mass error tolerance were colored blue (b-ion series) or red (y ion series). The lower graph shows mass error distribution of all matched fragment ions (represented by blue and red dots); Figure S7: Sequence coverage of mature BlatPII-c showing de novo sequenced tryptic peptides from HPLC Peak 1 with average local confidence (ALC) $\geq$99%, according to PEAKS 8 De Novo software. MS data were acquired using a Q Exactive Plus high resolution instrument. The similarity-based alignment was performed by the PepExplorer software, using the following search parameters: minimum identity = 75%; minimum number of residues per peptide = 6; substitution matrix = PAM30MS. Under these conditions, the sequence coverage obtained for BlatPII-c was 22.1%; Figure S8: Sequence coverage of mature BlatPII-c showing de novo sequenced tryptic peptides from HPLC Peak 2 with average local confidence (ALC) $\geq$99%, according to PEAKS 8 De Novo software. MS data were acquired using a Q Exactive Plus high resolution instrument. The similarity-based alignment was performed by the PepExplorer software, using the following search parameters: minimum identity = 75%; minimum number of residues per peptide = 6; substitution matrix = PAM30MS. Under these conditions, the sequence coverage obtained for BlatPII-c was 49.1%.

Acknowledgments: The authors thank Goran Neshich for support in the modeling studies and Paulo Carvalho (Fiocruz-Paraná, Brazil) for fruitful discussions on the analysis of de novo sequencing results. This study was supported by Vicerrectoría de Investigación, Universidad de Costa Rica (Projects 741-B0-528 and 741-B2-517), the International Foundation for Science (Grant F/4096-2) and Ministerio de Economía y Competitividad, Madrid, Spain (Grant BFU2013-42833-P). This work was performed in partial fulfillment of the requirements for the PhD degree of Erika Camacho at Universidad de Costa Rica.

Author Contributions: E.C., L.S., T.E., J.J.C., J.M.G. and A.R. conceived of and designed the experiments; E.C., L.S., T.E., F.V., A.G.C.N.-F., B.L., A.F., A.P., J.J.C., J.M.G. and A.R. performed the experiments; E.C., L.S., T.E., F.V., A.G.C.N.-F., B.L., J.J.C., J.M.G and A.R. analyzed the data; E.C., L.S., T.E., A.G.C.N.-F., B.L., J.J.C., J.M.G. and A.R. wrote the paper.

Conflicts of Interest: The authors declare no conflict of interest. The founding sponsors had no role in the design of the study; in the collection, analyses, or interpretation of data; in the writing of the manuscript, and in the decision to publish the results.

References

1. Fox, J.W.; Serrano, S.M.T. Structural considerations of the snake venom metalloproteinases, key members of the M12 reprolysin family of metalloproteinases. *Toxicon* **2005**, *45*, 969–985. [CrossRef] [PubMed]

2. Gutiérrez, J.M.; Rucavado, A.; Escalante, T. Snake Venom Metalloproteinases. Biological Roles and Participation in the Pathophysiology of Envenomation. In *Handbook of Venoms and Toxins of Reptiles*; Mackessy, S.P., Ed.; CRC Press: Boca Ratón, FL, USA, 2010; pp. 115–138.

3. Calvete, J.J. Proteomic tools against the neglected pathology of snake bite envenoming. *Expert Rev. Proteom.* **2011**, *8*, 739–758. [CrossRef] [PubMed]

4. Bode, W.; Grams, F.; Reinemer, P.; Gomis-Rüth, F.-X.; Baumann, U.; McKay, D.; Stöcker, W. The Metzincin-superfamily of zinc-peptidases. *Intracell. Protein Catabolism* **1996**, *389*, 1–11.

5. Fox, J.W.; Serrano, S.M. Snake venom metalloproteinases. In *Handbook of Venoms and Toxins of Reptiles*; Mackessy, S.P., Ed.; CRP Press: Boca Ratón, FL, USA, 2009; pp. 95–113.

6. Casewell, N.R.; Sunagar, K.; Takacs, Z.; Calvete, J.J.; Jacson, T.N.W.; Fry, B.G. Snake Venom Metalloprotease Enzymes. In *Venomous, Reptiles and Their Toxins. Evolution, Pathophysiology and Biodiscovery*; Fry, B.G., Ed.; Oxford University Press: Oxford, UK, 2015; pp. 347–363.

7.   Tanjoni, I.; Evangelista, K.; Della-Casa, M.S.; Butera, D.; Magalhães, G.S.; Baldo, C.; Clissa, P.B.; Fernandes, I.; Eble, J.; Moura-da-Silva, A.M. Different regions of the class P-III snake venom metalloproteinase jararhagin are involved in binding to alpha2beta1 integrin and collagen. *Toxicon* **2010**, *55*, 1093–1099. [CrossRef] [PubMed]

8.   Moura-da-Silva, A.M.; Serrano, S.M.; Fox, J.W.; Gutierrez, J.M. Snake Venom Metalloproteinases. Structure, function and effects on snake bite pathology. In *Animal Toxins: State of the Art. Perspectives in Health and Biotechonology*; de Lima, M.E., Monteiro de Castro, A., Martin-Euclaire, M.F., Benedeta, R., Eds.; UFMG: Belo Horizonte, Brasil, 2009; pp. 525–546.

9.   Casewell, N.R.; Wagstaff, S.C.; Harrison, R.A.; Renjifo, C.; Wüster, W. Domain loss facilitates accelerated evolution and neofunctionalization of duplicate snake venom metalloproteinase toxin genes. *Mol. Biol. Evol.* **2011**, *28*, 2637–2649. [CrossRef] [PubMed]

10. Moura-da-Silva, A.M.; Theakston, R.D.G.; Crampton, J.M. Evolution of disintegrin cysteine-rich and mammalian matrix-degrading metalloproteinases: Gene duplication and divergence of a common ancestor rather than convergent evolution. *J. Mol. Evol.* **1996**, *43*, 263–269. [CrossRef] [PubMed]

11. Fry, B. From genome to "venome": Molecular origin and evolution of the snake venom proteome inferred from phylogenetic analysis of toxin sequences and related body proteins. *Genome Res.* **2005**, *15*, 403–420. [CrossRef] [PubMed]

12. Casewell, N.R. On the ancestral recruitment of metalloproteinases into the venom of snakes. *Toxicon* **2012**, *60*, 449–454. [CrossRef] [PubMed]

13. Juarez, P.; Comas, I.; Gonzalez-Candelas, F.; Calvete, J.J. Evolution of snake venom disintegrins by positive Darwinian selection. *Mol. Biol. Evol.* **2008**, *25*, 2391–2407. [CrossRef] [PubMed]

14. Solórzano, A. *Snakes of Costa Rica Distribution, Taxonomy, and Natural History*; INBio: Heredia, Costa Rica, 2004.

15. Carbajo, R.J.; Sanz, L.; Perez, A.; Calvete, J.J. NMR structure of bitistatin—A missing piece in the evolutionary pathway of snake venom disintegrins. *FEBS J.* **2015**, *282*, 341–360. [CrossRef] [PubMed]

16. Serrano, S.M.T.; Jia, L.-G.; Wang, D.; Shannon, J.D.; Fox, J.W. Function of the cysteine-rich domain of the haemorrhagic metalloproteinase atrolysin A: Targeting adhesion proteins collagen I and von Willebrand factor. *Biochem. J.* **2005**, *391*, 69–76. [CrossRef] [PubMed]

17. Webb, B.; Sali, A. Comparative protein structure modeling using MODELLER. In *Current Protocols in Bioinformatics*; John Wiley & Sons, Inc: Hoboken, NJ, USA, 2002; Volume 47, pp. 5.6.1–5.6.32.

18. Lovejoy, B.; Hassell, A.M.; Luther, M.A.; Weigl, D.; Jordan, S.R. Crystal Structures of Recombinant 19-kDa Human Fibroblast Collagenase Complexed to Itself. *Biochemistry* **1994**, *33*, 8207–8217. [CrossRef] [PubMed]

19. Whittaker, M.; Floyd, C.D.; Brown, P.; Gearing, A.J.H. Design and Therapeutic Application of Matrix Metalloproteinase Inhibitors. *Chem. Rev.* **1999**, *99*, 2735–2776. [CrossRef] [PubMed]

20. Bates, E.E.M.; Fridman, W.H.; Mueller, C.G.F. The ADAMDEC1 (decysin) gene structure: Evolution by duplication in a metalloprotease gene cluster on chromosome 8p12. *Immunogenetics* **2002**, *54*, 96–105. [CrossRef] [PubMed]

21. Edwards, D.R.; Handsley, M.M.; Pennington, C.J. The ADAM metalloproteinases. *Mol. Aspects Med.* **2008**, *29*, 258–89. [CrossRef] [PubMed]

22. Frayne, J.A.N.; Jury, J.A.; Barker, H.L.; Hall, L.E.N. Rat MDC Family of Proteins: Sequence Analysis, Tissue Distribution, and Expression in Prepubertal. *Mol. Reprod. Dev.* **1997**, *167*, 159–167. [CrossRef]

23. Liu, H.; Shim, A.H.R.; He, X. Structural characterization of the ectodomain of a disintegrin and metalloproteinase-22 (ADAM22), a neural adhesion receptor instead of metalloproteinase: Insights on ADAM function. *J. Biol. Chem.* **2009**, *284*, 29077–29086. [CrossRef] [PubMed]

24. Waters, S.I.; White, J.M. Biochemical and Molecular Characterization of Bovine Fertilin a and (ADAM 1 and ADAM 2): A Candidate Sperm-Egg Binding/Fusion Complex. *Biol. Reprod.* **1997**, *56*, 1245–1254. [CrossRef] [PubMed]

25. Cornwall, G.A.; Hsia, N. ADAM7, A Member of the ADAM (A Disintegrin And Metalloprotease) Gene Family is Specifically Expressed in the Mouse Anterior Pituitary and Epididymis. *Endocrinology* **1997**, *138*, 4262–4272. [CrossRef] [PubMed]

26. Petretski, J.H.; Kanashiro, M.M.; Rodrigues, F.R.; Alves, E.W.; Machado, O.L.; Kipnis, T.L. Edema induction by the disintegrin-like/cysteine-rich domains from a *Bothrops atrox* hemorrhagin. *Biochem. Biophys. Res. Commun.* **2000**, *276*, 29–34. [CrossRef] [PubMed]

27. Kamiguti, A.S.; Zuzel, M.; Reid, A. Snake venom metalloproteinases and disintegrins: Interactions with cells. *Braz. J. Med. Biol. Res.* **1998**, *31*, 853–862. [CrossRef] [PubMed]
28. Calvete, J.J.; Marcinkiewicz, C.; Monleón, D.; Esteve, V.; Celda, B.; Juárez, P.; Sanz, L. Snake venom disintegrins: Evolution of structure and function. *Toxicon* **2005**, *45*, 1063–1074. [CrossRef] [PubMed]
29. Wijeyewickrema, L.C.; Berndt, M.C.; Andrews, R.K. Snake venom probes of platelet adhesion receptors and their ligands. *Toxicon* **2005**, *45*, 1051–1061. [CrossRef] [PubMed]
30. Lomonte, B.; Rangel, J. Snake venom Lys49 myotoxins: From phospholipases A2 to non-enzymatic membrane disruptors. *Toxicon* **2012**, *60*, 520–530. [CrossRef] [PubMed]
31. Kurtović, T.; Brgles, M.; Leonardi, A.; Lang Balija, M.; Sajevic, T.; Križaj, I.; Allmaier, G.; Marchetti-Deschmann, M.; Halassy, B. VaSP1, catalytically active serine proteinase from *Vipera ammodytes ammodytes* venom with unconventional active site triad. *Toxicon* **2014**, *77*, 93–104. [CrossRef] [PubMed]
32. Wu, J.; Jin, Y.; Zhong, S.; Chen, R.; Zhu, S.; Wang, W.; Lu, Q.; Xiong, Y. A unique group of inactive serine protease homologues from snake venom. *Toxicon* **2008**, *52*, 277–284. [CrossRef] [PubMed]
33. Lomonte, B.; Sanz, L.; Angulo, Y.; Calvete, J.J.; Escolano, J.; Fernández, J.; Sanz, L.; Angulo, Y.; Gutiérrez, J.M.; Calvete, J.J. Snake venomics and antivenomics of the arboreal neotropical pitvipers *Bothriechis lateralis* and *Bothriechis schlegelii*. *J. Proteome Res.* **2008**, *7*, 2445–2457. [CrossRef] [PubMed]
34. Camacho, E.; Villalobos, E.; Sanz, L.; Pérez, A.; Escalante, T.; Lomonte, B.; Calvete, J.J.; Gutiérrez, J.M.; Rucavado, A. Understanding structural and functional aspects of PII snake venom metalloproteinases: Characterization of BlatH1, a hemorrhagic dimeric enzyme from the venom of *Bothriechis lateralis*. *Biochimie* **2014**, *101*, 145–155. [CrossRef] [PubMed]
35. Gutiérrez, J.M.; Romero, M.; Diaz, C.; Borkow, G.A.D.; Ovadia, M. Isolation and characterization of a metalloproteinase with weak activity from the venom of the snake *Bothrops asper* (Terciopelo). *Toxicon* **1995**, *33*, 19–29. [CrossRef]
36. Laemmli, U.K. Cleavage of Structural Proteins during the Assembly of the Head of Bacteriophage T4. *Nature* **1970**, *227*, 680–685. [CrossRef] [PubMed]
37. Leprevost, F.V.; Valente, R.H.; Lima, D.B.; Perales, J.; Melani, R.; Iii, J.R.Y.; Barbosa, V.C.; Junqueira, M.; Carvalho, P.C. PepExplorer: A Similarity-driven Tool for Analyzing de Novo Sequencing Results. *Mol. Cell. Proteom.* **2014**, *13*, 2480–2489. [CrossRef] [PubMed]
38. Wang, W.; Shih, C.; Huang, T. A novel P-I class metalloproteinase with broad substrate-cleaving activity, agkislysin, from *Agkistrodon acutus* venom. *Biochem. Biophys. Res. Commun.* **2004**, *324*, 224–230. [CrossRef] [PubMed]
39. Herron, G.S.; Banda, M.J.; Clark, E.J.; Gavrilovic, J.; Werb, Z. Secretion of Metalloproteinases by Stimulated Capillary Endothelial Cells. *J. Biol. Chem.* **1986**, *261*, 2814–2818. [PubMed]
40. Rucavado, A.; Núñez, J.; Gutiérrez, J.M. Blister formation and skin damage induced by BaP1, a haemorrhagic metalloproteinase from the venom of the snake *Bothrops asper*. *J. Exp. Pathol. Int.* **1998**, *79*, 245–254.
41. Gutiérrez, J.; Gené, J.; Rojas, G.; Cerdas, L. Neutralization of proteolytic and hemorrhagic activities of Costa Rican snake venoms by a polyvalent antivenom. *Toxicon* **1985**, *23*, 887–893. [CrossRef]
42. Escalante, T.; Núñez, J.; Moura da Silva, A.M.; Rucavado, A.; Theakston, R.D.G.; Gutiérrez, J.M. Pulmonary hemorrhage induced by jararhagin, a metalloproteinase from *Bothrops jararaca* snake venom. *Toxicol. Appl. Pharmacol.* **2003**, *193*, 17–28. [CrossRef]
43. Lomonte, B.; Tarkowski, A.; Hanson, L.Å. Host response to *Bothrops asper* snake venom: Analysis of edema formation, inflammatory cells and cytokine release in a mouse model. *Inflammation* **1993**, *17*, 93–105. [CrossRef] [PubMed]
44. Gené, J.A.; Rojas, G.; Gutiérrez, J.M.; Cerdas, L. Comparative study on the coagulant, defibrinating, fibrinolytic and fibrinogenolytic activities of Costa Rican crotaline snake venoms and their neutralization by a polyvalent antivenom. *Toxicon* **1989**, *27*, 841–848. [CrossRef]
45. Kamiguti, A.; Hay, C.; Zuzel, M. Inhibition of collagen-induced platelet aggregation as the result of cleavage of alpha 2 beta 1-integrin by the snake venom metalloproteinase jararhagin. *Biochem. J.* **1996**, *320*, 635–641. [CrossRef] [PubMed]

toxins

MDPI

Article

Viperid Envenomation Wound Exudate Contributes to Increased Vascular Permeability via a DAMPs/TLR-4 Mediated Pathway

Alexandra Rucavado [1,†], Carolina A. Nicolau [2,†], Teresa Escalante [1], Junho Kim [3], Cristina Herrera [1,4], José María Gutiérrez [1,*] and Jay W. Fox [5,*]

[1]   Instituto Clodomiro Picado, Facultad de Microbiología Universidad de Costa Rica, San José 11501-2060, Costa Rica; alexandra.rucavado@ucr.ac.cr (A.R.); teresa.escalante@ucr.ac.cr (T.E.); cristina.herreraarias@gmail.com (C.H.)
[2]   Laboratório de Toxinologia, Instituto Oswaldo Cruz, Rio de Janeiro CEP 21040-360, Brazil; carolnicolau.bio@gmail.com
[3]   Department of Fine Chemistry & New Materials, Sangji University, Wonju-si, Kangwon-do 220-702, Korea; jhokim@sangji.ac.kr
[4]   Facultad de Farmacia, Universidad de Costa Rica, San José 11501-2060, Costa Rica
[5]   Department of Microbiology, Immunology and Cancer Biology, University of Virginia School of Medicine, P.O. Box 800734, Charlottesville, VA 22908, USA
*   Correspondence: jose.gutierrez@ucr.ac.cr (J.M.G.); jwf8x@eservices.virginia.edu (J.W.F.); Tel.: +506-2511-7865 (J.M.G.); +1-434-4924-0050 (J.W.F.)
†   These authors contributed equally to this work.

Academic Editor: R. Manjunatha Kini
Received: 4 October 2016; Accepted: 17 November 2016; Published: 24 November 2016

Abstract: Viperid snakebite envenomation is characterized by inflammatory events including increase in vascular permeability. A copious exudate is generated in tissue injected with venom, whose proteomics analysis has provided insights into the mechanisms of venom-induced tissue damage. Hereby it is reported that wound exudate itself has the ability to induce increase in vascular permeability in the skin of mice. Proteomics analysis of exudate revealed the presence of cytokines and chemokines, together with abundant damage associated molecular pattern molecules (DAMPs) resulting from both proteolysis of extracellular matrix and cellular lysis. Moreover, significant differences in the amounts of cytokines/chemokines and DAMPs were detected between exudates collected 1 h and 24 h after envenomation, thus highlighting a complex temporal dynamic in the composition of exudate. Pretreatment of mice with Eritoran, an antagonist of Toll-like receptor 4 (TLR4), significantly reduced the exudate-induced increase in vascular permeability, thus suggesting that DAMPs might be acting through this receptor. It is hypothesized that an "Envenomation-induced DAMPs cycle of tissue damage" may be operating in viperid snakebite envenomation through which venom-induced tissue damage generates a variety of DAMPs which may further expand tissue alterations.

Keywords: snake venom metalloproteinases (SVMPs); TLR4; damage associated molecular pattern molecules (DAMPs); exudate; increased vascular permeability

---

1. Introduction

Envenomation by viperid snakes causes local and systemic pathological and pathophysiological manifestations [1,2]. Many of these pathologies, such as hemorrhage and necrosis, have been associated with the action of the snake venom metalloproteinases (SVMPs) [3–7] which elicit these effects by both direct and indirect mechanisms [5,8]. One of the best described outcomes of the direct action of the

SVMPs is hemorrhage, which is the result of proteolytic degradation of key extracellular matrix proteins in the host stroma and capillaries, allowing extravasation of capillary contents into the stroma [9–12]. In addition, viperid envenomation triggers an inflammatory response which likely contributes to these pathological features, as well as being involved in tissue repair and regeneration [13,14].

Several years ago we described in animal models of envenomation that there is a copious volume of wound exudate at the site of tissue damage [5,8]. Markers for disease and various pathological conditions are often best observed in proximity to the site of tissue damage. As such we utilized venom-induced wound exudate for proteomic exploration of the effects of whole venom and isolated toxins, as well as neutralization by antivenoms and toxin inhibitors, to gain insight into the biological mechanisms by which these agents in vivo give rise to the symptoms and pathologies observed during snakebite envenomation [15–17]. These studies, in addition to illuminating many aspects of toxin action, also demonstrated that venom-induced wound exudate is a very rich reservoir of various inflammatory mediators and of damage-associated (or danger-associated) molecular pattern molecules (DAMPs). Some of the DAMPs are derived from the proteolysis of host proteins by the SVMPs and activated endogenous host proteinases and some the result of cell lysis and the escape of cellular proteins into the exudate due to cytotoxic phopholipases A_2 present in the venom [15].

DAMPs, like the pathogen-associated molecular pattern molecules (PAMPs), generate most of their biological activities via engagement of toll-like receptors (TLR) [18,19]. As their name implies DAMPs are the result of cellular and extracellular damage and serve to provide a host response to cellular injury by launching innate immunity inflammatory responses [20]. Prolonged or excessive DAMP response rather than being only beneficial to the host can in some situations also be deleterious. The microvascular endothelium is intimately related to microbial sepsis-induced organ failure by impacting the vascular barrier following engagement of endothelial TLRs by PAMPs [21,22]. Furthermore DAMP engagement with TLRs on endothelial surfaces has been shown to exacerbate the action of PAMPs-associated microbial sepsis [23]. In addition, DAMPs themselves have been implicated in ischemic disease, pulmonary disease, cancer and metastasis, ocular disease [24], and kidney injury [25]. Noteworthy many of these diseases share a fascinating level of overlapping pathophysiology with viperid envenomation, including increased vascular permeability. Thus, the possible role of DAMPs in the inflammatory scenario in snakebite envenomation requires exploration.

Previous works with snake venoms have shown a role for TLR2 as well as MYD88-dependent TLR signaling in venom-induced inflammation [26–28]. Those studies suggest a potential for increases in vascular permeability being mediated through these and other pathways. Given the critical role of increase in vascular permeability in envenomation pathophysiology we hypothesized that DAMPs and other envenomation-derived and inflammatory products found in wound exudate may play a functional role in envenomation. Here we report that wound exudate generated in a mouse model of viperid envenomation contain multiple cytokines and chemokines, as well as DAMPs, and induces increase in vascular permeability. Moreover, we show that this effect is likely to be mediated at least in part by TLR4, as demonstrated by its reduction by pretreatment with Eritoran, a specific TRL4 antagonist. These findings shed new light in the complex mechanisms involved in the inflammatory responses of tissue in snakebite envenomation and novel routes for therapeutic intervention to attenuate envenomation mortality and morbidity.

2. Results

2.1. Exudates Collected from Mice Injected with B. asper Venom Increase Vascular Permeability

When exudates collected from animals treated with *B. asper* venom were injected intradermally in the skin of mice, they induced an increase in vascular permeability, as reflected by the extravasation of Evans Blue (Figure 1). In order to assess whether this effect was due to the action of venom components present in the exudate, samples of exudate were incubated with polyvalent antivenom before testing in the mouse skin. As depicted in Figure 1, a large reduction in the effect was observed after incubation with antivenom in exudate of 1 h, but not in the neutralized exudate of 24 h, indicating that venom components play a role in the effect only in 1 h exudate samples. However, even in the 1 h exudate, there was a residual effect after neutralization by antivenom, indicating a venom-independent effect of exudate on permeability (Figure 1). As controls, normal mouse plasma and polyvalent antivenom did not induce an increase in vascular permeability (Figure 1). In order to attenuate concern that the observed effect was due to the presence of bacterial lipopolysaccharides, exudate collected at 24 h was incubated with polymyxin B before injection in mice. No reduction in the effect was observed, indicating that it is not due to the action of bacterial endotoxins. On the basis of these findings, the composition of the exudates in terms of inflammatory mediators was investigated.

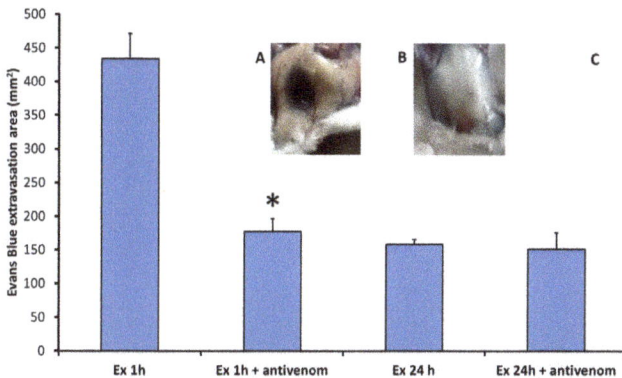

Figure 1. Wound exudate induces an increase in vascular permeability. Upper figures show samples of skin of mice injected intradermally with (**A**) wound exudate collected from mice 1 h after intramuscular injection of *B. asper* venom; or (**B**) blood plasma from untreated mice. Both groups of mice received an intravenous injection of Evans Blue solution before the injection of exudate or plasma (see Materials and Methods for details). Note the absence of Evans Blue extravasation in the control (**B**), whereas a clear extravasation was observed after injection of exudate (**A**); (**C**) Quantitative analysis of the increase in vascular permeability in mouse skin after injection of exudates collected 1 h and 24 h after injection of *B. asper* venom. In both types of exudates experiments were also performed with samples previously incubated with antivenom to neutralize the venom toxins present. One hour exudate induced a higher increase in vascular permeability than 24 h exudate. * A significant reduction in the activity by antivenom ($p < 0.05$) was observed only with 1 h exudate. Controls injected with mouse plasma alone or with antivenom alone did not show increase in vascular permeability.

2.2. Exudates Contain High Concentrations of Inflammatory Mediators

Given the clear capability of exudate to induce an increase in vascular permeability, it was necessary to investigate its molecular composition. When the cytokine and chemokine subproteome in exudates was analyzed by the Luminex technology, abundant inflammatory mediators were detected (Table 1). A dynamic development of the composition of the exudate was observed when comparing these subproteomes in exudates collected at 1h and 24 h, since a higher concentration of cytokines and chemokines was observed in the 24 h exudate as compared to the 1 h exudate. Among 32 mediators quantified, 14 of them presented more than 10-fold increase in concentration in the 24 h samples (Table 1). The highest increases were observed in IL-1β, CCL3, and CCL4. Thus, exudates, particularly the one collected at 24 h, contain abundant cytokines and chemokines.

2.3. Abundant DAMPs Are Identified in the Proteomes of Exudates

In order to ascertain whether exudates contained proteins that have been categorized as DAMPs, and which could play a role in the inflammatory event described, the full proteomes of exudates were analyzed (Table S1) vis-à-vis the information collected concerning the identity of DAMPs in exudates. As shown in Table 2, many proteins identified as DAMPs in the literature are observed present in both 1 h and 24 h exudates. When comparing the quantitative values of DAMPs in the exudates at the two time intervals, there is a clear trend towards higher abundance of many of these in the 24 h samples. Interesting exceptions are basement membrane-specific heparan sulfate proteoglycan core protein, 60 kDa heat shock protein (mitochondrial), and heat shock protein beta 2, whose quantitative values were higher in the 1 h exudate (Table 2). Thus, exudates contain a wide range of DAMPS, some of which are notably abundant at 24 h.

Table 1. Cytokine profile (subproteome) of wound exudates collected at 1 h and 24 h.

Analytes (pg/mL)	Exudate 1 h	Exudate 24 h	Fold change *
CCL11 (EOTAXIN)	220.0	982.4	4.5
CSF-3 (G-CSF)	1670.0	>11,610.0	>6.9
CSF-2 (GM-CSF)	21.2	219.0	10.3
IFNy	3.2	37.8	11.8
IL-10	436.3	3419.0	7.8
IL-12p40	10.3	37.5	3.6
IL-12p70	5.2	23.8	4.6
IL-13	268.4	1217.0	4.5
IL-15	25.0	117.3	4.7
IL-17	<2.9	12.0	>4.1
IL-1a	228.3	5952.0	26.1
IL-1b	8.0	843.1	105.4
IL-2	4.8	9.2	1.9
IL-3	<2.4	10.2	>4.2
IL-4	<1.4	4.7	>1.9
IL-5	15.9	68.1	4.3
IL-6	7901.0	>17,536.0	>2.2
IL-7	3.8	10.5	2.8
IL-9	324.0	509.9	1.6
CXCL10 (IP-10)	52.6	2424.0	46.1
CXCL1/GRO alpha (KC)	4514.0	15,957.0	3.5
LIF	24.6	2252.0	91.5
CXCL5 (LIX)	1494.0	3817.0	2.5
CCL2 (MCP-1)	938.8	>18,874.0	>20.1
CSF-1 (M-CSF)	26.1	529.8	20.3
CXCL9 (MIG)	228.6	3034.0	13.3
CCL3 (MIP-1a)	27.9	>14,741.0	>528.3
CCL4 (MIP-1b)	80.7	>14,663.0	>181.7
CXCL2 (MIP-2)	4623.0	12,954.0	2.8
CCL5 (RANTES)	5.0	307.3	61.4
TNF-a	9.0	799.2	88.8
VEGF	<1.3	89.3	>68.7

Analyses were performed by using the Luminex quantitative analysis (see Materials and Methods for details). * Proteins showing a difference higher than 10-fold between exudates collected at the two times are highlighted.

Table 2. DAMPs identified in wound exudates collected 1 and 24 h after injection of *B. asper* venom.

Identified Proteins	Accession Number	Molecular Weight	Quantitative Value 1 h	Quantitative Value 24 h	Fold Change *
Hemoglobin subunit beta-2	P02089	16 kDa	745	1329	1.8
Fibronectin	P11276	273 kDa	274	290	1.0
Fibrinogen gamma chain	Q8VCM7	49 kDa	49	145	2.9
Heat shock cognate 71 kDa protein	P63017	71 kDa	50	17	2.9
Fibrinogen beta chain	Q8K0E8	55 kDa	12	107	8.9
Heat shock protein HSP 90-beta	P11499	83 kDa	41	26	1.6
Basement membrane-specific heparan sulfate proteoglycan core protein	B1B0C7	469 kDa	83	0	>83
Serum amyloid P-component	P12246	26 kDa	27	65	2.4
Histone H4	P62806	11 kDa	55	46	1.2
Histone H2B type 1-M	P10854	14 kDa	46	29	1.6
Proteoglycan 4	E9QQ17	111 kDa	18	12	1.5
Protein S100-A9	P31725	13 kDa	1	21	21
Myosin light chain 1/3, skeletal muscle isoform	P05977	21 kDa	10	59	5.9
Myosin-9	Q8VDD5	226 kDa	64	30	2.1
Serum amyloid A-4 protein	P31532	15 kDa	13	11	1.1
Myosin-10	Q3UH59	233 kDa	1	23	23

Table 2. *Cont.*

Identified Proteins	Accession Number	Molecular Weight	Quantitative Value 1 h	Quantitative Value 24 h	Fold Change *
60 kDa heat shock protein	P63038	61 kDa	18	0	>18
40S ribosomal protein S19	Q9CZX8	16 kDa	0	34	>34
Decorin	P28654	40 kDa	1	22	22
Chondroitin sulfate proteoglycan 4	Q8VHY0	252 kDa	0	11	>11
Isoform 2 of Myosin-11	O08638-2	223 kDa	1	34	34
Myosin regulatory light chain 12B	Q3THE2	20 kDa	18	57	3.1
Endoplasmin	P08113	92 kDa	1	45	45
Heat shock protein beta-1	P14602	23 kDa	37	45	1.2
Calreticulin	P14211	48 kDa	0	22	>22
Protein S100-A8	P27005	10 kDa	0	80	>80
Isoform Smooth muscle of Myosin	Q60605-2	17 kDa	1	12	12
Myosin light chain 3	P09542	22 kDa	0	14	>14
Heat shock protein beta-2	Q99PR8	20 kDa	27	0	>27
Biglycan	P28653	42 kDa	0	22	>22
Serum amyloid A-1 protein	P05366	14 kDa	0	16	>16

* Proteins showing a difference higher than 10-fold between exudates collected at the two times are highlighted.

2.4. Eritoran, an Inhibitor of TLR4, Inhibits the Vascular Permeability Effect Induced by Exudate

The presence of abundant DAMPs in exudates prompted us to assess whether the effect of exudate in vascular permeability could be due to the action of DAMPs. Since many DAMPs act in the cells of the innate immune system through TLR4, the effect on vascular permeability after blocking this receptor with Eritoran was assessed. To this end, Eritoran was administered to mice before the injection of exudates collected from mice injected with venom. As shown in Figure 2, treatment with Eritoran significantly reduced the effect of exudates on vascular permeability, but only in the case of 24 h exudates. Thus, the increase in vascular permeability induced by 24 h exudate, but not by 1 h exudate, is mediated by TLR-4. We suggest this may be due to the presence of venom components in the 1 h exudate directly giving rise to permeability swamping out the TLR4 permeability axis and thus inhibition by Eritoran. Control mice receiving Eritoran and then injected intradermally with either normal mouse plasma or antivenom alone did not develop any extravasation of Evans blue in the skin.

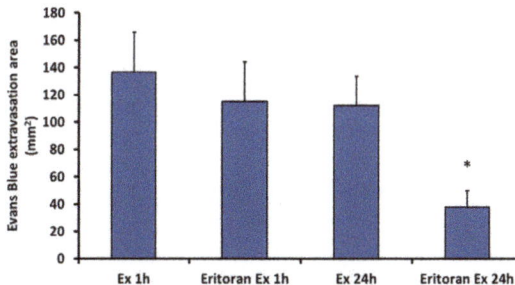

Figure 2. Effect of Eritoran in the increase of vascular permeability induced by exudates (Ex). Exudates were collected at 1 h and 24 h from mice injected with venom. Then, a separate group of mice were pretreated with either Eritoran or saline solution. Afterwards, these mice were injected intradermally in the skin with either 1 h exudate or 24 h exudate previously incubated with antivenom to neutralize venom toxins, as described in the legend of Figure 1. The increase in vascular permeability was assessed by extravasation of Evans blue, as described in Materials and Methods. The following experimental groups were used: Ex 1h: Mice injected with 1 h exudate; Eritoran Ex 1h: Mice pretreated with Eritoran and then injected with 1 h exudate; Ex 24h: Mice injected with 24 h exudate; Eritoran Ex 24h: Mice pretreated with Eritoran and then injected with 24 h exudate. Control mice pretreated with Eritoran and then injected intradermally with either mouse plasma or antivenom did not develop any extravasation of Evans blue. * Eritoran significantly reduced the effect induced by 24 h exudate ($p < 0.05$) but not by 1 h exudate.

3. Discussion

Snakebite envenomation involves highly complex and interrelated pathological and pathophysiological alterations which result from both the direct action of venom components on the host as well as a variety of tissue responses, in a dynamic interplay [7,29]. Many studies have assessed the direct action of venom toxins in the tissues, particularly of myotoxic PLA$_2$s, serine proteinases, and hemorrhagic SVMPs [11,12,30]. Inflammation in venom-affected tissues is associated with edema, hyperalgesia, and recruitment of inflammatory cells. These effects are induced by a variety of mediators released in the tissues, including histamine, eicosanoids, nitric oxide, complement anaphylatoxins, bradykinin, and cytokines, among others [14,31]. The present investigation explores an aspect of this pathophysiology that to date has received little attention, i.e., the possible role of the exudate, generated in the tissue, as a reservoir of potent mediators in the inflammatory events characteristic of these envenomations.

The ability of exudates collected from venom-damaged tissue to induce increases in vascular permeability was used in our studies as an index of pro-inflammatory action of exudates. We deemed the increase in vascular permeability is one of the landmarks of envenomation-induced inflammation. Significantly, our results indicated that exudates collected at early and late time intervals after envenomation, i.e., 1 h and 24 h, induced an increase in vascular permeability when injected in the skin of mice. However, the effect induced by the 1 h exudate was significantly reduced when exudate was incubated with antivenom, whereas such inhibition did not occur in the case of the 24 h exudate. This indicates that the effect of 1 h exudate was predominantly due to residual venom components present in the tissues and in the exudate. In contrast, the effect of 24 h exudate does not seem to be caused by the direct action of venom toxins, but very likely due to tissue-derived and/or endogenously released inflammatory components. In agreement with our observations, it has been shown that *B. asper* venom concentration in the tissue, after an intramuscular injection in mice, is high at 1 h, being largely reduced by 24 h probably as a consequence of venom diffusion and systemic distribution [32].

In order to identify possible components in the exudate responsible for this inflammatory action, a subproteome analysis of cytokines and chemokines in exudates was performed. Our results revealed the presence of abundant inflammatory mediators. Interestingly, much higher abundance of these components was observed in the 24 h exudate, as compared to the 1 h exudate, indicating that the composition of the exudate varies in concordance with the extent of tissue damage and subsequent repair processes. This agrees with the variations observed in the overall proteomes of exudates at different time intervals in the same experimental model of envenomation by *B. asper* venom [33]. Hence, although an abundant volume of exudate is present in venom-affected tissue even at the early stages of envenomation, its composition varies significantly as a result of the complex dynamics of tissue damage and response to venom deleterious effects.

Exploring the composition of this complex inflammatory milieu in the form of the exudate proteome and function may provide clues for a deeper understanding of the tissue dynamics in snakebite envenomation. The observation that 24 h exudate has a higher concentration of cytokines and chemokines predicts that this exudate would be more active at stimulating inflammatory cells and processes. The high concentration of many of these mediators in the 24 h exudate is likely to depend, at least partially, on the abundant population of inflammatory cells in the damaged tissue at that time interval, as previously described [34]. These cells, in particular monocytes/macrophages, are known to synthesize many of these mediators [35,36]. These findings suggest that the recruitment of inflammatory cells to the site of tissue damage is associated with the synthesis of mediators which, in turn, recruit additional cells and, at the same time, stimulate these cells to produce more mediators and set up a cycle of expansion of the inflammatory response. This may explain the reparative and regenerative processes that follow the acute tissue damage, but also may exacerbate tissue damage, a delicate balance that needs to be further investigated.

In addition to cytokines and chemokines, the exudate generated in venom-damaged tissue is known to contain many proteins of various types derived from the affected cells and extracellular matrix [35]. We were particularly interested in the identification of DAMPs, which are endogenous molecules, or fragments of molecules, released in the tissues as a consequence of damage of cells and extracellular matrix [19,37,38]. After binding to pattern-recognition receptors in cells of the innate immune system, DAMPs stimulate the synthesis of inflammatory mediators and, therefore, participate in the overall response of tissues to cellular and extracellular damage [39,40]. DAMPs act in concert with chemokines to regulate the recruitment and trafficking of leukocytes in inflammation [35]. Since snake venoms induce drastic pathological events in tissues, it is likely that abundant DAMPs are generated upon venom injection. It has been shown that mitochondrial DNA, cytochrome C, and ATP are released in tissue affected by venoms of *Bothrops asper* and *Crotalus durissus* [41,42]. To further explore this phenomenon, we performed a complete proteomic analysis of exudates collected after injection of *B. asper* venom, and identified DAMPs in these exudates, on the basis of information available in the literature. As expected, many DAMPs were identified in exudates, strongly supporting the concept that venom-induced tissue damage results in the release of DAMPs, which add to the complexity of the local milieu that develops in snakebite envenomation.

The inflammatory effects exerted by many DAMPs are mediated through their recognition by TLR4, a pattern-recognition receptor present in the membrane a various types of inflammatory cells [43]. To assess whether exudate-induced vascular permeability might be mediated by DAMPs, we used Eritoran, a specific blocker of TLR4. Pretreatment of mice with Eritoran significantly reduced the increase in vascular permeability, thus supporting the view that signals mediated through TLR4, probably generated by DAMPs, contribute to the pro-inflammatory activity of 24 h exudate. TLR4 mediates the action of a number of DAMPs, such as hyaluronan fragments [44,45], S100A9 [46], heat shock protein 60 [47], soluble heparan sulfate [48], and fibronectin fragments [49]. Fibrinogen stimulates secretion of chemokines by macrophages through TLR4 [50]. Interestingly, TLR4 has been shown to exert a protective role in muscular tissue damage induced by the venom of *Bothrops jararacussu* [51]. Thus, TLR4 is likely to be a centerpiece in the detection of venom-induced tissue damage and in the stimulus to inflammation in this pathology. Previous studies have demonstrated the involvement of TLR2 in the modulation of the inflammatory response after injection of *Bothrops atrox* venom [28], and the participation of MyD88 adaptor protein in this inflammatory scenario. MyD88 is involved in cellular activation after binding of DAMPs to TLRs [26,52]. Thus, various pattern recognition receptors are likely to participate in the tissue responses in snakebite envenomation.

Among the DAMPs identified in the exudates collected from the tissue of mice injected with *B. asper* venom, several are known to play roles in inflammation. Examples are fibrinogen [15,17,33], which is known to stimulate chemokine secretion by macrophages through TLR4 [50]. In addition, fibrinogen products transmit activating signals to leukocytes through interactions with integrins, inducing cytokine secretion by these cells [53]. Fibronectin fragments, also identified in the exudates, are known to induce expression of matrix metalloproteinases (MMPs) [54], and fragments of extracellular matrix proteins, and of the glycosaminoglycan hyaluronic acid, as a consequence of hydrolysis by venom or tissue proteinases and hyaluronidases, exert a variety of pro-inflammatory roles [55,56]. Decorin, found in our proteomic analysis, stimulates the expression of TNF-α and IL-1β, and reduces the expression of the anti-inflammatory IL-10 [56]. Therefore, many of the DAMPs detected in exudates from venom-affected tissue are known to play a variety of pro-inflammatory roles and are probably involved in the tissue responses to venom toxins.

The relationship of various DAMPs with the observed increase in vascular permeability deserves consideration. Some DAMPs might increase vascular permeability indirectly, by stimulating inflammatory cells to synthesize cytokines or chemokines which in turn act on the microvasculature. However, some DAMPs may also directly interact with the endothelial cells in venules. This is the case of S100 proteins, which have been repeatedly detected in exudates collected from tissues affected by venom and toxins of *B. asper* [15,17,33]. S100 A8 and S100 A9 increase monolayer permeability in a human endothelial cell line [57]. Direct stimulation of endothelial cells by DAMPs is therefore likely to also contribute to an increase in vascular permeability, trafficking of inflammatory cells and additionally generate a procoagulant phenotype which might contribute to hemostatic alterations in envenomations [23]. Moreover, TLR4 is known to mediate the disruption of the vascular endothelial barrier in the lungs [58].

In conclusion, our results show that exudates collected from tissue damaged by *B. asper* snake venom toxins contain abundant cytokines and chemokines, as well as DAMPs, and is able to increase vascular permeability. In this context, early and late events take place in the tissue after venom injection, associated with the direct and indirect effects of venom components on muscle fibers and the microvasculature. This complex interplay of direct and indirect effects and early and late phenomena are hypothetically summarized in Figure 3. The key role of TLR4 in this phenomenon is centrally illustrated, suggesting that DAMP-mediated signals to inflammatory cells in the damaged tissue environment may be a significant and cyclic component in the overall inflammatory scenario giving rise to a "DAMPs derived tissue damage cycle" (DDTD cycle). As such we are further examining the role of DAMPs in snake venom-induced tissue damage in order to gain a more complete understanding of this complex pathological phenomenon, and to identify novel therapeutic avenues to attenuate envenomation mortality and morbidity.

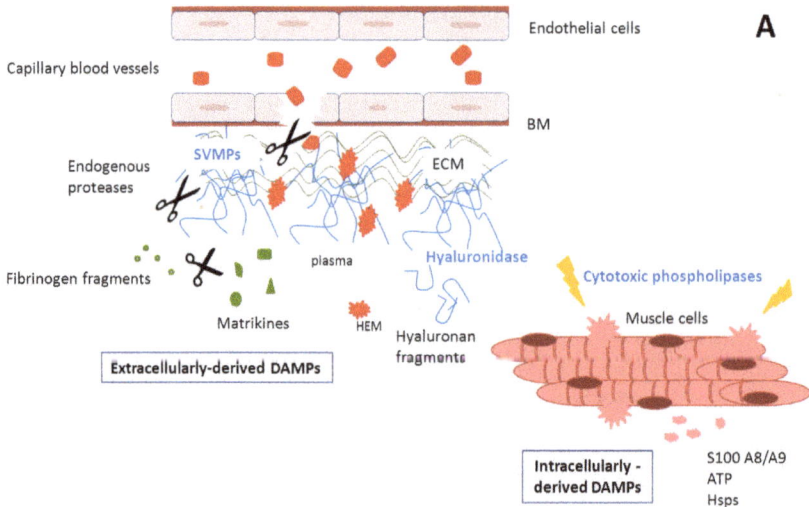

Figure 3. *Cont.*

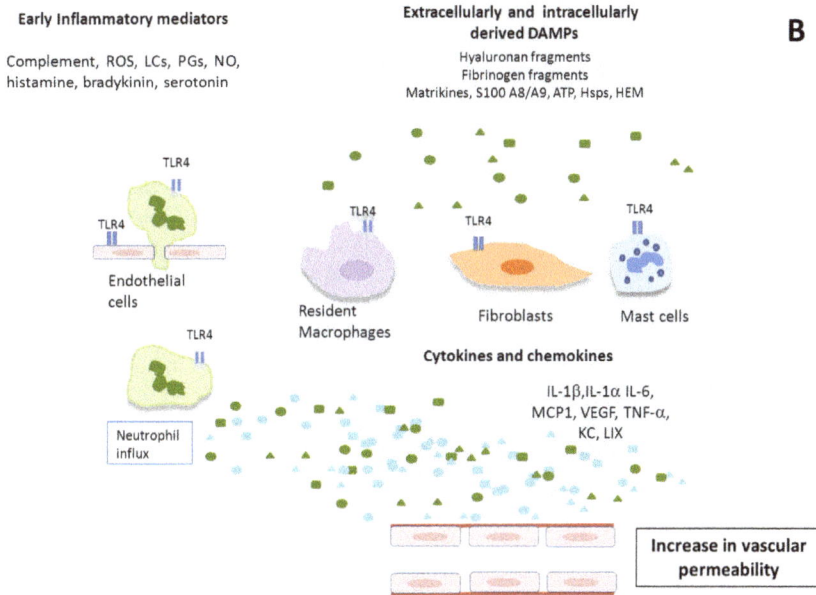

Figure 3. Hypothetical summary of the proposed events occurring in tissue injected with *B. asper* venom. (**A**) Venom toxins, particularly snake venom metalloproteinases (SVMPs), PLA$_2$s and hyaluronidases, induce direct damage to the tissue, especially acute muscle fiber necrosis and degradation of extracellular matrix components, such as those of the basement membrane of capillary vessels, and other matrix molecules, including hyaluronic acid. Acute inflammation ensues, with the release of many types of mediators that promote an increase in vascular permeability, recruitment of inflammatory cells, and pain. Such acute tissue damage is also associated with the release of multiple damage associated molecular pattern molecules (DAMPs), both intracellular and extracellular; (**B**) DAMPs act on a variety of cells, including endothelial cells, other resident cells, and incoming inflammatory leucocytes, to generate diverse tissue responses, such as increase in vascular permeability, and the synthesis of a variety of cytokines and chemokines, which further contribute to the inflammatory scenario in a highly complex interplay. ROS: Reactive oxygen species; LCs: leukotrienes; PGs: prostaglandins; NO: Nitric oxide.

4. Materials and Methods

4.1. Venom

B. asper venom was obtained from more than 40 adult specimens collected in the Pacific region of Costa Rica and kept at the serpentarium of the Instituto Clodomiro Picado. After collection, venoms were pooled, lyophilized, and stored at −20 °C until used.

4.2. Exudate Collection

Groups of five CD-1 mice were injected intramuscularly (i.m.) with 50 µg *B. asper* venom, dissolved in 50 µL of apyrogenic 0.15 M NaCl (saline solution, SS). One and 24 h after venom injection, animals were sacrificed by CO$_2$ inhalation, and an incision was made in the skin overlying the injected muscle, with care taken to avoid contamination. Wound exudate was collected from each animal individually into heparinized capillary tubes, pooled, and centrifuged to eliminate erythrocytes and other cells [15]. For some experiments, exudates were lyophilized and stored at −70° C until use. Exudates were reconstituted in the original volume with apyrogenic saline solution or water for further analyses.

All experiments involving the use of mice were approved by the Institutional Committee for the Care and Use of Laboratory Animals (CICUA) of the University of Costa Rica (CICUA-025-15).

4.3. Increase in Vascular Permeability

Groups of 5 of CD-1 mice (18–20 g) received an intravenous (i.v.) injection of 200 μL of an Evans blue (EB; Sigma-Aldrich, St. Louis, MO, USA) solution (6 mg/mL; 60 mg/kg). Twenty min after EB injection, mice were injected intradermally (i.d.) with 50 μL of exudate samples collected 1 h and 24 h after venom injection in mice. In some groups, 40 μL of polyvalent antivenom (anti-*Bothrops*, *Crotalus* and *Lachesis* antivenom, Instituto Clodomiro Picado, Vázquez de Coronado, Costa Rica) was added to 250 μL of exudate and incubated for 20 min before injection, in order to neutralize the venom toxins that might be present in the exudates. Control groups of mice received either normal mouse plasma or antivenom. One hour after exudate injection or, in the case of controls, plasma or antivenom injection, animals were sacrificed, their skin was removed, and the areas of EB extravasation in the inner side of the skin were measured. To rule out the possibility that an increase in vascular permeability was due to the presence of bacterial lipopolysaccharide, exudate samples were incubated with polymyxin B (15 μg/mL) before testing in the mouse model as previously described [50].

4.4. Effect of Eritoran in Exudate-Induced Vascular Permeability

CD-1 mice (18–20 g) were injected i.m. with 50 μg *B. asper* venom, and exudate was collected 1 h and 24 h after envenomation, as described above. After that, two groups of 5 mice each were pretreated with either Eritoran (E5564; Eisai Co, Ltd., Woodcliff Lake, NJ, USA; 200 μg/100 μL i.v.) or SS (100 μL i.v.). After 1 h, mice received an intradermal injection of 50 μL of exudates collected at either 1 h or 24 h, previously incubated with antivenom to neutralize venom toxins, as described in Section 4.3. Control groups of mice were injected with Eritoran, as described, and then received an intradermal injection of 50 μL of either normal mouse plasma or antivenom. Increase in vascular permeability was assessed, as described above.

4.5. Quantification of Inflammatory Mediators in Exudates by Luminex Assays

For the analysis of the exudate subproteome associated with inflammatory mediators (i.e., cytokines and chemokines), 20 μL of each exudate (1 h or 24 h) were used for Luminex quantitative analysis of 32 analytes (Mouse Premixed Multi-Analyte Kit, R & D systems, Minneapolis, MN, USA) following the methodology recommended by the manufacturer.

4.6. Complete Proteomic Analysis of Wound Exudates and Identification of DAMPs

Lyophilized wound exudate samples were dissolved in water and protein quantification was performed using micro BCA protein assay kit (Thermo Scientific, Waltham, WA, USA). Twenty micrograms of protein from each sample was precipitated with acetone, resuspended in Laemmli buffer under reducing conditions and electrophoresed in a 5%–20% precast acrylamide gel (Bio-Rad, Hercules, CA, USA). The gel was stained with Coomassie Brilliant Blue and lanes were cut into 8 equal sized slices. Gel slices were destained for 3 h and the proteins in the gels were reduced (10 mM dithiothreitol, DTT) and alkylated (50 mM iodoacetamide) at room temperature. Gel slices were then washed with 100 mM ammonium bicarbonate, dehydrated with acetonitrile and dried in a speed vac, followed by in-gel digestion with a solution of Promega modified trypsin (20 ng/μL) in 50 mM ammonium bicarbonate for 30 min on ice. Excess trypsin solution was removed and the digestion continued for 18 h at 37 °C. The resulting tryptic peptides were extracted from gel slices with two 30 μL aliquots of a 50% acetonitrile/5% formic acid solution. These extracts were combined and evaporated to 15 μL for mass spectrometric (MS) analysis.

LC/MS/MS was performed using a Thermo Electron Orbitrap Velos ETD mass spectrometer system. Analytical columns were fabricated in-house by packing 0.5 cm of irregular C18 Beads (YMC Gel ODS-A, 12 nm, I-10-25 um) followed by 7.5 cm Jupiter 10 μm C18 packing material

(Phenomenex, Torrance, CA, USA) into 360 × 75 μm fused silica (Polymicro Technologies, Phoenix, AZ, USA) behind a bottleneck. Samples were loaded directly onto these columns for the C18 analytical runs. 7 μL of the extract was injected, and the peptides were eluted from the column at 0.5 μL/min using an acetonitrile/0.1 M acetic acid gradient (2%–90% acetonitrile over 1 h). The instrument was set to Full MS (*m/z* 300–1600) resolution of 60,000 and programmed to acquire a cycle of one mass spectrum followed by collision-induced dissociation (CID) MS/MS performed in the ion trap on the twenty most abundant ions in a data-dependent mode. Dynamic exclusion was enabled with an exclusion list of 400 masses, duration of 60 s, and repeat count of 1. The electrospray voltage was set to 2.4 kV, and the capillary temperature was 265 °C.

The data were analyzed by database searching using the Sequest search algorithm in Proteome Discoverer 1.4.1 against the Uniprot Mouse database from July 2014. Spectra generated were searched using carbamidomethylation on cysteine as a fixed modification, oxidation of methionine as a variable modification, 10 ppm parent tolerance and 1 Da fragment tolerance. All hits were required to be fully tryptic. The results were exported to Scaffold (version 4.3.2, Proteome Software Inc., Portland, OR, USA) to validate MS/MS based peptide and protein identifications, and to visualize multiple datasets in a comprehensive manner. Proteins shown were identified in Scaffold with a confidence of 95%. The relative abundance of proteins was determined in Scaffold. This is a normalization algorithm that gives a unit less output of Quantitative Value as defined by the software provider, Proteome Software [59] based on averaging the unweighted spectral counts for all of the samples and then multiplying the spectrum counts in each sample by the average divided by the individual sample's sum. The Quantitative Value allows a relative abundance comparison between a specific protein from different samples and relative abundance between proteins for a particular exudate sample.

DAMP proteins were identified in the exudate proteomic analysis, on the basis of the characterization of DAMPs in the literature. A list of DAMPs present in the proteomics analysis of exudate was prepared and the quantitative values of these proteins in 1 h and 24 h exudates were compared.

4.7. Statistical Analyses

The significance of the differences of mean values between experimental groups was assessed by analysis of variance (ANOVA), followed by Tukey-Kramer test to compare pairs of means. A p value < 0.05 was considered significant.

Supplementary Materials: The following are available online at www.mdpi.com/2072-6651/8/12/349/s1, Table S1: List of all proteins identified by proteomics analysis in exudates collected at 1 h and 24 h after injection of *B. asper* venom. Quantitative values for all proteins are included.

Acknowledgments: Thanks are due to Alvaro Segura for performing some of the animal injecting procedures. This study was supported by Vicerrectoría de Investigación, Universidad de Costa Rica (projects 741-B6-125). C.A.N. received a Ph.D. fellowship from CAPES (2012) and a visiting graduate student fellowship from Programa Nacional em Áreas Estratégicas e INCT from CAPES (process number: BEX 2832/15-1).

Author Contributions: J.W.F., J.M.G., T.E. and A.R. conceived and designed the experiments; C.A.N., A.R., T.E., J.K. performed the experiments; C.A.N., J.K., A.R., T.E., C.H., J.M.G. and J.W.F. analyzed the data; J.M.G. and J.W.F. wrote the paper.

Conflicts of Interest: The authors declare no conflict of interest. The founding sponsors had no role in the design of the study; in the collection, analyses, or interpretation of data; in the writing of the manuscript, and in the decision to publish the results.

References

1. Warrell, D.A. Snakebites in Central and South America: Epidemiology, clinical features, and clinical management. In *The Venomous Reptiles of the Western Hemisphere*; Campbell, J.A., Lamar, W.W., Eds.; Cornell University Press: Ithaca, NY, USA, 2004; pp. 709–761.
2. Warrell, D.A. Snake bite. *Lancet* **2010**, *375*, 77–88. [CrossRef]

3. Fox, J.W.; Bjarnason, J.B. New proteases from *Crotalus atrox* venom. *J. Toxicol. Toxin Rev.* **1983**, *2*, 161–204. [CrossRef]
4. Bjarnason, J.B.; Fox, J.W. Hemorrhagic toxins from snake venoms. *J. Toxicol. Toxin Rev.* **1988**, *7*, 121–209. [CrossRef]
5. Gutiérrez, J.M.; Rucavado, A. Snake venom metalloproteinases: Their role in the pathogenesis of local tissue damage. *Biochimie* **2000**, *82*, 841–850. [CrossRef]
6. Fox, J.W.; Serrano, S.M.T. Structural considerations of the snake venom metalloproteinases, key members of the M12 reprolysin family of metalloproteinases. *Toxicon* **2005**, *45*, 969–985. [CrossRef] [PubMed]
7. Gutiérrez, J.M.; Rucavado, A.; Escalante, T.; Lomonte, B.; Angulo, Y.; Fox, J.W. Tissue pathology induced by snake venoms: How to understand a complex pattern of alterations from a systems biology perspective? *Toxicon* **2010**, *55*, 166–170. [CrossRef] [PubMed]
8. Gallagher, P.G.; Bao, Y.; Serrano, S.M.T.; Kamiguti, A.S.; Theakston, R.D.G.; Fox, J.W. Use of microarrays for investigating the subtoxic effects of snake venoms: Insights into venom-induced apoptosis in human umbilical vein endothelial cells. *Toxicon* **2003**, *41*, 429–440. [CrossRef]
9. Baramova, E.N.; Shannon, J.D.; Bjarnason, J.B.; Fox, J.W. Identification of the cleavage sites by a hemorrhagic metalloproteinase in type IV collagen. *Matrix* **1990**, *10*, 91–97. [CrossRef]
10. Baramova, E.N.; Shannon, J.D.; Fox, J.W.; Bjarnason, J.B. Proteolytic digestion of non-collagenous basement membrane proteins by the hemorrhagic metalloproteinase Ht-e from *Crotalus atrox* venom. *Biomed. Biochim. Acta* **1991**, *50*, 763–768. [PubMed]
11. Gutiérrez, J.M.; Rucavado, A.; Escalante, T.; Díaz, C. Hemorrhage induced by snake venom metalloproteinases: Biochemical and biophysical mechanisms involved in microvessel damage. *Toxicon* **2005**, *45*, 997–1011. [CrossRef] [PubMed]
12. Escalante, T.; Ortiz, N.; Rucavado, A.; Sanchez, E.F.; Richardson, M.; Fox, J.W.; Gutiérrez, J.M. Role of collagens and perlecan in microvascular stability: Exploring the mechanism of capillary vessel damage by snake venom metalloproteinases. *PLoS ONE* **2011**, *6*. [CrossRef] [PubMed]
13. Gallagher, P.; Bao, Y.; Serrano, S.M.T.; Laing, G.D.; Theakston, R.D.G.; Gutiérrez, J.M.; Escalante, T.; Zigrino, P.; Moura-Da-Silva, A.M.; Nischt, R.; et al. Role of the snake venom toxin jararhagin in proinflammatory pathogenesis: In vitro and in vivo gene expression analysis of the effects of the toxin. *Arch. Biochem. Biophys.* **2005**, *441*, 1–15. [CrossRef] [PubMed]
14. Teixeira, C.; Cury, Y.; Moreira, V.; Picolo, G.; Chaves, F. Inflammation induced by *Bothrops asper* venom. *Toxicon* **2009**, *54*, 67–76. [CrossRef] [PubMed]
15. Escalante, T.; Rucavado, A.; Pinto, A.F.M.; Terra, R.M.S.; Gutiérrez, J.M.; Fox, J.W. Wound exudate as a proteomic window to reveal different mechanisms of tissue damage by snake venom toxins. *J. Proteome Res.* **2009**, *8*, 5120–5131. [CrossRef] [PubMed]
16. Rucavado, A.; Escalante, T.; Shannon, J.D.; Ayala-Castro, C.N.; Villalta, M.; Gutiérrez, J.M.; Fox, J.W. Efficacy of IgG and F(ab')2 antivenoms to neutralize snake venom-induced local tissue damage as assessed by the proteomic analysis of wound exudate. *J. Proteome Res.* **2012**, *11*, 292–305. [CrossRef] [PubMed]
17. Rucavado, A.; Escalante, T.; Shannon, J.; Gutiérrez, J.M.; Fox, J.W. Proteomics of wound exudate in snake venom-induced pathology: Search for biomarkers to assess tissue damage and therapeutic success. *J. Proteome Res.* **2011**, *10*, 1987–2005. [CrossRef] [PubMed]
18. Newton, K.; Dixit, V.M. Signaling in innate immunity and inflammation. *Cold Spring Harb. Perspect. Biol.* **2012**, *4*, 829–841. [CrossRef] [PubMed]
19. Piccinini, A.M.; Midwood, K.S. DAMPening inflammation by modulating TLR signalling. *Mediat. Inflamm.* **2010**. [CrossRef] [PubMed]
20. Seong, S.-Y.; Matzinger, P. Hydrophobicity: An ancient damage-associated molecular pattern that initiates innate immune responses. *Nat. Rev. Immunol.* **2004**, *4*, 469–478. [CrossRef] [PubMed]
21. Aird, W.C. The role of the endothelium in severe sepsis and multiple organ dysfunction syndrome. *Blood* **2003**, *101*, 3765–3777. [CrossRef] [PubMed]
22. Kumar, P.; Shen, Q.; Pivetti, C.D.; Lee, E.S.; Wu, M.H.; Yuan, S.Y. Molecular mechanisms of endothelial hyperpermeability: Implications in inflammation. *Expert Rev. Mol. Med.* **2009**, *11*. [CrossRef] [PubMed]
23. Khakpour, S.; Wilhelmsen, K.; Hellman, J. Vascular endothelial cell Toll-like receptor pathways in sepsis. *Innate Immun.* **2015**, *21*, 827–846. [CrossRef] [PubMed]

24. Park-Windhol, C.; D'Amore, P.A. Disorders of vascular permeability. *Annu. Rev. Pathol. Mech. Dis.* **2016**, *11*, 251–281. [CrossRef] [PubMed]

25. Allam, R.; Scherbaum, C.R.; Darisipudi, M.N.; Mulay, S.R.; Hägele, H.; Lichtnekert, J.; Hagemann, J.H.; Rupanagudi, K.V.; Ryu, M.; Schwarzenberger, C.; et al. Histones from dying renal cells aggravate kidney injury via TLR2 and TLR4. *J. Am. Soc. Nephrol.* **2012**, *23*, 1375–1388. [CrossRef] [PubMed]

26. Moreira, V.; Teixeira, C.; Borges da Silva, H.; D'Império Lima, M.R.; Dos-Santos, M.C. The crucial role of the MyD88 adaptor protein in the inflammatory response induced by *Bothrops atrox* venom. *Toxicon* **2013**, *67*, 37–46. [CrossRef] [PubMed]

27. Giannotti, K.C.; Leiguez, E.; Moreira, V.; Nascimento, N.G.; Lomonte, B.; Gutiérrez, J.M.; Lopes de Melo, R.; Teixeira, C. A Lys49 phospholipase A2, isolated from *Bothrops asper* snake venom, induces lipid droplet formation in macrophages which depends on distinct signaling pathways and the C-terminal region. *Biomed. Res. Int.* **2013**. [CrossRef]

28. Moreira, V.; Teixeira, C.; Borges da Silva, H.; D'Império Lima, M.R.; Dos-Santos, M.C. The role of TLR2 in the acute inflammatory response induced by *Bothrops atrox* snake venom. *Toxicon* **2016**, *118*, 121–128. [CrossRef] [PubMed]

29. Gutiérrez, J.M.; Rucavado, A.; Chaves, F.; Díaz, C.; Escalante, T. Experimental pathology of local tissue damage induced by *Bothrops asper* snake venom. *Toxicon* **2009**, *54*, 958–975. [CrossRef] [PubMed]

30. Gutiérrez, J.M.; Ownby, C.L. Skeletal muscle degeneration induced by venom phospholipases A 2: Insights into the mechanisms of local and systemic myotoxicity. *Toxicon* **2003**, *42*, 915–931. [CrossRef] [PubMed]

31. Teixeira, C.F.P.; Zamunér, S.R.; Zuliani, J.P.; Fernandes, C.M.; Cruz-Hofling, M.A.; Fernandes, I.; Chaves, F.; Gutiérrez, J.M. Neutrophils do not contribute to local tissue damage, but play a key role in skeletal muscle regeneration, in mice injected with *Bothrops asper* snake venom. *Muscle Nerve* **2003**, *28*, 449–459. [CrossRef] [PubMed]

32. Saravia-Otten, P.; Robledo, B.; Escalante, T.; Bonilla, L.; Rucavado, A.; Lomonte, B.; Hernández, R.; Flock, J.I.; Gutiérrez, J.M.; Gastaldello, S. Homogenates of skeletal muscle injected with snake venom inhibit myogenic differentiation in cell culture. *Muscle Nerve* **2013**, *47*, 202–212. [CrossRef] [PubMed]

33. Herrera, C.; Macêdo, J.K.A.; Feoli, A.; Escalante, T.; Rucavado, A.; Gutiérrez, J.M.; Fox, J.W. Muscle tissue damage induced by the venom of *Bothrops asper*: Identification of early and late pathological events through proteomic analysis. *PLoS Negl. Trop. Dis.* **2016**, *10*. [CrossRef] [PubMed]

34. Gutiérrez, J.M.; Chaves, F.; Cerdas, L. Inflammatory infiltrate in skeletal muscle injected with *Bothrops asper* venom. *Rev. Biol. Trop.* **1986**, *34*, 209–214. [PubMed]

35. Mahdavian Delavary, B.; van der Veer, W.M.; van Egmond, M.; Niessen, F.B.; Beelen, R.H.J. Macrophages in skin injury and repair. *Immunobiology* **2011**, *216*, 753–762. [CrossRef] [PubMed]

36. Brancato, S.K.; Albina, J.E. Wound macrophages as key regulators of repair: Origin, phenotype, and function. *Am. J. Pathol.* **2011**, *178*, 19–25. [CrossRef] [PubMed]

37. Schaefer, L. Complexity of danger: The diverse nature of damage-associated molecular patterns. *J. Biol. Chem.* **2014**, *289*, 35237–35245. [CrossRef] [PubMed]

38. Vénéreau, E.; Ceriotti, C.; Bianchi, M.E. DAMPs from cell death to new life. *Front. Immunol.* **2015**, *6*, 422. [CrossRef] [PubMed]

39. Yang, D.; Wei, F.; Tewary, P.; Howard, O.M.Z.; Oppenheim, J.J. Alarmin-induced cell migration. *Eur. J. Immunol.* **2013**, *43*, 1412–1418. [CrossRef] [PubMed]

40. Turner, N.A. Inflammatory and fibrotic responses of cardiac fibroblasts to myocardial damage associated molecular patterns (DAMPs). *J. Mol. Cell. Cardiol.* **2016**, *94*, 189–200. [CrossRef] [PubMed]

41. Zornetta, I.; Caccin, P.; Fernandez, J.; Lomonte, B.; Gutierrez, J.M.; Montecucco, C. Envenomations by *Bothrops* and *Crotalus* snakes induce the release of mitochondrial alarmins. *PLoS Negl. Trop. Dis.* **2012**, *6*. [CrossRef] [PubMed]

42. Cintra-Francischinelli, M.; Caccin, P.; Chiavegato, A.; Pizzo, P.; Carmignoto, G.; Angulo, Y.; Lomonte, B.; Gutiérrez, J.M.; Montecucco, C. *Bothrops* snake myotoxins induce a large efflux of ATP and potassium with spreading of cell damage and pain. *Proc. Natl. Acad. Sci. USA* **2010**, *107*, 14140–14145. [CrossRef] [PubMed]

43. Deguchi, A.; Tomita, T.; Ohto, U.; Takemura, K.; Kitao, A.; Akashi-Takamura, S.; Miyake, K.; Maru, Y. Eritoran inhibits S100A8-mediated TLR4/MD-2 activation and tumor growth by changing the immune microenvironment. *Oncogene* **2016**, *35*, 1445–1456. [CrossRef] [PubMed]

44. Taylor, K.R.; Trowbridge, J.M.; Rudisill, J.A.; Termeer, C.C.; Simon, J.C.; Gallo, R.L. Hyaluronan fragments stimulate endothelial recognition of injury through TLR4. *J. Biol. Chem.* **2004**, *279*, 17079–17084. [CrossRef] [PubMed]
45. Voelcker, V.; Gebhardt, C.; Averbeck, M.; Saalbach, A.; Wolf, V.; Weih, F.; Sleeman, J.; Anderegg, U.; Simon, J. Hyaluronan fragments induce cytokine and metalloprotease upregulation in human melanoma cells in part by signalling via TLR4. *Exp. Dermatol.* **2008**, *17*, 100–107. [CrossRef] [PubMed]
46. Tsai, S.-Y.; Segovia, J.A.; Chang, T.-H.; Morris, I.R.; Berton, M.T.; Tessier, P.A.; Tardif, M.R.; Cesaro, A.; Bose, S. DAMP molecule S100A9 acts as a molecular pattern to enhance inflammation during influenza A virus infection: Role of DDX21-TRIF-TLR4-MyD88 pathway. *PLoS Pathog.* **2014**, *10*. [CrossRef] [PubMed]
47. Ohashi, K.; Burkart, V.; Flohé, S.; Kolb, H. Cutting edge: Heat shock protein 60 is a putative endogenous ligand of the toll-like receptor-4 complex. *J. Immunol.* **2000**, *164*, 558–561. [CrossRef] [PubMed]
48. Johnson, G.B.; Brunn, G.J.; Kodaira, Y.; Platt, J.L. Receptor-mediated monitoring of tissue well-being via detection of soluble heparan sulfate by Toll-like receptor 4. *J. Immunol.* **2002**, *168*, 5233–5239. [CrossRef] [PubMed]
49. Okamura, Y.; Watari, M.; Jerud, E.S.; Young, D.W.; Ishizaka, S.T.; Rose, J.; Chow, J.C.; Strauss, J.F. The extra domain A of fibronectin activates Toll-like receptor 4. *J. Biol. Chem.* **2001**, *276*, 10229–10233. [CrossRef] [PubMed]
50. Smiley, S.T.; King, J.A.; Hancock, W.W. Fibrinogen stimulates macrophage chemokine secretion through toll-like receptor 4. *J. Immunol.* **2001**, *167*, 2887–2894. [CrossRef] [PubMed]
51. Paiva-Oliveira, E.L.; Ferreira da Silva, R.; Correa Leite, P.E.; Cogo, J.C.; Quirico-Santos, T.; Lagrota-Candido, J. TLR4 signaling protects from excessive muscular damage induced by Bothrops jararacussu snake venom. *Toxicon* **2012**, *60*, 1396–1403. [CrossRef] [PubMed]
52. Leiguez, E.; Giannotti, K.C.; Moreira, V.; Matsubara, M.H.; Gutiérrez, J.M.; Lomonte, B.; Rodríguez, J.P.; Balsinde, J.; Teixeira, C. Critical role of TLR2 and MyD88 for functional response of macrophages to a group IIA-secreted phospholipase A2 from snake venom. *PLoS ONE* **2014**, *9*. [CrossRef] [PubMed]
53. Davalos, D.; Akassoglou, K. Fibrinogen as a key regulator of inflammation in disease. *Semin. Immunopathol.* **2012**, *34*, 43–62. [CrossRef] [PubMed]
54. Saito, S.; Yamaji, N.; Yasunaga, K.; Saito, T.; Matsumoto, S.; Katoh, M.; Kobayashi, S.; Masuho, Y. The fibronectin extra domain A activates matrix metalloproteinase gene expression by an interleukin-1-dependent mechanism. *J. Biol. Chem.* **1999**, *274*, 30756–30763. [CrossRef] [PubMed]
55. Kelsh, R.M.; McKeown-Longo, P.J. Topographical changes in extracellular matrix: Activation of TLR4 signaling and solid tumor progression. *Trends Cancer Res.* **2013**, *9*, 1–13. [PubMed]
56. Järveläinen, H.; Sainio, A.; Wight, T.N. Pivotal role for decorin in angiogenesis. *Matrix Biol.* **2015**, *43*, 15–26. [CrossRef] [PubMed]
57. Wang, L.; Luo, H.; Chen, X.; Jiang, Y.; Huang, Q. Functional characterization of S100A8 and S100A9 in altering monolayer permeability of human umbilical endothelial cells. *PLoS ONE* **2014**, *9*. [CrossRef] [PubMed]
58. Tauseef, M.; Knezevic, N.; Chava, K.R.; Smith, M.; Sukriti, S.; Gianaris, N.; Obukhov, A.G.; Vogel, S.M.; Schraufnagel, D.E.; Dietrich, A.; et al. TLR4 activation of TRPC6-dependent calcium signaling mediates endotoxin-induced lung vascular permeability and inflammation. *J. Exp. Med.* **2012**, *209*, 1953–1968. [CrossRef] [PubMed]
59. Proteome Software. Available online: http://www.proteomesoftware.com (accessed on 18 April 2016).

toxins

MDPI

Article

Snake Venom Metalloproteinases and Their Peptide Inhibitors from Myanmar Russell's Viper Venom

Khin Than Yee [1], Morgan Pitts [2], Pumipat Tongyoo [1], Ponlapat Rojnuckarin [1,*] and Mark C. Wilkinson [2,*]

[1] Faculty of Medicine, Chulalongkorn University, Bangkok 10330, Thailand; khinthanyee@gmail.com (K.T.Y.); kpsppto@ku.ac.th (P.T.)
[2] Institute of Integrative Biology, University of Liverpool, Liverpool L69 7ZB, UK; morganpitts@hotmail.co.uk
*   Correspondence: rojnuckarinp@gmail.com (P.R.); Mwilk@liverpool.ac.uk (M.C.W.);
    Tel.: +66-2-256-4564 (P.R.); +44-151-795-4464 (M.C.W.)

Academic Editors: Jay Fox and José María Gutiérrez
Received: 23 October 2016; Accepted: 23 December 2016; Published: 30 December 2016

Abstract: Russell's viper bites are potentially fatal from severe bleeding, renal failure and capillary leakage. Snake venom metalloproteinases (SVMPs) are attributed to these effects. In addition to specific antivenom therapy, endogenous inhibitors from snakes are of interest in studies of new treatment modalities for neutralization of the effect of toxins. Two major snake venom metalloproteinases (SVMPs): RVV-X and Daborhagin were purified from Myanmar Russell's viper venom using a new purification strategy. Using the Next Generation Sequencing (NGS) approach to explore the Myanmar RV venom gland transcriptome, mRNAs of novel tripeptide SVMP inhibitors (SVMPIs) were discovered. Two novel endogenous tripeptides, pERW and pEKW were identified and isolated from the crude venom. Both purified SVMPs showed caseinolytic activity. Additionally, RVV-X displayed specific proteolytic activity towards gelatin and Daborhagin showed potent fibrinogenolytic activity. These activities were inhibited by metal chelators. Notably, the synthetic peptide inhibitors, pERW and pEKW, completely inhibit the gelatinolytic and fibrinogenolytic activities of respective SVMPs at 5 mM concentration. These complete inhibitory effects suggest that these tripeptides deserve further study for development of a therapeutic candidate for Russell's viper envenomation.

Keywords: snake venom metalloproteinases; snake venom metalloproteinase inhibitors; Russell's viper; viper venom

---

1. Introduction

Russell's viper (*Daboia russelii*) is a medically important snake, variants of which are distributed throughout East and Southeast Asia. A Russell's viper bite has a 60% morbidity rate and the fatality rate is 8.2% in Myanmar [1]. The cause of death includes shock, massive bleeding and renal failure. Snake venom metalloproteinases (SVMPs) play a major role in the local and systemic clinical manifestations: blistering, necrosis and bleeding from the fang marks and incoagulable blood, thrombocytopenia, spontaneous systemic bleeding, hypotension, increased permeability and reduced urine output [2]. Although Russell's viper antivenoms are available, their efficacy in reversal tissue damage, such as acute renal failure, is limited [3]. Novel treatment modalities are required.

Snake venom metalloproteinases (SVMPs) play major roles in pathogenesis of Russell's viper bites [4,5]. SVMPs are categorised into P-I to P-III classes according to their domain organization with different molecular weights [6]: Class I (P-I) contains only a prodomain and a metalloproteinase domain (20–30 kDa); Class II (P-II) contains a prodomain, metalloproteinase domain followed by disintegrin domain (30–60 kDa); Class III (P-III) contains a pro, metalloproteinase, disintegrin-like and cysteine-rich domain (60–100 kDa). There are subclasses in P-II and P-III depending on post-translational modifications. The variation in domain composition between SVMP classes contributes to a wide spectrum of substrate specific proteolytic activity. The active site of the metalloproteinase domain has a consensus $H^{142}EXXHXXGXXH^{152}$ sequence. The catalytic zinc-ion is located at the bottom of the active-site cleft, and tetrahedrally coordinated by His^{142}, His^{146}, His^{152}, and a water molecule anchored to Glu^{143} [7]. The degradation of endothelial cell membrane proteins (integrin and cadherin), basement membrane components (fibronectin, laminin and collagen) and blood coagulation proteins (fibrinogen, factor X and prothrombin) leads to haemorrhage. Generally, P-III SVMPs have more potent haemorrhagic activity than P-I and P-II SVMPs [8].

In order to protect against auto-digestion by SVMPs, snake venom of several species are found to contain natural protease inhibitors: citrate and small peptides. The latter bind selectively to SVMPs in the venom glands to protect glandular tissues and venom factors from self-digestion by SVMPs [9]. Three endogenous peptides: pyroGlu-Lys-Trp (pEKW), pyroGlu-Asn-Trp (pENW) and pyroGlu-Gln-Trp (pEQW) isolated from venom of Taiwan habu (*Trimeresurus mucrosquamatus*) showed an inhibitory action on proteolytic activity of metalloproteinases present in the crude venom [10]. It is reported that these peptide inhibitors regulate the proteolytic activities of their SVMPs in a reversible manner under physiological conditions [11]. Other pit vipers, such as *Bothrops asper* [12] and some rattlesnakes [13], also have venoms containing endogenous tripeptides: pEQW and pENW. African vipers, *Echis ocellatus* and *Cerastes cerastes cerastes*, have pEKW tripeptides. These tripeptides are encoded by tandemly repeating elements from the transcripts which also contain a CNP (C-type natriuretic peptide) homologous sequence at the C-terminus [14]. Two peptides: PtA (pENW) and PtB (pEQW) isolated from venom liquor of *Deinagkistrodon acutus* (Hundred-pacer viper) showed anti-human platelet aggregation activity in vitro and protection effects on ADP-induced paralysis and formation of pulmonary thrombosis in mice [15].

We hypothesized that Myanmar Russell's viper venom might contain endogenous peptides to neutralise its own potent SVMPs. The goal of this research was to purify and identify specific SVMP inhibitors (SVMPIs) from the venom as well as from venom glands and to determine their inhibitory action on purified SVMPs from same source of venom. From the transcriptome of the snake, novel SVMPI transcripts containing tripeptide motifs and ANP (atrial natriuretic peptide) sequences were found. Two tripeptides were purified from the venom and identified as pERW and pEKW. Their effect on biological activities of two SVMPs: RVV-X and Daborhagin from the same venom, purified through newly developed strategy, were examined. Both synthetic peptides showed complete inhibitory action on the gelatinolytic activity of RVV-X and fibrinogenolytic activity of Daborhagin at 5 mM concentration (approximate protease to inhibitor molar ratio of 1:500). The results might contribute to the development of complementary candidates for current antivenom therapy of Russell's viper bites, as well as for novel therapeutic agents for cardiovascular diseases.

2. Results

2.1. Purification and Identification of SVMPs from Myanmar Russell's Viper Venom

2.1.1. Purification of SVMPs

The crude venom of Myanmar Russell's viper (MRV) was initially separated on a Superdex 200 column. Of the three major protein-containing peaks, only the first possessed caseinolytic activity (Figure 1). These fractions were pooled and further purified on a Resource Q anion-exchange column. The proteins resolved into two peaks and the first peak (Q1) exhibited caseinolytic activity (Figure 2a). The purity of proteins in Q1 was determined on both reducing and non-reducing SDS-PAGE. Non-reducing SDS-PAGE of this fraction showed it to contain two bands at 85 kDa and 67 kDa. Under reducing conditions, the main protein bands ran at approximately 67 kDa band and low molecular weight (15–20 kDa) bands were evident.

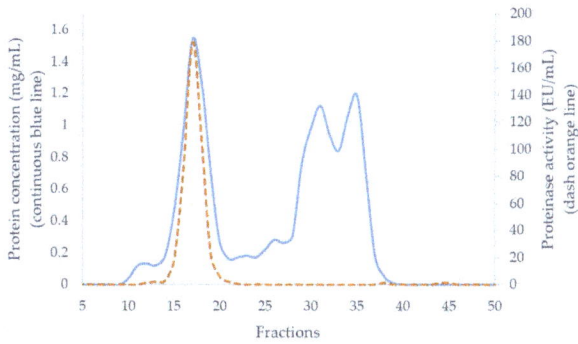

Figure 1. Fractionation of Myanmar Russell's viper crude venom through Superdex 200 gel filtration column (5 × 160 cm). Crude venom was separated in 0.01 M phosphate buffered saline (pH 7.4) at 2 mL/min. Each fraction was 6 mL in volume. The blue continuous line shows the protein concentration (mg/mL) and the orange dashed line shows protease activity (EU/mL) in collection fractions.

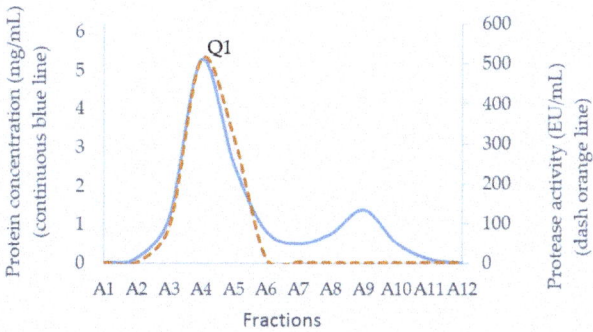

(a)

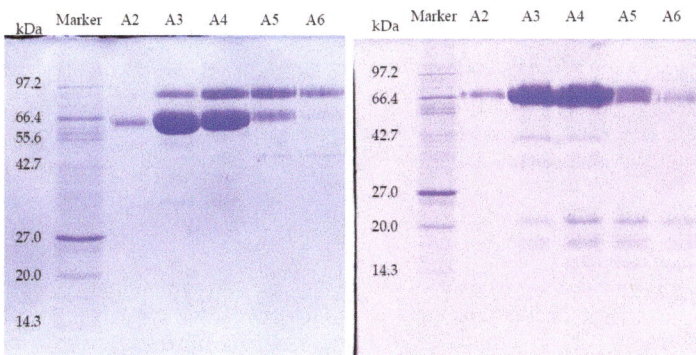

(b)   (c)

Figure 2. Separation of fractions 15–18 from GFC on a Resource Q anion-exchange column (**a**) Chromatography trace showing protein concentration and caseinolytic activity. Peak one (Q1) contained fractions with protease activity; SDS-PAGE of the purified proteins under (**b**) non-reducing; and (**c**) reducing conditions.

This material (Q1) was then subjected to further separation on either HIC for activity studies, or RP-HPLC when proteins were prepared for mass spectrometry. A Phenyl Superose column was used for HIC during which the protein fraction resolved into 2 peaks: H1 (eluted at 13 min), and H2 (eluted at 29 min), respectively (Figure 3a–c). For RP-HPLC, a Phenomenex Luna C4 column was used and again the proteins were separated into 2 peaks (R1 and R2) (Figure 3d–f). SDS-PAGE analysis and activity studies showed H1 to be the same protein as R1 running at 85 kDa under non-reducing conditions, but at 67 kDa with several subunits at 15–20 kDa when reduced. H2 is the same as R2, with a single band at 68 kDa under both reducing and non-reducing conditions.

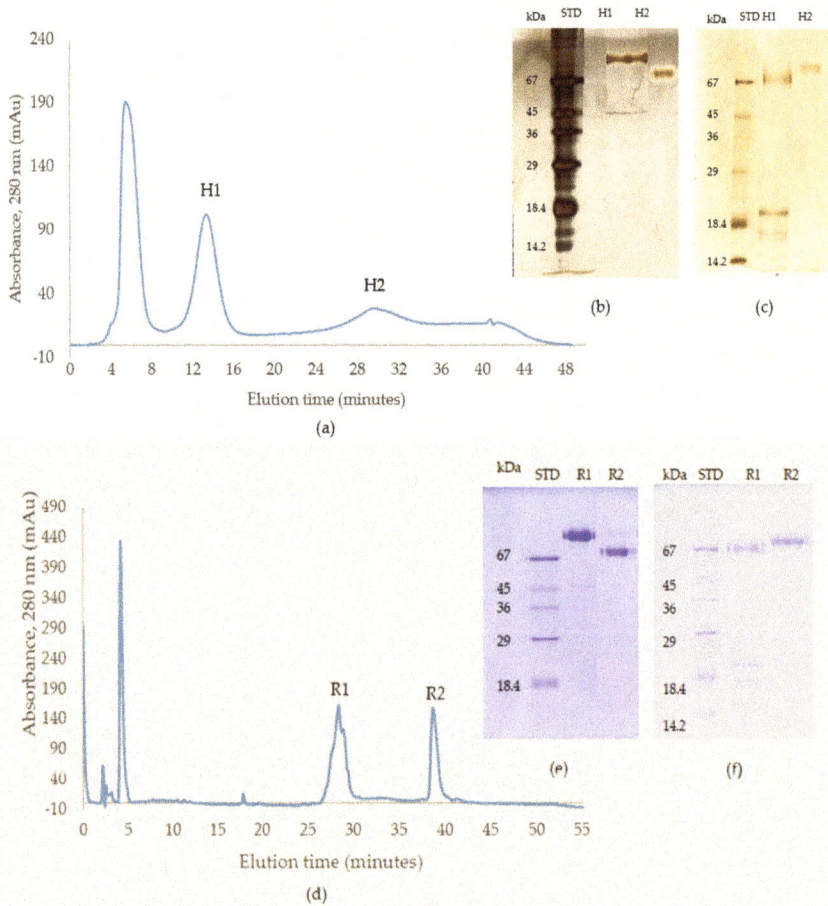

Figure 3. (a) Chromatography of fraction Q1 from the Resource Q column on Phenyl Superose column (HIC) showing protein-containing peaks (H1 and H2); (b) non-reducing; and (c) reducing SDS-PAGE of purified proteins (silver-stained); (d) Chromatography of fraction Q1 from Resource Q column on a C4 RP-HPLC column; Two protein peaks were observed: R1 and R2; SDS-PAGE of purified proteins R1 and R2 (e) under non-reducing conditions; and (f) reducing conditions.

2.1.2. Identification of SVMPs

Both proteins with protease activity were purified by C4 RP-HPLC in preparation for mass spectrometric analysis (see R1 and R2 in Figure 3d). For R1, the protein was reduced and treated with iodoacetamide and digested with trypsin in the presence of 2 M urea and the digest was analysed using LC-ESI-MS/MS (Figure 4). LC-ESI-MS/MS analysis of the tryptic peptides provided sufficient sequence coverage to match to the mature sequence (residues 189–615) of the Eastern Russell's viper (*Daboia russelii siamensis*) RVV-X H chain VM3CX_DABSI (Q7LZ61) (Figure 4). In the same digest mixture we also found matches to RVV-X light chain proteins LC1 SLLC1_DABSI (Q4PRD1) and LC2 SLLC2_DABSI (Q4PRD2) from the same species (data not shown).

```
1   MMQVLLVTISLAVFPYQGSSIILESGNVNDYEVVYPQKVTALPKGAVQQPEQKYEDTMQY   60

61  EFEVNGEPVVLHLEKNKILFSEDYSETHYYPDGREITTNPPVEDHCYYHGRIQNDAHSSA  120

121 SISACNGLKGHFKLRGEMYFIEPLKLSNSEAHAVYKYENIEKEDEIPKMCGVTQTNWESD  180

181 KPIKKASQLVSTSAQFNKIFIELVIIVDHSMAKKCNSTATNTKIYEIVNSANEIFNPLNI  240

241 HVTLIGVEFWCDRDLINVTSSADETLNSFGEWRASDLMTRKSHDNALLFTDMRFDLNTLG  300

301 ITFLAGMCQAYRSVEIVQEQGNRNFKTAVIMAHELSHNLGMYHDGKNCICNDSSCVMSPV  360

361 LSDQPSKLFSNCSIHDYQRYLTRYKPKCIFNPPLRKDIVSPPVCGNEIWEEGEECDCGSP  420

421 ANCQNPCCDAATCKLKPGAECGNGLCCYQCKIKTAGTVCRRARDECDVPEHCTGQSAECP  480

481 RDQLQQNGKPCQNNRGYCYNGDCPIMRNQCISLFGSRANVAKDSCFQENLKGSYYGYCRK  540

541 ENGRKIPCAPQDVKCGRLFCLNNSPRNKNPCNMHYSCMDQHKGMVDPGTKCEDGKVCNNK  600

601 RQCVDVNTAYQSTTGFSQI  619
```

Figure 4. Data from LC-ESI-MS/MS analysis of the tryptic peptides from purified RVV-X. The data was obtained by digesting the R1 fraction from RP-HPLC with trypsin. The prepro-sequence of RVV-X H chain (VM3CX_DABSI; Q7LZ61) annotated to show the peptides (underlined) identified in this analysis. All matched peptides were found within the sequence of the processed protein (residues 189–615).

The protein R2 from RP-HPLC was digested in the same way as R1. In this case MALDI-MS analysis was used to identify tryptic peptides that matched the mass [M + H$^+$] of those predicted from the sequence of Daborhagin-K (Indian Russell's viper) (VM3DK_DABRR) (B8K1W0) (Figure 5). The majority of the most abundant peptides matched the mass of expected tryptic peptides and notably we found many of the same tryptic peptides as did Chen et al. (Table 2 in ref. [16]) for Daborhagin-M.

As a result of this work we can identify R1, H1 as the Myanmar Russell's viper RVV-X and R2, H2 as Myanmar Russell's viper Daborhagin and will to refer them as such from hereon.

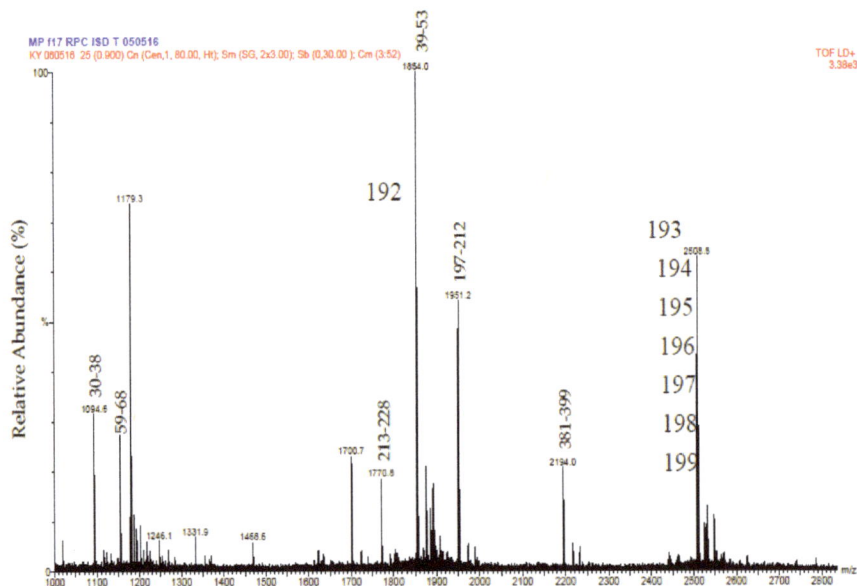

Figure 5. MALDI-MS spectrum of tryptic peptides from purified Daborhagin. The data was obtained by digesting the R2 fraction from RP-HPLC with trypsin. The numbers above the peptides masses indicates the residue numbers for the peptides matched to the sequence of Daborhagin K (VM3DK_DABRR) (B8K1W0). All *m/z* values are for the M + H⁺ ions. The ions at 1854 and greater have been labeled with the *m/z* value for the ion containing one carbon as the C¹³ isotope.

2.2. Analysis of SVMPI Transcripts from Myanmar Russell's Viper Transcriptome

From the transcriptome of Myanmar Russell's viper venom glands, a total of 4 contigs were annotated as the Snake Venom Metalloproteinase Inhibitors (SVMPI). The conceptually translated proteins were aligned with those transcripts of African vipers, *Echis ocellatus* (A8YPR6) and *Cerastes cerastes cerastes* (A8YPR9). The signal peptides are highly similar and a new tripeptide QRW motif in addition to a QKW motif was found in the MRV transcripts. The tripeptides were flanked by the conserved PXXQ(K/R)WXXP motifs. The SVMPI transcripts of MRV also contained a conserved poly-Gly (pG) motif instead of the poly-His poly-Gly (pHpG) seen in *E. ocellatus* SVMPI transcripts. Moreover, the C-terminal portion of the SVMPI transcripts of MRV have an atrial natriuretic peptide (ANP) domain in place of the C-type natriuretic peptide (CNP) domain in the two African viper SVMPI transcripts (Figure 6).

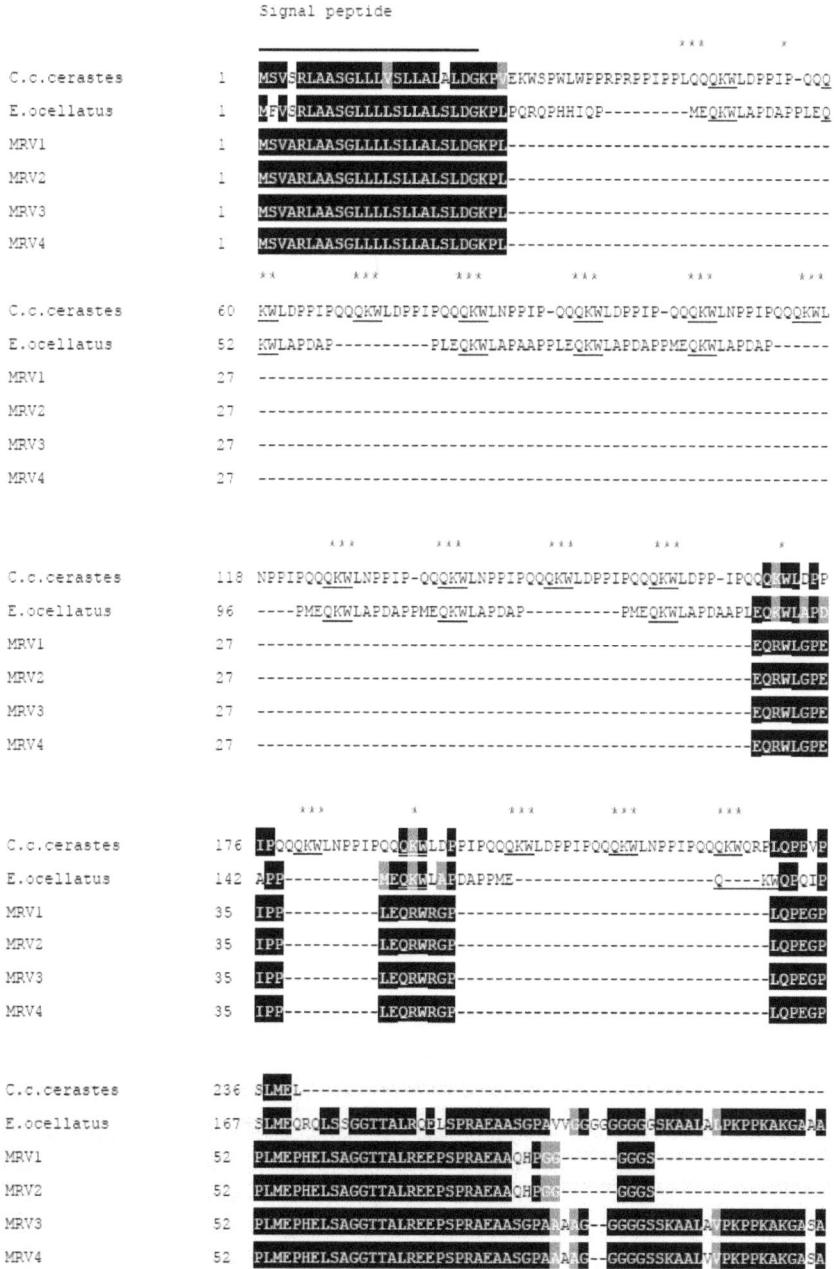

Signal peptide

```
                                                        * * *           *
C.c.cerastes    1   MSVSRLAASGLLLVSLLAIALDGKPVEKWSPWLWPPRPRPPIPPLQQQKWLDPPIP-QQQ
E.ocellatus     1   MFVSRLAASGLLLLSLLALSLDGKPLPQRQPHHIQP---------MEQKWLAPDAPPLEQ
MRV1            1   MSVARLAASGLLLLSLLALSLDGKPL-----------------------------------
MRV2            1   MSVARLAASGLLLLSLLALSLDGKPL-----------------------------------
MRV3            1   MSVARLAASGLLLLSLLALSLDGKPL-----------------------------------
MRV4            1   MSVARLAASGLLLLSLLALSLDGKPL-----------------------------------

                    * *       * * *        * * *          * * *        * * *         * * *
C.c.cerastes   60   KWLDPPIPQQQKWLDPPIPQQQKWLNPPIP-QQQKWLDPPIP-QQQKWLNPPIPQQQKWL
E.ocellatus    52   KWLAPDAP----------PLEQKWLAPAAPPLEQKWLAPDAPPMEQKWLAPDAP------
MRV1           27   ------------------------------------------------------------
MRV2           27   ------------------------------------------------------------
MRV3           27   ------------------------------------------------------------
MRV4           27   ------------------------------------------------------------

                        * * *         * * *          * * *         * * *            *
C.c.cerastes  118   NPPIPQQQKWLNPPIP-QQQKWLNPPIPQQQKWLDPPIPQQQKWLDPP-IPQQQKWLDPP
E.ocellatus    96   ----PMEQKWLAPDAPPMEQKWLAPDAP----------PMEQKWLAPDAAPLEQRWLAPD
MRV1           27   ---------------------------------------------------EQRWLGPE
MRV2           27   ---------------------------------------------------EQRWLGPE
MRV3           27   ---------------------------------------------------EQRWLGPE
MRV4           27   ---------------------------------------------------EQRWLGPE

                        * * *         *          * * *         * * *        * * *
C.c.cerastes  176   IPQQQKWLNPPIPQQQKWLDPPIPQQQKWLDPPIPQQQKWLNPPIPQQQKWQRPLQPEVP
E.ocellatus   142   APP---------MEQKWLAEDAPPME-----------------Q----KWQPQIP
MRV1           35   IPP---------LEQRWRGP--------------------------------LQPEGP
MRV2           35   IPP---------LEQRWRGP--------------------------------LQPEGP
MRV3           35   IPP---------LEQRWRGP--------------------------------LQPEGP
MRV4           35   IPP---------LEQRWRGP--------------------------------LQPEGP

C.c.cerastes  236   SLMPL-------------------------------------------------------
E.ocellatus   167   SLMEQRQLSSGGTTALRQELSPRAEAASGPAVVGGGGGGGGSKAAIALPKPPKAKGAAA
MRV1           52   PLMEPHELSAGGTTALREEPSPRAEAAQHPGG------GGGS-----------------
MRV2           52   PLMEPHELSAGGTTALREEPSPRAEAAQHPGG------GGGS-----------------
MRV3           52   PLMEPHELSAGGTTALREEPSPRAEAASGPAAAG--GGGGSSKAALAVPKPPKAKGASA
MRV4           52   PLMEPHELSAGGTTALREEPSPRAEAASGPAAAAG--GGGGSSKAALVVPKPPKAKGASA
```

Figure 6. *Cont.*

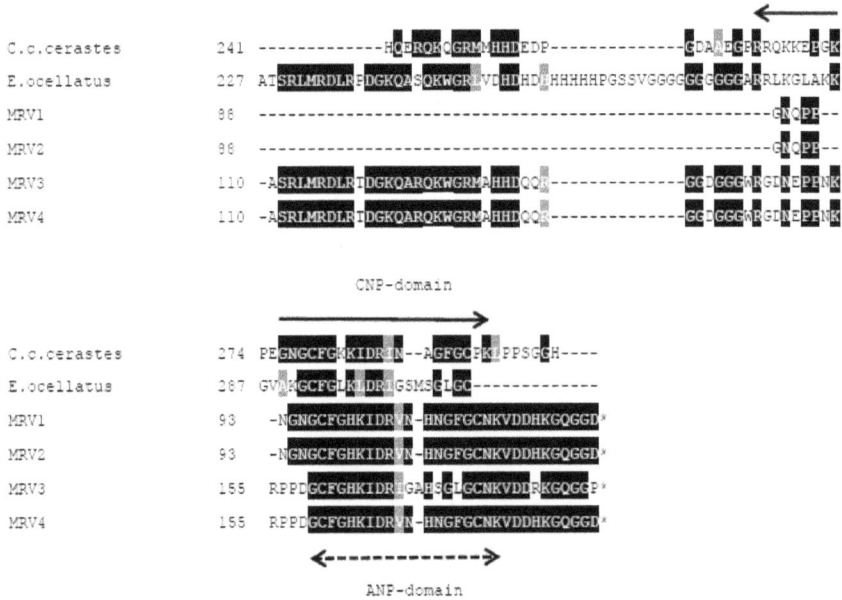

Figure 6. Multiple sequence alignment of the polypeptide encoded by Myanmar Russell's viper SVMPI transcripts (MRV1-4) with those of two African vipers [*C. c. cerastes* (A8YPR9) and *E. ocellatus* (A8YPR6)]. The signal peptides are denoted by a solid line. The active tripeptides are underlined and identified with three asterisks. The varied residue is identified by a single asterisk. The CNP domains are indicated with a solid arrowed line and ANP domains with a dashed arrowed line.

2.3. Purification and Identification of Tripeptides

The low molecular fractions from Superdex 200 chromatography were analysed using C18 RP-HPLC. Fraction 48 was found to contain the highest concentration of the tripeptides. Upon RP-HPLC analysis of this fraction, two peaks (A_p and B_p) eluted close together at 31–33 min (Figure 7). These peaks possessed the same elution time as that of two synthetic peptides pEKW (peak A_s) and pERW (peak B_s), respectively. RP-HPLC analysis of mixtures of natural and synthetic tripeptides showed perfect co-chromatography. The purified endogenous tripeptides were then analysed using ESI-MS. The resultant spectra of peak A_p showed a strong M + H$^+$ ion at m/z 444.2, (the predicted monoisotopic mass of pEKW is 443.2). Analysis of peak B_p, also showed a strong M + H$^+$ ion at m/z 472.2(the predicted monoisotopic mass of pEKW is 471.2) (Figure 8). MS/MS analysis of these tripeptides produced a set of fragment ions consistent with their expected amino acid sequence (data not shown).

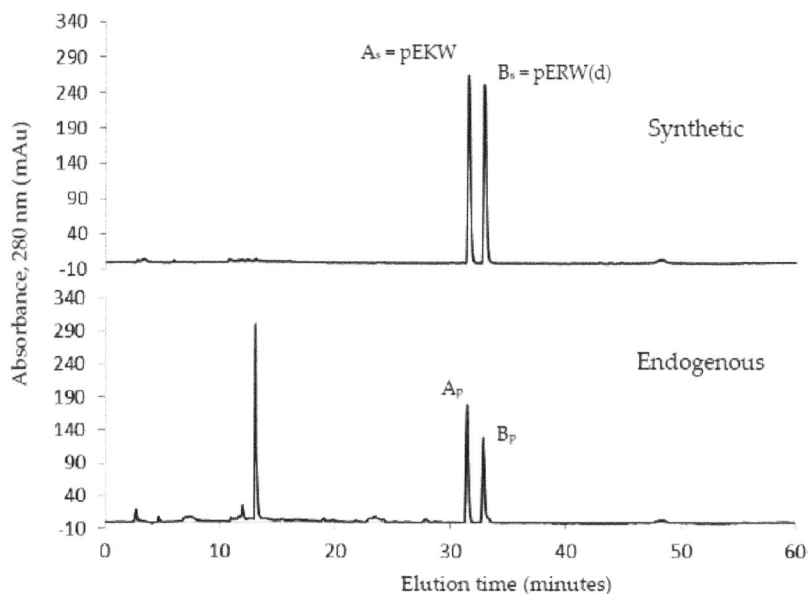

Figure 7. C18 RP-HPLC analysis of synthetic tripeptides, pEKW and pERW (1 µg of each) (**upper panel**); and fraction 48 (200 µL) (**lower panel**) obtained from gel filtration chromatography of crude venom. Peak A_s and peak B_s represent the two synthetic tripeptides. Peak A_p and peak B_p represent the two tripeptides from Fraction 48.

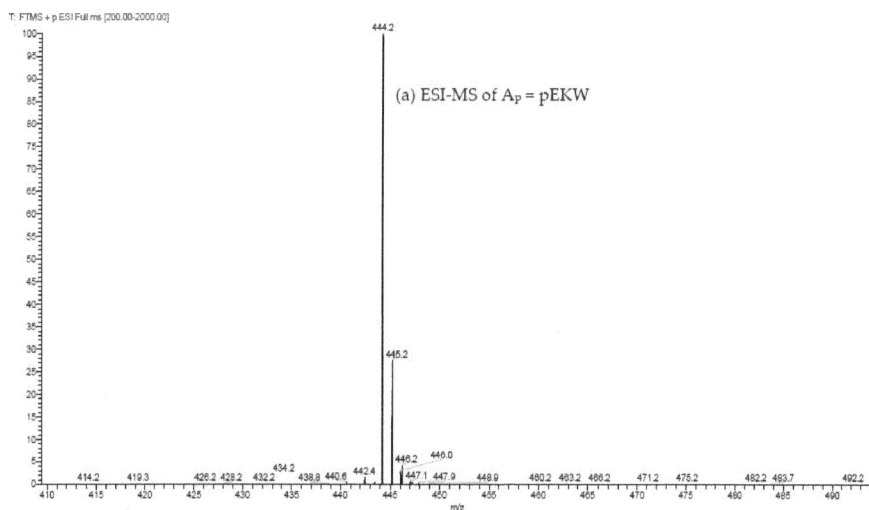

Figure 8. *Cont.*

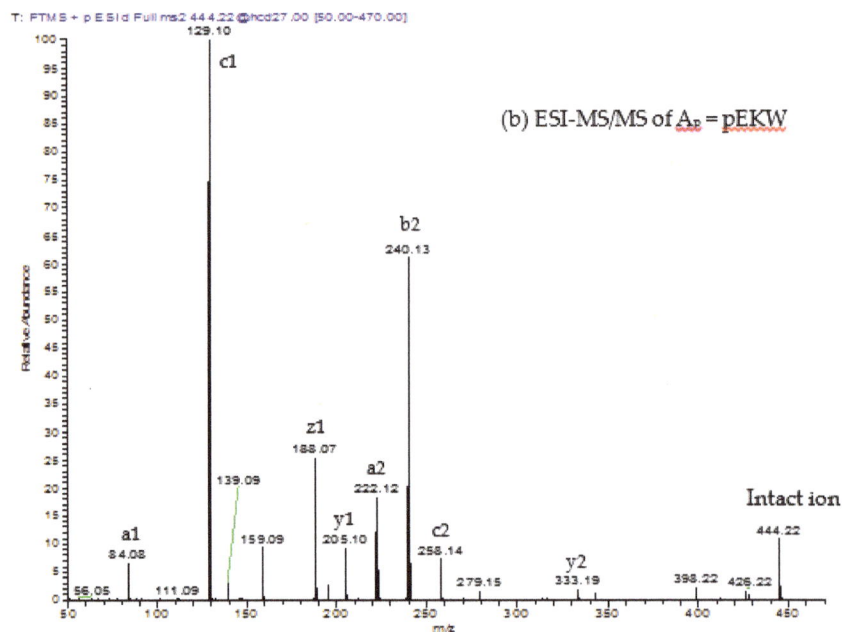

(b) ESI-MS/MS of A_p = pEKW

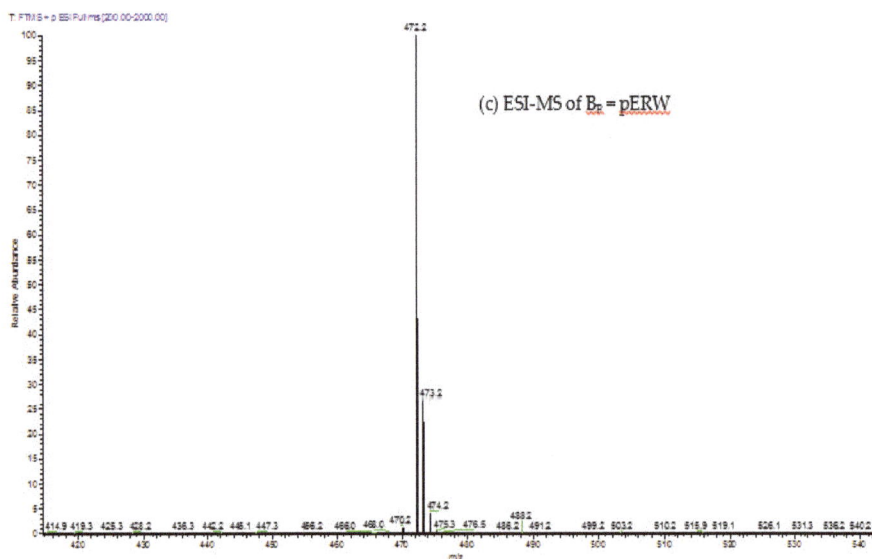

(c) ESI-MS of B_p = pERW

Figure 8. *Cont.*

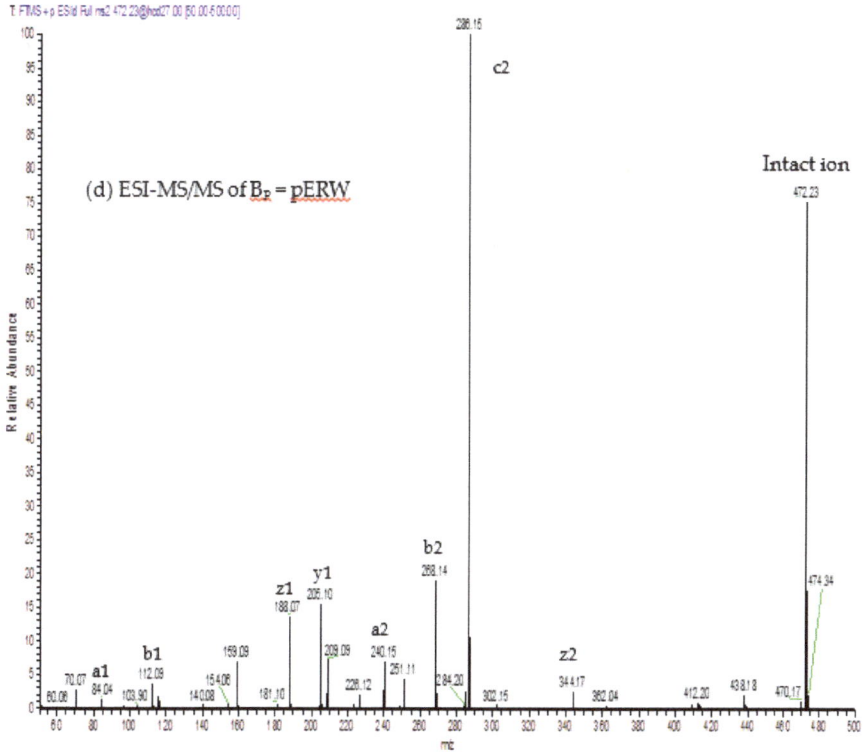

Figure 8. ESI-MS and ESI-MS/MS spectra of (**a,b**) peak A_p; and (**c,d**) peak B_p isolated via RP-HPLC of low molecular material obtained from GFC of crude MRV venom; (**a,c**) ESI-MS spectra. The values indicated are for the M + H$^+$ ions. These are within 0.05 Da of the predicted values for pEKW and pERW (monoisotopic masses are 443.2 and 471.2 respectively); (**b,d**) ESI-MS/MS spectra. The predicted a, b, c, y and z ions are indicated above the mass values.

2.4. Characterization of RVV-X and Daborhagin

The purified proteins RVV-X and Daborhagin from HIC were used for characterization of their gelatinolytic and fibrinogenolytic activities. Using a caseinolytic assay, both proteins were shown to be completely inhibited with metal chelators such as EDTA, 1,10-phenanthroline and citrate.

The gelatinolytic activity was analysed by zymography. On the gelatin zymogram (0.25% gelatin), RVV-X showed a clear band but Daborhagin did not show any gelatin degradation (Figure 9).

The fibrinogenolytic activity of the two proteins was determined using 12% SDS-PAGE after incubation with fibrinogen solution for different times at 37 °C. Daborhagin digested the α-chain of human fibrinogen within 1 h of incubation. RVV-X only revealed fibrinogenolytic activity after an overnight incubation (Figure 10).

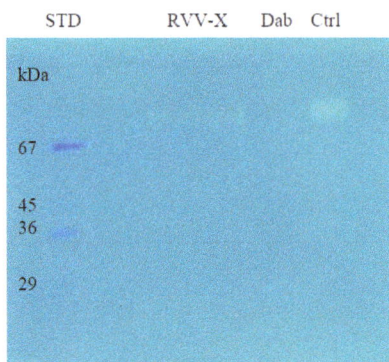

Figure 9. Gelatinolytic activity of RVV-X and Daborhagin on 0.25% gelatin zymogram after 48 h-incubation at 37 °C. Dab: Daborhagin; Ctrl: combined sample of two purified proteins. RVV-X, but not Daborhagin, showed gelatinolytic activity.

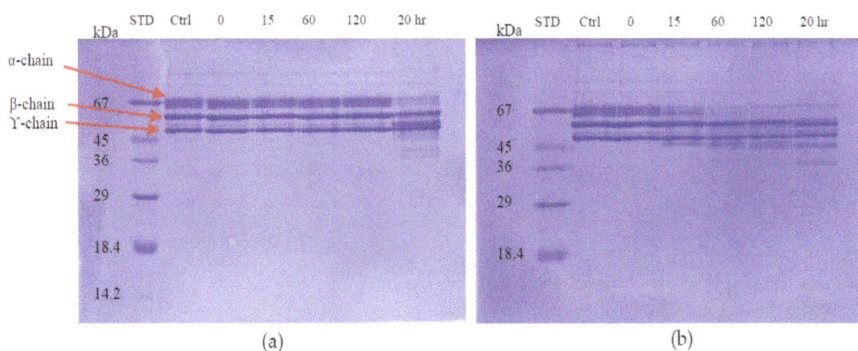

Figure 10. Fibrinogenolytic activity of RVV-X and Daborhagin. 10 µg/mL purified enzyme was incubated with 1 mg/mL fibrinogen solution at 0, 15, 60, 120 min and 20 h-incubation. Sample: (**a**) RVV-X; (**b**) Daborhagin. Ctrl: fibrinogen control; STD: molecular weight markers.

2.5. Inhibitory Assay with Synthetic Tripeptides

2.5.1. Effect of Synthetic Tripeptides on the Gelatinolytic Activity of RVV-X

The gelatinolytic activity of RVV-X was completely inhibited by both synthetic tripeptides pEKW and pERW at 5 mM concentration when incubated with 1 mg/mL gelatin solution at 37 °C (Figure 11). The α-chains (100 kDa & 130 kDa), β-chain (200 kDa) and γ-chain (300 kDa) of gelatin were totally degraded by RVV-X in a 20 h-incubation, whereas these gelatin subunits were still intact in samples containing tripeptides or EDTA after 20 h of incubation. The tripeptide pEEW was included in the assay to test the specificity of amino acid residue in the second position of the tripeptides.

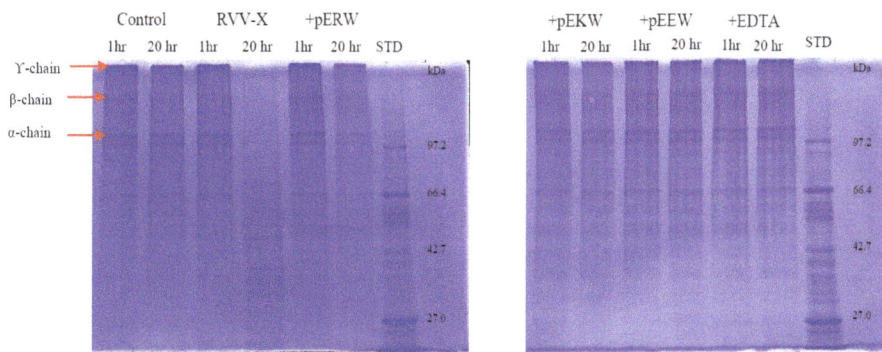

Figure 11. Gelatinolytic activity of RVV-X. Gelatin (1 mg/mL) was incubated with 10 μg/mL RVV-X for 1 h and 20 h at 37 °C, either with or without EDTA or synthetic tripeptides: pERW, pEKW, pEEW. Control = reduced gelatin, STD = molecular weight markers.

2.5.2. Effect of Synthetic Tripeptides on the Fibrinogenolytic Activity of Daborhagin

The fibrinogenolytic activity of Daborhagin was completely inhibited by both synthetic tripeptides pEKW and pERW at 5 mM concentration when incubated with 1 mg/mL fibrinogen at 37 °C (Figure 12).

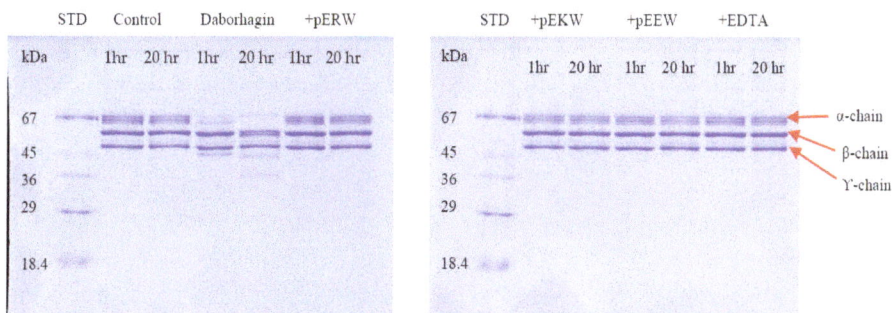

Figure 12. Fibrinogenolytic activity of Daborhagin. Fibrinogen (1 mg/mL) was incubated with 10 μg/mL Daborhagin for 1 h and 20 h at 37 °C with or without EDTA or synthetic tripeptides: pERW, pEKW, pEEW. Control = reduced fibrinogen, STD = molecular weight markers.

3. Discussion

In the current study, we have developed a new method which can be used to simultaneously isolate the two SVMPs, RVV-X and Daborhagin, from Myanmar Russell's viper venom. The relative amounts of these enzymes in the venom were determined. In addition, four novel RNA sequences of SVMP inhibitor (MRV1-4) were derived from the venom gland transcriptome. These sequences are different from those of previously reports in other snakes. For the first time in studies of Russell's viper venom, two tripeptide SVMP inhibitors, pERW and pEKW have been isolated. Evidence for the complete inhibition of RVV-X and Daborhagin activities by these tripeptides is presented to support our hypothesis.

Russell's viper is a venomous species of the South-East Asian region. The clinical manifestations of its bites reflect the high content of proteases such as snake venom serine proteases and snake venom metalloproteinases (SVMPs). It has been shown that the SVMPs comprise approximately 11% to 65% of the total protein in the Viperidae venoms [17]. In Myanmar Russell's viper, SVMPs contribute to 20% of the crude venom (data not shown) and Class III SVMPs are found to be the major component.

In comparison with other species, the Myanmar species have 6–7 times more Daborhagin than Indian species [16] and SVMPs, mainly RVV-X, in Sri Lankan species comprise just 6.9% of the crude venom [18]. Thus, it can be noted that the Myanmar venom contains greater amounts of SVMPs than that of the Indian and Sri Lankan species. The variations in types and amounts of SVMPs in venom among different subspecies of Russell's viper might be due to diversity in their prey at different locations and this could lead to the dissimilar severity or clinical presentations of snakebite patients.

Russell's viper venom factor X activator (RVV-X) is a well-characterised Class III metalloproteinase (formally known as Class IV) which specifically activates coagulation factor X by hydrolysis of an Arg-Ile bond in factor X. It is a glycoprotein consisting of a heavy chain (α-chain, 57.6 kDa) and two light chains (β- and γ-chains, 19.4 kDa and 16.4 kDa) linked by disulfide bonds [19]. In addition to proteolytic activity on factor X and IX, RVV-X also inhibits collagen- and ADP-stimulated platelet aggregation [20] and has a strong affinity for protein S [21]. Factor X activators are also found in *Vipera lebetina* (blunt-nosed viper) in which it exhibits specific proteolytic activity towards human factor X and also factor IX, but it is not active against prothrombin nor fibrinogen [22].

In the present study, the purified RVV-X was shown to be composed of a heavy chain (67 kDa) and two light chains (20 kDa and 15 kDa). The two thin bands on SDS-PAGE at around 15 kDa level suggested that the γ-light chain in Myanmar species might exist as 2 forms, likely due to either amino acid variation or differences in N-glycosylation. Our experiments showed that MRV RVV-X possesses hydrolytic activity to gelatin (Type I collagen, bovine), which had not been characterised before for RVV-X.

Another potent Class III SVMP, Daborhagin, composed of metalloproteinase, disintegrin and cysteine-rich domains, was also purified from MRV venom. The Daborhagin-M from Myanmar Russell's viper venom specifically digested the α-chain of fibrinogen, fibronectin and type IV collagen in vitro and exhibited haemorrhagic [16], edema inducing and myonecrotic activity in mice [23]. In our studies, a 67 kDa metalloproteinase was isolated and matched to Daborhagin-K from Indian species using mass spectrometric analysis of tryptic peptides. This MRV Daborhagin exhibited potent α-fibrinogenolytic activity, but did not digest gelatin.

In the current purification strategy, the two SVMPs were co-purified initially, but then could be separated from each other using either hydrophobic interaction chromatography or RP-HPLC. Better resolution was evident on RP-HPLC, and the presence of multiple forms of RVV-X was indicated by the irregularity of the RVV-X RP-HPLC peak, suggesting heterogeneity of the protein (R1, Figure 3d). Two isoforms of the heavy chain and 6 isoforms of the light chain from RVV-X have been revealed on 2-D electrophoresis in the proteomic study of Risch, M et al. in the same species [24].

New SVMPI transcripts from Myanmar Russell's viper were discovered containing novel two inhibitory tripeptides, QKW and QRW. The tripeptide sequences are found in the same transcript as natriuretic peptide sequences, as is the case in African vipers. This assortment of different peptide sequences in the same transcript could be related to independent evolution of toxin genes in snakes. The conserved proline residues in the consensus sequence PXXQ(K/R)WXXP might be a signal point for cleavage of tripeptides from transcripts. The mechanism for release of tripeptides from their transcripts is still unknown. These tripeptides and natriuretic peptides are observed separately in venom, although they are encoded from the same transcript. Since ANP is homologous to hormone, it might be processed near the effective cells. The release and modification of tripeptides [25] might probably occur during the exocytosis process at an earlier stage than the natriuretic peptides [26].

The aforementioned inhibitory tripeptides were purified from the MRV venom as their pyroglutamate forms, pEKW and pERW. These were identified using RP-HPLC (co-chromatography with synthetic peptides), LC-ESI-MS analysis of intact mass and LC-ESI-MS/MS sequencing. Although the tripeptide pEKW purified from MRV venom has been found in other snake species, such as *Trimeresurus mucrosquamatus* [10], *Echis ocellatus*, and *Cerastes cerastes cerastes* [14], the tripeptide pERW purified here has not been found in the venom of any other snake species.

The synthetic tripeptides pERW, pEKW and pEEW showed complete inhibition of the gelatinolytic activity of RVV-X and of the fibrinogenolytic activity of Daborhagin at 5 mM concentration of each inhibitor. Non-selective inhibition of all three synthetic peptides on biological activities of SVMPs reflects the importance of the first pyroGlu and the final tryptophan residue in the blocking mechanism at the active site of SVMP. The crystal structure of TM-3 (a SVMP from *Trimeresurus mucrosquamatus*) bound to tripeptide inhibitors (a proteinase and inhibitors model from Taiwan habu) revealed that the inhibitor Trp residue deeply inserts into the S-1 pocket of the protease and provides a greater inhibition than other smaller amino acids. Similarly, the pyro-ring of the inhibitor is required for fitting into the S-3 position of the protease and the activity of inhibitor becomes weaker in the absence of pyro-ring. The native middle residue is also position-specific to the S-2 site [11]. Tripeptides from different species share same first (pyroGlu-) and third (tryptophan) residues. The variability of the middle residue may be dependent on species variation of the SVMPs.

4. Conclusions

In summary, we have isolated and identified two major SVMPs and two endogenous tripeptides from Myanmar Russell's viper venom. The two synthetic tripeptides showed specific inhibition against the fibrinogenolytic and gelatinolytic activities of the SVMPs. These findings may provide a means to explore potential drug design in using these tripeptide inhibitors, or analogues of theses as alternative or additional tools in treating the toxic effects of envenomation, as well as in thrombosis and related diseases.

5. Materials and Methods

5.1. Venoms and Venom Glands

Lyophilised crude venom was obtained from No. 1, Myanmar Pharmaceutical Factory, Yangon, Myanmar. The salivary glands from Myanmar Russell's viper (*Daboia russelii siamensis*) were dissected in the Snake Farm, MPF, Yangon, Myanmar and used for RNA-Seq (RNA sequencing). The experimental plan was approved by the Animal Care and Use Committee, Chulalongkorn University (CU-ACUC) (No. 17/2558). Synthetic tripeptides (98% purity) were purchased from Severn Biotech Ltd., Worcestershire, UK and supplied with data from MS analysis to confirm their masses to be within 0.4% of the predicted values.

5.2. Transcripts Analysis of SVMPIs

Next-generation sequencing of mRNA from Myanmar Russell's viper (2 adult males and 2 adult females) venom glands was performed on an Illumina HiSeq2000 platform. De novo assembly was performed using Trinity (r20140717). Annotation of SVMPI transcripts were archived through Blastn searches against the collected NCBI nucleotide database with search words "venom" and "serpents". The annotations with a high score at the top hit list were picked up.

The SVMPI transcripts were further analysed with Blastx and ORF finder for final best annotation and identification of the full-length transcript. The alignment of translated SVMPI sequences with those from other snakes were performed by using Clustal Omega followed by shading with BOXSHADE 3.21 (K. Hofmann, Koeln, Germany & M. Baron, Surrey, UK).

5.3. Protein Concentration

Protein concentration was determined by using Bradford reagent (BioRad, Hemel Hempstead, UK). The absorbance was measured at 595 nm and the calibration curve was prepared with a bovine gamma globulin standard (0–1.5 mg/mL).

5.4. Purification of SVMPs

All chromatographic procedures were performed on either a Bio-Cad Vision Workstation (GFC) or a GE Healthcare AKTA System (anion-exchange, HIC and RP-HPLC).

5.4.1. Gel Filtration Chromatography (GFC)

Crude venom (100 mg) dissolved in 5 mL of 0.01 M phosphate buffered saline (pH 7.4) was applied to a Superdex 200 column (5 × 160 cm) pre-equilibrated with the same buffer. Elution was carried out with the same buffer. The flow rate was 2 mL/min and 6 mL fractions were collected. The fractions having metalloproteinase activity (fractions 15–18) were combined for further purification. The fractions eluting near the total volume were analysed for tripeptides using RP-HPLC with subsequent MS analysis.

5.4.2. Anion-Exchange Chromatography

The SVMP-containing sample obtained from GFC was applied to a Resource Q anion-exchange column (6 mL) pre-equilibrated with 0.05 M Tris-Cl buffer (pH 8.0). Elution was achieved with a linear NaCl gradient from 0 to 0.5 M in the same buffer at a flow rate of 0.6 mL/min and 1.8 mL fractions were collected. Elution was monitored at 280 nm.

5.4.3. Hydrophobic Interaction Chromatography (HIC)

To further purify the SVMPs for activity measurements, fractions from Resource Q were loaded onto a Phenyl Superose column (1 mL) equilibrated in 2.5 M NaCl, 50 mM Tris-Cl, pH 7.8. Samples in Tris-Cl were adjusted to 2.5 M in NaCl and were centrifuged at 10,000× *g* for 5 min before loading onto the column. Separation was achieved by a 30 min-gradient of 2.5–0 M NaCl in 50 mM Tris-Cl, pH 7.0, using a flow rate of 0.25 mL/min. Elution was monitored at 280 nm and 0.25 mL fractions were collected.

5.4.4. Reversed Phase High Performance Liquid Chromatography (RP-HPLC)

For MS analysis, SVMPs were purified using RP-HPLC rather than HIC. Fractions from Resource Q chromatography were made up to 0.2% (*v*/*v*) in TFA, centrifuged at 10,000× *g* for 5 min and then applied to Phenomenex Aeris C4 column (150 × 2.1 mm, 5 micron). The proteins were separated in a two-part acetonitrile gradient in 0.08% TFA: 0%–40% over 25 min then 40%–65% over 5 min and elution was monitored at 280 nm. The flow rate was 0.15 mL/min and 0.25 mL fractions were collected.

5.5. Purification of Tripeptides

Fractions from GFC suspected to contain small molecular weight components were made up to 0.2% (*v*/*v*) in TFA, centrifuged at 10,000× *g* for 5 min and applied to Phenomenex Luna C18 RP-HPLC column (100 × 2.1 mm) equilibrated in 0.08% TFA. The components were separated at 0.15 mL/min with a three-part acetonitrile gradient in 0.08% TFA: 0%–12% over 5 min, 12%–28% over 50 min and then 28%–65% over 10 min. Elution was monitored at 280 nm.

5.6. Mass Spectrometric Analyses

Putative RVV-X (10 µg of R1 from RP-HPLC) was reduced, treated with iodoacetamide and digested with 1.0 µg trypsin in the presence of 2 M urea. The resulting peptides were analysed by LC-ESI-MS/MS using an Acquity UPLC CSH Peptide C18 RP column (Waters, Milford, MA, USA) connected to a Q-Exactive (ThermoFisher, Northumberland, UK) MS instrument. Peaks Studio 8.0 (BSI, Waterloo, Canada) was used to analyse the resulting MS/MS data against the sequences for Eastern Russell's viper RVV-X H chain VM3CX_DABSI (Q7LZ61) and light chains LC1 SLLC1_DABSI (Q4PRD1) and LC2 SLLC2_DABSI (Q4PRD2).

Putative Daborhagin (5 μg of R2 from RP-HPLC) was reduced, treated with iodoacetamide and digested with 0.5 μg trypsin in the presence of 2 M urea. The resulting peptides were desalted and mass spectrometric analysis was performed using a MALDI-TOF instrument (Waters-Micromass, Milford, MA, USA). Samples were analysed by mixing a 1 μL solution of the tryptic peptides with an equal volume of 5.7 mg/mL α-cyano-4-hydroxycinnamic acid in 60% acetonitrile/0.1% trifluoroacetic acid and laying this onto a dried bed of 1 μL of 25 mg/mL α-cyano-4-hydroxycinnamic acid. Laser energy was set at 25% and detector voltage 1800 V. Ion spectra were collected in the mass range of 1000–3000 Da. Data analysis was performed using MassLynx (Waters, Milford, MA, USA). The tryptic peptide masses obtained were matched manually with those predicted (using ExPASy Peptide Mass) of a sequence for Daborhagin-K (VM3DK_DABRR; B8K1W0) retrieved from UniprotKB [27] using search word 'Daborhagin'.

The purified tripeptides were analysed by ESI-MS and ESI-MS/MS using the same instrument and conditions as used to analyse the tryptic peptides from RVV-X.

5.7. Analysis by SDS-PAGE

Protein purity was determined by SDS-PAGE [28] on a 12% or 15% resolving gel and 4% stacking gel using a Mini-PROTEAN 3 electrophoresis system (BioRad, Hemel Hemstead, UK). Samples were loaded in either reduced or non-reduced form. Gels were run at 200 V, 30 mA per gel, for 50 min. Proteins were visualised with Coomassie Brilliant Blue R250 V followed by destaining with methanol: water: acetic acid (30:60:10). Alternatively, proteins were visualised by silver staining as performed by method of Heukeshoven & Dernick [29].

5.8. Caseinolytic Activity

The proteolytic activity was estimated by hydrolysis of heated casein using the Anson method [30]. The reaction mixture, consisting of 500 μL casein (20 mg/mL) in 0.1 M Tris-Cl (pH 8.0), 20 μL venom was incubated for 30 min at 37 °C. The reaction was quenched by the addition of 500 μL of 5% trichloroacetic acid (TCA) at room temperature. After centrifugation at $10,000\times g$ for 5 min, the hydrolysed substrate un-precipitated with TCA was determined by Folin Ciocalteau method [31]. Thus, 400 μL of the supernatant was mixed with 1 mL of 0.5M Na_2CO_3 and 200 μL of diluted (1:5) Folin & Ciocalteau's phenol reagent. The mixture was then incubated at 37 °C for 30 min and the absorbance was measured at 660 nm. One enzyme unit is defined as the amount of enzyme which hydrolyses casein to produce color equivalent to 1.0 μmole of tyrosine per minute at pH 8.0 at 37 °C.

5.9. Gelatinolytic Activity

The gelatinolytic activity of the purified enzyme was analysed by zymography [32]. The purified metalloproteinase was diluted in SDS sample buffer under non-reducing conditions and run on 10% SDS-polyacrylamide gels (0.75 mm) co-polymerised with 0.5 mg/mL of gelatin. After electrophoresis, the gels were washed in 2.5% Triton X-100 for 30 min and then washed three times in distilled water to remove any Triton. Gels were then incubated in developing buffer (50 mM Tris-Cl, pH 7.8, 200 mM NaCl, 5 mM $CaCl_2$, 0.02% Brij 35) for 18 h at 37 °C. The gels were stained with 0.5% Coomassie blue R-250 in methanol: acetic acid: water (5:10:85) solution and subsequently destained in methanol: acetic acid: water (10:5:85). The presence of gelatinolytic activity was defined as clear bands on the dark blue background.

5.10. Fibrinogenolytic Activity

The fibrinogenolytic activity was assayed by SDS-PAGE (4% stacking/12% resolving gel) as described by Ouyang & Teng [33]. Equal volumes of fibrinogen (1 mg/mL in 0.05 M Tris-Cl, pH 8.5) and 20 μg/mL of enzyme were incubated at 37 °C for various times intervals. At 0, 5, 15, 30, 60 and 120 min, 200 μL of the incubated solution was mixed with 400 μL of denaturing buffer containing 0.2 M Tris-Cl (pH6.8), 20% glycerol, 10% sodium dodecyl sulfate (SDS), 0.05% bromophenol blue and

10 mM β-mercaptoethanol and heated at 100 °C for 10 min to stop the digestion. Proteolytic activity was determined on the Coomassie blue-stained gel after electrophoresis by observing the cleavage patterns of purified fibrinogen chains.

5.11. Inhibition of Gelatinolytic Activity

The effect of synthetic tripeptides and EDTA on purified protein was assayed using SDS-PAGE (4% stacking/10% resolving gel) to determine gelatin degradation. The purified protein (10 ng/μL) was incubated firstly with synthetic tripeptide (5 mM) or EDTA (100 μM) at 37 °C for 10 min. Then, 10 μL of gelatin solution (2 mg/mL in distilled water) was added and 20 μL of this incubated solution was taken out at 1 h and 20 h, mixed with 5 μL of 5× denaturing buffer and heated at 95 °C for 2 min. The cleavage patterns on gelatin by the enzyme was observed on Coomassie blue-stained gels after electrophoresis.

5.12. Inhibition of Fibrinogenolytic Activity

The effect of synthetic tripeptides or EDTA on purified protein was assayed using SDS-PAGE (4% stacking/12% resolving gel) to determine fibrinogen degradation. The purified protein (32 ng/μL) was incubated firstly with synthetic tripeptide (5 mM) or EDTA (100 μM) at 37 °C for 10 min. Then, 10 μL of fibrinogen solution (2 mg/mL in distilled water) was added and 20 μL of this incubated solution was taken out at 1 h and 20 h, mixed with 5 μL of 5× denaturing buffer and heated at 95 °C for 2 min. The cleavage effect on fibrinogen chains by the SVMPs were observed on Coomassie blue-stained gel following electrophoresis.

Acknowledgments: This study was supported by the Thailand Research Fund (TRF) and Chulalongkorn University.

Author Contributions: K.T.Y., P.R., and M.C.W. conceived and designed the experiments; K.T.Y. and M.C.W. performed the experiments; K.T.Y., M.P., P.R. and P.T. analysed the data; P.R. and M.C.W. contributed reagents/materials/analysis tools; K.T.Y. wrote the paper.

Conflicts of Interest: The authors declare no conflict of interest. The founding sponsors had no role in the design of the study; in the collection, analyses, or interpretation of data; in the writing of the manuscript, and in the decision to publish the results.

Abbreviations

The following abbreviations are used in this manuscript:

SVMPs	Snake Venom Metalloproteinases
RVV-X	Russell's Viper Venom factor X activator
NGS	Next Generation Sequencing
SVMPI	Snake Venom Metalloproteinase Inhibitor
pERW	pyroglutamate-arginine-tryptophan
pEKW	pyroglutamate-lysine-tryptophan
pENW	pyroglutamate-asparagine-tryptophan
pEQW	pyroglutamate-glutamine-tryptophan
ANP	Atrial Natriuretic Peptide
CNP	C-type Natriuretic Peptide
MRV	Myanmar Russell's Viper
CVO	*Crotalus viridis oreganus*

References

1. Myint, A.A.; Pe, T.; Maw, T.Z. An epidemiological study of snakebite and venomous snake survey in Myanmar. In *Management of Snakebite and Research; Proceedings of the Report and Working Papers of a Seminar, Yangon, Myanmar, 11–12 December 2001*; WHO, Regional Office for South-East Asia: New Delhi, India, 2002; pp. 12–16.

2. Phillips, R.; Warrell, D. Bites by Russell's viper (*Vipera russelli siamensis*) in Burma: Haemostatic, vascular, and renal disturbances and response to treatment. *Lancet* **1985**, *326*, 1259–1264.

3. Hung, D.Z.; Yu, Y.J.; Hsu, C.L.; Lin, T.J. Antivenom treatment and renal dysfunction in Russell's viper snakebite in Taiwan: A case series. *Trans. R. Soc. Trop. Med. Hyg.* **2006**, *100*, 489–494. [CrossRef] [PubMed]

4. Suntravat, M.; Yusuksawad, M.; Sereemaspun, A.; Perez, J.C.; Nuchprayoon, I. Effect of purified Russell's viper venom-factor X activator (RVV-X) on renal hemodynamics, renal functions, and coagulopathy in rats. *Toxicon* **2011**, *58*, 230–238. [CrossRef] [PubMed]

5. Mitrmoonpitak, C.; Chulasugandha, P.; Khow, O.; Noiprom, J.; Chaiyabutr, N.; Sitprija, V. Effects of phospholipase A_2 and metalloprotease fractions of Russell's viper venom on cytokines and renal hemodynamics in dogs. *Toxicon* **2013**, *61*, 47–53. [CrossRef] [PubMed]

6. Fox, J.W.; Serrano, S.M. Insights into and speculations about snake venom metalloproteinase (SVMP) synthesis, folding and disulfide bond formation and their contribution to venom complexity. *FEBS J.* **2008**, *275*, 3016–3030. [CrossRef] [PubMed]

7. Markland, F.S.; Swenson, S. Snake venom metalloproteinases. *Toxicon* **2013**, *62*, 3–18. [CrossRef] [PubMed]

8. Gutierrez, J.M.; Escalante, T.; Rucavado, A.; Herrera, C. Hemorrhage caused by snake venom metalloproteinases: A journey of discovery and understanding. *Toxins* **2016**, *8*. [CrossRef] [PubMed]

9. Kato, H.; Iwanaga, S.; Suzuki, T. The isolation and amino acid sequences of new pyroglutamanylpeptides from snake venoms. *Experientia* **1966**, *22*, 49–50. [CrossRef] [PubMed]

10. Huang, K.F.; Hung, C.C.; Wu, S.H.; Chiou, S.H. Characterization of three endogenous peptide inhibitors for multiple metalloproteinases with fibrinogenolytic activity from the venom of Taiwan habu (*Trimeresurus mucrosquamatus*). *Biochem. Biophys. Res. Commun.* **1998**, *248*, 562–568. [CrossRef] [PubMed]

11. Huang, K.F.; Chiou, S.H.; Ko, T.P.; Wang, A.H.J. Determinants of the inhibition of a Taiwan habu venom metalloproteinase by its endogenous inhibitors revealed by x-ray crystallography and synthetic inhibitor analogues. *Eur. J. Biochem.* **2002**, *269*, 3047–3056. [CrossRef] [PubMed]

12. Francis, B.; Kaiser, I.I. Inhibition of metalloproteinases in *Bothrops asper* venom by endogenous peptides. *Toxicon* **1993**, *31*, 889–899. [CrossRef]

13. Munekiyo, S.M.; Mackessy, S.P. Presence of peptide inhibitors in rattlesnake venoms and their effects on endogenous metalloproteases. *Toxicon* **2005**, *45*, 255–263. [CrossRef] [PubMed]

14. Wagstaff, S.C.; Favreau, P.; Cheneval, O.; Laing, G.D.; Wilkinson, M.C.; Miller, R.L.; Stocklin, R.; Harrison, R.A. Molecular characterisation of endogenous snake venom metalloproteinase inhibitors. *Biochem. Biophys. Res. Commun.* **2008**, *365*, 650–656. [CrossRef] [PubMed]

15. Ding, B.; Xu, Z.; Qian, C.; Jiang, F.; Ding, X.; Ruan, Y.; Ding, Z.; Fan, Y. Antiplatelet aggregation and antithrombosis efficiency of peptides in the snake venom of *Deinagkistrodon acutus*: Isolation, identification, and evaluation. *Evid. Based Complement. Alternat. Med.* **2015**. [CrossRef] [PubMed]

16. Chen, H.-S.; Tsai, H.-Y.; Wang, Y.-M.; Tsai, I.-H. P-III hemorrhagic metalloproteinases from Russell's viper venom: Cloning, characterization, phylogenetic and functional site analyses. *Biochimie* **2008**, *90*, 1486–1498. [CrossRef] [PubMed]

17. Calvete, J.J.; Juarez, P.; Sanz, L. Snake venomics. Strategy and applications. *J. Mass Spectrom.* **2007**, *42*, 1405–1414. [CrossRef] [PubMed]

18. Tan, N.H.; Fung, S.Y.; Tan, K.Y.; Yap, M.K.; Gnanathasan, C.A.; Tan, C.H. Functional venomics of the Sri Lankan Russell's viper (*Daboia russelii*) and its toxinological correlations. *J. Proteom.* **2015**, *128*, 403–423. [CrossRef] [PubMed]

19. Gowda, D.C.; Jackson, C.M.; Hensley, P.; Davidson, E.A. Factor X-activating glycoprotein of Russell's viper venom. *J. Biol. Chem.* **1994**, *269*, 10644–10650. [PubMed]

20. Takeya, H.; Nishida, S.; Miyata, T.; Kawada, S.-I.; Saisaka, Y.; Morita, T.; Iwanaga, S. Coagulation factor X activating enzyme from Russell's viper venom (RVV-X). *J. Biol. Chem.* **1992**, *267*, 14109–14117. [PubMed]

21. Chen, H.S.; Chen, J.M.; Lin, C.W.; Khoo, K.H.; Tsai, I.H. New insights into the functions and N-glycan structures of factor X activator from Russell's viper venom. *FEBS J.* **2008**, *275*, 3944–3958. [CrossRef] [PubMed]

22. Siigur, E.; Tonismagi, K.; Trummal, K.; Samel, M.; Vija, H.; Subbi, J.; Siigur, J. Factor X activitor from *Vipera lebetina* snake venom, molecular characterization and substrate specificity. *Biochim. Biophys. Acta* **2001**, *1568*, 90–98. [CrossRef]

23. Yee, K.T.; Khow, O.; Noiphrom, J.; Kyaw, A.M.; Maw, L.Z.; Kyaw, M.T.; Chulasugandha, P. Purification and characterization of metalloproteinase from Myanmar Russell's viper (*Vipera russelii*) venom. *Myanmar Health Sci. Res. J.* **2014**, *26*, 93–102.

24. Risch, M.; Georgieva, D.; von Bergen, M.; Jehmlich, N.; Genov, N.; Arni, R.K.; Betzel, C. Snake venomics of the Siamese Russell's viper (*Daboia russelli siamensis*)—Relation to pharmacological activities. *J. Proteom.* **2009**, *72*, 256–269. [CrossRef] [PubMed]

25. Fischer, W.H.; Spiess, J. Identification of a mammalian glutaminyl cyclase converting glutaminyl into pyroglutamyl peptides. *Proc. Natl. Acad. Sci. USA* **1987**, *84*, 3628–3632. [CrossRef] [PubMed]

26. Varro, A. Posttranslational processing: Peptide hormones and neuropeptide transmitters. *eLS* **2007**. [CrossRef]

27. UniprotKB. Available online: http://www.uniprot.org (accessed on 15 November 2016).

28. Laemmli, U.K. Cleavage of structural proteins during the assembly of the head of bacteriophage T4. *Nature* **1970**, *227*, 680–685. [CrossRef] [PubMed]

29. Heukeshoven, J.; Dernick, R. Simplified method for silver staining of proteins in polyacrylamide gels and the mechanism of silver staining. *Electrophoresis* **1985**, *6*, 103–112. [CrossRef]

30. Anson, M.L. The estimation of pepsin, trypsin, papain, and cathepsin with hemoglobin. *J. Gen. Physiol.* **1938**, *22*, 79–89. [CrossRef] [PubMed]

31. Folin, O.; Ciocalteu, V. On tyrosine and tryptophane determinations in proteins. *J. Biol. Chem.* **1927**, *73*, 627–650.

32. Toth, M.; Fridman, R. Assessment of gelatinases (MMP-2 and MMP-9) by gelatin zymography. In *Metastasis Research Protocols*; Brooks, S.A., Schumacher, U., Eds.; Humana Press: Totowa, NJ, USA, 2001; Volume 57, pp. 163–173.

33. Ouyang, C.; Teng, C.-M. Fibrinogenolytic enzymes of *Trimeresurus macrosquamatus* venom. *Biochim. Biophys. Acta* **1976**, *420*, 298–308. [CrossRef]

Article

Insights into the Evolution of a Snake Venom Multi-Gene Family from the Genomic Organization of *Echis ocellatus* SVMP Genes

Libia Sanz * and Juan J. Calvete *

Laboratorio de Venómica Estructural y Funcional, Instituto de Biomedicina de Valencia,
Consejo Superior de Investigaciones Científicas, Jaume Roig 11, 46010 València, Spain
* Correspondence: libia.sanz@ibv.csic.es (L.S.); jcalvete@ibv.csic.es (J.J.C.); Tel.: +34-96-339-1760 (L.S.);
+34-96-339-1778 (J.J.C.)

Academic Editors: Jay Fox and José María Gutiérrez
Received: 12 June 2016; Accepted: 6 July 2016; Published: 12 July 2016

Abstract: The molecular events underlying the evolution of the Snake Venom Metalloproteinase (SVMP) family from an A Disintegrin And Metalloproteinase (ADAM) ancestor remain poorly understood. Comparative genomics may provide decisive information to reconstruct the evolutionary history of this multi-locus toxin family. Here, we report the genomic organization of *Echis ocellatus* genes encoding SVMPs from the PII and PI classes. Comparisons between them and between these genes and the genomic structures of *Anolis carolinensis* ADAM28 and *E. ocellatus* PIII-SVMP EOC00089 suggest that insertions and deletions of intronic regions played key roles along the evolutionary pathway that shaped the current diversity within the multi-locus SVMP gene family. In particular, our data suggest that emergence of EOC00028-like PI-SVMP from an ancestral PII(e/d)-type SVMP involved splicing site mutations that abolished both the 3′ splice AG acceptor site of intron 12* and the 5′ splice GT donor site of intron 13*, and resulted in the intronization of exon 13* and the consequent destruction of the structural integrity of the PII-SVMP characteristic disintegrin domain.

Keywords: Snake venom toxin multi-gene family; snake venom metalloproteinase; genomic organization of SVMP genes; PII-SVMP; PI-SVMP; gene duplication; intronic retroelements; intronization

---

1. Introduction

The ADAM (A Disintegrin-like And Metalloproteinase) family of transmembrane type 1 proteins belongs to the MEROP database M12 family of Zn^{2+}-dependent metalloendopeptidases [1] and PFAM family PF01421 [2]. Members of the ADAM family play important roles in cell signaling and in regulating cell-cell and cell-matrix interactions [3,4]. The ADAM family comprises ancient proteins whose origin extends back >750 My [5,6]. To date, close to 40 ADAM genes have been identified in vertebrate and invertebrate bilaterian animals, both in deuterostomes, from the basal chordate, *Ciona intestinalis*, to higher vertebrates, and in protostome, such as arthropods, nematodes, platyhelminths, rotifers, molluscs, and annelids. The evolutionary history of vertebrate ADAM genes is punctuated by gene duplication and retroposition events [7,8], followed by neo- or subfunctionalization [7]. Gene duplications are an essential source of genetic novelty that can lead to evolutionary innovation if the new function has no deleterious effects to its host organism or provides selective advantages. For example, in mammalian species, including marsupials and monotremes, except the platypus, ADAM28, ADAMDEC1 (decysin, a soluble ADAM-like protein), and ADAM7 form a cluster, likely as a result of tandem duplication of ADAM28 [9]. Instead, in most non-mammalian vertebrate genomes investigated, including those of aves, reptiles, and fishes, a single ADAM28 locus is present in this region [7,10]. The data suggest that ADAM7 and ADAMDEC1 were duplicated from

ADAM28, probably only in mammals [7]. On the other hand, as described below in more detail, it is thought that ADAM28 played a starring role in the emergence of toxic metalloproteinases in the superfamily Colubroidea of Caenophidian snakes (viperids, elapids, and colubrids).

The concept that gene duplication plays a major role in evolution has been around for over a century [11]. In his classic and influential book *"Evolution by Gene Duplication"* [12] Susumo Ohno argued that gene duplication is the most important evolutionary force since the emergence of the universal common ancestor. Common sources of gene duplications include ectopic homologous recombination, retrotransposition event, aneuploidy, polyploidy, and replication slippage [13]. Duplication creates genetic redundancy, where the second copy of the gene is often free from selective pressure. Thus, over generations of the organism, duplicate genes accumulate mutations faster than a functional single-copy gene, making it possible for one of the two copies to develop a new and different function. Duplicated genes may switch their transcription to other tissues by localizing closely to, and utilizing the regulatory elements of, a neighboring gene [14–16]. Examples of this are (i) the formation of toxin gene families during the evolution of the venom system of advanced snakes by co-option, multiplication, and weaponization in the venom gland of paralogs of genes encoding for normal body proteins [17–20], and (ii) the finding of 309 distinct widow spider genes exhibiting venom gland biased expression [21], suggesting that the switching of genes to venom gland expression in numerous unrelated gene families has been a dominant mode of evolution [21–23].

Because of its functional importance for prey capture, predator defense, and competitor deterrence, venom represented a key innovation that has underpinned the explosive radiation of toxicoferan reptiles in the Late Jurassic period of the Mesozoic era, ~150 million years before present (MYBP) [24–28]. Toxicofera [18] (Greek for "those who bear toxins") is the term coined for the clade of squamate reptiles that includes the Serpentes (snakes), Anguimorpha (monitor lizards, gila monster, and alligator lizards) ,and Iguania (iguanas, agamas, and chameleons) lizards. One of the founding families of advanced snake venom comprises the Zn^{2+}-dependent metalloendopeptidases (SVMPs) [17–19,29–32]. SVMPs are key enzymes contributing to toxicity of vipers and pitvipers venoms. Hemorrhage is one of the most significant effects in envenomings induced by viperid and crotalid snakebites. Damage to the microvasculature, induced by SVMPs, is the main event responsible for this effect. In addition to hemorrhagic activity, members of the SVMP family also have fibrin(ogen)olytic activity, act as prothrombin activators, activate blood coagulation factor X, possess apoptotic activity, inhibit platelet aggregation, are proinflammatory, and inactivate blood serine proteinase inhibitors [33–36].

The closest non-venom ancestors of SVMPs was likely an ADAM28 precursor gene [37]. The origin of SVMPs has been inferred to have occurred following the split of the Pareatidae from the remaining Caenophidians, approximately 60 MYBP around the Cretaceous–Paleocene boundary of the Cenozoic Era [18,19,29,31,38]. SVMPs are found in the venoms of all advanced snakes and are classified into different classes depending upon their domain structure [39–41]. The ancestral multidomain PIII form, which is found in all snake venoms, derives from the extracellular region (metalloproteinase domain with disintegrin-like and cysteine-rich domains at the C-terminus) of a duplicated ADAM28 precursor gene that lost the C-terminal epidermal-growth-factor-(EGF-)-like, transmembrane, and cytoplasmic domains [31,32,41–43]. On the other hand, the derived PII-SVMPs, comprising the metalloproteinase and C-terminal disintegrin domain, have been only found in venoms of vipers and rattlesnakes (Viperidae). This strongly suggests that they emerged, subsequently to the separation of Viperidae and Elapidae, ~37 million years ago, in the Eocene epoch of the Cenozoic era, but before the separation of the Viperidae subfamilies Viperinae and Crotalinae 12–20 MYBP, from a duplicated PIII-SVMP gene that lost its cysteine-rich domain (see Figures 1 and 8 in [43] and Figure 18.1 in [44]). The disintegrin domain has been lost from the PII-SVMP structure on multiple occasions, resulting in the formation of the PI class of SVMPs [45] made only by the catalytic Zn^{2+}-metalloproteinase domain [39–41].

Details on the mechanisms of co-option and the molecular events underlying the transformation of an ADAM28 precursor gene copy into the SVMP multi-gene family of extant snake venoms remain elusive. In previous works, we described a family of RPTLN genes that exhibit a broad and reptile-specific distribution, for which we hypothesize may have played a key role in the recruitment and restricted expression of SVMP genes in the venom gland of Caenophidian snakes [46]. We have also reported the genomic organization of *Echis ocellatus* PIII-SVMP gene EOC00089, and compared it to those of its closest orthologs from *Homo sapiens* and the lizard, *Anolis carolinensis* [47]. Now, we fit two new pieces in the puzzle: the genomic structures of *E. ocellatus* PII—(EOC00006-like) and PI—(EOC00028-like) SVMP genes. Insights into post-duplication events gained from the structural comparison of the three classes of SVMP genes are discussed.

2. Results and Discussion

2.1. The Genomic Structure of Pre-Pro EOC00006-Like PII-SVMP and Pre-Pro EOC00028-Like Genes

Genomic sequences encoding full-length pre-pro EOC00006-like PII-SVMP (17828 nt) [KX219964] (Figure A1) and EOC00028-like PI-SVMP (21605 nt) [KX219965] (Figure A2) genes were assembled from overlapping PCR-amplified fragments (Appendix A, Figures A1 and A2). The pre-pro PII-SVMP gene consists of 15 exons interrupted by 14 introns (Figure 1A), whereas the pre-pro PI-SVMP gene contains 13 exons and 12 introns (Figure 1B).

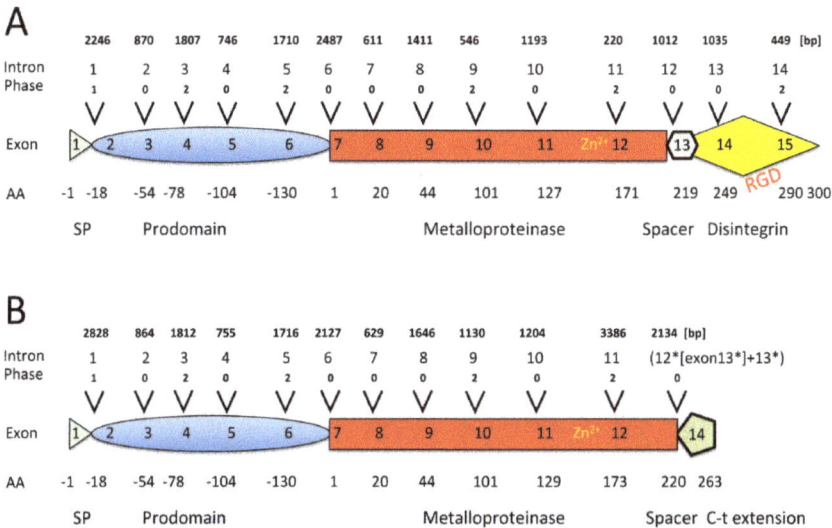

Figure 1. Scheme of the genomic organization of pre-pro EOC00006-like PII-SVMP (**A**) and pre-pro EOC00028-like PI-SVMP (**B**) genes. The distribution, phase, and size of the 14 (PII) and 12 (PI) introns and the boundaries of the protein-coding regions are highlighted. SP, signal peptide. Homologous exons and introns have identical numbering. Intron 12 of the PI-SVMP gene corresponds to the fusion of the genomic segment spanning intron12*-exon13*-intron13*. Mature PII- and PI-SVMP amino acid sequences span 299 and 263 amino acid residues, respectively. Zn^{2+}, relative location of the catalytic Zn^{2+}-binding environment; RGD, integrin-binding arginine-glycine-aspartic acid tripeptide motif.

The translated 494 (PII) and 457 (PI) pre-pro-SVMP amino acid sequences exhibit identical distribution and features (in terms of codon location and phase) for their first 11 introns and 12 exons, which code for the signal peptide (SP), prodomain (PD), metalloproteinase (MP) domain, and the short tetrapeptide (ELLQ) "spacer" sequence (Appendix A, Figures A1 and A2). These 413 (PII)/414 (PI) amino acid sequences show 85% identity, strongly suggesting that both SVMPs have a shared ancestry. It is also worth noting that the protein-coding positions interrupted by each of the introns of the PII- and PI-SVMP genes are entirely conserved in *Anolis carolinensis* [XP_008118058] (and also in human [NG_029394]) ADAM28 gene. Introns are inserted after or between secondary structure elements, supporting the "introns-added-late" model, which proposes that during the evolution of the eukaryotic branch, introns were added at the boundaries of structural modules coded for by ancestral continuous genes [48]. In addition, as will be analyzed in detail below, pairwise alignment of topologically equivalent PII- and PI-SVMP introns show that homologous intronic nucleic acid sequences share 88%–99% identity (Figure 2). This clearly indicates that EOC00006-like PII-SVMP and EOC00028-like PI-SVMP represent paralog genes.

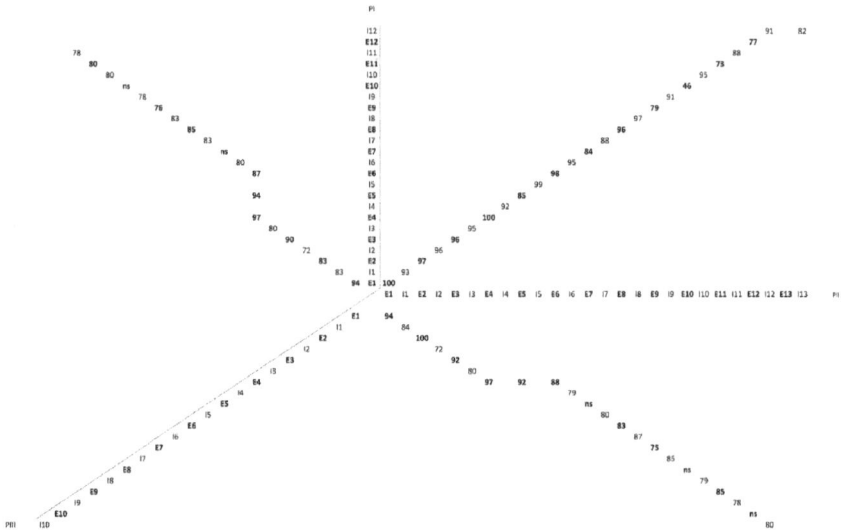

Figure 2. Pairwise comparisons of the sequence identities between the exonic and intronic nucleic acid sequences of pre-pro EOC00089-like PIII-SVMP, EOC00006-like PII-SVMP, and EOC00028-like PI-SVMP genes.

New genes can arise through four mechanisms: gene duplication, retroposition, horizontal gene transfer, and de novo origination from non-coding sequences [49]. Available evidence strongly suggests that gene duplication has played a pivotal role in the origin of venom multi-gene families [20–23,50,51]. Although the fate of many new genes may be to lose their function and become pseudogenes, some can be fixed through evolution of redundancy, subfunctionalization, or neofunctionlization. Several models have been proposed to explain functional divergence following venom toxin gene duplications [52–55]. However, this issue remains controversial and is the subject of vivid debates. The family portrait of SVMPs shows a complicated picture. SVMPs belong to different "generations", that in the canonical model for the evolutionary expansion of this multi-gene family are hierarchically related, being PIII-SVMPs the most ancient and the PII- and PI- SVMPs the succesively most recently derived family members [31,32,42]. However, due to the limited genomic information available, this model can be confounded by high rates of protein amino acid sequence divergence [56], and the occurrence of alternative routes (e.g., PIII > PI) can not be presently ruled out. The only other full-length viperid SVMP gene sequenced to date is *E. ocellatus* EOC00089-like PIII-SVMP [47] [KX219963]. The ORF encoding the pre-pro-metalloproteinase domains of this gene exhibits 63% amino acid sequence identity with the homologous coding regions of the PII- and PI-SVMPs here reported, and 72%–83% nucleotide sequence identity between topologically equivalent PIII-, PII-, and PI-SVMP introns (Figure 2). Although these figures clearly point to a common origin, it is not possible to infer whether they belong to the same or to a different PIII > PII > PI hierarchical lineage. Nonetheless, the fact that the PIII-SVMP gene has lost introns 5 and 6 (ADAM28 numbering), with the consequence that exons 4, 5, and 6 have merged into a single exon, suggests that either these events occurred after the duplication that gave rise to the PII-SVMP ancestor, or that the PIII-SVMP EOC00089-like gene does not lay in the direct line of descent of the EOC00006-like PII-SVMP and EOC00028-like PI-SVMP genes. Refinement of the family tree of the multi-gene family of *E. ocellatus* SVMPs will surely emerge from future comparative genomic analysis of the carpet viper and other viperid species.

2.2. Role of Introns in the Evolution of the SVMP Multi-Gene Family

Since their discovery in 1977 [57,58], introns have been the subject of considerable debate. It is now generally accepted that introns represent more than merely junk DNA that must be pruned from pre-mRNAs to yield mature, functional mRNAs prior to their translation. Mounting evidence indicates that while introns do not encode protein products, they play essential roles in a wide range of gene expression regulatory functions such as non-sense mediated decay [59], mRNA export [60], and regulation of the amount of recombination between the flanking exons [61], or they serve as locations for nonhomologous recombination that would allow for exon shuffling [62,63]. As discussed below, most of the structural divergence between the EOC00006-like PII-SVMP and EOC00028-like PI-SVMP genes is due to the different size of their topologically equivalent eleven (1–11) introns (Supplementary Figure S1). The role of introns in the evolution of snake venom gene families remains elusive. However, in other biological systems, i.e., *Arabidopsis* and *Drosophila*, intron features, such as sequence and length, have been shown to function in maintaining pre-mRNA secondary structure, thus influencing temporal and spatial patterns of gene expression by modulating transcription efficiency and splicing accuracy [64–67].

Most PII- and PI-SVMP introns belong to phase 0, followed by phase 2; and, in both genes, only intron 1, separating the monoexonic signal peptide from the start of the prodomain, is a phase 1 intron (Figure 1). Analysis of the exon–intron structures of a large number of human genes has revealed a statistically highly significant enrichment of phase 1 introns flanking signal peptide cleavage sites [68]. Phase 1 introns most frequently split the four GGN codons encoding glycine. A plausible explanation for the correlation between signal peptide domains and the intron phase is that the base preferences of proto-splice sites [69,70] mirrors the amino acid preference for glycine in the signal peptidase consensus cleavage site [71].

The signal peptide is the most conserved structural element between pre-pro EOC00006-like PII-SVMP and EOC00028-like PI-SVMP is (Figure 2). In both genes, it is encoded by identical exon 1 amino acid sequences (Figures A1 and A2), which is also highly conserved in present-day SVMPs [46]. These findings support the view that co-option of this signal peptide may have played a role in the restricted expression of SVMP genes in the venom gland of Caenophidian snakes, some 60–50 Mya [46].

Nucleotide sequence comparison of the topologically equivalent introns of the *E. ocellatus* PII- and PI-SVMPs (Supplementary Figure S1) provide insights into the events underlying the conversion of a PII-SVMP into a PI-SVMP gene. In this regard, some introns differ in the number and location of intronic retroelements (Table 1). Thus, insertions in introns PI-SVMP 1 and 9 introduced complete and truncated SINE/Sauria elements in positions 1764–2101 (Figure S1, panel A) and 321–502 (Figure S1, panel I), respectively. The inserted nucleic acid sequence in intron 9 retains the GT-AG splicing sites, indicating that this insertion event created a twintron, an intron within an intron. PII-SVMP intron 6 (Figure S1, panel F) and PI-SVMP introns 11 (Figure S1, panel K) and 12 (Figure S1, panel L) are also twintrons. Compared to its topologically equivalent PII-SVMP intron, a large insertion in intron 11 of the EOC00028-like PI-SVMP gene replaced the first 66 nucleotides for a longer stretch of 3281 nucleotides; region 2461–2561 of the inserted nucleic acid sequence is 97% identical to *Hyla tsinlingensis* Hts-35 [KP204922], a microsatellite sequence that is also partly present in intron 61 of *Podarcis* reelin (RELN) genes [GU181006-13] (positions 554–623) [72]. Microsatellites are simple nucleotide sequence repeats (SSR) ranging in length from two to five base pairs that are tandemly repeated, typically 5–50 times (reviewed in [73]). These non-coding elements are abundant in major lineages of vertebrates. Mammalian, fish, and squamate reptile genomes appear to be relatively microsatellite rich [74]. However, besides Hts35, RepeatMasker only identified few SSR tracks in introns 1 (5× GTTT; 28× TC) and 2 (13× ATTT; 4× TAA) of the PII-SVMP gene (Figure A1), and introns 1 (11× GTTT; 21× AG) and 2 (9× GTTT; 4× TAA) of the PI-SVMP gene (Figure A2).

Table 1. Comparison of type and location of retroelements identified in introns of *E. ocellatus* PII-SVMP EOC00006-like and PI-SVMP EOC00028-like genes.

Intron	PII-SVMP	PI-SVMP
	Inserted Retroelement	
1	SINE/Sauria	2 SINE/Sauria, LTR/ERV1, DNA/hAT-Ac
3	LINE/L2/CR1	LINE/L2/CR1
5	LINE/L2/CR1	LINE/L2/CR1
6	SINE/Sauria	-
8	LINE/L2/CR1	-
9	-	SINE/Sauria
10	DNA transposon	DNA transposon

Growing evidence supports that repetitive intronic elements, such as the long interspersed elements (LINEs) and the short interspersed elements (SINEs) contained in several introns of both PII- and PI-SVMP genes (Table 1) can influence genome stability and gene expression (reviewed in [75]). Thus, these interspersed repeats may alter genome recombination structure and rates, through a number of mechanisms, including replication slippage and unequal crossover [76,77], potentially impacting regulation of gene expression [78], recombination events leading to tandem duplication of segments of the genome [79,80], gene conversion [81], and chromosomal organization [79]. Moreover, the insertion of interspersed repeats into a new genomic position may introduce promoter or enhancer sequence motifs for transcription of nearby genes [82,83], and alternative splicing sites or polyadenylation sites [84], thereby resulting in a change of overall level of gene expression. Interspersed repetitive elements have also played an important role in expanding the repertoire of transcription factor binding sites in eukaryotic genomes [85]. However, whether these elements have contributed to the genomic context that facilitated the evolution and radiation of venom *loci* in snakes deserves future detailed comparative genomic studies.

2.3. A Fusion Event Led to the Conversion of a PII(e/d)-Type SVMP into EOC00028-like PI-SVMP

PI-SVMP intron 12 is a twintron resulting from the fusion of the genomic region spanning ancestral introns 12* and 13* and exon 13* (homologous to identical numbered elements in the genomic structure PII) (Figures 1 and 3A). Splicing site mutations affecting both the 3' splice AG acceptor site of intron 12* and the 5' splice GT donor site of intron 13* led to the retention, and subsequent intronization, of exon 13* within a fused (12* + 13*) twintron (Figure 3A). Intronization of exon-coding nucleic acid sequences has been proposed as a major contributor to intron creation [86]. Intron 13* encoded part of the N-terminal region of a disintegrin domain, most likely, as discussed below, an eventual subunit of dimeric disintegrin. In addition to the disruption of the structural integrity of the disintegrin domain, a stop codon after exon 14 removed intron 14 and exon 15 from the PII(e/d)-type SVMP (Fox & Serrano's nomenclature [40]) precursor gene structure, thereby completing the conversion of the PII-SVMP into present EOC00028-like PI-SVMP gene (Figure 3A).

A PII-SVMP

↦ Disintegrin

Exon 12	Intron 12	Exon 13	Intron 13	Exon 14
.... TPVSENELLQ	GT 1012 nt AG	NSVNPC ... DNCK	GT 1035 nt AG	FLKEG ... PMEW

PI-SVMP

Exon 12	Intron 12 (2134 nt) [*Intron 12* + Exon 13* + Intron 13**]	Exon 14
... TPVSENELLQ	GT at *° YDPCC ... LVYF °* ac	AG FLRAR ... WNDLQ STOP

B

```
Exon 14   TTT TTG AGA GCA CGA ACA GTA AGC AAG AGA GCA GTG AGT GAT GAC ATG GAT GAT TAC TGC
PI-SVMP    F   L   R   A   R   T   V   S   K   R   A   V   S   D   D   M   D   D   Y   C

Exon 2    TTT TTG AGA GCA GGA ACA GTA TGC AAG AGA GCA GTG GGT GAT GAC ATG GAT GAT TAC TGC
ML-G1      F   L   R   A   G   T   V   C   K   R   A   V   G   D   D   M   D   D   Y   C


Exon 14   TCT GGC ATA ACT TCT GAC TGT GCC AGA AAT CCT ACA AAG GCT AAG CAA CAG AGA TGG AAC
PI-SVMP    S   G   I   T   S   D   C   A   R   N   P   T   K   A   K   Q   Q   R   W   N

Exon 2    ACT GGC ATA TCT TCT GAC TGT CCC AGA AAT CCC TAC AAA GAC TAA
ML-G1      T   G   I   S   S   D   C   P   R   N   P   Y   K   D   STOP


Exon 14   GAT CTG CAG TAG
PI-SVMP    D   L   Q   STOP
```

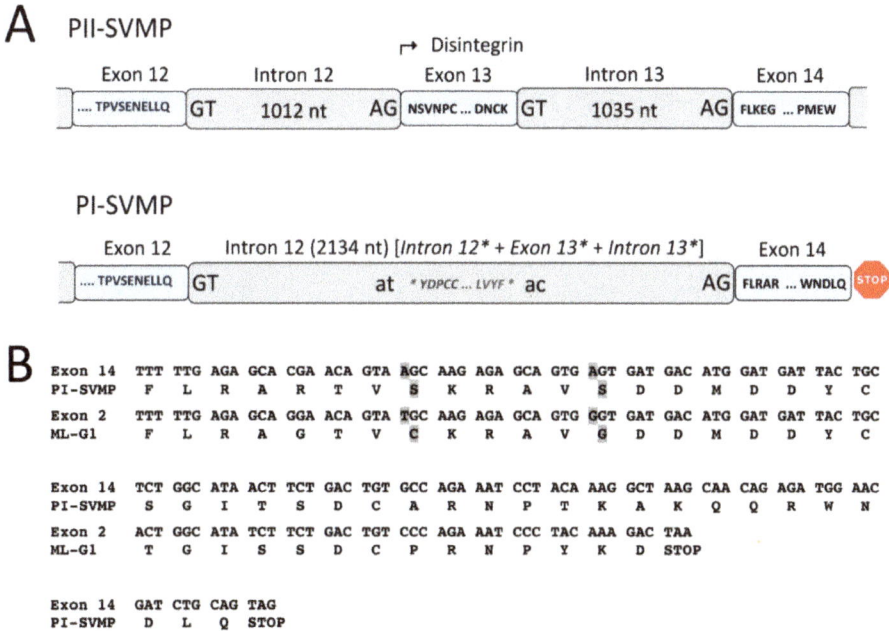

Figure 3. Panel **A**, cartoon comparing the 3′ regions of the PII-SVMP and PI-SVMP genes and highlighting the processes (intronization of ancestral exon 13* inside twintron 12 resulting from the fusion of introns 12* and 13*, and creation of a stop codon after exon 14) that destroyed the integrity of the disintegrin domain, converting an ancestral PII(e/d)-type SVMP into extant EOC00028-like PI-SVMP. Panel **B**, alignment of the amino acid sequences encoded by exon 14 of EOC00028-like PI-SVMP and exon 2 of the dimeric disintegrin subunit ML-G1 [AM261811] [87]. Degeneration of PI-SVMP's conserved functional and structural amino acid residues in dimeric disintegrins are highlighted in boldface and grey background.

Region 1013–2134 of PI-SVMP intron 12 exhibits 91% nucleotide sequence identity with range 14 to 1135 of *Macrovipera lebetina* gene encoding part of exon 1 and full-length intron 1 of the VGD-containing dimeric disintegrin subunit precursor, ML-G1 [AM261811] [87]. PI-SVMP exon 14 (mature protein amino acid residues 221–263, Figure A2) exhibits strong homology (79% identity) to exon 2 of the same VGD-bearing dimeric disintegrin subunit. The PI-SVMP exon 14 shows the consequences of genetic drift (Figure 3B): the conseved $\alpha_5\beta_1$ integrin-inhibitory VGD tripeptide motif [44] of the PII-SVMP precursor gene has been replaced by a VSD motif (generated by a G > A mutation: GTG AGT GAT > GTG GGT GAT), and the absolutely conserved tenth cysteine residue of dimeric disintegrin subunits has degenerated (TGC) to a serine residue (AGC) (Figure 3B).

3. Concluding Remarks and Perspectives

The event that gave birth to the family of SVMPs was the generation of a STOP codon at the 3′ end of exon 16 of a duplicated ADAM28 gene (Figure 4). This mutation produced an ORF truncated at the N-terminal part of the EGF-like domain, which encoded a precursor of an ancestral PIII-SVMP lacking this domain and the C-terminal membrane anchoring and cytoplasmic polypeptides (Figure 4). On the other hand, our results comparing the available genomic structures of SVMP genes, e.g., EOC00089-like PIII-SVMP [47] [KX219963], EOC00006-like PII-SVMP [KX219964], and EOC00028-like PI-SVMP [KX219965] (this work), suggest that the evolutionary history of SVMPs is marked with events of insertions and deletions of intronic regions. This scenario points to introns as key players in the formation of the multi-locus SVMP gene multifamily. Thus, comparison of the genomic structures of EOC00089-like PIII-SVMP and EOC00006-like PII-SVMP (Figure 5) indicates that replacement of the PIII-specific cysteine-rich domain by a non-homologous region encoding intron 14-exon 15 followed by a STOP codon may represent a step in the conversion of a PIII-SVMP into a PII-SVMP.

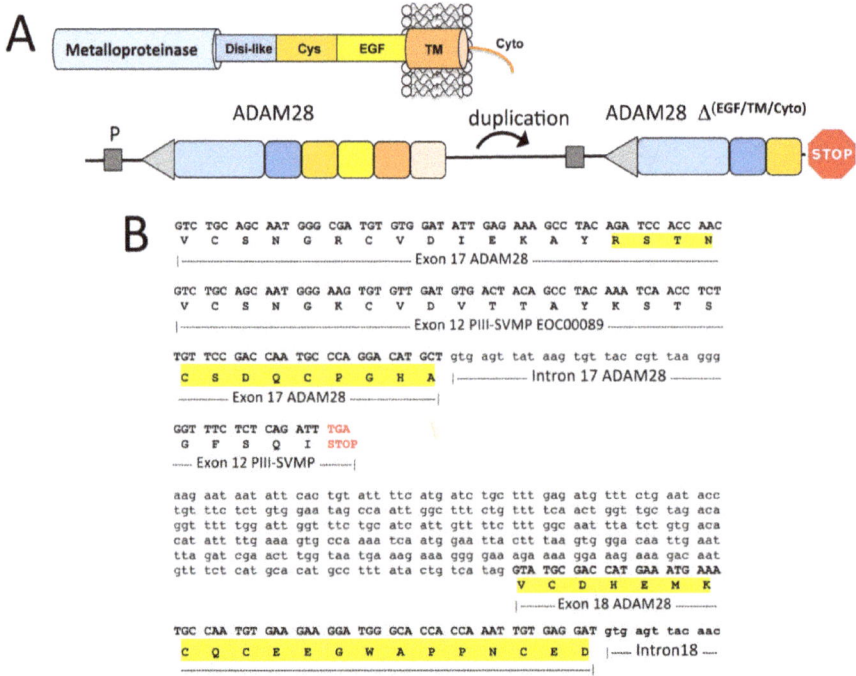

Figure 4. Comparison of the genomic region encompassing exons 17 through 18 of *Anolis carolinensis* ADAM28 [XP_003226913] and the homologous amino acid sequence of *E. ocellatus* SVMP EOC00089 [ADW54351], highlighting the STOP codon after exon 12 of the latter generating a C-terminally truncated molecule, which eventually gave rise to the ancestor of the PIII-SVMPs.

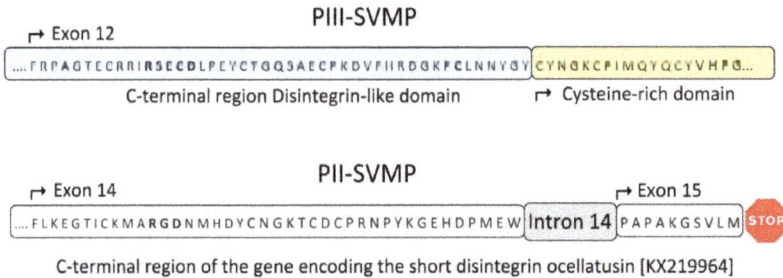

Figure 5. Comparison of the genomic region encoding the C-terminal domains of *E. ocellatus* EOC00089-like PIII-SVMP and EOC00006-like PII-SVMP, suggesting that 3′ genomic remodeling represents a seminal step in the generation of PII-SVMPs.

This view is consistent with structural evidence suggesting that the loss of the cysteine-rich domain represents an early seminal event that facilitated the formation of PII class SVMPs [43].

The PII subfamily of SVMPs is characterized by the diversity of disintegrin domains exhibited by different family members [39,40], ranging from the more ancestral long disintegrin domains (~84 amino-acid-residue polypeptide cross-linked by 7 disulfide linkages) to the more recently evolved short disintegrin (41–51 amino-acid-residues crosslinked by 4 disulfide bonds) [42]; for a scheme of the evolutionary path of the disintegrin domains, see Figure 1 in [43]. EOC00006-like is an example of a PII-SVMP with short disintegrin domain. Given the structural diversity of PII-SVMPs, genomic sequences from the different members of the subfamily are required for a more accurate glimpse of the genomic mechanisms operating in the generation and subsequent diversification of PII-SVMPs.

Comparison of the EOC00006-like PII-SVMP and EOC00028-like PI-SVMP gene structures also points to genomic remodeling of the 3′ region of a PII(e/d)-type SVMP precursor gene [39,40] as the EOC00028-like PI-SVMP gene generator mechanism. The PII > PI conversion involved the generation of twintron 12 (by fusion of introns 12* and 13*) and the loss, by intronization, of exon 13*, thereby destroying the consistency of the region coding for the disintegrin domain. This elaborated mechanism indicates that the structural diversification of SVMPs is not due to a random mutation generating a STOP codon before the disintegrin domain, but follows a well orchestrated sequence of events imprinted in the genome of snake species sometime after the split of Viperidae and Elapidae, 37 million years ago, but before the separation of the Viperidae subfamilies Viperinae and Crotalinae 12–20 MYBP. The mechanisms underlying loss or gain of spliceosomal introns are still poorly understood. The most widely accepted hypothesis is that intron insertion may occur via a process similar to group II intron retrotransposition [88,89]. According to this view, the spliceosomal components remain transiently associated with a recently excised intron and then attach at a potential splice site of a non-homologous pre-mRNA, where they catalyze the reverse reaction [90,91]. The modified pre-mRNA is reverse-transcribed and the resulting cDNA participates in a recombination with its parent gene, thereby inserting a novel intron into the target gene [90–93]. An attractive feature of this mechanism is that it ensures that the inserted nucleic acid sequence has the full complement of intron signature sequences required for efficient splicing [94].

Studies of multi-gene protein families are crucial for understanding the role of gene duplication and genomic exon-intron organization in generating protein diversity. For example, full-length genomic sequences of Crotalinae group II PLA$_2$ isogenes from *P. flavoviridis* (Tokunoshima and Amami-Oshima islands, Japan) [95], and *T. gramineus* (Taiwan) [96] have been reported. All these genes exhibit four coding regions and conserved exon-intron structures spanning about 1.9 kb. A cluster of five tandemly arranged PLA$_2$ genes have been located in a 25 kb 3′ segment of a 31 kb fragment of the Amami-Oshima *P. flavoviridis* genome [97], which in addition harbors a PLA$_2$ pseudogene in its 6 kb 5′ region [98]. Genomic sequence comparisons between the pancreatic PLA$_2$ gene of *P. elegans*, group IB pancreatic PLA$_2$ gene of *L. semifasciata*, and the *L. semifasciata* group IA venom PLA$_2$ gene, suggest that Crotalinae group II venom PLA$_2$ genes emerged before the divergence of Elapinae and Crotalinae, whereas groups of IB and IA PLA$_2$ genes appeared after Elapinae was established as a taxonomic lineage [99].

Duplicated structures found in eukaryotic genomes may result from complex interplays between different mechanisms [100]. Mitotic and meiotic non-allelic homologous recombination (NAHR) events, resolved as unequal crossing-over, have been traditionally invoked to account for segmental duplications within genomes [101,102]. Duplicated regions can be organized as direct tandems (e.g., the cluster of tandem snake venom PLA$_2$ genes), but also be separated by hundreds of kb [100]. Our present and previous work [47] inaugurate a line of research that will allow the depiction of a more precise characterization of the genomic context in which the SVMP multi-gene family has emerged. This goal demands populating the current databases with genomic sequences of genes representing the different members of the SVMPs. Although the variety of structural forms comprising the PII family may be considered a challenge for this purpose, this circumstance can be also regarded as a valuable opportunity for the step-by-step description of the molecular pathways that led to the formation of this multi-gene family. Without a doubt, ongoing Viperidae snake genome sequencing projects will mark the beginning of comparative snake genomics, and will be key to revealing not only the topology and copy number of the genes encoding SVMPs, but also to provide decisive information to reconstruct the evolutionary history of this multilocus gene family.

4. Materials and Methods

4.1. Genomic DNA

Genomic DNA was extracted from the fresh liver of *E. ocellatus* (Kaltungo, Nigeria) maintained at the herpetarium of the Liverpool School of Tropical Medicine. *Echis ocellatus* liver was ground to a fine powder under liquid nitrogen and the genomic DNA extracted using a Roche DNA isolation kit for cells and tissue containing SDS (2% final concentration) and proteinase K (400 μg/mL final concentration). The homogenates were incubated at 55 °C overnight. Thereafter, 300 μL of 6 M NaCl (NaCl-saturated H_2O) was added to each sample, and the mixture was vortexed for 30 s at maximum speed and centrifuged for 30 min at 10,000 *g*. An equal volume of isopropanol was added to each supernatant, and the sample mixed, incubated at −20 °C for 1 h, and centrifuged for 20 min at 4 °C and 10,000 *g*. The resulting pellets were washed with 70% ethanol, dried, and, finally, resuspended in 300–500 μL sterile distilled H_2O.

4.2. Strategy for PCR Amplification of Overlapping Genomic DNA Fragments

For sequencing *E. ocellatus* genes encoding PII-SVMP EOC00006 [Q14FJ4] and PI-SVMP EOC00028 [Q2UXQ3] we employed a similar iterative process as described in [47]. Full-length cDNA-deduced amino acid sequences of disintegrin domains [103] and of the genomic organization of dimeric disintegrin domains [AM286800] [87] and PIII-SVMP EOC00089 [47] from the same species were used as templates to design primers for the PCR-amplification of protein-specific genomic sequences (Table 2).

PI-SVMP stretch ^{72}AREILNS.....QRWNDLQ^{263} was amplified on an Eppendorf Mastercycle® epgradient S instrument in a 50 μL reaction mixture containing 17.5 μL of H_2O, 25 μL Master-Mix (Thermo Scientific, Waltham, MA USA) including buffer, dNTPs, and Phusion High-Fidelity DNA polymerase, 2.5 μL of each primer (10 μM) Met1PIRv and Met5PIFw, 1.5 μL of DMSO (100%), and 1 μL of genomic DNA (50 ng/μL). PCR conditions included an initial denaturation step at 98 °C for 30 s followed by 35 cycles of denaturation (20 s at 98 °C), annealing (15 s at 63 °C), extension (300 s at 72 °C), and a final extension for 5 min at 72 °C. All other PCR amplifications were carried out in the same thermocycler using iProof High Fidelity polymerase (BioRad, Hercules, CA, USA). The 50 μL reaction mixture contained 10 μL of 5×buffer, 1 μL of 10 mM (each) dNTPs, 2 μL of $MgCl_2$ 50 mM, 1.5 μL of DMSO (100%), 1 μL of each Fw and Rv primer (10 μM), 1 μL of genomic DNA (50 ng/μL), and 32.5 μL of water. PCR conditions included an initial denaturation step at 98 °C for 120 s followed by 35 cycles of denaturation (10 s at 98 °C), annealing (15 s at the lower melting temperature of the primers), extension (60 s per Kb at 72 °C), and a final extension for 5 min at 72 °C.

4.3. Purification and Cloning of PCR Products

PCR-amplified DNA fragments were purified from agarose electrophoretic bands using the GENECLEAN Turbo kit (MP Biomedicals). The purified fragments were inserted into pJET_1.2 (Thermo Scientific, Waltham, MA USA) using phage T4 ligase and cloned into *E. coli* DH5α by electroporation at 1700 V. Transformed cells, resuspended in 200 μL LB medium, were incubated at 37 °C for 1 h, and were subsequently plated on LB agar/ampicilline to select positive clones. The presence of the inserted DNA fragments was verified by PCR amplification or digestion of the expression vector with the restriction enzyme Bgl II. The inserted DNA fragments were sequenced in-house on an Applied Biosystems model 377 DNA sequencing system (Foster City, CA, USA) using pJETFw and pJETRv primers.

Table 2. Forward (Fw) and reverse (Rv) primers used to PCR-amplify genomic DNA stretches from *E. ocellatus* PII-SVMP EOC00006-like (**right**) and PI-SVMP EOC00028-like (**left**) genes.

Primer	DNA sequence	Primer	DNA sequence
Sp35_Eo Fw	ATGATCCAAGTTCTCTTGGTAACTATATGCTTAGC	5′ PS-Disi Fw	ATGATCCAAGTTCTCTTGG
Met14PI Fw	CTATATGCTTAGCAGTTTTTCCATATC	Intr4 Fw	ATGACACTGACCTCTAGAGTTGG
Intr1F1PI Fw	CTAGTCATTCCGGCCATATGAC	IntrB9_4-2 Fw	AAGCTTGCTTGCTAGTAGGTGG
Intr2F1PI Fw	ATCAGTCTGAGAGGATGCATTTCC	Intr4 Rv	TGGACATTGTATGGTCACCTG
Intr3F1PI Fw	GTGACCATGCAATGTCCATATG	Prodom 3 Fw	GGAGCTTTTAAGCAGCCAGAG
Met15PI Fw	GTTGCCTGTAGGAGCTGTTAAG	Prodom 3 Rv	CTCTGGCTGCTTAAAAGCTCC
Prodom 2 Fw	GACGCTGTGCAATATGAATTTG	Prodom 2 Fw	GACGCTGTGCAATATGAATTTG
Prodom 2 Rv	CAAATTCATATTGCACAGCGTC	Prodom 2 Rv	CAAATTCATATTGCACAGCGTC
Intr3 Rv	GCACCAACTCTGTATCTCAGTC	Intr3 Fw	CACAGGTAAATAAGCCACAAACACC
Pro2 Rv	CAGTGAGACTCATTATTCCCCTGATGGCAG	Intr3 Rv	GCACCAACTCTGTATCTCAGTC
Pro3 Rv	CTGCCATCAGGGGAATAATGAGTCTCACTG	Pro2-SVMP_Fw	CAGAAGATTACAGTGAGACTCATTA TTCCCWGATGG
IntrB13-1 Fw	CTTGCCTCCCTATAGGATCACTGC	Pro3-SVMP_Rv	CTGCCATCAGGGGAATAATGAGTCTCACT
Met16PI Rv	GATGCGTCCATAATAATAGCAGTG	IntrB13-1 Fw	CTTGCCTCCCTATAGGATCACTGC
Prodom 1 Fw	GATGCCAAAAAAAAGGATGAGG	Prodom 1 Fw	GATGCCAAAAAAAAGGATGAGG
Prodom 1 Rv	CCTCATCCTTTTTTTTGGCATC	Prodom 1 Rv	CCTCATCCTTTTTTTTGGCATC
IntronB7PI Fw	TGGAACAACAGCTGTTGTTATGACG	Intr2 Fw	ACAATGGGAAACTGAGGAACAG
IntronB7PI Rv	TGAGAGACATGCTGATGTGGTC	Intr2 Rv	GGGAACTCTGACTTAGAGAAAGTC
Met4 PI Fw	GACCCAAGATACATTCAGCTTGTC	Met1PII Fw	CAACAGCATTTTCACCCAAGATAC
Met4 PI Rv	GACAAGCTGAATGTATCTTGGGTC	Met1PII Rv	GTATCTTGGGTGAAAATGCTGTTG
Met8PI Rv	TATCCATGTTGTTATAGCAGTTAAATC	Met 1-2 Fw	CATGGATACATCAAATTGTCAACG
Intron B16 Fw	TGTGCTTACCCAACACTGAGCC	Met 1-3 Rv	TGTACATCTGTCAGGTGGACATG
Met5 PI Fw	GCACGTGAAATTTTGAACTCA	Met2PII Fw	GCCGTTCACCTTGATAACCTTATAGG
Met5PI Rv	GAGTTCAAAATTTCACGTGCTG	Met2PII Rv	CCTATAAGGTTATcAAGGTGAACGGC
Met9PI Rv	AGCATTATCATGCGTTATGCG	Met 6 PII Fw	CCACAATCGTCTGTAGCAATTACTGA
Met3 PI Fw	GGAAGAGCTTACATGGAGAG	Met 6 PII Rv	TCAGTAATTGCTACAGACGATTGTGG
Met3PI Rv	CTCTCCATGTAAGCTCTTCC	Met3 PII Fw	GATCATAGCACAGATCATCTTTGG
Met2PI Rv	GCTCCCCAGACATAACGCATC	Met3PII Rv	CCAAAGATGATCTGTGCTATGATCc
IntrB23PI Fw	CTGACTATGACTCACTTAACAACTGG	Met 4 Fw	ATGATCCAGGTTCTCTTGGTAACTATATG
IntrF2PI Fw	GGCCGCGTGAATGCATCTGCTTC	Met 4 Rv	TGAACTGATAGGAACGGTATTGTG
Intr2F2PI Fw	GCATCAGTTTGTTCGCACTCAATAAAG	Fw_Ocella NcoI	ATCCATGGTAGACTGTGAATCTGGACC
Intr3F2PI Fw	GAGCATAATCTGGAACTAAGATCAAG	IntrDis1 Rv	ATACGGCTAGTATGGAGCAAG
Met7PI Fw	GCACAAGATTCCTATCACTTCAG	Dis PII Rv	TCACATCAACACACTGCCTTTTGC
Met13PI Rv	TCCTACCTGCAAAAGTTCATTTTC	-	-
Intron B10PI Rv	CTGACTCAGGGCACCAATCTC	-	-
Met1PI Rv	CTACTGCAGATCGTTCCATCTCTG	-	-

4.4. Sequence Analysis

Exon-intron boundaries were localized by visual inspection and corroborated using Wise2 [104]. Amino acid and nucleotide sequence similarity searches were done using BLAST [105]. Multiple sequence alignments were performed using ClustalW2 [106]. The occurrence of retrotransposable elements and simple nucleotide sequence repeats (SSRs) were assessed using RepeatMasker (version rm-20110920) [107], a program that screens DNA sequences for interspersed repeats and low complexity DNA sequences included in the Repbase database [108].

4.5. Sequence Availability

Pre-pro EOC00006-like PII-SVMP and EOC00028-like PI-SVMP gene sequences have been deposited with the NCBI GeneBank [109] and are accessible under accession codes KX219964 and KX219965, respectively.

Supplementary Materials: The following are available online at www.mdpi.com/2072-6651/8/7/216/s1, Figure S1: Pairwise nucleotide sequence alignments of topologically equivalent paralog introns 1–12 from Pre-pro EOC00006-like PII-SVMP and 1–13 from Pre-pro EOC00028-like PI-SVMP gene sequences.

Acknowledgments: This work has been financed by Grant BFU2013-42833-P from the Ministerio de Economía y Competitividad, Madrid (Spain).

Author Contributions: J.J.C. and L.S. conceived and designed the experiments; L.S. performed the experiments; J.J.C. and L.S. analyzed the data; J.J.C. wrote the paper.

Conflicts of Interest: The authors declare no conflict of interest.

Appendix A

Genomic sequence of *E. ocellatus* EOC00006-like PII-SVMP gene. The locations and identities of the primers used to PCR-amplify genomic sequences (listed in Table 2) are indicated. Protein-coding DNA regions are in upper letters and boldface, and the encoded amino acid sequence is displayed below the DNA sequence. Start of introns are labelled EoPII-X, where "X" corresponds to intron number. The beginning and the signal peptide, propeptide, metalloproteinase and the short-disintegrin domains are specified. Numbers at the right correspond to amino acid numbering of the DNA-deduced pre-pro-PII-SVMP relative to the mature SVMP. The N-terminal glutamine of the metalloproteinase domain has been assigned residue 1. The extended Zn^{2+}-binding environment (HEXXHXXGXXH) and the RGD integrin inhibitory motif stand on yellow background. The only two amino acids (−70I/V, and −111H/R) that distinguish this sequence from that of PII-SVMP EOC00006 (Q14FJ4) are shown in bold and red. The remains of a disintegrin-like domain transformed into intron EoPII-12 are underscored in italics and on cyan background. SINE/Sauria, LINE/L2/CR1 and DNA transposon retroelements are highlighted on a gray background. Simple sequence repeats (SSR, microsatellites) are shown in light green background.

```
ATG ATC CAA GTT CTC TTG GTA ACT ATA TGC TTA GCA GTT TTT CCA TTT CAA Ggt aag atg
    5'PS-Disi Fw                                                      |--- EoPII-1 →
 M   I   Q   V   L   L   V   T   I   C   L   A   V   F   P   F   Q                     -17
|------------------------------- Signal peptide -------------------------------|
ttc tct tta gtt ccc ttg ttc aga atc tta ctg cta aga cta ttg cac cca aca gat tgc
tat gtt gtt ggc ttt ggt ttt att ttt gac aat taa cta aag ttt gct tca ctt cag ttt
cca ctg att aag caa aag aag gtt ctc aag gat cgc att tgt tct aaa gtt aaa taa atg
gtt tat tgg aat ttg ctt aaa ttt tgt aaa gcc agg aga aaa tac taa gga aga aaa tcc
taa gat ttt cag aaa aaa aaa aaa acc ctc att cat ttt aga ggg aac gtt ttc caa tga
tgt gag tta aaa aaa aag aat taa aac act gaa gag gac aga gag aaa gtg ttt gtt gca
aga aaa ttt cag aaa gca aaa gaa aaa aat gct att gtt ttt aat ttt ctt tca ata ggt
cag tta gac atc ctg aaa tta aac atg cat aat att tag ccc ggg aca cga tgg ctc agt
aga tta gga tgg tga gct tat tga cca gca att tgg cgg tca aat acc tag tgc ctg tta
gct att tgt tag gta taa caa agt gag tgc ctg tta ctc atc cca gct tct gcc aac caa
|--------------------------------- SINE/Sauria →
gca gtt caa aag cat gta aaa atc caa gta gaa aaa tat ggg cca cta cag tgg gaa ggt
cac agc att cta tgt gct ttt ggt gtc tag tca tgc tag cca ttt gac cat aga gat gtc
ctt gga caa aca ctg gct ctt tga ctt tga aat aga aAT GAC ACT GAC CTC TAG AGT TGG
                                                                    Intr4 Fw
gaa atg agt agt atg cat atg tgg ggg act tta ctt tta cct aat att tag cca aat aaa
agc tag tgc ttt tga tgt tac aga gac tct tca gac tgt act taa tat aat atg aca cta
gct ttt ctc aat gca aga ttc cgt att atc tta aag tat ttc ttt cct cct tcc ctt tta
ctc ttc aaa aaa tag agc agt ccc ttc tga aat tct tcc caa gtt ctc ctt ctg tga gct
aag ata cat ttt aca tgc cag tta gcc aat cta tgg cta ttc ttc ctg tat ccc aag tca
ttc cca cat atc att aca aac aag aat act caa tag att cag aag ttg cca aac aaa ttt
aag ggt gag agg ctc act gca gta gcc ctg ttt gtt tgt ttg ttt gtt tgt tta aat agc
aaa gaa aga tga gaa tgc cct tat cag gag taa ttt atg ggg aac aat gta gga aag aag
aaa ggg aga tgg agc aga aat tgc tgc tta atc tgt gct cta agc aag aga gga ata gtA
AGC TTG CTT GCT AGT AGG TGG aag tga acc cct ccc atg ttt ctg gga gat cag gta caa
      IntrB9_4-2 Fw
ata ggt agg tac atc aat agg tat gag gaa gat gca ttc ttc tgt ctg gga ctg ccc atc
aat ctg aaa gga tga gtc att tcc agg gta aaa tgc tgg tat ctc tgt ttc tcc ctg tct ctc
tct ctc tgc ctc tct gtc tct ctc tct ctt tct ctc tct ctc tct ctc tca cac aca cac
aca gag act ata cat tta cag tat aag tta aaa aac aaa tat CAG GTG ACC ATA CAA TGT
                                                              Intr4 Rv
CCA tat ggg gaa caa ctc taa ttt ttt cca cta ttt ttc taa ttc tga aat cat tga att
atg aaa atg tta tta ggt gct tca gtg ata act atg ata cta tgg act ctc aat agc ttc
ttt tga aag tga aag aaa aga aag tca gca aaa gta tca gga tac cag tga agt agc att
tca gct gac agg cat tcc aga taa gat gca ttt gca tag aaa tct aag ttt gcc tag att
tta gac taa tgg tta tgc taa acc ata acg cag tat ggt tta ctg aat tgt ctg aat cag
att att gtg att ctc atc tta tca ctc agt gca caa act tgc aca ttg gtt gaa atc tat
tta ttt ttc taa ttt ttc aac ctc tga aat gtt tcc ttt gca cct gca caa aga aat gtc
tcc tgt tag act tcc atc ttt cag aca tca ttc agc aaa tta cat tcc cag aga aaa tct
gga ttt ctg tgc tgt gtc cca ctg gga aaa gat tcc att gtt cac tga taa ctg aat aac
aaa ttg tgt tgc aga tgt gaa att gat aga ata aca tcc caa aga aaa cca aat ctc ctt
ttc ttt ttc ctt aca aag GG AGC TCT AAA ACC CTG AAA TCT GGG AAT GTT AAT GAT TAT
                         G   S   S   K   T   L   K   S   G   N   V   N   D   Y      -31
|----------------------------------------- Propeptide →
GAA GTA GTG AAT CCA CAA AAA ATC ACT GGG TTG CCT GTA GGA GCT TTT AAG CAG CCA GAG
                                                     Prodom 3 Fw/Prodom 3 Rv
```

Figure A1. *Cont.*

```
  E     V     V     N     P     Q     K     I     T     G     L     P     V     G     A     F     K     Q     P     E          -51
AAA AAG gta aga tat ttc ttt cag caa caa att att ttt gtc agc cca tag aaa gtt tga
  K     K   |———————————— EoPII-2 →                                                                                            -53
tat tcc ttt cct gct att taa tgg tta ttt gga ttt tgc att gca atc tat gtt cct gtt
tga aat att tat tta ttt att tat tta ttt att tat tta ttt att tat tta ttt att tag
caa att tac act gcc tcc cag ttt aca caa aat tga gga taa tcc tgg caa ttt ata aac
aga att tcc taa tat taa tag gtt aat ata act aat caa gtg ttg gga gaa agc cag cca
tac agt gaa gag gaa tca agc agt gaa ata acc att agt cat ttc ata ctg caa cag atc
cag caa aga cat ccc ctc cct cca gga aat ccc att aaa gtt tat ctg atc tgc aga
gaa tgg aag cag tct agg gag agt cct gta tta cag gca aaa aat aat aaa ctg ctt aat
tta aac ttt gta tat cta cat gcg tca cat aga aat ata aaa gat ata ttt cta gat atg
ata gta tag tgg tta cag ctc aag gtt agt tct ata tac cag att cat gta tag cgt ggg
atg act ttt gag tca atc tgt ctg tcc cag atc aac ctc att cag caa tat gtt cag tta
gag aat gag atc ttg aac cta tgc agg aaa aat aaa taa aga tct tat cat tca aag cac
cag gtg aaa tac tct aat aat aat aat cta ata aat atg act atc aca aat tat tca
tta gat tta aat taa tac aga tgt gta tta aga aat gac aca ttt tat ttg cct aaa
ttt tga gaa gta aaa cat aaa ctc ttt gtt ttt cag TAT GAA GAC GCT GTG CAA TAT GAA
                                                 Prodom 2 Fw/Prodom 2 Rv
                                                   Y     E     D     A     V     Q     Y     E          -61
TTT GAA GTG AAT GGA GAG CCA GTG ATC CTT CAT CTG GAA AAA AAT AA    g tat gtt aac
  F     E     V     N     G     E     P     V     I     L     H     L     E     K     N     K   |———————————— EoPII-3 →     -77
tca gaa att ttc tta act tta cta aac aat gtg gaa aat gta ttt tcc tgg aca caa tct
gac aga aat aat aat tgc att cct ttg ttt gct ctt aaa tta aat gtt act ata tag gaa
aag tgg ata tgt taa tta tgt ctt acc ttg gaa ctg aaa ttt ttt tac tgc tgt tct cct
atg gga cac tgc tga act ata aca att tta gac tca gtt gaa cca ttc aag gct cat aaa
cgt cta tca ctt gaa tta act agg tta gtt atg aaa tat tgc aga gat ttc aaa atg ttt
ctt tat gct gct tca tag aca ctc cat caa tct gaa taa cat ttt ctg tga gac ctg tag
agc tgc tca atg cac atc aaa att tat atg tta aat cat gaa agg cag taa atg tca tct
cct att aag gcc cga aag ggg aca gga cat aaa taa gcc cag att tca caa gcc ttt ttt
aaa aga agt cct tgt gag tct tct cag aat tgt tcc att tga act gta aat caa gat att
aat aaa aat att ata caa ata tga acc aca aaa atg gaa tga aat tat ttt tca gca caa
aag aca aac ata gtt gga agg gac ctt gga ggc ctt cta gtc cac ccc ctg ctg aag cag
       |———————————————————————————————— LINE/L2/CR1 →
gag act ata tca ttc cat ggg tgt cca att gtt cct tga aaa cct cca gtg atg gga tac
cca cag ctt ctg aaa gct agc cat tcc acc gat taa ttg ttc tca ttc tca gga aat ttc
ttt tag ttc tag gtt gaa ttt ttC ACA GGT AAA TAA GCC ACA AAC ACC att aaa aga aag
                                      Intr3 Fw
tgt gcc att ttc att acc cta aaa act gta aat gtc ata aca cag ttt aaa tta gtc tgg
cag tca cta aac tgc aaa cac gta cct cag cct ggt tca gtt taa tac tat agt taa GAC
TGA GAT ACA GAG TTG GTG Cat tgg gaa aga ctg cct gcc ttt gat ttt agc atg tat ctt
Intr3 Rv
ggt ttt gat cat tct gca tgt acc ttt aaa agt ata ttg tct tta tta tta aac ata tgc
tac aag aaa ttg cac tat atg tta gta tct gtg tgt cac ttc cta ttc cac cgg aga
tat ttc atg cta aat tct aat gtg tgc ttt aga cgt tct agt gtc ccc ttt tgt ttt gtt
taa ata tga agg cta tac aga aat tca agg gta gca tat gga tgg tct tct ttt gtc ata
gaa gca aag cag ggc tac agg ggg aag acc agc aag gat ttc tag att ggc aaa aat gga
aga cga tga caa gtt tat ttt gac ttc agg att gga atg tga ctc cat ata ctg gtt
taa agc ttc aaa ata tat ctt tct gta tga atg ttg aag gca cac att tcc ttc agg ttg
ttt cat ttt acc ctt tag taa ata gaa cta tac cgt ttt cct tta gag aca aag cag ttt
cat cag att gtt gca tag aga aga aga ata aaa gga aga cgc aat aca aac tgt cct ctc
ttt gtc cat cac aat cct ctg agc att gat gaa gag aat tgc aca gct tca ttt ggt gaa
tct agt atg cac tgc ctt cct ctt aaa aag cca tgg gat aga aat acc tgc tcc att ctt
ttg ata cca aag aag ata ctt aaa ttg cca tta tac ggt tgt ttt aaa ctt ctt ata aaa
tta tct ctg tgc tac ata att cct gat ata gat atc ctt tct ttg cat tct ttc cag    A
GGA CTT TTT TCA GAA GAT TAC AGT GAG ACT CAT TAT TCC CCT GAT GGC AGC GAA ATT ACA
         Pro2-SVMP_Fw/Pro3_SVMP_Rv
  G     L     F     S     E     D     Y     S     E     T     H     Y     S     P     D     G     S     E     I     T          -97

ACA AAC CCT CCT GTT GAG gta ggg tct cac ttt tat gag cct ttt ttt tag gaa gta aac
  T     N     P     P     V     E   |———————————— EoPII-4 →                                                                   -103
tga aac aaa tgt ttg tgc aca ata tta caa ata tac aag aat gag acc agg cta ctc aaa
caa agt gta tat atg tat gaa gta ttt tat att gat atg tac gta caa gga tgc ctg gat
tgt taa acc ctg gtt aaa agc caa cat att tgg gag gtg agt ttc aca aat aga ttt att
atg aga aca tca ggt ttg taa gat tat att ttc att ttt aaa cca gac tac agg gat aaa
tgc aaa gtc ttt tat aaa tga taa caa ttc act ttg ctc cta tac aga aat cca ttt aat
atc ttt cat aat aaa atg gtg cca aaa atg gct cta tca gat gta aaa caa tta gag cac
taa cat gct tca tgt tgg ctc cat gct ccc aaa ttg att taa aag tgc att ctg tgt cta
ttt ctg gtt tat tat ctt caa ggt tca cac aaa tta ctc ttt tgt gtg atc agt ggg agt
ccc ctg cag tgt gac ttg att tat gga gac ttg cat tta tcc tat gtt cat ttt gca aca
atc aat att aag aag gtc ttc ggt tct cct gaa tca aaa ttt tct gga aaa att ctg tct
aaa tat ttc att gat gtt ctg gaa tac att gga act gta ctt ctc ctc atc aaa tca caa
tac aaa caa gct tta acc agt gta gtc ctC TTG CCT CCC TAC AG  GAT CAC TGC TAT TAT
                                          IntrB13-1 Fw
                                                             D     H     C     Y     Y          -108
CAT GGA CAC GTC CAG AAT GAT GCT GAC TCA ACT GCA AGC ATC AGC ACA TGC AAT GGT TTG
```

Figure A1. *Cont.*

```
   H   G   H   V   Q   N   D   A   D   S   T   A   S   I   S   T   C   N   G   L      -128
   AA  gta aga tag tct cta atc ttt tat ttg ttt att aat aat aat ata gtg ctc ttg gag
   K   |------------------------------------ EoPII-5 →                               -129
       ttc taa ttg tta aaa tga agg aca tcc tca gtt ttt cat gga aat tag ttg ggt gtg atc
       cag gat ttc ggc aga att aag aca tac ttt ggt tga aaa cca aga aga gct gct gcc agc
       cag gag aaa aac tat gga gca aaa tca cat aag tct aaa gga gct tcc aag ccc cgg tct
       cct ttc cca ggg tga ggt gat att aca ggt aga tga gat tag tag gtt tca aat tgg aga
       cct tgc tag aaa gtg tac agg aag agg caa gaa gtt tca gtt cta ccc aga aac act ttc
       ttg agt cac tct gca cac ttt ctt cag cca act aga tat gtt aac tac ata aag atc cca
       gaa ttc aga agg tcc cta tca ata gta aga atg aac atc acc tca aca tct ttt act gaa
       aaa aga cac tga aac tca cct ttg aac aga gac tgt gtc cat gga gtg gag gaa taa atg
       aaa agc tgg aac aga gca gaa taa caa cag aaa aat aaa gga aaa aca gaa tga cag aat
                                                                                    |--
       aat agc att gga agg gac ttt gag gtc ttc tag tcc aac ttc ctg ctc aag tag gag acc
       ------------------------- LINE/L2/CR1 →
       tat atc atc cta gac aaa tag ctg tca atc ttc tct taa aaa gca gta gtg atg gaa cac
       cca caa tgt ctg aat agg tta att gtt cca ttt gtg aga aaa tta ctc ctt agt tct aac
       tta ttt ctc tct ttg gtt act ttc cac gca ttg ctt ctc ctg cca tca ggt gaa gaa
       tag gtt gtc cca cat ttt tta tga cag cct ctt aaa tac tta aag att atc aag tca tct
       cta ccc ctt ctt gtc act agc atg agt ata ctc att gtc tgc agc cat tct aac cct cca
       gtt agt atg cat tct tat tcc ttt cat tgt tac tcc ctg ttg ttc tgc att gac ttc tct
       atg aga tgc ttg cca aga atc tat ttc att att tat aaa tat cct gtc atc tga ctc tat
       cta aat tgc tat caa act aat ctg att tta ttt cct tga cca cag aca aat att gtt cta
       tac ttg ttt aaa gta aat tgc agt acc tat aac tct ttt tag ata ttt tag cag tta
       tat ttt tcc ttt ttt atc cta ctt agt tgt gat tct tga gct tta tca gta ata tat atg
       ata aat ata aag tat ttt acc ctt atg aaa taa agt ttt aca caa agc aga atg tta caa
       ttg gct tta gtg ttg tat tta tgt agc tag aat ctt att ttt tta aca tcc tgg aaa tat
       aca ata ttg ggt tcc atg cca aaa tat ttc caa aca aaa ctg tac acc tat ttt gtg gct
       gca ctg agt ttg tga aat ctc tca tat ctt tct gat cat aac tgc atc tat gaa aag tat
       gag aaa gtg att tga gtg ctg agg aaa gaa tat aaa ata ttc act cat tgt taa gaa gga
       att caa aaa cat gag gtt agt tga aaa tgg gtc tca gag ccg agt ttc att acc caa cta
       ggt aac atc atc agt gca gtt ctt ctc tga act aac aat att ctc ttc ttt tgc ttc tcc
       atc tct gat cat cct ttt cac att gtt tta cag   A GGA TTT TTT ACG CTT CGT GGG GAG
                                                       G   F   F   T   L   R   G   E     -137
   ACG TAC TTA ATT GAA CCC TTG AAG GTT CCC GAC AGT GAA TCC CAT GCA GTC TAC AAA TAT
     T   Y   L   I   E   P   L   K   V   P   D   S   E   S   H   A   V   Y   K   Y      -157
   GAA GAT GCC AAA AAA AAG GAT GAG GCC CCC AAA ATG TGT GGG GTA ACC CTG ACT AAT TGG
     E   D   A   K   K   K   D   E   A   P   K   M   C   G   V   T   L   T   N   W      -177
              Prodom 1 Rv /Prodom 1 Fw
   GAA TCA GAT GAG CCC ATC AAA AAG GCT TCT CAT TTA GTT GCT ACT TCT GAA gta agt ctc
     E   S   D   E   P   I   K   K   A   S   H   L   V   A   T   S   E   |--- EoPII-6 →  -194
       ata gta aac ata gtt taa gat cac ata ctc att tgc ttg ttt aga aaa tat aaa gta aga
       gag aaa ttc ctt tgg gga gag gtg ata gat aga att taa aat gga gaa gcc ccc att tct
       ata ttt tta ttg taa agg taa agg taa agg ttc ccc cac aca tat gtg cta gtc gtt tct
       |------------------------------------------ SINE/Sauria →
       ggt ccg gtg ctc atc tcc gtt tca aag ccg aag ggc cag tgc ttg tct gag gac ata tcc
       gtg gtc atg tgg ctg gca tga cta gac acc aaa ggc gta cgg aac gct ttt tcc ttc cca
       ccg tag tgg tcc cta ttt ttc tac ttg cat ttt tac atg gtt ttg aat tgc tag ggt ggc
       aga agc tgg gac gag taa cag gcg ctc act ctg tta cac agc act cgg gat ttg aac cgc
       caa act gct gac ctt cag atc aac aag ctc agt gtc cta acc cac tga gcc act gcg tcc
       ctc att ttt att gta gcc aag gta taa aag aaA CAA TGG GAA ACT GAG GAA CAG aaa ata
                                                    Intr2 Fw
       aat ctt cca gtt gtt tga tca aac aaa ctt agt ttg gag att tga atc aaa aat gga ttt
       aaa tga gtt tct aaa tca tct cca ctt tct aag tca att ttg aaa agt aat taa att atc
       aat ttg gat gcg tct ttt atg gat gca gaa aga aat tga gat ggg gga gaa agt ggt ttg
       aaa tat tta atg gtt tta aga tgt ctg ata agg cca tta cat aat tgt gac tcc att ttc
       cag ttt gat ttg aat cat caa gtt gga ttg atg tga tga tgc aat aac tgg atg gaa agt
       ggc aat gtg aac cta gtc aca att ggc cct tat gct ctc aat gtt ttc ctc ctt tac tgg
       agg cac aaa aat tag aaa aca aaa tat tgc atc caa agt gac agt tcc tta cat ctt ttt
       ggt ggc aaa agt tga aac tgg ctc aaa aat ctc aac tgt ttt tat tag aat gtt aaa att
       gac atg gaa caa cag ctg ttg tta tga cgg aat acc aaa aca caa gtg aag acg cca aat
       gaa gtc ggg ttt gtc ttt tgg ctt ctt tca ttc tgg caa ttc aag att ctt tat cct cag
       caa ttt gtg gtt ata cgt tac att taa ctc ctt aat tgg ttc ttt atc ttg aat gtt
       ctt gtc tca aag tca gtc atg atc act tta ttc gga ctc gat ctt gtg cag cag aat taa
       gaa agt ggc tgt gag agg aga agg aga gaa att gca ttc tag aat tgc aac ttg gct tgg
       tag ttc tgt aca ctc tca tag gga aag aag caa tta atg cac act cac aca ttc agt ctt
       agc ata agt gaa ctt cct gag caa cct ggg ttc tct agt gga gga atc tcc ctt tgg aat
       gca gag agg ggc tct ttc cta gag aaa taa gag gca att cta gaa tga tca gaa aga ttc
       ttg cag aaa gat gcc cgg aag gga tcc tgc cta aag cct tca cag gat gta tct cat ggt
       tac acg gag tgt gct tta gtc tgg caa gat tcc tgt cta cag gag aag agc aat aaa gaa
       act act ctg agt aat taa cca cag ctt ctt gtt cct tgt gct caa gaa gtc tga aac aat
       att tca gta atc att tta aaa ata cat cac tga aaa ggc ata att ctt tgg cca tta aga
       gtg agt tca gtg ggt caa aat gtc tat ttg ctt ctc ctt acc ttg tct ttg tta cat agt
       gtg ata tga ttc cag gcc agg tct aat tgc atg acc aca tca gca tgt ctc tca cgt gat
```

Figure A1. *Cont.*

```
tgg ttg gca tcc ttt gtt tga gaa agg gaa gga aag ttg aga aag tca ttg aag cat cat
ttt gac agg gtg aaa aac acg tca aag aga aca gtt tcc tca tat gcc tat taa att ctt
ttc aga gtt agg tat tca tat ata cca tta tct tga caa tcc att gaa taa cgt act ttt
ttc ttc aaa act tta tca tGA CTT TCT CTA AGT CAG AGT TCC Caa acc ttt cca gct ttg
                          Intr2 Rv
```

```
ggg ata ggt gag gga gag ggg atg gtt cca cgt gaa cag tgg ggt cag gtg tgt acc cag
ctc tat ttg tgt gag gag tgg gca cac ata ccc act cgt gta aac aga gca cac cca cct
atg ctt gtt cac tgg tca tac aag tag aga tgc agc tgc tca cct gcc att tcc atg gcc
cag ttc tga agg gct gca ggc cca ggg cta aaa ttt taa caa gct gtc tcc ctg taa ata
tct tct tga aag aac tga tat ttc tgg aag ttg acc aga gag taa aac aag cat ttt tct
gat tat ctg agg ttg acc aac gtt ctt gtt ggc tgc tgg gag taa cat gtt taa aca gcc
att tac ttt tcc gtt tag CAA CAG CAT TTT CAC CCA AGA TAC GTT CAG CTT GTC ATA GTT
                        Met1PII Fw/Met1PII Rv
```

```
                        Q   Q   H   F   H   P   R   Y   V   Q   L   V   I   V        14
                        |------------------------------------- Metalloproteinase →
GCA GAC CAC TCA ATG gta agt atc ttg gat atc ttt cta ttt act ttt tgc att gag cgc
A   D   H   S   M   |---------- EoPII-7 →                                            19
agc att tcg ttt tgg cct ttt tta atg tgg gca ttt ttc aag aga tta tcc taa ctg aac
ttt ctc tta aaa tgc gct tta tca taa act ttg ata ttt ttt gtt att gga ccc aac caa
aat tta caa agc taa aag tca ttt gta aat ata ttt taa ttt gca cac ttg tta ttc ctg
gat acc att taa gtt tat ttt tat ccc aca ctg gtt aaa aag ttc tgt agg ttt ctt tga
aat gtt ccc acg ctc ttt ttt ttg tcc aag ttg gca aca tac aaa gaa aaa agt gaa gaa
tgc tta tct cac aca tct ctg aaa gag gaa ata ttt tct cta acc agg aaa aag gcc cat
gta tgt tgc tga aaa gtt aaa aat ctt aat ata tta atg gca caa atg tag att aaa aaa
aat agc tca aaa gaa ttc ttt gga ctg ctt gga taa aaa tta tta cat caa gaa att caa
aga tcc ttt cac taa tat att ctt ttt ctc ctc cct tct ttc cct tct tat aat tat caa
ctt gtc tta act ttt ttt ttt tgt ag GTC ACG AAA AAC AAC AAT GAT TTA ACT GCT TTA
                                    V   T   K   N   N   N   D   L   T   A   L        30
ACA ACA TGG ATA CAT CAA ATT GTC AAC GAT ATG ATT GTG gta aga aca aat gct tgt tca
    Met 1-2 Fw                                        |------------------- EoPII-8 →
T   T   W   I   H   Q   I   V   N   D   M   I   V                                     43
ttt taa act tca ctt agg ccc agc cga gat ttt gat tgt gtt aag ata aca aac ata atc
agg taa ata aag tag atg gat ttc taa atg caa acc tct gct ccg cac gct gca ttg gct
                                                                  |—— LINE/L2/CR1 →
ccc agt tgc tct ccg ggt gag att cag tgc tgg taa tga cct ata aag ccc tac atg gct
tgg gtc cag aat atc tga ggg aac acc tgc agc caa gtt ctc atc gtc cgg tac gct ccc
aca ggg agg ggc tcc tta gag tac cgc cgg caa agg att gcc ggc ggg tga ctc cta gag
aga ggg cct tct ctg tgg cac ccg tcc ttt gga acc agc tcc cag tgg agt tga gga
ctg ccc ccg acc tgc gtt ttt ttc gga gga acc tga aaa cat ggc tgt tta atc tga ccc
agg ctg gtg ttt tta gat ttg ggg tta ctt ggt ttt aat ttt gag gat tgt gtt taa
                         |——————— LINE/L2/CR1 →
tgt att ttt agc tgt ttt tta att ttt gta ata atg tct ttt aaa ttc ctg tac acc tcc
ctg agt cct tcg gga aaa ggg tgg ttt aaa aat aga att aaa taa ata aaa ata aat aaa
taa atg aat ggt ttc ttt tgt ggc ttt gag tga tct aca act atg tta ttt ggc cag ctg
gtg atc cag cat gtg agg cca tcc cag aaa atg ggt taa ttc agt aac taa att aat aag
cct cta atg gaa ata gag tca ttg ggt ttg gag gat gca aat cca aaa gtg ttg caa gaa
ttc agt cag aag tat atc tgt agt tgg att ctt tca ggg ctg tgg tta atg gtt ctg agc
ttg gag att aaa aaa tga tgg aca agg tca ggc tat tcc aag ctc agt tga tta tga aaa
tga tct ctg agt aga aac ctt gag aga agc att caa ttt gac ttg gat tgt ggt act tag
gag gag ata cag tca tat tgc att ctc cga tta gcc ttg ttt aca ttt tca cta ttg atg
atc agt tag aaa cag gga gag aac agg aag atg gaa gaa tca act ctc ctg tta gtc ctc
cta ctc ttc tgt tcc ttt tta cag aaa ata atc acc tgc ttt ttt tat cat gtt att tat
tag agt cct tag tac tgt cta agc ttg gtg gtt acc ttg cag aca ttt cat tat ctg act
act cta cat gag taa cca aag cag caa gct cat agc acc aag gac ttc aac ctt gag ctt
cta ttg atg tct agg tct att ttt ttt tca tct aca aaa agg gcc caa tgc cag ttg ctt
ttc aat att gaa atg ttt ctc aag gtt tac ttt gtt tga gtt tct gac tga tgc CAT GTC
CAC CTG ACA G   ATG TAC AGA ATT CTG AAT ATT CAT ATA ACA CTG GCT AAC GTA GAA ATT
Met 1-3 Rv
                  M   Y   R   I   L   N   I   H   I   T   L   A   N   V   E   I        59
TGG TCC AGT GGA GAT TTG ATT GCT GTG ACA TCA TCA GCA CCT ACT ACT TTG AGG TCA TTT
W   S   S   G   D   L   I   A   V   T   S   S   A   P   T   T   L   R   S   F        79
GGA GAA TGG AGA GCG AGA AAT TTG GTG AAT CGC ATA ACG CAT GAT AAT GCT CAA TTA ATC
G   E   W   R   A   R   N   L   V   N   R   I   T   H   D   N   A   Q   L   I        99
AC  gta tgt ctc att gtg ggg aaa ggg agt gag agt ggc tgg gag tgg agg att atg gaa
T   |--------------------------- EoPII-9 →                                          100
agg tta atg ctt gcc tag agc ttc tgt tct atg ctg tat gct tta aac cat gca tgt agt
aca ttt ctg ggt caa agt caa cca ctt ata tta tag atg aga cct ggc ttt gag aaa tct
ttg aat gat tgc agg tga aaa atc cat taa ata aat atg tat ttg ggg ttt gca tag tta
atg gaa tta aat taa ccc aaa tgg gtt gga ttt gac aac atg cta acc cct ctc ccc aac
ctc att cat gct aat gcc agc caa aca caa cat ata gtg aaa aat ata tcc tga ata ttt
tta acg gac ctg ctt aac taa atg gct ttt gag act aga gcc taa aat gaa ttc tag cat
tct gaa aac tgg tag taa ctg agg acg ggt caa agg gtt tac aga aat cca tat tta tgt
atg acc taa gac tac ata aca tgc ttc tat ctt cta tca att tta tcc ccc tcc cct tct
```

Figure A1. *Cont.*

```
ttt tta tag   A GCC GTT CAC CTT GAT AAC CTT ATA GGA TAC GGT TAC TTA GGT ACT ATG
              Met2PII Rv/ Met2PII Fw
              A   V   H   L   D   N   L   I   G   Y   G   Y   L   G   T   M      116
TGC GAT CCA CAA TCG TCT GTA GCA ATT ACT GAG gtt agt aga aag gat act tta tta tct
         Met6PII Fw/Met6PII Rv              |------------------------ EoPII-10 →
 C   D   P   Q   S   S   V   A   I   T   E                                      127
att tgt act caa gtg aaa cct tac ata cag aca aaa cat ctt ttc aaa taa agt ctc ttt
ctt att ttt gag cca cgt cat ttt cac cca tat tta ttt gca gat ttg aca tct cca ggt
cct gcg tca act aat ggc att ttg aca cag tgc att cta gaa caa gct ttt tta atg caa
tga gct ata tgt caa gga tga gaa tat att ata atg ttt atg gtt cag tca aac tgt act
ctg att ggc aaa tga aca ggt caa agc atg tta cac cac ttc caa ata atg ctt ctg aac
aat agt ctt agc aat ccc aaa gac aaa cat gaa ttc att cca aga aat tta gtg tct aga
ttg cat atg att gaa ttc tag tac att gag aaa aca aaa aat tac taa atc tac tca aaa
aga aaa aaa acc ctc tag ata tta gtt aag gtg atg cta tgc att tat tga gaa aga gta
aac tta gct ttt tgt tca cat aga aag aat gga gag aca tgg taa taa aca aaa gtt ata
caa caa aac tca taa agt ttt gtt tct taa taa gca gga tcc tgg tag tag tag
gta ctc ata agc cta ctt gct caa gaa ggt tat ttt att cag aaa gag caa ctc att cta
agt ctg ttt agg atg gct acc ttc aat att ctg aaa atg caa gat tgt agc aaa gga cac
tga gta gtt ttt ttc gac tga agt ttc ctg taa gtc agg gct gtc aaa ctc aat ttc att
     |---------- DNA transposon  →
gag ggc cac atc agc att gcg gtt gcc ctc aag ggg gtg gtt ggg tgt ggc cag ggt ggg
cac agg cca cag gca tgg ctg gaa tgt ata tgg cta agt ttt agt aac tga ata agt gca
gac agc aaa tgg atg cat aca ttt tga tct tat tct gtg ctg tag ctt ctg gct ata aag
ttt cct tct gga tgt att tgt gta tgt tct gga gtc ttg gtg ggc aca gat act ttc aga
gga gct aga aga atc ctg aga tgg tat cct caa cct aaa att ggt cac ttg gtc acc agt
ttt agc cac tta gtg gta ata ata att gga ttc act ttc agt ttc ttg gca gag taa caa
taa aaa aag tat tct tat ttc ttc aG GAT CAT AGC ACA GAT CAT CTT TGG GTT GCA GCT
                                     Met3PII Rv/ Met3 PII Fw
              D   H   S   T   D   H   L   W   V   A   A                          138
ACA ATG GCC CAT GAG ATG GGT CAT AAT CTG GGT ATG AAT CAT GAT GGA AAT CAG TGT AAT
 T   M   A   H   E   M   G   H   N   L   G   M   N   H   D   G   N   Q   C   N   158
TGT GGT GCT GCC GGA TGC ATT ATG TCT GCG ATC ATA TC gta agt att gag gaa tat gct
 C   G   A   A   G   C   I   M   S   A   I   I   S   |------------- EoPII-11 →   171
taa tgg ctt tcc aat caa gtt att ttt aaa tgg ttg caa aaa tga ata aag tat tct ctt
atc cat tct gtt agc ttt aga aga aaa caa atc att aca ttt ctt cat tag caa ttc ctt
ttc ctt ata tgt ttt tgc aat gaa att ctg ctc cta gtc caa agt tgg agg atg tca tga
tct ttt ttc ata tct aca g  A CAA TAC CGT TCC TAT CAG TTC AGT GAT TGT AGT ATG
                             Met4 Rv / Met4 Fw
              Q   Y   R   S   Y   Q   F   S   D   C   S   M                      183
AAT GAA TAT CGC AAC TAT ATT ACT ACT CAT AAC CCA CCA TGC ATT CTC AAT CAA GCC CTG
 N   E   Y   R   N   Y   I   T   T   H   N   P   P   C   I   L   N   Q   A   L   203
AGA ACA GAT ACT GTT TCA ACT CCA GTT TCT GAA AAT GAA CTT TTG CAG gta aga gaa gaa
 R   T   D   T   V   S   T   P   V   S   E   N   E   L   L   Q   V   R   E   E   219
                                    |------- Spacer ------| |------- EoPII-12 →

tgt gac tgt ggt tct cct gca tta agt ctt ttt ttt taa tca aca aaa gta att tga aga
 C   D   C   G   F   P   A   L   S   L   F   F   Stop
ata ttc tca gaa atg aga atc ctt gaa aaa tca tct agc ttt cta agt ggt ttg agc cat
cca aga gct tgg ctt gtg aat ggc tga ggt ttg tgc ctt tca tgt aca tgc atg tat gaa
gtg gtt tct tgg gtt gta gag gaa tgg aga act ggt atc tca cta cta ttt tgg gaa gat
ggt gaa ttt tta aaa acg ggt gat tga cca ctc caa gaa aat ctt tcc ctc ctg aaa ccc
cct att ttg ttg ata tag cca cat tat ctt gta cca cga ttt tct cga act gct ccc tcc
cat atc tga tta tct tta atc tat gct ctg atc cta ata ata ttt tta taa gaa cag taa
tat agt gtt ttt atg ttg tta aat aca cct gtg atg gtc tgt gag aat gtc ctt aag aga
caa aay aay gac gaa aca tcc ayt taa tyy tcy tat aay aag yay att aac cty cay aay
caa tgg caa aaa tct caa gat gga cac ttc cca ccc att ctc ttg gtc cgt aaa gat gag
gtg gta caa ata gat ttt cag tat tga aag att ctg cta ctg taa cct tac aat cat gtc
gca tta ata ctc aag gtt gct cct tct tct cta gac taa ctc aaa ggc tgg cat gat gag
tag aaa atc cat cat gaa taa gaa agg aat ggg gct gta ggt tat gtg ggc ttc aca att
agg aga tga gga tat ttt tgt ttt att ctt ttc acg tag gaa ata tca gat aag gct ctt
tcg cag aga aat gcc ttt agc tgt ttt caa taa caa aca att tgt gca tct cct agc atg
aac tca taa gag gga aca tat cgc aga aat gtt cct ctt caa aat aga cca att aaa aaa
gaa aat tct atg cca tca ttc gat atg ttt tgg ttt tca g  AAT TCT GTA AAT CCA TGC
                                                       N   S   V   N   P   C    225
TAT GAT CCT GTA ACA TGT CAA CCA AAA GAA AAG GAA GAC TGT GAA TCT GGA CCA TGT TGT
                                                      Fw_Ocella NcoI
 Y   D   P   V   T   C   Q   P   K   E   K   E   D   C   E   S   G   P   C   C   245
                                    |------- Short  Disintegrin Ocellatusin→
GAT AAC TGC AAA gta aga ctt att tat ttt taa cac caa gag aaa ttt tac cct gct cca
 D   N   C   K   |------------------------- EoPII-13 →                          249
tac tag ccg tat aga aat ata ata ttt ctt ggc tgt tta cta tga taa aaa cat ttc agc
IntrDis1 Rv
tct att tcc tat ccc ttc ttc cag ttt att tga ccc tta tga aca taa gca aag gga aga
taa ttt aac aaa att tct ccc tta ttt caa ttt caa atg cac tct ttc agc atg cta aat
```

Figure A1. *Cont.*

```
cat atc tgt gaa aat aat aca ttt gta gtt tga ctg aaa tta cat gga aac taa gtt taa
aca agg gtg agc agt gta tga gat tgg tgc cct tac tca gct tcc tga gtt tct gga agg
ttc taa gag ttt cct ggt aat gct gtg aca ttt ttt tct ctg agc ctt tta aga agg aaa
tca atg cac aga ctt ctg gaa gta aaa tcg cct ttt ttc ccc att aag ttc tct tcc tac
tct cta aag cac taa att cag gta ttt tgg tgg tac att ctg gaa gtg ctg cag cac cat
gaa aag aga ggt gca cgt tgc cca ttc ctt ctt tct atc tgg cat cac att tga ctc ttt
tga gca gaa tgg ccc aaa aca ttt tgt tat tac cat att tcc atc aca agc cta gct tcc
gag caa gaa aag ggg gcc atg tgt ttt cag caa gtg ata gaa aat tct aca aat gct tcc
tcc aat gta aag aaa taa aaa tag atc aga ata agt tca gca ttt aaa ttt ttg ctg ctt
ttt caa ggc agc tca act gat ttt cac ttt atg gtc agc caa cat gta gaa gtt ctg ttt
cag gaa ttg agc ctt tca ttg caa cca ttt ccc caa agc aaa caa gtt gga ctg gga ctt
cta ggc aac aca cac agt tgt agc agg gca ggg atg cct ttc ttg gtg atc ctc aag aca
gat gaa gag gag gtt ttg aaa tgt gtc cct ctt tga tct ctg cta ctg aag aat gat agc

tgg agt att ttt tat tct cac cca cag TTT CTG AAG GAA GGA ACA ATA TGC AAG ATG GCA
                                     F   L   K   E   G   T   I   C   K   M   A    260
AGG GGT GAT AAC ATG CAT GAT TAC TGC AAT GGC AAA ACT TGT GAC TGT CCC AGA AAT CCT
 R   G   D   N   M   H   D   Y   C   N   G   K   T   C   D   C   P   R   N   P    280
TAC AAA GGC GAA CAT GAT CCG ATG GAA TG  gtg agt aaa aga tta cct cta acc tgt gtg
 Y   K   G   E   H   D   P   M   E   W  |------------------------- EoPII-14 →       290
ttc taa agt ctg att cca agg ggt aat acc taa aaa aaa gaa gta atc ttt caa tac taa
aag ctg tga atc tac ttg aaa gaa aaa gta tcc atc tac cct tct ttt ggt tgt tgt tat
ttt gat ttt tct tca gac aac aac cac aac aaa tgc ggt caa tgt cca gga ctg ttc ctt
tct tgc aag aac aaa atg ctt ggc ctt ctc agg gcc ttg tgc tta ggt gga aga gag aaa
tga gaa aaa tgg ggc aga tct agt tgt gac cta aca atg aag caa acc caa atc tta cct
taa aga atc agg aat tgc tga atc ccc ttg att ttt ata caa tag aac ctg aaa gaa gtt
tgg gtt agt ttg gaa agt gct gtc tta cac cat tga aaa tct ctt tct ttg act ttc   ag
G CCT GCA CCA GCA AAA GGC AGT GTG TTG ATG TGA
         Dis PII Rv
 P   A   P   A   K   G   S   V   L   M  STOP                                      300
```

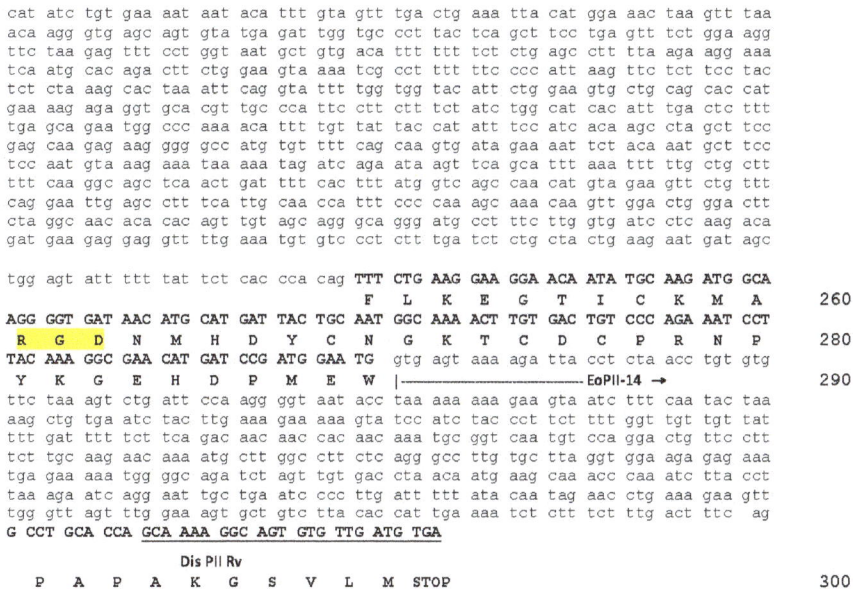

Figure A1. Genomic organization of *E. ocellatus* EOC00006-like PII-SVMP gene.

Genomic sequence of *E. ocellatus* EOC00028-like PI-SVMP gene. The locations and identities of the primers used to PCR-amplify genomic sequences (listed in Table 2) are indicated. Protein-coding DNA regions are in upper letters and boldface, and the encoded amino acid sequence is displayed below the DNA sequence. Start of introns are labelled EoPI-X, where "X" corresponds to intron number. The beginning and the signal peptide, propeptide, metalloproteinase domains and the C-terminal extension are specified. Numbers at the right correspond to amino acid numbering of the DNA-deduced pre-pro-PII-SVMP relative to the mature SVMP. The extended Zn^{2+}-binding environment (HEXXHXXGXXH) stands on yellow background. The only two amino acids (-124T/A and 15T/A) that distinguish this sequence from that of PI-SVMP EOC00028 (Q2UXQ3) are shown in bold and red. The remains of a disintegrin-like domain and a dimeric disintegrin domain transformed into intron EoPI-12 are underscored in italics and on cyan background. The *N*-terminal glutamine of the metalloproteinase domain has been assigned residue 1. SINE/Sauria, LINE/L2/CR1, LTR/ERV1, DNA/hAT-Ac and DNA transposon retroelements are highlighted on a gray background. Inserted nucleotide sequences in introns 1 (582 nucleotides between positions 1534–1582, including a SINE/Sauria element); 9 (between nucleotides 194–195 of the topologically equivalent intron of PII); 11 (replacing nucleotides 1-66 of PII intron 11 for a stretch of 3281 nucleotides); and 12 (after nucleotide 999 of the homologous PII intron) are underlined. Simple sequence repeats (SSR, microsatellites) are shown in light green background.

```
ATG ATC CAA GTT CTC TTG GTA ACT ATA TGC TTA GCA GTT TTT CCA TAT CAA G gta aga tgt tct gtt tag ttc cct tgt
      Sp35_Eo Fw                                                      |------------------------- EoPI-1 →
   M   I   Q   V   L   L   V   T   I   C   L   A   V   F   P   Y   Q                                          – 17
   |------------------------------ Signal peptide ------------------------------|
tca gaa tct tac tgc taa aag act att gca ccc aaa aga ctg cta tgt tgg tag ttt tgg
ttt tat ttt tga caa tta acc aaa gtt tac ttc act tca gtt tct aaa gat taa gca aaa
gaa tgt tct caa gga tca cat ttg ttc taa agt tac ata aat ggt tta ttg gta ttt gtt
taa att ttc tca agt aag gag caa atc cta agg aaa aaa gtc cta aga ttt tca tta aaa
aag cac att cat gtt gga ggt aat ttt ttc caa tga taa gat taa cac att gaa gag gac
aga gag aaa gtg ttt gtt gca aaa aaa att cag aaa gca aaa gaa aaa aat gtt ttt att
ttt aat ttt ctt tca ata ggt caa tta gac atc ctg aaa tta agc atg cat aat att tag
cca ggg aca caa tgg ctc agt agg ttc gga taa tga att tgt taa cca gat ggt gag cag
   |------------------------------------- SINE/Sauria → 
act ggc ggg tca aat ccc aag tgc cac gta aca gag tga gtg cct gtt act tgt ccc agt
ttc tgc caa cct agt aat tca aaa gca tgt aaa aat cca agt aga aaa ata agg acc act
aca gtg gga atg taa cag cat tct atg tgc ttt tgg cat CTA GTC ATT CCG GCC ATA TGA
                                                     Intr1F1PI Fw
cca ctg aga tgt cct tgg aca aac act ggc tct ttg act ttg aat gga gat gag cac caa
ccc cta gag ttg gaa atg agt agt atg cat gtg tgg ggg aac ctt tac ttt tac cta ata
ttt agc caa ata aaa gct agt gct ttt gat gtt aca gag aat ctt cag act gta ctc aat
ata atg tga cag tag ctt ttc tca atg caa gat tcc gta tta tct taa agt att tct ttc
ctc ctt ccc ttt tac tct gca aaa aat aga gca gtc cct tct gaa att ctt ccc aag ttc
tcc ttc tgt gag cta aga tac att tta cat gcc agt tag cca atc tat ggc tat tct tcc
tgt atc cca agt cat tcc cac ata tca tta caa aca aga ata ctc aat aga ttc aga agt
tgc caa aca aat tta agg gtg aga ggc tca ctg cag tgg ccc tgt ttg ttt gtt tgt ttg
ttt gtt tgt ttg ttt gtt tgt ttg ttt aaa tag caa aga aag atg aga atg ccc tta tca
gga gta att tat ggg gaa caa tgt agg aaa gaa gaa agg gag atg gaa gag aaa ttg ctg
ctt aat ctg tgc tct aag caa gag agg aat agt aag ctt gct agt agg tgg agg tga acc
cct ccc atg ttt ctg gga caa tca ggt aca aat agg tag gta cat caa tag ata taa gga
aga tgt att ctt ctg tct ggg gct gcc cAT CAG TCT GAG AGG ATG CAT TTC Cag ggt aaa
                                      Intr2F1PI Fw
atg ctg gtt tct ctg ttt ctc cct ctc tcc ctc tct ctg ttt ctc ttt cac tct ctc tct
   |------------------------------- LTR/ERV1 and DNA/hAT-Ac →
ctc tct cac aca cac aca cac aca aac aca aac aca cac aga cac aga cac aca cag aga
gag aga gag aga gag aga gag aga gag aga gag aga gtc tgt ctc tct agc tcc ttc cct
ctc ttt ctc tct ctc tct gtc tgt ctc tgt ctc aat tgt tta ggt gtt tgg gta gca gtt
tct ttc cta taa ggt aaa ggt aaa gtt tcc ccc ata cac acg tgc tag tca ttt cca act
   |------------------------------- SINE/Sauria →
cta gga gct gat gct cat cgc cgt ttc aaa gct gaa gag cca gtg ctt gtc cat gga cat
ctc cgg gat cat gtg acc agc atg act aaa tgc cag aga tac atg gaa aac tgt tat ctt
```

Figure A2. *Cont.*

```
ccc act gca gta gtc cct att ttt cta ctt gca ttt tta tgt gct ttc aaa ctg cta ggt
ggg cag aag ctg gga caa gtt aac aag tta act cac tct gtt acg ctg cac tgg gga ttc
aac cca cca atc tgc cga cct tct gac cga caa gct cga tat cct aag cta ccg tgt ctc
ttt tct ttc cta tac cct gtg gtt att gat att tga att tat tgt ttt gtt tct gct tgt
aat tat gag cac aac ata ccc tgt cct ttc cat taa tct gct aac aaa tct tgg tct cag
ttc tga ttg tac atc caa gca tac ctg cag gtg tca gac tgc ctc att tac aat ata agt
taa aca aca aat atc aGT GAC CAT GCA ATG TCC ATA TGG gga cca act cta aat ttt tca
                                                      Intr3F1PI Fw
cta ttg gtc taa ttc tga atc att gaa tta tgg atg tta tta gat aat tat gat gct gtg
aac tct cca tag att att tgg aaa gtg aaa gaa aag aaa gtc agc aaa agt att agg ata
gca gta aag tag cat ttc agc tga aag gaa ttc caa ata atc tgc att tgc ata gaa atc
taa gtt tgc cta ggt ttt aga cta atg gat atg cta aac cat aac aca ata tgg ttt act
gaa ttg tat gaa tca att att gtg att tac agg ttc tca tct tat cac tca ggg cac aaa
ctt gca cat tgg ttg aaa tct att tat ttt tct aat tat tca agc tct gaa atg ttt cct
ttg cac ctg cac aaa gaa atg tct cct gtt aga cta cca ggt ctg tca gac cat cat gca gca
aat tgc att ccc aga gaa aat ctg gat ttc tgt gct gtg tcc cac tgg aaa aag att cca
ttc ttc act gat aac tga ata aca aat tgt gtt gtg gat gtg aaa tcg ata gaa taa cat
ccc aaa gaa aac cga atc ttc ttt tct ttt tcc tta tga ag  GG AGC TCT AAA ACC CTG             -23
                                                         G  S   S   K   T   L
                                                         |------- Propeptide  →
AAA TCT GGG AAT GTT AAT GAT TAT GAA GTA GTG AAT CCA CAA AAA ATC ACT GGG TTG CCT             -43
  K   S   G   N   V   N   D   Y   E   V   V   N   P   Q   K   I   T   G   L   P
GTA GGA GCT GTT AAG CAG CCT GAG AAA AAG gta aga tat ttc ttt cat caa caa att att
    Met15PI Fw                           |-------------------------- EoPI-2 →
  V   G   A   V   K   Q   P   E   K   K                                                      -53
ttt tgt cag tcc ata gaa ggt ttg ata ttc ctt tcc tgc cat tta atg gtt att tgg att
ttt cat tgc aat cca tgt tcc tgt ttt att tat tta ttt gtt tgt ttg ttt gtt tgt ttg
ttt gtt tgt ttg ttt gtt agc aaa ttt tac tgc ctc cca gtt tac aca aaa ttg agg aga
atc ttg gca act tac aaa cag aat ttc cta ata tta ata gtt tag tat aac taa tca agt
gtt ggg aga aag cca gcc ata cag tga aga gga atc aag cag tga aat aac cat tag tca
ttt cat act gca aca gat cca aca aag aca tct ccc tcc tcc agg aaa tcc cat taa
agt tta tct gat ctg cag aga atg gaa gca gtc tag gga gag tcc tgt att aca ggc aaa
aaa taa taa act gct taa ttt aaa ctt tgt ata tct aca tgc gtc aca tag aaa tat aaa
aga tat att tct aga tat gat agt ata gtg att aca gct aaa gat tag ttc tat ata cca
gat tca att ctg gcg tgg gac gac ttt tga gtc aat ctc tct gtc cca gat caa cct cat
tca gca ata tgt tca gtt aga gaa tga gat ctt gaa cct atg cag gaa aaa taa ata aag
atc tta tca ttc aaa gca cca ggt gaa ata cct aat aat aat cta ata aat aat ttg
act atc aca aat tat tca tta gat tta aat taa tac aga tgt aga gta tta aga aat gac
aca ttt tat ttg cct aaa ttt tga gaa gta aaa cat aaa ctc ttt gtt ttt cag TAT GAA            -55
                                                                          Y   E
GAC GCT GTG CAA TAT GAA TTT GAA GTG AAT GGA GAG CCA GTG GTC CTT CAT CTG GAA AAA
     Prodom 2 Fw/Prodom 2 Rv
  D   A   V   Q   Y   E   F   E   V   N   G   E   P   V   V   L   H   L   E   K            -75
AAT AA gta tgt taa ctc aga att ttt ttt aac ttt act aaa caa tgt gga aaa tgt ata
  N   K  |-------------------------------- EoPI-3 →                                          -77
ttc ctg gac aca atc tga gag aaa taa taa ttg cat tcc ttt gtt tgg aaa tta aaa tta
aat taa atg tta cta tat aga aaa agt gga tat aga tat taa gta tgt taa tta tgt ctt
acc atg aaa ctg aat ttt ttt tac tgc tgt ttt cct atg gaa cat tgc tga acc ata aca
att tta gac tca gtt gaa cca ttc aag gct cat aac ctt cta tca ctt gaa tta act agg
tta gtt atg aaa tat tgc agt gat ttc aaa atg ttt gtt tat gct gct tca tag aca ctc
cat caa tct gaa taa aat ttt cta tga gac ctg tag agc tgt tca acg tac ata aaa aaa
att ata tta aat cat aaa agg cag gaa atg tca tct cct att aag acc cga aag gga cag
gac cta aat aac cct gga ttt cac aag cct ttt tta aaa tgc ctt gtg agt ctt ctc
aga gtt gtt cca ttt aaa ctg taa atc aag ata tta ata aaa ata tta tac aaa tac gaa
cca caa aaa tgg aat gaa att att ttt cag cac aaa aga caa aca tag ttg gaa ggg acc
                                                                 |------------------- LINE/L2/CR1 →
ttg gag gcc ttc tag tcc acc ccc tgc tga agc agg aga cta tat cat tcc atg ggt gtc
caa ttg ttc ctt gaa aac ttc cca cag ctt ctg aaa gct acc cat tcc acc gat taa ttg
ttc tca ttc tca gga att ttt tta tag ttc tga gtt gga ttt ttc aca ggt aaa taa gcc
aca aac acc cct aaa aga aag tgt gcc att aac att acc cta aaa act gta aat gtc ata
aca cag ttt aaa tta gtc tgg cag tca cta aac tgc aaa cac gta cct cag cct ggt tca
gtt taa tac tat agt taa GAC TGA GAT ACA GAG TTG GTG Cat tgg gaa aga ctg cct gcc
                        Intr3 Rv
ttt gat ttt agc atg tat ctt ggt ttt gat cat tct gca tgt acc ttt aaa agt ata ttg
tct tta tta tta aac atg tgc tac aag aaa ttg cac tat atg tta gta tct gtg tgt cac
ttc ctc cta ttc cac cgg aga tat ttc atg cta aat tct aat gtg tgc ttt aga cgt tct
agt gtc ccc ttt tgt ttt gtt taa ata tga agg cta tac aga aat tca agg gta gca tat
gga tgg tct tct ttt gtc ata gaa gca aag cag ggc tac agg ggg aag acc agc aag gat
ttc tag att ggc aaa aat gga aga cga tga caa gtt tat ttt tct gac atc agg act gga
atg tga ctc cat ata ctg gtt taa agc ttc aaa ata tat ctt tct gta tga atg ttg aag
gca taa att tcc ttg agg ttg ctt cat ttt act ctt tag taa ata gaa cta tat cgt tct
ctt tta gag aca atg cag ttt cat cag att gtt gca tag act aaa aga ata aaa gga aga
cgc aat aca aac tgt cct ctc ttt gtc cat cac aat cct ctg agc att gat gaa gag aat
tgc aca gct tca ttt ggt gaa tct agt atg cac tgc ctt cct ctt aaa aag cca tgg gat
aga aat acc tgc tcc att ctt ttg ata cca aag aag ata ctt aaa ttg cca tta tac ggt
```

Figure A2. *Cont.*

```
tgt ttt aaa ctt ctt ata aaa tta tct ctg tgc tac ata att cct gat ata gat atc tct
tct ttg cat tct ttc cag  A GGA CTT TTT TCA GAA GAT TAC AGT GAG ACT CAT TAT TCC
                                                      Pro2 Fw/Pro3 Rv
                         G   L   F   S   E   D   Y   S   E   T   H   Y   S          -90
CCT GAT GGC AGC GAA ATT ACA ACA AAC CCT CCT GTT GAG gta ggg tct cac ttt tat gag
 P   D   G   S   E   I   T   T   N   P   P   V   E  |--------------------- EoPI-4 →  -103
cct ttt ttt aag gaa gta aat tga aac aaa tgt ttg tgc act ata tta caa ata tac aaa
aat gag acc agg cta ctc aaa caa agt gta tat aag tat aaa gta tct tat att gat atg
tac tta caa aga tgc ctg gat tgt taa tcc ttg gtt aaa agc caa cat att tgg gag gtg
agt ttc aca aat aga ttt att atg aga aca tca ggt ttg taa gat tat att ttc att ttt
aaa cca gac tac agg gat aaa tgc aaa gtc ttt tat ctg taa tac caa aag tga taa caa
ttc act ttg ctc cta tac aga aat cca ttt aac atc ttt cat att aaa atg gtg cca aaa
atg gct cta tca gag gtt aaa aaa tta cag cac taa tat gct tca tgt tgg ctc cat ttc
ccc aaa ttg att taa aag tgc att ctg tgt cta ttt ctg gtt tag cat ctt cat ggg ttg
cac aaa tta ctc ctt tgt gcc atc agt ggc act ctc cca tag tgt gac ttg att tat gga
gac ttg cat tta tcc tat gtt cct ttt gca ata gtc agt att aag aag gtt ttc tgt cct
cct gaa tca aaa ttt tct gga aaa ctg ctg tct aaa tat ttc att gat gtt atg gaa tac
att gga act gta ctt ctg ctc atc aaa tca caa tca acc agt gta gtc ctC
TTG CCT CCC TAT AG  GAT CAC TGC TAT TAT TAT GGA CGC ATC CAG AAT GAT GCT GAC TCA
     IntrB13-1 Fw/Met16PI Rv
                     D   H   C   Y   Y   Y   G   R   I   Q   N   D   A   D   S      -118
ACT GCA AGC ATC AGC ACA TGC AAT GGT TTG AA gta aga tag tct cta atc ttt tat ttg
 T   A   S   I   S   T   C   N   G   L   K |------------------------ EoPI-5 →        -129
ttt att aat aat aat ata gtg ctc ttg gag ttc taa ttg tta aaa tga agg aca tcc tca
gtt ttt cat gga aat tag ttg ggt gtg atc cag att ggc aga att aag aca tac ttt
ggt tga aaa cca aga aga gct gct gcc agc cag gag aaa aac tat gga gct aaa tca cat
aag tct aaa gga gct tcc aag ccc cgg tct cct ttc cca ggg tga ggt gat att aca ggt
aga aga gat tag tag gtt tca aat tgg aga cct tgc tag aaa gtg tac agg aag agg caa
gaa gtt tca gtt cta ccc aga aac act ttc ttg agt cac tct gca cac ttt ctt cag cca
act aga tat gtt aac tac ata aag atc cca gaa ttc aga agg tcc cta tca ata gta aga
atg aac atc acc tca aca tct ttt act gaa aga cac tga aac tca cct ttg aac aga
gac tgt gtc cat gga gtg gag gaa taa atg aaa agc tgg aac aga gca gaa taa caa cag
aaa aat aaa gga aaa aca gaa tga cag  aat aat agc att gga agg gac ttt gag gtc ttc
                                    |--------------------- LINE/L2/CR1 →
tag tcc aac ttc ctg ctc aag tag gag acc tat atc atc cta gac aaa tag ctg tca atc
ttc tct taa aaa gca gta gtg atg gaa cac cca caa tgt ctg aat agg tta att gtt cca
ttt gtg aga aaa tta ctc ctt agt tct aaa tta ttt ctc tct ttg gtt act ttc cac gca
ttg ctt ctt ctc ctg cca tca ggt gaa gaa tag gtt gtc cca cat ttt tta tga cag cct
ctt aaa tac tta aag att atc aag tca tct cta ccc ctt ctt gtc act agc atg agt ata
ctc att gtc agc cat tct aac ctc cag tta gta tgc att ctt att cct tca ttg tta
ctc ctg ttg ttc tgc att gac ttc tct atg aag atg ctt gcc aag aat tct tat ttt cat
tat tta tta aat atc ctg gtc atc ctg act ctt atc tta aat tgc tat caa act aat ctg
att tta ttt cct tga cca cag aca aat att gtt cta tac ttg ttt aaa gta aat tgc agt
att acc tat aac tct ttt tag atta ttt tag cag tta tat ttt tcc ttt ttt atc cta ctt
agt tgt gat tct tga gct tta tca gta ata tat atg ata aat ata aag tat ttt acc ctt
atg aaa taa agt ttt aca caa agg atg tta caa ttg gct tta gtg ttg tat tta tgt
agc tag aaa ctt att ttt tta aca tcc tgg aaa tat aca ata ttg ggt tcc atg cca aaa
tat ttc caa aca aaa ctg tac acc tat ttt gtg gct gca ctg agt ttg tga aat ctc tca
tat ctt tct gat cat aac tgc atc tat gaa aag tat gag aaa gtg att tga gtg ctg agg
aaa gaa tat aaa ata ttc act cat tgt taa gaa gga att caa aaa cat gag gtt agt tga
aaa tgg gtc tca gag ccg agt ttc att acc caa cta ggt aac atc atc agt gca gtt ttt
ctc tga act aac aat att ctc ttc ttt tgc ttc tcc atc tct gat cat cct ttt cac att
gtt tta cag  A GGA TTT TTT ACG CTT CGT GGG GAG ACG TAC TTA ATT GAA CCC TTG AAG
             G   F   F   T   L   R   G   E   T   Y   L   I   E   P   L   K          -145
GTT CCC GAC AGT GAA TCC CAT GCA GTC TAC AAA TAT GAA GAT GCC AAA AAA AAG GAT GAG
                                                      Prodom 1 Rv/Prodom 1 Fw
 V   P   D   S   E   S   H   A   V   Y   K   Y   E   D   A   K   K   K   D   E      -165
GCC CCC AAA ATG TGT GGG GTA AAC CTG ACT AAT TGG GAA TCA GAT AAG CCC ATC AAA AAG
 A   P   K   M   C   G   V   T   L   T   N   W   E   S   D   K   P   I   K   K      -185
GCT TCT CAT TTA GTT GCT ACT TCT GAA gta agt ctc ata ata aac ata gtt taa gat tac
 A   S   H   L   V   A   T   S   E  |------------------------ EoPI-6 →               -194
ata cta att tcc ttg tct tga aaa tat aaa gta aga agg aat ttc ctt tgg gaa ggg gtg
ata gat aga att caa aag gga gaa gcc ccc att tct ata ttt tta ttg tag cca tgg cat
aaa aga aag aat gga aac ttg agg aac aga aaa tac att ttc cag gct tat agc att ttc
ttt ggt cat tca aac tta gtt tag gga ttt gaa tca aaa tct att taa atg agt ttc taa
att atc tct agt ttc taa gtc aat gtt gaa aag taa tta aat tat caa ttt gga ttc ctc
ttt tat gca tgc aga gag gat ggg gga caa agt ggt ttg aaa tat taa atg gtt tta aga
tgt ctg ata agg cca tta cat aat tgt tac tcc att atc caa ttt gat ttg aat cat caa
gtt gga ttg atg caa tga atg gat gaa aag tga caa tgt gaa cct agt cac aat tga ccc
tta tgc tct caa tat ttt cct cct tta ttg gac gca caa aaa tta gaa aac aaa ata ttg
cat cca aag tga cag ttc ctt tcc gtt gtg gca aaa gtt gaa act ggc tga aaa
atc tct act gtt ttt att aga atg tta aaa ttg aca TGG AAC AAC AGC TGT TGT TAT GAC
                                                  IntronB7PI Fw
Gga ata cca aaa cac aag tga aga cgc caa atg aag cct ggt ttg tct ttt ggc ttc ttt
cat tct ggc aat tca aga ttc ttt atc ctc agc aat ttg tgg tta tac gtt aca ttt aac
```

Figure A2. *Cont.*

```
tgg tcc tta att ggt tct cta tct tga atg ttc ttg tct caa agt cag tca tga tca ctt
tat tcg gac tcg atc ttg tgc agc aga att aag aaa gtg gct gtg aga ggg gaa gga gag
aaa ttg cat tct aga att gca act tgg cct ggt agt tct gta cac ttt cat agg gaa aga
agc aat taa tgc aca ctc aca cat tca gtc tta gca aaa gtg tac ttc ctg agc aac ctg
ggt act cta gtg gag gaa tct ccc ttt gga atg cag gaa gga ggg gct ctt tcc tag aga
aat aag agg cag atc aga aat gaa tcc ttg gat tgc aga aag atg ccc gga agg gat cct
gcc taa agc ctt cac agg atg tat ctc atg gtt aca cag agt gtg ctt tag tct ggc aag
gtt cct gtc tac agg aga aga gca ata aag aaa cta ctc taa gta att aac cac agt ttc
ttg ttc ctt gtg ctc aag aag tct gaa aca ata ttt cag taa tca ttt taa aat tac atc
act gaa aag aca taa ttc ttt ggc cat taa gag tga gtt tag tgg gtc aaa atg tct att
tgc ttc tcc tta cct tgt ctt tga tgc ata gtg tga tat gat tcc agg cca ggt cta att
gca tGA CCA CAT CAG CAT GTC TCT CAc gtg att ggt tgg cat cct ttg ttt gag aag ggg
        IntronB7PI Rv
aag gaa agt tga gaa agt cat tga agc atc att ttg aca ggg tga aaa aca ggt caa aga
gaa cag ttt cct cat atg cct att aaa ttc ttt tca gag tta ggt att cat ata tac cat
tat ctt gac aat cca ttg aat aac gta ctt ttt tct tca aaa ctt tgt cag cac ttt ctc
taa gtc aga gtt ccc aaa cct ttc cag ctt tgg gga tag gtg agg gag agg gga tgg ttc
cac atg aac agt ggg gtc agg tgt gta ccc agc tct att tgt gcg agc agt ggg cag caca
tac cca ctc gtg taa aca gaa cac att cac cta tgc ttg ttc act ggt cat aca agt aga
gat gca gct gct cac ctg cca ttt cca tgg ccc agt tct aaa ggg atc aag gcc cag ggc
taa aac tct aac aaa ctg tct ccc tgt aaa tat ctt ctt gaa aga act gat att tct gga
agt tga cca gag agt aaa aca agc att ttt ctg att atc tga ggt tga cca aca ttc ttg
ttg gct gct ggg agt aac atg ttt aaa cat cca ttt ttt aat tgt tct gtt tag CAA CAA
                                                                          Q   Q        2
                                                                          |--------
CAT TTT GAC CCA AGA TAC ATT CAG CTT GTC ATA GTT ACA GAC CAC GCA ATG gta agt atc
         Met4 PI Rv/ Met4 PI Fw                                     |— EoPI-7 →
  H   F   D   P   R   Y   I   Q   L   V   I   V   T   D   H   A   M                    19
-----------------------------------   Metalloproteinase   →
tta aat acc ttt cca ttt act ttc tgc att gag tac agc att ttg ttt ttg aat ttt tta
atg cgg gca ttt ttc aag aga tta ata cac atc cta act gac ttt ctc tta taa tgt gcc
tta tca taa act ttg ata ttt ttt gtt att gat cca acc aat att tac aaa gtt aag agt
cat ttg taa ata tat tct aat ttg cac att ttt tat ttc tgg ata cca tct aag ttt att
ttt atc cca cta tag tta aaa att tct gta ggt ttc ttt gaa atg ttc cca cac tct ttt
ttc atc caa gct ggc aac aca cac aca cac aaa aaa aga tgt aca gtg aag aat gct ttt
ctc aca cat ctc gga aca tgg aaa tat ttt ctc caa cca aaa aaa agg ccc atg tat gct
gct gaa aag tta aaa atc tta ata tat taa cgg cac aaa tgt aga tta aaa aat cag cac
aaa aga att att tgg act gct tgg ata aaa att att aca tca aca aat tca aag atc ctt
tca cta ata tgt tct ttt tct cct ccc ttc ttt ccc ttc tta taa tta tca act tgt ctt
aac ttt ttt ttt ttt tgt ag  GTC ACG AAA AAC AAC AAT GAT TTA ACT GCT ATA ACA ACA
                                                        Met8PI Rv
                            V   T   K   N   N   N   D   L   T   A   I   T   T        32
TGG ATA CAT CAA ATT GTC AAC GAT ATG ATT GTG gta aga aca aat gct tgt tca ttt taa
W   I   H   Q   I   V   N   D   M   I   V   |-------------------------------------   43
                                           |----------------- EoPI-8 →
act tca ctt agg ccc agc cga gat ttt gat tgt gtt aag ata aca aac tta atc agg taa
aga aag tag atg gat ttc taa atg aat ggt ttc ttt tgt ggc ttt gag tga tgt aaa act
aag tta ttt gac cag ctg gtg atc cag cat gtg agg cca tcc cag aaa atg ggt taa ttc
agt aac taa att aat aag cct cta atg gaa ata gag tca ttg ggt tgg gag gat gca aat
cca aaa gtg ttg caa gaa ttc agt cag aag tat acc tgt agt tgg att ctg tca ggg ctg
tgg tta atg gtt ctg aac ttg gag att aga aag tga tgg aca gag tca gtg tca ggc gcc
gtg cct gac aca tgc act ggg ggt gga gga gga gta gcc gcg cgc ccc tcg ttc att gga
ggg gcc ggt aca agg acg cgg agc tcg ctc gca gct gga ttg tgg cgg caa ggc aac tga
cag ctg cag agg gtg gga gcg gcc ttt cca gtg gct ggt gga cac tgc gga aga gag gag
acc ggc gga caa act aag cca cag aac ttt gga gtg gtt cca gca cag aag gag gag cca
cga ggg cgg TGT GCT TAC CCA ACA CTG AGC Ctg gaa aaa agg acg ata tgt tta cgt ttg
            Intron B16 Fw
gct gtt tgg agt tca ttc ccc ccc cct tgt ctg cca gat gct gct gcg ggg ctg tat ttt
ggg tgg tgt tag gga ggg gct agg ata ggg gta aag gct aag tgt gga gaa ggt gat ggg
aat aca tac ttg gtg gtg gat tag gaa gaa ggg cgg gtg aaa tga agg gtg gtg att ttg
tgt cac cat gct ggg agg ggt gga gcc cag gcg gag ggg tgt ggc tgg gtg gtg taa caa tgt
att taa tag tgt ggg agg gat atg taa gcc aac gct ggc ttt ttc cca ctt cta cgt tga
gtt ttg ttg ctg aat aaa gcg tta ttt ctt ttt gga tac ttc ccg tgc ctg tga gac tgc
tca ttg gtg agt aac tgg acg gga ggg act gac agt cag cct att cca agc tca gtt gat
tat gaa aat gat ctc tta gaa gaa acc ttg aga gaa gaa ctc aat ttg act tgg att gtg
gta ctt agg agg aga tac agt cat act gaa ttc tct gat tag cct tat tta cat ttt cac
tat tga tga tca gtt aga aac agg gag aga aca gga aga ttg agg aga taa ctc tcc tgt
tag tcc acc tac tct tct gtt cct gtt tac aga aaa taa tca cct gct tct ttt tat cat
gtt att tat ttc tct tgc tta ata cat tgt tat tat tcc ctg ata tat aga gtc ctt agt
act atc tca gct tgg tta cct tgc aga cgt ttc att att tga gta ctc tac atg aat
aac caa aca gca agc tca gag cac caa gga ctt caa cct tga gct tct att gat acc tag
gtc tat ttt ttt ttt taa atc tac aaa aag agg cca atg tca gtt gct ttt taa tat tga
aat gtt tct caa ggt tta ctt tgt ttg agt ttc tga atg cca tgt cca cct gac ag
ATG TAC ATA GAT TTG AAT ATT CAT ATA ACA CTG GCT GCC GTA GAA ATT TGG TCC AAT GGA
M   Y   I   D   L   N   I   H   I   T   L   A   A   V   E   I   W   S   N   G        63
GAT TTG ATT ACT GTG ACA TCA TCA GCA CGT GAA ATT TTG AAC TCA TTT GGA GAA TGG AGA
            Met5 PI Rv/Met5 PI Fw
```

Figure A2. *Cont.*

```
tgt ttt aaa ctt ctt ata aaa tta tct ctg tgc tac ata att cct gat ata gat atc tct
tct ttg cat tct ttc cag  A GGA CTT TTT TCA GAA GAT TAC AGT GAG ACT CAT TAT TCC
                                                           Pro2 Fw/Pro3 Rv
                         G   L   F   S   E   D   Y   S   E   T   H   Y   S        -90
CCT GAT GGC AGC GAA ATT ACA ACA AAC CCT CCT GTT GAG gta ggg tct cac ttt tat gag
 P   D   G   S   E   I   T   T   N   P   P   V   E  |--------------------- EoPI-4 →  -103
cct ttt ttt aag gaa gta aat tga aac aaa tgt ttg tgc act ata tta caa ata tac aag
aat gag acc agg cta ctc aaa caa agt gta tat aag tat aaa gta tct tat att gat atg
tac tta caa aga tgc ctg gat tgt taa tcc ttg gtt aaa agc caa cat att tgg gag gtg
agt ttc aca aat aga ttt att atg aga aca tca ggt ttg taa gat tat att ttc att ttt
aaa cca gac tac agg gat aaa tgc aaa gtc ttt tat ctg taa tac caa aag tga taa caa
ttc act ttg ctc cta tac aga aat cca ttt aac atc ttt cat att aaa atg gtg cca aaa
atg gct cta tca gag gtt aaa aaa tta cag cac taa tat gct tca tgt tgg ctc cat ttc
ccc aaa ttg att taa aag tgc att ctg tgt cta ttt ctg gtt tag cat ctt cat ggg ttg
cac aaa tta ctc ctt tgt gcc atc agt ggc act ctc cca tag tgt gac ttg att tat gga
gac ttg cat tta tcc tat gtt cct ttt gca ata gtc agt att aag aag gtt ttc tgt cct
cct gaa tca aaa ttt tct gga aaa ctg ctg tct aaa tat ttc att gat gtt atg gaa tac
att gga act gta ctt ctg ctc atc aaa tca caa tac aaa gtc cta acc agt gta gtc ctC
TTG CCT CCC TAT AG  GAT CAC TGC TAT TAT TAT GGA CGC ATC CAG AAT GAT GCT GAC TCA
      IntrB13-1 Fw/Met16PI Rv
                    D   H   C   Y   Y   Y   G   R   I   Q   N   D   A   D   S      -118
ACT GCA AGC ATC AGC ACA TGC AAT GGT TTG AA gta aga tag tct cta atc ttt tat ttg
 T   A   S   I   S   T   C   N   G   L   K |--------------------- EoPI-5 →        -129
ttt att aat aat aat ata gtc ctc ttg gag ttc taa ttg tta aaa tga agg aca tcc tca
gtt ttt cat gga aat tag ttg ggt gtg atc ggt gat ttc ggc aga att aag aca tac ttt
ggt tga aaa cca aga aga gct gct gcc agc cag gag aaa aac tat gga gct aaa tca cat
aag tct aaa gga gct tcc aag ccc cgg tct cct ttc cca ggg tga ggt gat att aca ggt
aga gaa gat tag tag gtt tca aat tgg aga cct tgc tag aaa gtg tac agg aag agg caa
gaa gtt tca gtt cta ccc aga aac act ttc ttg agt cac tct gca cac ttt ctt cag cca
act aga tat gtt aac tac ata aag atc cca gaa ttc aga agg tcc cta tca ata gta aga
atg aac atc acc tca aca tct ttt act gaa aaa aga cac tga aac tca cct ttg aac aga
gac tgt gtc cat gga gtg gga gaa taa atg aaa agc tgg aac aga gca gaa taa caa cag
aaa aat aaa gga aaa aca gaa tga cag aat aat agc att gga agg gac ttt gag gtc ttc
                                             |------------------ LINE/L2/CR1 →
tag tcc aac ttc ctg ctc aag tag gag acc tat atc atc cta gac aaa tag tgt tca atc
ttc tct taa aaa gca gta gtg atg gaa cac cca caa tgt ctg aat agg tta att gtt cca
ttt gtg aga aaa tta ctc ctt agt tct aac tta ttt ctc tct ttg gtt act ttc cac gca
ttg ctt ctt ctc ctg cca tca ggt gaa gaa tag gtt gtc cca cat ttt tta tga cag cct
ctt aaa tac tta aag att atc aag tca tct cta ccc ctt ctt gtc act agc atg agt ata
ctc att gtc tgc agc cat tct aac ctc cag tta gta tgc att ctt att cct tca ttg tta
ctc ctg ttg ttc tgc att gac ttc tct atg aag atg ctt gcc aag aat tct tat ttt cat
tat tta tta aat atc ctg gtc atc ctg act ctt atc tta aat tgc tat caa act aat ctg
att tta ttt cct tga cca cag aca aat att gtt cta tac ttg ttt aaa gta aat tgc agt
att acc tat aac tct ttt tag ata ttt tag cag tta tat ttt tcc ttt ttt atc cta ctt
agt tgt gat tct tga gct tta tca gta ata tat atg ata aat ata aag tat ttt acc ctt
atg aaa taa agt ttt aca caa agc aga atg tta caa ttg gct tta gtg ttg tat tta tgt
agc tag aaa ctt att ttt tta aca tcc tgg aaa tat aca ata ttg ggt tcc atg cca aaa
tat ttc caa aca aaa ctg tac acc tat ttt gtg gct gca ctg agt ttg tga aat ctc tca
tat ctt tct gat cat aac tgc atc tat gaa aag tat gag aaa gtg att tga gtg ctg agg
aaa gaa tat aaa ata ttc act cat tgt taa gaa gga att caa aaa cat gag gtt agt tga
aaa tgg gtc tca gag ccg agt ttc att acc caa cta ggt aac atc atc agt gca gtt ttt
ctc tga act aac aat att ctc ttt tgc ttc tcc atc tct gat cat cct ttt cac att
gtt tta cag  A GGA TTT TTT ACG CTT CGT GGG GAG ACG TAC TTA ATT GAA CCC TTG AAG
             G   F   F   T   L   R   G   E   T   Y   L   I   E   P   L   K        -145
GTT CCC GAC AGT GAA TCC CAT GCA GTC TAC AAA TAT GAA GAT GCC AAA AAA AAG GAT GAG
                                                          Prodom 1 Rv/Prodom 1 Fw
 V   P   D   S   E   S   H   A   V   Y   K   Y   E   D   A   K   K   K   D   E     -165
GCC CCC AAA ATG TGT GGG GTA AAC CTG ACT AAT TGG GAA TCA GAT AAG CCC ATC AAA AAG
 A   P   K   M   C   G   V   T   L   T   N   W   E   S   D   K   P   I   K   K     -185
GCT TCT CAT TTA GTT GCT ACT TCT GAA gta agt ctc ata ata aac ata gtt taa gat tac
 A   S   H   L   V   A   T   S   E  |--------------------- EoPI-6 →               -194
ata cta att tcc ttg tct tga aat aat aaa gta aga gag aat ttc ctt tgg gaa ggg gtg
ata gat aga att caa aag gga gaa gcc ccc att tct ata ttt tta ttg tag cca tgg cat
aaa aga aag aat gga aac ttg agg aac aga aaa tac att ttc cag gct tat agc att ttc
ttt ggt cat tca aac tta gtt tag gga ttt gaa tca aaa tct att taa atg agt ttc taa
att atc tct agt ttc taa gtc aat gtt gaa aag taa tta aat tat caa ttt gga ttc ctc
ttt tat gca tgc aga gag gat ggg gga caa agt ggt ttg aaa tat taa atg gtt tta aga
tgt ctg ata agg cca tta cat aat tgt tac tcc att atc caa ttt gat ttg aat cat caa
gtt gga ttg atg caa tga atg gat gaa aag tga caa tgt gaa cct agt cac aat tga ccc
tta tgc tct caa tat ttt cct cct tta ttg gac gca caa aaa tta gaa aac aaa ata ttg
cat cca aag tga cag ttc ctt tcc gtg gtg aaa gtt gaa act ggc tga aaa
atc tct act gtt ttt att aga atg tta aaa ttg aca TGG AAC AAC AGC TGT TGT TAT GAC
                                                   IntronB7PI Fw
Gga ata cca aaa cac aag tga aga cgc caa atg aag cct ggt ttg tct ttt ggc ttc ttt
cat tct ggc aat tca aga ttc ttt atc ctc agc aat ttg tgg tta tac gtt aca ttt aac
```

Figure A2. *Cont.*

```
 D   L   I   T   V   T   S   S   A   R   E   I   L   N   S   F   G   E   W   R        83
CAG AGA GAT TTG GTG AAT CGC ATA ACG CAT GAT AAT GCT CAG TTA CTC AC gta cgt ctc
                        Met9PI Rv                                  |— EoPI-9 →
 Q   R   D   L   V   N   R   I   T   H   D   N   A   Q   L   L   T                    100
act gtg ggg aat ggg agt tag ggt gtc tga gag tgg agg gtt atg gga agg tta ctg ctt
gca tag agc ttc tgt tct atg ctg tat gct tga aac cat gca tgt act aca ttt ctg ggt
caa agt cag cca ctt ata gat gag agc tgg ctt tga gaa gtc ctt gaa tga tta cag gtg
aaa aac ctg gat tta aaa ctc aaa ctt tta tga cta tcc tgc tat act ttt gca gtt ctt
ata cat aca ttt taa ttc aat taa ata atg ttt tgg ttc cca ttc tgc acc aag taa atg
aac act tgt att ata caa aag cat aca gaa tgc tat tac ctt ctc tga aat ggt gcc tat
             |————————————————————————— SINE/Sauria →
tta att act tgt gtt tga act gct agg gta gca gga gct gga gca agt aac agg agc tca
ttg cat cag gca aaa ctt agg tct caa ct att cac ttt cca gtc aac tag tcc act gta
tta act gct cag ttc cac acc tca gcc tat ata ttg tca aac aat tta atg agt gag
aaa ctg cta tgt tta gtg tct tat gtc tgt aac ata ctc agg aaa taa caa ttg tgt tgt
gag caa tcc tac agg atg ctt gca taa caa gta ggt ctt atg aaa cag gga tta cat tga
agg cag att taa gtg caa ggg aga ata tta aaa aaa tac cca cag atg gtg gtt tga ctt
gca att caa tca ctc ttc cca aaa cca tct agt cca ata acc cca caa cta aat tga tag
aca aaa tct agt aaa taa ata tgt att tgt taa ttt aat gga att aaa tta acc caa atg
ggt tga att ctg aca aca cct gc taa ccc tct ccc caa cct cat tca tat taa tgc cat cca
aac aca act tat agt gaa aaa tag atc ctg cat att ttt aac ggg cct gct tag tta aat
gac agg tcg agg agt tta cag aaa tcc ata ctt atg tat gcc aga cta cat aac atg
ctt cta tat tca att tta tcc ccc tcc cct tct tct tat ag   A GCC GTT AAC CTC AAT
                                                          A  V   N   L   N          105
GGT GAT ACT ATA GGA AGA GCT TAC ATG GAG AGC ATG TGT GAT CCA AAG AAA TCT GTA GGA
             Met3PI Rv/Met3PI Fw
 G   D   T   I   G   R   A   Y   M   E   S   M   C   D   P   K   K   S   V   G        125
ATT AAT CAG gtt agt aga aag gat att cta tta tct att tgt act caa gcg aaa cgt ggc
 I   N   Q  |——————————————————————————— EoPI-10 →                                  128
ata cag aca aaa cat ctt tac caa taa agt ctc ttt ctt att ttt gag cca cgt cat ttt
cac cca tat tta ttt gca gat ttg aca tct cca ggt cct gcg tca act aat ggc att ttg
aca cag tgc att cta gaa caa gct ttt tta atg caa tga gct ata tgt caa gga tga gaa
tat att ata atg ttt atg gtt cag tca aac tgt act ctg att ggc aaa tga aca ggt caa
agc atg tta caa cac ttc caa ata atg ctt ctg aac aat agt ctt agc aat ccc aaa gac
aaa cat gaa ttc att cca aga aat tta gtg tct aga ttg cat atg att gaa ttc tag tac
att gag aaa aca aaa aaa tac taa atc tac tca aaa aga aaa aaa acc ctc tag tat ttt
aag aaa acc ata tta gtt aag tga tgc tat gca ttt att gag aaa gag taa act tag ctt
ttt gtt cac ata gaa aga atg gag aga cat ggt aat gaa caa aag tta tac aac aaa act
cat aaa gtt ttg ttt ctt aat aag cag agt tag gat cct ggt agt agt agg tac tca taa
gcc tac ttg ctc aag aag gtt att tta ttc aga aag agc agc tca ttc tta agt gtg ttt
agg atg gct acc ttc att att ctg aaa atg taa gct tgt agc aaa gga cac tga gta gtt
                                                                            |————————
ttt ttc aac tga cgt ttc ctg taa gtc agg gct gtc aaa ctc aat ttc att gag gga cgc
————————————————— DNA Transposon →
atc agc att gcg gtt gcc ctc atg ggg gca gtc ggg tat ggc cag ggt ggg cac agc cca
cag gca tgg ctg gag tgg gta tgg cta agt ttt agt aac tga ata agt gca cat agc aaa
tgg atg cat aca ttt tga tct tat cct gtg ctg tag ctt ctg ggc ttt aaa gtt tcc ttc
tgg atg tat ttg tgt atg ttc tgg agt cat ggt ggg cac aca tac ttt cag agg ag tag
agg aat cct gag atg gta tcc tca act taa aat tgg tca tcc ggt cac cag ttt agc cac
tta gtt cta ata att gaa gtt cac tta cgt ttc tgg cag agt aac aat aaa aaa agt att
ctt att tct tca g  GAT CAT AGC ACA GTA CAT CTT TTG GTT GCA GTT ACA ATG GCC CAT
                    D   H   S   T   V   H   L   L   V   A   V   T   M   A   H        143
GAG CTG GGT CAT AAT CTG GGT ATG GAT CAT GAT GAA AAT CAG TGT AAT TGT GGT GCT TCC
 E   L   G   H   N   L   G   M   D   H   D   E   N   Q   C   N   C   G   A   S        163
GGA TGC GTT ATG TCT GGG GAG CTA AG gta agt act gag gaa tat gct taa tgg gtt ttg
   Met2PI Rv                         |————————————————————————— EoPI-11 →
 G   C   V   M   S   G   E   L   R                                                   172
aat caa ctt att ttt aaa tgg tta caa aaa tga aaa ggt cag ttt agt tac aaa aaa gag
gtc att tgg tca ttt gtt tgt cac ttt atc atg tgt agt taa aat gtt tca tat tta aaa
agg tag aaa tgt ttt cag caa ttg aat agc cat cat tca cag ctt cct cca aaa cac cag
aat tta aag aaa cag cat agt cag agc gaa cta ata tta atc gtg atc aat aaa aga aga
gga aac caa gct aaa tgt taa aga ttt tcc taa agt ggg aag cag aat ggg gaa aat tat
tct gaa cca ggc aac acc agg aaa aaa aga taa att cgc atc ctt ggg gtt ttc cta ctt
tgt cct ccc atc aag gta caa tat gtg aca tct ctt tca tat act gca gtc ttc tga aat
aaa cag ata gca atg tac agg aag tca ttg act tac agt tac ttt gaa gtt aca acg tcc
atg aga act atg act tat gac tgg tcc tcg cag tta caa cca tca tag act ccc ttc agc
aac aaa atc aaa att tgg gca gtt ggt cac acc caa gtg ggt tca tga tgc cta cat cat
cct gta ttc atg gga tcc tca ttt gta cat tcc aag gtg tct tcc cat atg caa agc caa
tag acc aaa ctg ggt acc tta aag act gtg tga ttc act taa caa agt ggc aaa act ggt
ttt aaa aCT GAC TAT GAC TCA CTT AAC AAC TGG ctt ccc ttg gaa agt gga aat tct ggt
            IntrB23PI Fw
cct agt tgt ggc agc gag tca aga act agc aat att ttc ttc gag gtg aaa gtt tag gcc
tta ctt cat caa gaa gga aaa aga aaa gga aaa agg aag caa aat ctt ctc agc aat tgt
att ttg caa aac tct cca taa ctt ttg tgt tca gtg ttt tgt tta ggc tca gtg agc atc
```

Figure A2. *Cont.*

```
tgt ttt aaa ctt ctt ata aaa tta tct ctg tgc tac ata att cct gat ata gat atc tct
tct ttg cat tct ttc cag  A GGA CTT TTT TCA GAA GAT TAC AGT GAG ACT CAT TAT TCC
                                                      Pro2 Fw/Pro3 Rv
                         G   L   F   S   E   D   Y   S   E   T   H   Y   S        -90
CCT GAT GGC AGC GAA ATT ACA ACA AAC CCT CCT GTT GAG gta ggg tct cac ttt tat gag
 P   D   G   S   E   I   T   T   N   P   P   V   E  |---------------- EoPI-4 →    -103
cct ttt ttt aag gaa gta aat tga aac aaa tgt ttg tgc act ata tta caa ata tac aag
aat gag acc agg cta ctc aaa caa agt gta tat aag tat aaa gta tct tat att gat atg
tac tta caa aga tgc ctg gat tgt taa tcc ttg gtt aaa agc caa cat att tgg gag gtg
agt ttc aca aat aga ttt att atg aga aca tca ggt ttg taa gat tat att ttc att ttt
aaa cca gac tac agg gat aaa tgc aaa gtc ttt tat ctg taa tac caa aag tga taa caa
ttc act ttg ctc cta tac aga aat cca ttt aac atc ttt cat att aaa atg gtg cca aaa
atg gct cta tca gag gtt aaa aaa tta cag cac taa tat gct tca tgt tgg ctc cat ttc
ccc aaa ttg att taa aag tgc att ctg tgt cta ttt ctg gtt tag cat ctt cat ggg ttg
cac aaa tta ctc ctt tgt gcc atc agt ggc act ctc cca tag tgt gac ttg att tat gga
gac ttg cat tta tcc tat gtt cct ttt gca ata gtc agt att aag aag gtt ttc tgt cct
cct gaa tca aaa ttt tct gga aaa ctg ctg tct aaa tat ttc att gat gtt atg gaa tac
att aga act gta ctt ctg ctc atc aaa tca caa tca acc agt gta gtc ctC
TTG CCT CCC TAT AG  GAT CAC TGC TAT TAT TAT GGA CGC ATC CAG AAT GAT GCT GAC TCA
                    IntrB13-1 Fw/Met16PI Rv
                     D   H   C   Y   Y   Y   G   R   I   Q   N   D   A   D   S    -118
ACT GCA AGC ATC AGC ACA TGC AAT GGT TTG AA gta aga tag tct cta atc ttt tat ttg
 T   A   S   I   S   T   C   N   G   L   K |---------------- EoPI-5 →           -129
ttt att aat aat aat ata gtg ctc ttg gag ttc taa ttg tta aaa tga agg aca tcc tca
gtt ttt cat gga aat tag ttg ggt gtg atc cga att ggc aga att aag aca tac ttt
ggt tga aaa cca aga aga gct gct gcc agc cag gag aaa aac tat gga gct aaa tca cat
aag tct aaa gga gct tcc aag ccc cgg tct cct ttc cca ggg tga ggt gat att aca ggt
aga gaa gat tag gtt tca aat tgg aga cct tgc tag aaa gtg tac agg aag agg caa
gaa gtt tca gtt cta ccc aga aac act ttc ttg agt cac tct gca cac ttt ctt cag cca
act aga tat gtt aac tac ata aag atc cca gaa ttc aga agg tcc cta tca ata gta aga
atg aac atc acc tca aca tct ttt act gaa aaa gca cac tga aac tca cct ttg aac aga
gac tgt gtc cat gga gtg gag gaa taa atg aaa agc tgg aac aga gca gaa taa caa cag
aaa aat aaa gga aaa aca gaa tga cag **aat aat agc att gga agg gac ttt gag gtc ttc**
                                  |----------------- LINE/L2/CR1 →
**tag tcc aac ttc ctg ctc aag tag gag acc tat atc atc cta gac aaa tag ctg tca atc**
**ttc tct taa aaa gca gta gtg atg gaa cac cca caa tgt ctg aat agg tta att gtt cca**
**ttt gtg aga aaa tta ctc ctt agt tct aaa tta ttt ctc tct ttg gtt act ttc cac gca**
**ttg ctt ctt ctc ctg cca tca ggt gaa gaa tag gtt gtc cca cat ttt tta tga cag cct**
**ctt aaa tac tta aag att atc aag tca tct cta ccc ctt ctt gtc act agc atg agt ata**
**ctc att gtc tgc agc cat tct aac ctc cag tta** gta tgc att ctt att cct tca ttg tta
ctc ctg ttg ttc tgc att gac ttc tct atg aag atg ctt gcc aag aat tct tat ttt cat
tat tta tta aat atc ctg gtc atc ctg act ctt atc tta aat tgc tat caa act aat ctg
att tta ttt cct tga cca cag aca aat att gtt cta tac ttg ttt aaa gta aat tgc agt
att acc tat aac tct ttt tag att ttt tag cag tta tat ttt tcc ttt ttt atc cta ctt
agt tgt gat tct tga gct tta tca gta ata tat atg ata aat ata aag tat ttt acc ctt
atg aaa taa agt ttt aca caa gga atg tta caa ttg gct tta gtg ttg tat tta tgt
agc tag aaa ctt att ttt tta aca tcc tgg aaa tat aca ata ttg ggt tcc atg cca aaa
tat ttc caa aca aaa ctg tac acc tat ttt gtg gct gca ctg agt ttg tga aat ctc tca
tat ctt tct gat cat aac tgc atc tat gaa aag tat gag aaa gtg att tga gtg ctg agg
aaa gaa tat aaa ata ttc act cat tgt taa gaa gga att caa aaa cat gag gtt agt tga
aaa tgg gtc tca gag ccg agt ttc att acc caa cta ggt aac atc atc agt gca gtt ttt
ctc tga act aac aat att ctc ttc ttt tgc ttc tcc atc tct gat cat cct ttt cac att
gtt tta cag  A GGA TTT TTT ACG CTT CGT GGG GAG ACG TAC TTA ATT GAA CCC TTG AAG
             G   F   F   T   L   R   G   E   T   Y   L   I   E   P   L   K        -145
GTT CCC GAC AGT GAA TCC CAT GCA GTC TAC AAA TAT GAA GAT GCC AAA AAA AAG GAT GAG
                                                      Prodom 1 Rv/Prodom 1 Fw
 V   P   D   S   E   S   H   A   V   Y   K   Y   E   D   A   K   K   K   D   E    -165
GCC CCC AAA ATG TGT GGG GTA AAA CTG ACT AAT TGG GAA TCA GAT AAG CCC ATC AAA AAG
 A   P   K   M   C   G   V   T   L   T   N   W   E   S   D   K   P   I   K   K    -185
GCT TCT CAT TTA GTT GCT ACT TCT GAA gta agt ctc ata ata aac ata gtt taa gat tac
 A   S   H   L   V   A   T   S   E  |---------------- EoPI-6 →                  -194
ata cta att tcc ttg tct tga aaa tat aaa gta aga gag aat ttc ctt tgg gaa ggg gtg
ata gat aga att caa aag gga gaa gcc ccc att tct ata ttt tta ttg tag cca tgg cat
aaa aga aag aat gga aac ttg agg aac aga aaa tac att ttc cag gct tat agc att ttc
ttt ggt cat tca aac tta gtt tag aga ttt gaa tca aaa tct att taa atg agt ttc taa
att atc tct agt ttc taa gtc aat gtt gaa aag taa tta aat tat caa ttt gga ttc ctc
ttt tat gca tgc aga gag gat ggg gga caa agt ggt ttg aaa tat taa atg gtt tta aga
tgt ctg ata agg cca tta cat aat tgt tac tcc att atc caa ttt gat ttg aat cat caa
gtt gga ttg atg caa tga atg gat gaa aag tca caa tgt gaa cct agt cac aat tga ccc
tta tgc tct caa tat ttt cct cct tta ttg gac gca caa aaa tta gaa aac aaa ata ttg
cat cca aag tga cag ttc ctt tcc atc ttt ttg gtg gca aaa gtt gaa act ggc tga aaa
atc tct act gtt ttt att aga gta tta aaa ttg aca TGG AAC AAC AGC TGT TGT TAT GAC
                                                IntronB7PI Fw
Gga ata cca aaa cac aag tga aga cgc caa atg aag cct ggt ttg tct ttt ggc ttc ttt
cat tct ggc aat tca aga ttc ttt atc ctc agc aat ttg tgg tta tac gtt aca ttt aac
```

Figure A2. *Cont.*

```
cca gct atg gat cag ttt tgg att tct tct gct aaa gcc tga aga ctt tgt tgc ctc cta
ttt cat gca atg aat agg agc cca gta aat atg gag aat atc aca tag cca ttc ctg cag
tgg cgt agc atg ggg gtg cag ggg ggg cag ccg cac cgg gca caa cat ctg ggg ggg cgc
gct cgc act cgc agc tct ctg ccc ctg cct ggc tca ctc att ctc tct cca ctg aga aac
cac gcc gga ttc ccc tca cac gac cac tca ccc ggg aaa gcc gag cga gct cgc ccc acc
ttt tcg agc ctt ttc tct cat ctc cag cct gtt ggc aac cgc aaa ttg ttt tga gcc ctg
ttc tct tct ctc ccc ccc cgg ccg tgt taa gcc aag gac aaa ctt tgc aag aaa ttg cag
ttt tgc ttt ctc ttc ttc ccc ctc ccc tct ccc gta gag tag tgg ggg aaa ggg aat gtg
gga gat ttg cca gcg gac aca gac ttt cca cta aac tcc ccc cgc ctg gca tct cca cct
cat ttt GGC CGC GTG AAT GCA TCT GCT TCt ttc tct ctt ctc acc cca ccc cac cat cca
        IntrF2Pl Fw
ctc gtg aaa agg gag ggg gag gtg cta ata cct gga aag aaa cta act ttc att tgc caa
ttt cat aaa tgg tgg agt taa aga gag ttg ctg gaa aat tat ata tag tta cac gtt cgg
ttt gtg tga gga aaa caa agt gaa tgc taa ttc ctt caa ggg ggt aat ttc ttt cca gct
gaa ctg act agc cta tct atg agc cag tgt tgt tta aag aaa act aaa ttt aca aag aaa
tct gtt gag aaa ttc tat tga ttc tga gca tat ttc atg ggg gca aag agg aga aat tag
atc tct tga ctc ttt cag aat ctt gct cct ttg tac act tct tta taa cag cac tgg gcc
tgt gaa agc agg cga gaa gtc cta gag aag cta cca ttt caa tgc aga att cgc gac cca
tta aac cta ttc ttt tat ggc agc ttc acc aca aac agc agc ttc tcc att ctc tac agg
cag aaa aaa aat ggg aga ggg gca ttc atg cat cat ttc ctg aaa gaa tct ata cta aag
aaa gta tgt aaa gcg ttg tct atc aaa aac tta ttc aaa agt tac gta acc aag gga tta
tga att tgc agc aaa ttt gcc tat gaa ttt gga gta aat ttc ttt gta ggg tag caa aat
gtg aca cca cta tat aca tat ata cag ggt tgg cca aaa tct gaa aag gaa tat agt cta
tga aga gtt ata gac aaa tta aGC ATC AGT TTG TTC GCA CTC AAT AAA Gtg ttt gaa aat
                           Intr2F2Pl Fw
aaa ctt gta ttt aga tgc att tta ctt taa tta cat cag tat ttt cac aac aaa caa tac
atg tgc tta ggg ggt aag ggt ttt ttt aac taa tct agt gga aga gac ttg agt gct aaa
                                          |------------------ microsatellite ------→
atc cac ggg tta ggg ggc gca aat tac ttg cct tgc ccc agg tgc tga caa ccc atg cta
cgc cac tgc att cct gtt ctt cct gaa gaa tgt ctg ggc atc acg ctt act ctt taa tag
ttc tag aca ctt tgt att gta ttg tac aga gga gtt tgc tta gaa aac aat ttt tct caa
tta acc cca cca aaa ggg tct gct gca act ttg act ttg gaa gaa gac cgc att gtt tgt
tta atg gcc acc agc aga ttc tct ttg cat tct tgt ttc tac ccc tta agt tga ggg cca
atg atg ata ttc cta gta gtt tga att aag tta gaa tgc cat gtt tgc atc tgt tta agt
att cag ctt cag agt tca gtg gca tgt cgc tga tat tat cac ttt tca acc aaa tga atc
aat tga aaa tcc ttt aag gta aat aga aat tta gca tat tta gct ctc tct tca tcc tta
taa aag gtc cta tcc ctt caa aag aag aaa aga ata act gga tca aaa tta cac agt ttt
ccc ttc aaa tat att att gtc ctc cca cag aga ggc tga cta ata atg aaa ata aaa ggc
agc cca ttc aaa cat aag acg ctc gat caa cta att tca cat gga ttt aaa tgt aaa
tgg att tgc taa aaa aag aaa aga aag tta gag aaa tta aaa ctc agg GAG CAT AAT CTG
GAA CTA AGA TCA AGg ggt ata aag tat tct ctt atc ttt ttt gtt agc ttt aga aga aaa
     Intr3F2Pl Fw
aaa aat cac tgc att tct tct tta gca att cct ttt cct tat atg ttt ttg aaa tga att
ttg ttc cta gtc tga att tgg agg atg tca tga tct ttt ttc cac ttc tac ag   G GCA
                                                                           A        173
CAA GAT TCC TAT CAC TTC AGT GAT TGT AGT AAG AAT GAA TAT CAG AGC TAT ATT GCT ACT
     Met7Pl Fw
 Q   D   S   Y   H   F   S   D   C   S   K   N   E   Y   Q   S   Y   I   A   T    193
TAT AAC CCA CAG TGC ATT CTC AAT CAA CCC TTG AGA ACA GAT ACT GTT TCA ACT CCA GTT
 Y   N   P   Q   C   I   L   N   Q   P   L   R   T   D   T   V   S   T   P   V    213
TCT GAA AAT GAA CTT TTG CAG gta gga gaa gaa tgt gac tgt ggc ttt cct gca tta agt
     Met13Pl Rv              V   G   E   E   C   D   C   G   F   P   A   L   S
 S   E   N   E   L   L   Q   |---------------------------------- EoPI-12 --------→   220
                     |----- Spacer -----|
ctt ttt ttt taa tca aca aaa gta att tga aga ata ttc tca gaa atg aga atc ctt gaa
 L   F   F   STOP
aaa tca tct agc ttt cta agt ggt ttg agc cat cca aga ggt tgg ctt gtg aat ggc tga
ggt ttg tgc ctt tca tgt aca tgc atg tat gaa atg gtt tct tgg gtt gta gag gaa tag
aga aat ggt atc tca cta cta ttt ggg gaa gat ggt gaa ttt tta aaa agg ggt gat tga
cca ttc cat gaa aat ctt tcc ctc ctg aaa acc cct att ttg ttg ata tag cca cat tat
cct gtc cca caa ttt tct cga act gct cct tcc cat atc tga tta tct tta atc tat gct
ctg atc cta ata ata ttt tta taa gaa cag taa tat agt gtt ttt atg ttg tta aat aca
cct gta atg gtc tgt gag aat gtc ctt aag aga caa aag aag gag gaa aca tcc agt cag
tgg tca tat aaa aag gag att aac ctg cag aaa caa agg ggc ata gca aaa atc tca aga
ggg aca cct cct acc cat tct ctt ggt ccg taa agta tgg ggt aga aat gga ctt tca
gta ttg aaa gat tct gct act gta act gta caa tca agg tag tgt taa tgc tca tgg ttg
gtg ctt ctt ctc tgg att acc tca aaa gct ggc atg atg agt aga aaa tct ctc atg aat
aag aaa gga atg ggg ctg tag gtt atg tgg gct tca caa tta gga gat gag gat att ttt
gtt tta ttc ttt tca cgt agg aaa tat cag ata agg ctc ttt cgc aga gaa atg cca tta
tct gtt ttc aat aac aaa caa ttt ttg cat ttg cta gca tga acc cat aaa agg gaa cac
att gca gaa att tcc ctc ttc aaa ata gac caa tta aaa aag aaa att cta tgc cat cat
ttg ata tga tat gat ccg tgc tgt gat cct ata acg tgt aaa cca aga caa ggg aaa cat
     *   Y   D   P   C   C   D   P   I   T   C   K   P   R   Q   G   K   H
tgt gta tct gga ctg tgt tgt tgt agc tac aaa gta aga ctt gtt tat ttt taa cac cag
 C   V   S   G   L   C   C   C   S   Y   K   V   R   L   V   Y   F   *
gag aaa ttt tac cct gct cca tac tag gct tgt aga aat gta ata ttt ctt ggc ttt tta
```

Figure A2. *Cont.*

```
tgt ttt aaa ctt ctt ata aaa tta tct ctg tgc tac ata att cct gat ata gat atc tct
tct ttg cat tct ttc cag  A GGA CTT TTT TCA GAA GAT TAC AGT GAG ACT CAT TAT TCC
                                         Pro2 Fw/Pro3 Rv
                            G   L   F   S   E   D   Y   S   E   T   H   Y   S        -90
CCT GAT GGC AGC GAA ATT ACA ACA AAC CCT CCT GTT GAG gta ggg tct cac ttt tat gag
 P   D   G   S   E   I   T   T   N   P   P   V   E  |---------------------- EoPI-4 →  -103
cct ttt ttt aag gaa gta aat tga aac aaa tgt ttg tgc act ata tta caa ata tac aag
aat gag acc agg cta ctc aaa caa agt gta tat aag tat aaa gta tct tat att gat atg
tac tta caa aga tgc ctg gat tgt taa tcc ttg gtt aaa agc caa cat att tgg gag gtg
agt ttc aca aat aga ttt att atg aga aca tca ggt ttg taa gat tat att ttc att ttt
aaa cca gac tac agg gat aaa tgc aaa gtc ttt tat ctg taa tac caa aag tga taa caa
ttc act ttg ctc cta tac aga aat cca ttt aac atc ttt cat att aaa atg gtg cca aaa
atg gct cta tca gag gtt aaa aaa tta cag cac taa tat gct tca tgt tgg ctc cat ttc
ccc aaa ttg att taa aag tgc att ctg tgt cta ttt ctg gtt tag cat ctt cat ggg ttg
cac aaa tta ctc ctt tgt gcc atc agt ggc act ctc cca tag tgt gac ttg att tat gga
gac ttg cat tta tcc tat gtt cct ttt gca ata gtc agt att aag aag gtt ttc tgt cct
cct gaa tca aaa ttt tct gga aaa ctg ctg tct aaa tat ttc att gat gtt atg gaa tac
att gga act gta ctt ctg ctc atc aaa tca caa gcc tta acc agt gta gtc ctC
TTG CCT CCC TAT AG  GAT CAC TGC TAT TAT TAT GGA CGC ATC CAG AAT GAT GCT GAC TCA
      IntrB13-1 Fw/Met16PI Rv
                      D   H   C   Y   Y   Y   G   R   I   Q   N   D   A   D   S      -118
ACT GCA AGC ATC AGC ACA TGC AAT GGT TTG AA gta aga tag tct cta atc ttt tat ttg
 T   A   S   I   S   T   C   N   G   L   K |---------------------- EoPI-5 →          -129
ttt att aat aat aat ata gtg ctc ttg gag ttc taa ttg tta aaa tga agg aca tcc tca
gtt ttt cat gga aat tag ttg ggt gtg atc cag att ttc ggc aga att aag aca tac ttt
ggt tga aaa cca aga aga gct gct gcc atc cag gag aaa aac tat gga gct aaa tca cat
aag tct aaa gga gct tcc aag ccc cgg tct cct ttc cca ggg tga ggt gat att aca ggt
aga gaa gat tag tag gtt tca aat tgg aga cct tgc tag aaa gtg tac agg aag agg caa
gaa gtt tca gtt cta ccc aga aac act ttc ttg agt cac tct gca cac ttt ctt cag cca
act aga tat gtt aac tac ata aag atc cca gaa ttc aga agg tcc cta tca ata gta aga
atg aac atc acc tca aca tct ttt act gaa aaa aga cac tga aac tca cct ttg aac aga
gac tgt gtc cat gga gtg gag gaa taa atg aaa agc tgg aac aga gca gaa taa caa cag
aaa aat aaa gga aaa aca gaa tga cag aat aat agc att gga agg gac ttt gag gtc ttc
                                         |---------------- LINE/L2/CR1 →
tag tcc aac ttc ctg ctc aag tag gag acc tat atc atc cta gac aaa tag ctg tca atc
ttc tct taa aaa gca gta gtg atg gaa cac cca caa tgt ctg aat agg tta att gtt cca
ttt gtg aga aaa tta ctc ctt agt tct aac tta ttt ctc tct ttg gtt act ttc cac gca
ttg ctt ctt ctc ctg cca tca ggt gaa gaa tag gtt gtc cca cat ttt tta tga cag cct
ctt aaa tac tta aag att atc aag tca tct cta ccc ctt ctt gtc act agc atg agt ata
ctc att gtc tgc agc cat tct aac ctc cag tta gta tgc att ctt att cct tca ttg tta
ctc ctg ttg ttc tgc att gac ttc tct atg aag atg ctt gcc aag aat tct tat ttt cat
tat tta tta aat atc ctg gtc atc ctg act ctt atc tta aat tgc tat caa act aat ctg
att tta ttt cct tga cca cag aca aat att gtt cta tac ttg ttt aaa gta aat tgc agt
att acc tat aac tct ttt tag ata ttt tag tag tta tat ttt tcc ttt ttt atc cta ctt
agt tgt gat tct tga gct tta tca gta ata tat atg ata aat ata aag tat ttt acc ctt
atg aaa taa agt ttt aca caa agc aga att tta caa ttg gct tta gtg ttg tat tta tgt
agc tag aaa ctt att ttt tta aca tcc tgg aaa tat aca ata ttg ggt tcc atg cca aat
tat ttc caa aca aaa ctg tac acc tat ttt gtg gct gca ctg agt ttg tga aat ctc tca
tat ctt tct gat cat aac tgt att tct gaa aag tat gag aaa gtg att tga gtg ctg agg
aaa gaa tat aaa ata ttc act cat tgt taa gaa gga att caa aaa cat gag gtt agt tga
aaa tgg gtc tca gag ccg agt ttc att acc caa cta ggt aac atc atc agt gca gtt ttt
ctc tga act aac aat att ctc ttc ttt tgc ttc tcc atc tct gat cat cct ttt cac att
gtt tta cag  A GGA TTT TTT ACG CTT CGT GGG GAG ACG TAC TTA ATT GAA CCC TTG AAG
             G   F   F   T   L   R   G   E   T   Y   L   I   E   P   L   K          -145
GTT CCC GAC AGT GAA TCC CAT GCA GTC TAC AAA TAT GAA GAT GCC AAA AAA AAG GAT GAG
 V   P   D   S   E   S   H   A   V   Y   K   Y   E   D   A   K   K   K   D   E      -165
                                                       Prodom 1 Rv/Prodom 1 Fw
GCC CCC AAA ATG TGT GGG GTA ACC CTG ACT AAT TGG GAA TCA GAT AGC CCC ATC AAA AAG
 A   P   K   M   C   G   V   T   L   T   N   W   E   S   D   K   P   I   K   K      -185
GCT TCT CAT TTA GTT GCT ACT TCT GAA gta agt ctc ata ata aac ata gtt taa gat tac
 A   S   H   L   V   A   T   S   E  |---------------- EoPI-6 →                      -194
ata cta att tcc ttg tct tga aaa tat aaa gta aga gag aat ttc ctt tgg gaa ggg gtg
ata gat aga att caa aag gga gaa gcc ccc att tct ata ttt tta ttg tag cca tgg cat
aaa aga aag aat gga aac ttg agg aac aga aaa tac att ttc cag gct tat agc att ttc
ttt ggt cat tca aac tta gtt tag aga ttt caa tca aaa tct att taa atg agt ttc taa
att atc tct agt ttc taa gtc aat gtt gaa aag taa tta aat tat caa ttt gga ttc ctc
ttt tat gca tgc aga gag gat ggg gga caa agt ggt ttg aaa tat taa atg gtt tta aga
tgt ctg ata agg cca tta caa tat tgt tac tcc att atc caa ttt gat ttg aat cat caa
gtt gga ttg gtc caa tga atg gat aaa agt tga caa tgt gaa cct agt cac aat tga ccc
tta tgc tct caa tat ttt cct cct tta ttg gac gca caa aaa tta gaa aac aaa ata ttg
cat cca aag tga cag ttc ctc tct tcc atc ttt ttg ctg gga act ggt gaa act ggc tga aaa
atc tct act gtt ttt att aga atg tta aaa ttg aca TGG AAC AAC AGC TGT TGT TAT GAC
                                                  IntronB7PI Fw
Gga ata cca aaa cac aag tga aga cgc caa atg aag cct ggt ttg tct ttt ggc ttc ttt
cat tct ggc aat tca aga ttc ttt atc ctc agc aat ttg tgg tta tac gtt aca ttt aac
```

Figure A2. *Cont.*

```
cta tga tca aaa cat ttc aac cct att tcc tat cct ttc agt tta tct gac cct tat gaa
cat acg cat agg gaa gat agt tta gca ata ttt cag cct tgt ctc aac ccc aga ctc aca
ctt tca gca tgt taa atc atg tct gtg aaa ata ata tat ttg ttc ttt tag ata gtt tgc
atg gaa acc cag ttt aaa taa ggg tga gta atg ttt GAG ATT GGT GCC CTG AGT CAG ctt
                                                  Intron B10PI Rv
cct ggc ttt ctg gaa ggt tct aag agg tcc ctg gta atg ctg tga cat ttt tct ccc aga
gac ttt tag gat gga aat tgg tgc agg aga aaa aag ttt ctt ttt cca ccc ctt aag tta
tct acc tgc tct gta aag ttc taa att cag gtg ttt tgg tgg cac att ctg gaa gtg ttt
caa gac cat gaa aat gga ggt gca agt tcc tca ttc ttt ctt tct acg tag gaa ccc agt
tga ctt tta atg aac ctt ttg agc aga atg gcc caa aac att ttg tta ttt tca tgt ttc
cct cac aag cct agt ttc aca gga aga gaa gag agc cat gag ttt ttc agc acg tga cag
aaa att cta tga atg ctt ctt ccc atg taa aga aat atc agg aga agt tca gca att cac
ttt ttg ttg ctt ttt cat ggc agc cca att gat ttt aac ttt acg gtc aac caa cat ata
gaa ctt ctg ttt cag gaa ctg agc ctt tca ttg cag tca ttt ccc tgt agc aaa taa gat
ggt ttg gga ctt cta ggc acc aca cac agt tgt aac agg gca ggg atg cct tgc ttg gtg
atc ctc aag aga gat gaa gag gag gtt ttg aaa tgt gta tca aat cat ggt ttg act ctt
tga tct ctg ctg ctg atg aat gat agc tga gaa tat ttt tga ttc tca ccc aca g    TTT
                                                                              F         221
                                                                              |-
TTG AGA GCA CGA ACA GTA AGC AAG AGA GCA GTG AGT GAT GAC ATG GAT GAT TAC TGC TCT
 L   R   A   R   T   V   S   K   R   A   V   S   D   D   M   D   D   Y   C   S        241
-------------------------------- C-terminal extension -------------------->
GGC ATA ACT TCT GAC TGT GCC AGA AAT CCT ACA AAG GCT AAG CAA CAG AGA TGG AAC GAT
                                                                 Met1PI Rv
 G   I   T   S   D   C   A   R   N   P   T   K   A   K   Q   Q   R   W   N   D        261
CTG CAG TAG
 L   Q   STOP                                                                         263
```

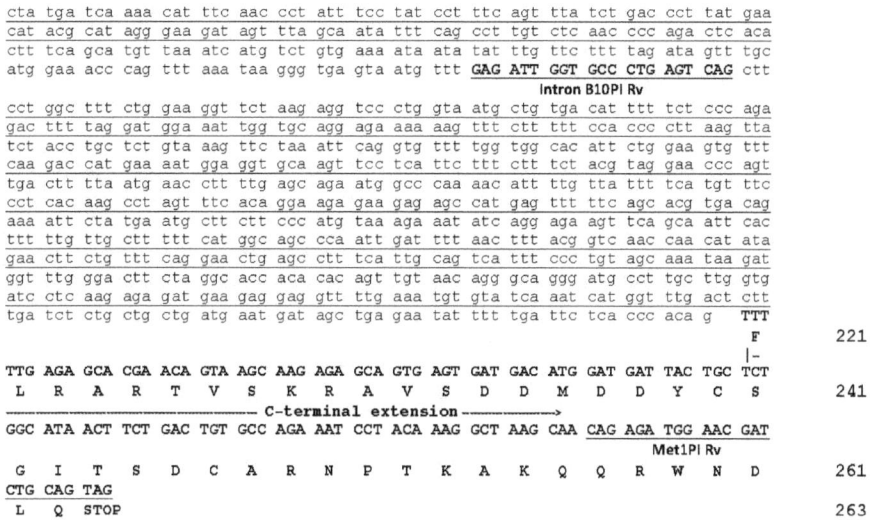

Figure A2. Genomic organization of *E. ocellatus* EOC00028-like PI-SVMP gene.

References

1. MEROPS- the Peptidase Database: Family M72. Available online: https://merops.sanger.ac.uk/cgi-bin/famsum?family=M12 (accessed on 8 May 2016).
2. PFAM Family: REprolysin. Available online: http://pfam.xfam.org/family/PF01421 (accessed on 8 May 2016).
3. Seals, D.F.; Courtneidge, S.A. The ADAMs family of metalloproteases: Multidomain proteins with multiple functions. *Genes Dev.* **2003**, *17*, 7–30. [CrossRef] [PubMed]
4. Giebeler, N.; Zigrino, P. A disintegrin and metalloprotease (ADAM): Historical overview of their functions. *Toxins* **2016**, *8*. [CrossRef] [PubMed]
5. Arendt, D.; Technau, U.; Wittbrodt, J. Evolution of the bilaterian larval foregut. *Nature* **2001**, *409*, 81–85. [CrossRef] [PubMed]
6. Tucker, R.P.; Adams, J.C. Adhesion Networks of Cnidarians: A Postgenomic View. *Int. Rev. Cell Mol. Biol.* **2014**, *308*, 323–377. [PubMed]
7. Bahudhanapati, H.; Bhattacharya, S.; Wei, S. Evolution of Vertebrate Adam Genes; Duplication of Testicular Adams from Ancient Adam9/9-like Loci. *PLoS ONE* **2015**, *10*, e0136281. [CrossRef] [PubMed]
8. Cho, C. Testicular and epididymal ADAMs: Expression and function during fertilization. *Nat. Rev. Urol.* **2012**, *9*, 550–560. [CrossRef] [PubMed]
9. Bates, E.E.; Fridman, W.H.; Mueller, C.G. The ADAMDEC1 (decysin) gene structure: Evolution by duplication in a metalloprotease gene cluster on chromosome 8p12. *Immunogenetics* **2002**, *54*, 96–105. [CrossRef] [PubMed]
10. Wei, S.; Whittaker, C.A.; Xu, G.; Bridges, L.C.; Shah, A.; White, J.M.; Desimone, D.W. Conservation and divergence of ADAM family proteins in the Xenopus genome. *BMC Evol. Biol.* **2010**, *10*. [CrossRef] [PubMed]
11. Taylor, J.S.; Raes, J. Duplication and divergence: The evolution of new genes and old ideas. *Annu. Rev. Genet.* **2004**, *38*, 615–643. [CrossRef] [PubMed]
12. Ohno, S. *Evolution by Gene Duplication*; Springer-Verlag: Berlin-Heidelberg, Germany, 1970.
13. Zhang, J. Evolution by gene duplication: An update. *Trends Ecol. Evol.* **2003**, *18*, 292–298. [CrossRef]
14. True, J.R.; Carroll, S.B. Gene co-option in physiological and morphological evolution. *Annu. Rev. Cell Dev. Biol.* **2002**, *18*, 53–80. [CrossRef] [PubMed]

15. Kaessmann, H.; Vinckenbosch, N.; Long, M. RNA-based gene duplication: Mechanistic and evolutionary insights. *Nat. Rev. Genet.* **2009**, *10*, 19–31. [CrossRef] [PubMed]
16. Vinckenbosch, N.; Dupanloup, I.; Kaessmann, H. Evolutionary fate of retroposed gene copies in the human genome. *Proc. Natl. Acad. Sci. USA* **2006**, *103*, 3220–3225. [CrossRef] [PubMed]
17. Fry, B.; Wüster, W. Assembling an Arsenal: Origin and evolution of the snake venom proteome inferred from phylogenetic analysis of toxin sequences. *Mol. Biol. Evol.* **2004**, *21*, 870–883. [CrossRef] [PubMed]
18. Fry, B.G.; Vidal, N.; Norman, J.A.; Vonk, F.J.; Scheib, H.; Ramjan, S.F.; Kuruppu, S.; Fung, K.; Hedges, S.B.; Richardson, M.K.; et al. Early evolution of the venom system in lizards and snakes. *Nature* **2006**, *439*, 584–588. [CrossRef] [PubMed]
19. Fry, B.G.; Casewell, N.R.; Wüster, W.; Vidal, N.; Young, B.; Jackson, N. The structural and functional diversification of the Toxicofera reptile venom system. *Toxicon* **2012**, *60*, 434–448. [CrossRef] [PubMed]
20. Casewell, N.R.; Wüster, W.; Vonk, F.J.; Harrison, R.A.; Fry, B.G. Complex cocktails: The evolutionary novelty of venoms. *Trends Ecol. Evol.* **2013**, *28*, 219–229. [CrossRef] [PubMed]
21. Haney, R.A.; Clarke, T.H.; Gadgil, R.; Fitzpatrick, R.; Hayashi, C.Y.; Ayoub, N.A.; Garb, J.E. Effects of gene duplication, positive selection, and shifts in gene expression on the evolution of the venom gland transcriptome in widow spiders. *Genome Biol. Evol.* **2016**, *8*, 228–242. [CrossRef] [PubMed]
22. Wong, E.S.; Belov, K. Venom evolution through gene duplications. *Gene* **2012**, *496*, 1–7. [CrossRef] [PubMed]
23. Vonk, F.J.; Casewell, N.R.; Henkel, C.V.; Heimberg, A.M.; Jansen, H.J.; McCleary, R.J.; Kerkkamp, H.M.; Vos, R.A.; Guerreiro, I.; Calvete, J.J.; et al. The king cobra genome reveals dynamic gene evolution and adaptation in the snake venom system. *Proc. Natl. Acad. Sci. USA* **2013**, *110*, 20651–20656. [CrossRef] [PubMed]
24. Hedges, S.B.; Vidal, N. Lizards, snakes, and amphisbaenians (Squamata). In *The Timetree of Life*; Hedges, S.B., Kumar, S., Eds.; Oxford University Press: Oxford, UK, 2009; pp. 383–389.
25. Jones, M.E.; Anderson, C.L.; Hipsley, C.A.; Müller, J.; Evans, S.E.; Schoch, R.R. Integration of molecules and new fossils supports a Triassic origin for Lepidosauria (lizards, snakes, and tuatara). *BMC Evol. Biol.* **2013**, *13*. [CrossRef] [PubMed]
26. Pyron, R.A.; Burbrink, F.T.; Wiens, J.J. A phylogeny and revised classification of Squamata, including 4161 species of lizards and snakes. *BMC Evol. Biol.* **2013**, *13*. [CrossRef] [PubMed]
27. Reeder, T.W.; Townsend, T.M.; Mulcahy, D.G.; Noonan, B.P.; Wood, P.L., Jr.; Sites, J.W., Jr.; Wiens, J.J. Integrated analyses resolve conflicts over squamate reptile phylogeny and reveal unexpected placements for fossil. *PLoS ONE* **2015**, *10*, e0118199. [CrossRef] [PubMed]
28. Hsiang, A.Y.; Field, D.J.; Webster, T.H.; Behlke, A.D.; Davis, M.B.; Racicot, R.A.; Gauthier, J.A. The origin of snakes: Revealing the ecology, behavior, and evolutionary history of early snakes using genomics, phenomics, and the fossil record. *BMC Evol. Biol.* **2015**, *15*. [CrossRef] [PubMed]
29. Reeks, T.A.; Fry, B.G.; Alewood, P.F. Privileged frameworks from snake venom. *Cell. Mol. Life Sci.* **2015**, *72*, 1939–1958. [CrossRef] [PubMed]
30. Hite, L.A.; Jia, L.G.; Bjarnason, J.B.; Fox, J.W. cDNA sequences for four snake venom metalloproteinases: Structure, classification, and their relationship to mammalian reproductive proteins. *Arch. Biochem. Biophys.* **1994**, *308*, 182–191. [CrossRef] [PubMed]
31. Moura-da-Silva, A.M.; Theakston, R.D.; Crampton, J.M. Evolution of disintegrin cysteine-rich and mammalian matrix-degrading metalloproteinases: Gene duplication and divergence of a common ancestor rather than convergent evolution. *J. Mol. Evol.* **1996**, *43*, 263–269. [CrossRef] [PubMed]
32. Casewell, N.R. On the ancestral recruitment of metalloproteinases into the venom of snakes. *Toxicon* **2012**, *60*, 449–454. [CrossRef] [PubMed]
33. Escalante, T.; Rucavado, A.; Fox, J.W.; Gutiérrez, J.M. Key events in microvascular damage induced by snake venom hemorrhagic metalloproteinases. *J. Proteomics* **2011**, *74*, 1781–1794. [CrossRef] [PubMed]
34. Markland, F.S., Jr.; Swenson, S. Snake venom metalloproteinases. *Toxicon* **2013**, *62*, 3–18. [CrossRef] [PubMed]
35. Herrera, C.; Escalante, T.; Voisin, M.B.; Rucavado, A.; Morazán, D.; Macêdo, J.K.; Calvete, J.J.; Sanz, L.; Nourshargh, S.; Gutiérrez, J.M.; et al. Tissue localization and extracellular matrix degradation by PI, PII and PIII snake venom metalloproteinases: Clues on the mechanisms of venom-induced hemorrhage. *PLoS Negl. Trop. Dis.* **2015**, *9*, e0003731. [CrossRef] [PubMed]
36. Gutiérrez, J.M.; Escalante, T.; Rucavado, A.; Herrera, C. Hemorrhage Caused by Snake Venom Metalloproteinases: A Journey of Discovery and Understanding. *Toxins* **2016**, *8*. [CrossRef] [PubMed]

37. Jia, L.G.; Shimokawa, K.; Bjarnason, J.B.; Fox, J.W. Snake venom metalloproteinases: Structure, function and relationship to the ADAMs family of proteins. *Toxicon* **1996**, *34*, 1269–1276. [CrossRef]

38. Pyron, R.A.; Burnbrink, F.T. Extinction ecological opportunity and the origins of global snake diversity. *Evolution* **2012**, *66*, 163–178. [CrossRef] [PubMed]

39. Fox, J.W.; Serrano, S.M. Structural considerations of the snake venom metalloproteinases, key members of the M12 reprolysin family of metalloproteinases. *Toxicon* **2005**, *45*, 969–985. [CrossRef] [PubMed]

40. Fox, J.W.; Serrano, S.M. Insights into and speculations about snake venom metalloproteinase (SVMP) synthesis, folding and disulfide bond formation and their contribution to venom complexity. *FEBS J.* **2008**, *275*, 3016–3030. [CrossRef] [PubMed]

41. Casewell, N.R.; Sunagar, K.; Takacs, Z.; Calvete, J.J.; Jackson, T.N.W.; Fry, B.G. Snake venom metalloprotease enzymes. In *Venomous Reptiles and Their Toxins: Evolution, Pathophysiology and Biodiscovery*; ISBN: 978-0-19-930939-9. Fry, B.G., Ed.; Oxford University Press: Oxford, UK, 2015; Chapter 23; pp. 347–363.

42. Juárez, P.; Comas, I.; González-Candelas, F.; Calvete, J.J. Evolution of snake venom disintegrins by positive Darwinian selection. *Mol. Biol. Evol.* **2008**, *25*, 2391–2407. [CrossRef] [PubMed]

43. Carbajo, R.J.; Sanz, L.; Pérez, A.; Calvete, J.J. NMR structure of bitistatin—A missing piece in the evolutionary pathway of snake venom disintegrins. *FEBS J.* **2015**, *282*, 341–360. [CrossRef] [PubMed]

44. Calvete, J.J. Brief History and Molecular Determinants of Snake Venom Disintegrin Evolution. In *Toxins and Hemostasis. From Bench to Bedside*; Kini, R.M., Markland, F., McLane, M.A., Morita, T., Eds.; Springer Science+Business Media B.V.: Amsterdam, The Netherlands, 2010; pp. 285–300.

45. Casewell, N.R.; Wagstaff, S.C.; Harrison, R.A.; Renjifo, C.; Wüster, W. Domain loss facilitates accelerated evolution and neofunctionalization of duplicate snake venom metalloproteinase toxin genes. *Mol. Biol. Evol.* **2011**, *28*, 2637–2649. [CrossRef] [PubMed]

46. Sanz-Soler, R.; Sanz, L.; Calvete, J.J. Distribution of *RPTLN* genes across Reptilia. Hypothesized role for *RPTLN* in the evolution of SVMPs. *Integr. Compar. Biol.* **2016**. [CrossRef] [PubMed]

47. Sanz, L.; Harrison, R.A.; Calvete, J.J. First draft of the genomic organization of a PIII-SVMP gene. *Toxicon* **2012**, *60*, 455–469. [CrossRef] [PubMed]

48. Endo, T.; Fedorov, A.; de Souza, S.J.; Gilbert, W. Do Introns Favor or Avoid Regions of Amino Acid Conservation? *Mol. Biol. Evol.* **2002**, *19*, 521–525. [CrossRef] [PubMed]

49. Zhou, Q.; Wang, W. On the origin and evolution of new genes-a genomic and experimental perspective. *J. Genet. Genomics* **2008**, *35*, 639–648. [CrossRef]

50. Kordis, D.; Gubensek, F. Adaptive evolution of animal toxin multigene families. *Gene* **2000**, *261*, 43–52. [CrossRef]

51. Cao, Z.; Yu, Y.; Wu, Y.; Hao, P.; Di, Z.; He, Y.; Chen, Z.; Yang, W.; Shen, Z.; He, X.; et al. The genome of Mesobuthus martensii reveals a unique adaptation model of arthropods. *Nat. Commun.* **2013**, *4*. [CrossRef] [PubMed]

52. Fry, B.G.; Wüster, W.; Kini, R.M.; Brusic, V.; Khan, A.; Venkataraman, D.; Rooney, A.P. Molecular evolution and phylogeny of elapid snake venom three-finger toxins. *J. Mol. Evol.* **2003**, *57*, 110–129. [CrossRef] [PubMed]

53. Reyes-Velasco, J.; Card, D.C.; Andrew, A.L.; Shaney, K.J.; Adams, R.H.; Schield, D.R.; Casewell, N.R.; Mackessy, S.P.; Castoe, T.A. Expression of venom gene homologs in diverse python tissues suggests a new model for the evolution of snake venom. *Mol. Biol. Evol.* **2015**, *32*, 173–183. [CrossRef] [PubMed]

54. Hargreaves, A.D.; Swain, M.T.; Logan, D.W.; Mulley, J.F. Testing the Toxicofera: Comparative transcriptomics casts doubt on the single, early evolution of the reptile venom system. *Toxicon* **2014**, *92*, 140–156. [CrossRef] [PubMed]

55. Sunagar, K.; Jackson, T.N.; Undheim, E.A.; Ali, S.A.; Antunes, A.; Fry, B.G. Three-fingered RAVERs: Rapid Accumulation of Variations in Exposed Residues of snake venom toxins. *Toxins* **2013**, *5*, 2172–2208. [CrossRef] [PubMed]

56. Chang, D.; Duda, T.F. Extensive and continuous duplication facilitates rapid evolution and diversification of gene families. *Mol. Biol. Evol.* **2012**, *29*, 2019–2029. [CrossRef] [PubMed]

57. Chow, L.T.; Gelinas, R.E.; Broker, T.R.; Roberts, R.J. An amazing sequence arrangement at the 5′ ends of adenovirus 2 messenger RNA. *Cell* **1977**, *12*, 1–8. [CrossRef]

58. Berget, S.M.; Moore, C.; Sharp, P.A. Spliced segments at the 5′ terminus of adenovirus 2 late mRNA. *Proc. Natl. Acad. Sci. USA* **1977**, *74*, 3171–3175. [CrossRef] [PubMed]

59. Bicknell, A.A.; Cenik, C.; Chua, H.N.; Roth, F.P.; Moore, M.J. Introns in UTRs: Why we should stop ignoring them. *BioEssays* **2012**, *34*, 1025–1034. [CrossRef] [PubMed]

60. Cenik, C.; Chua, H.N.; Zhang, H.; Tarnawsky, S.P.; Akef, A.; Derti, A.; Tasan, M.; Moore, M.J.; Palazzo, A.F.; Roth, F.P. Genome analysis reveals interplay between 5'-UTR introns and nuclear mRNA export for secretory and mitochondrial genes. *PLoS Genet.* **2011**, *7*, e1001366. [CrossRef] [PubMed]

61. Comeron, J.M.; Kreitman, M. The correlation between intron length and recombination in Drosophila: Dynamic equilibrium between mutational and selective forces. *Genetics* **2000**, *156*, 1175–1190. [PubMed]

62. De Souza, S.J.; Long, M.; Gilbert, W. Introns and gene evolution. *Genes Cells* **1996**, *1*, 493–505. [CrossRef] [PubMed]

63. Patthy, L. Exon shuffling and other ways of module exchange. *Matrix Biol.* **1996**, *15*, 301–310. [CrossRef]

64. Hughes, A.L.; Hughes, M.K. Small genomes for better flyers. *Nature* **1995**, *377*, 391. [CrossRef] [PubMed]

65. Lynch, M. Intron evolution as a population-genetic process. *Proc. Natl. Acad. Sci. USA* **2002**, *99*, 6118–6123. [CrossRef] [PubMed]

66. Haddrill, P.R.; Charlesworth, B.; Halligan, D.L.; Andolfatto, P. Patterns of intron sequence evolution in Drosophila are dependent upon length and GC content. *Genome Biol.* **2005**, *6*, R67. [CrossRef] [PubMed]

67. Zhao, M.; He, L.; Gu, Y.; Wang, Y.; Chen, Q.; He, C. Genome-wide analyses of a plant-specific LIM-domain gene family implicate its evolutionary role in plant diversification. *Genome Biol. Evol.* **2014**, *6*, 1000–1012. [CrossRef] [PubMed]

68. Tordai, H.; Patthy, L. Insertion of spliceosomal introns in proto-splice sites: The case of secretory signal peptides. *FEBS Lett.* **2004**, *575*, 109–111. [CrossRef] [PubMed]

69. Tomita, M.; Shimizu, N.; Brutlag, D.L. Introns and reading frames: Correlation between splicing sites and their codon positions. *Mol. Biol. Evol.* **1996**, *13*, 1219–1223. [CrossRef] [PubMed]

70. Long, M.; de Souza, S.J.; Rosenberg, C.; Gilbert, W. Relationship between proto-splice sites and intron phases: Evidence from dicodon analysis. *Proc. Natl. Acad. Sci. USA* **1998**, *95*, 219–223. [CrossRef] [PubMed]

71. Von Heijne, G. Patterns of amino acids near signal-sequence cleavage sites. *Eur. J. Biochem.* **1983**, *133*, 17–21. [CrossRef] [PubMed]

72. Pinho, C.; Rocha, S.; Carvalho, B.M.; Lopes, S.; Mourão, S.; Vallinoto, M.; Brunes, T.O.; Haddad, C.F.B.; Gonçalves, H.; Sequeira, F.; et al. New primers for the amplification and sequencing of nuclear *loci* in a taxonomically wide set of reptiles and amphibians. *Conserv. Genet. Resour.* **2010**, *2*, 181–185. [CrossRef]

73. Ellegren, H. Microsatellites: Simple sequences with complex evolution. *Nat. Rev. Genet.* **2004**, *5*, 435–445. [CrossRef] [PubMed]

74. Adams, R.H.; Blackmon, H.; Reyes-Velasco, J.; Schield, D.R.; Card, D.C.; Andrew, A.L.; Waynewood, N.; Castoe, T.A. Microsatellite landscape evolutionary dynamics across 450 million years of vertebrate genome evolution. *Genome* **2016**, *59*, 295–310. [CrossRef] [PubMed]

75. Liang, K.C.; Tseng, J.T.; Tsai, S.J.; Sun, H.S. Characterization and distribution of repetitive elements in association with genes in the human genome. *Comput. Biol. Chem.* **2015**, *57*, 29–38. [CrossRef] [PubMed]

76. Schlötterer, C.; Tautz, D. Slippage synthesis of simple sequence DNA. *Nucleic Acids Res.* **1992**, *20*, 211–215. [CrossRef] [PubMed]

77. Charlesworth, B.; Sniegowski, P.; Stephan, W. The evolutionary dynamics of repetitive DNA in eukaryotes. *Nature* **1994**, *371*, 215–220. [CrossRef] [PubMed]

78. Martin, P.; Makepeace, K.; Hill, S.A.; Hood, D.W.; Moxon, E.R. Microsatellite instability regulates transcription factor binding and gene expression. *Proc. Natl. Acad. Sci. USA* **2005**, *102*, 3800–3804. [CrossRef] [PubMed]

79. Li, Y.C.; Korol, A.B.; Fahima, T.; Beiles, A.; Nevo, E. Microsatellites: Genomic distribution, putative functions and mutational mechanisms: A review. *Mol. Ecol.* **2002**, *11*, 2453–2465. [CrossRef] [PubMed]

80. Shaney, K.J.; Schield, D.R.; Card, D.C.; Ruggiero, R.P.; Pollock, D.D.; Mackessy, S.P.; Castoe, T.A. Squamate reptile genomics and evolution. In *Handbook of Toxinology: Venom Genomics and Proteomics*; Gopalakrishnakone, P., Calvete, J.J., Eds.; Springer Science+Business Media: Dordrecht, The Netherlands, 2016; pp. 29–49.

81. Balaresque, P.; King, T.E.; Parkin, E.J.; Heyer, E.; Carvalho-Silva, D.; Kraaijenbrink, T.; de Knijff, P.; Tyler-Smith, C.; Jobling, M.A. Gene conversion violates the stepwise mutation model for microsatellites in Y-chromosomal palindromic repeats. *Hum. Mutat.* **2014**, *35*, 609–617. [CrossRef] [PubMed]

82. Eller, C.D.; Regelson, M.; Merriman, B.; Nelson, S.; Horvath, S.; Marahrens, Y. Repetitive sequence environment distinguishes housekeeping genes. *Gene* **2007**, *390*, 153–165. [CrossRef] [PubMed]

83. Sverdlov, E.D. Perpetually mobile footprints of ancient infections in human genome. *FEBS Lett.* **1998**, *428*, 1–6. [CrossRef]

84. Makalowski, W. Genomic scrap yard: How genomes utilize all that junk. *Gene* **2000**, *259*, 61–67. [CrossRef]

85. Bourque, G.; Leong, B.; Vega, V.B.; Chen, X.; Lee, Y.L.; Srinivasan, K.G.; Chew, J.L.; Ruan, Y.; Wei, C.L.; Ng, H.H.; et al. Evolution of the mammalian transcription factor binding repertoire via transposable elements. *Genome. Res.* **2008**, *18*, 1752–1762. [CrossRef] [PubMed]

86. Irimía, M.; Rukov, J.L.; Penny, D.; Vinther, J.; García-Fernández, J.; Roy, S.W. Origin of introns by 'intronization' of exonic sequences. *Trends Genet.* **2008**, *24*, 378–381. [CrossRef] [PubMed]

87. Bazaa, A.; Juarez, P.; Marrakchi, N.; Bel Lasfer, Z.; El Ayeb, M.; Harrison, R.A.; Calvete, J.J.; Sanz, L. Loss of introns along the evolutionary diversification pathway of snake venom disintegrins evidenced by sequence analysis of genomic DNA from *Macrovipera lebetina transmediterranea* and *Echis ocellatus*. *J. Mol. Evol.* **2007**, *64*, 261–271. [CrossRef] [PubMed]

88. Bonen, L.; Vogel, J. The ins and outs of group II introns. *Trends Genet.* **2001**, *17*, 322–331. [CrossRef]

89. Dibb, N.J.; Newman, A.J. Evidence that introns arose at proto-splice sites. *EMBO J.* **1989**, *8*, 2015–2021. [PubMed]

90. Logsdon, J.M., Jr. The recent origins of spliceosomal introns revisited. *Curr. Opin. Genet. Dev.* **1998**, *8*, 637–648. [CrossRef]

91. Lynch, M.; Richardson, A.O. The evolution of spliceosomal introns. *Curr. Opin. Genet. Dev.* **2002**, *12*, 701–710. [CrossRef]

92. Patel, A.A.; Steitz, J.A. Splicing double: Insights from the second spliceosome. *Nat. Rev. Mol. Cell Biol.* **2003**, *4*, 960–970. [CrossRef] [PubMed]

93. Rodríguez-Trelles, F.; Tarrío, R.; Ayala, F.J. Origins and evolution of spliceosomal introns. *Annu. Rev. Genet.* **2006**, *40*, 47–76. [CrossRef] [PubMed]

94. Luo, Y.; Li, C.; Gong, X.; Wang, Y.; Zhang, K.; Cui, Y.; Sun, Y.E.; Li, S. Splicing-related features of introns serve to propel evolution. *PLoS ONE* **2013**, *8*, e58547. [CrossRef] [PubMed]

95. Nakashima, K.; Ogawa, T.; Oda, N.; Hattori, M.; Sakaki, Y.; Kihara, H.; Ohno, M. Accelerated evolution of *Trimeresurus. flavoviridis* venom gland phospholipase A$_2$ isozymes. *Proc. Natl. Acad. Sci. USA* **1993**, *90*, 5964–5968. [CrossRef] [PubMed]

96. Nakashima, K.; Nobuhisa, I.; Deshimaru, M.; Nakai, M.; Ogawa, T.; Shimohigashi, Y.; Fukumaki, Y.; Hattori, M.; Sakaki, Y.; Hattori, S.; et al. Accelerated evolution in the protein-coding regions is universal in crotalinae snake venom gland phospholipase A$_2$ isozyme genes. *Proc. Natl. Acad. Sci. USA* **1995**, *92*, 5605–5609. [CrossRef] [PubMed]

97. Ikeda, N.; Chijiwa, T.; Matsubara, K.; Oda-Ueda, N.; Hattori, S.; Matsuda, Y.; Ohno, M. Unique structural characteristics and evolution of a cluster of venom phospholipase A$_2$ isozyme genes of *Protobothrops flavoviridis* snake. *Gene* **2010**, *461*, 15–25. [CrossRef] [PubMed]

98. Chijiwa, T.; Ikeda, N.; Masuda, H.; Hara, H.; Oda-Ueda, N.; Hattori, S.; Ohno, M. Structural characteristics and evolution of a novel venom phospholipase A$_2$ gene from *Protobothrops flavoviridis*. *Biosci. Biotechnol. Biochem.* **2012**, *76*, 551–558. [CrossRef] [PubMed]

99. Chijiwa, T.; Nakasone, H.; Irie, S.; Ikeda, N.; Tomoda, K.; Oda-Ueda, N.; Hattori, S.; Ohno, M. Structural characteristics and evolution of the *Protobothrops elegans* pancreatic phospholipase A$_2$ gene in contrast with those of *Protobothrops* genus venom phospholipase A$_2$ genes. *Biosci. Biotechnol. Biochem.* **2013**, *77*, 97–102. [CrossRef] [PubMed]

100. Koszul, R.; Fischer, G. A prominent role for segmental duplications in modeling eukaryotic genomes. *C. R. Biol.* **2009**, *332*, 254–266. [CrossRef] [PubMed]

101. Ohta, T. Simple model for treating evolution of multigene families. *Nature* **1976**, *263*, 74–76. [CrossRef] [PubMed]

102. Smith, G.P. Evolution of repeated DNA sequences by unequal crossover. *Science* **1976**, *191*, 528–535. [CrossRef] [PubMed]

103. Juárez, P.; Wagstaff, S.C.; Sanz, L.; Harrison, R.A.; Calvete, J.J. Molecular cloning of *Echis ocellatus* disintegrins reveals non-venom-secreted proteins and a pathway for the evolution of ocellatusin. *J. Mol. Evol.* **2006**, *63*, 183–193. [CrossRef] [PubMed]

104. GeneWise. Available online: http://www.ebi.ac.uk/Tools/Wise2/index.html (accessed on 31 May 2016).
105. Basic Local Alignment Search Tool (BLAST). Available online: http://blast.ncbi.nlm.nih.gov/Blast.cgi (accessed on 5 June 2016).
106. ClustalW2-Multiple Sequence Alignment. Available online: http://www.ebi.ac.uk/Tools/msa/clustalw2 (accessed on 5 June 2016).
107. RepeatMasker. Available on: http://www.repeatmasker.org (accessed on 30 December 2015).
108. Genetic Information Research Institute. Available on: http://www.girinst.org (accessed on 7 January 2016).
109. National Center for Biotechnology Information (NCBI) Database. Available on: http://www.ncbi.nlm.nih.gov (accessed on 10 January 2016).

MDPI AG
St. Alban-Anlage 66
4052 Basel, Switzerland
Tel. +41 61 683 77 34
Fax +41 61 302 89 18
http://www.mdpi.com

Toxins Editorial Office
E-mail: toxins@mdpi.com
http://www.mdpi.com/journal/toxins

www.ingramcontent.com/pod-product-compliance
Lightning Source LLC
Chambersburg PA
CBHW051722210326
41597CB00032B/5575